Semiparametric Odds Ratio Model and Its Applications

Semiparametric Odds Ratio Model and Its Applications

Hua Yun Chen

CRC Press
Taylor & Francis Group
Boca Raton London New York

CRC Press is an imprint of the
Taylor & Francis Group, an **informa** business
A CHAPMAN & HALL BOOK

First edition published 2022
by CRC Press
6000 Broken Sound Parkway NW, Suite 300, Boca Raton, FL 33487-2742

and by CRC Press
2 Park Square, Milton Park, Abingdon, Oxon, OX14 4RN

CRC Press is an imprint of Taylor & Francis Group, LLC

ISBN: 978-1-138-48532-7 (hbk)
ISBN: 978-1-032-15919-5 (pbk)
ISBN: 978-1-351-04975-7 (ebk)

DOI: 10.1201/9781351049757

Publisher's note: This book has been prepared from camera-ready copy provided by the authors.

In memory of my father, and to my mother, my wife and children.

Contents

Preface

Odds ratio parameter has long been used in epidemiology to measure the strength of the association of a risk factor and a disease outcome. One key feature making the odds ratio parameter popular is its invariance to retrospective sampling designs. As a result, studies of such association can be flexibly designed when the outcome is discrete. For continuous outcome with discrete or continuous exposures, the invariant property still holds. However, this property is not well explored due to the lack of a flexible modeling framework. This book systematically introduces such a modeling framework termed the semiparametric odds ratio model. The modeling framework can be applied to the discrete data, continuous data, or a mix of both. The unified modeling framework also extends the joint normal model to multivariate data without restrictions on the type of variables included, and generalizes the generalized linear models so that multivariate data with mixed variable types can be flexibly modeled. The scope of the applications of the modeling framework is much broadened comparing with either the joint normal model or the generalized linear models.

The introduction of semiparametric odds ratio model requires us to first generalize the odds ratio parameter to the odds ratio function. This generalization first appeared in Liang and Qin (2000) as far as we know. See also Satten and Carroll (2000) for the discrete case. Chen (2003, 2004) applied the concept of an odds ratio function to the biased sampling problem and missing data problem, respectively. Osuis (2005, 2009) characterized a bivariate density in terms of the odds ratio function and its marginal densities. The involvement of marginal density functions makes it difficult to flexibly model the odds ratio function because constraints on the odds ratio function are introduced by the marginal densities. This is similar to the difficulties encountered in modeling multivariate data by copula models. Chen (2007, 2010) took a different approach by characterizing a joint density in terms of the odds ratio function and the baseline densities at a fixed condition. This characterization has the important property of yielding variation-independent components, i.e., the odds ratio function and the baseline densities are variation-independent. This makes flexibly modeling the odds ratio function possible. The variation-independent property also enables us to perform doubly robust estimation of the association parameters (Chen, 2007; Tchetgen Techetgen, 2010). The semiparametric odds ratio model has a very close tie with the density ratio model (Qin, 1998) and the semiparametric likelihood ratio model (Lou and Tsai, 2012) that are developed separately. However, both models can be converted into special cases of the semiparametric odds ratio model. Applications of the semiparametric odds ratio model have been continuously evolved as computation challenges for parameter estimation and inference are tackled.

This book is composed of three parts: Chapters 1–4, Chapters 5–9, and Chapter 10. The first part (Chapters 1–4) introduces the semiparametric odds ratio model for multivariate data with possibly both discrete and continuous variables. Starting from the basic concept of an odds ratio parameter for modeling the association of exposure and disease in epidemiology, Chapter 1 examines the traditional loglinear models for contingency tables and reformulates them around odds ratio parameters to pave the way for the introduction of odds ratio function and semiparametric odds ratio model in Chapter 2. The semiparametric odds ratio model is based on the odds ratio representation of a density function that satisfies the positivity condition. The key feature of the representation is the variation-independent components that a density function can be decomposed. The semiparametric model generalizes the joint normal model and the generalized linear models so that it is more flexible to model both discrete and continuous data. Chapter 3 introduces estimation and inference methods for the semiparametric odds ratio model. Such methods include the pseudo-likelihood approaches, the permutation likelihood approach, and the semiparametric likelihood approach. The software package `SPORM` for model fitting and inference is also introduced. Chapter 4 studies estimation and inference on the conditional odds ratio parameters and examines the doubly robust approaches to estimation and inference on the conditional odds ratio parameters.

The second part (Chapters 5–9) discusses applications of the semiparametric odds ratio modeling framework to a number of statistical problems. Chapter 5 applies the model to the biased sampling designs to resolve issues with parameter identifiability, estimation, and inference. Chapter 6 introduces a set of likelihood-based tests for the conditional independence hypothesis and examines their ties to the traditional tests of conditional independence. Chapter 7 discusses the application of the modeling framework to network detection and estimation. Chapter 8 applies the model to missing data imputation, and to parameter identification and estimation for nonignorable missing data. Chapter 9 deals with additional applications including checking model compatibility and reconciling incompatible models, as well as estimating multivariate density semiparametrically. The third part (Chapter 10) develops some key theoretical results supporting the methods used in the previous chapters.

Acknowledgments

The writing of this book lasted a couple of years. I am very grateful to the patience and support of Mr. David Grubbs and his team from the Taylor & Francis Group over these years. I am also thankful to the continuous support and encouragement of my wife, Wei Qi, and my kids, Gary, John, and Andy. They made it possible for me to finally complete the book. A special thanks goes to my son, John Chen, for creating the digital cover image. I would also like to take this opportunity to thank Dr. Roderick Little, my dissertation advisor at the University of Michigan, for introducing me to the research topics on missing data, which directly or indirectly led to the development of much of this book. The part of my research presented in this book was partially supported by grants from the U.S. National Science Foundation (DMS1007726, DMS1512930) and the National Institutes of Health (R01ES028790).

1

Odds Ratio Parameter and Its Utilities

1.1 Relative risk and odds ratio parameter

Epidemiological study of the association of a disease with the exposure to a risk factor often generates data as shown in Table 1.1.

TABLE 1.1
Counts on disease and exposure status.

Disease	Exposure Yes	Exposure No	Total
Yes	n_{11}	n_{12}	n_{1+}
No	n_{21}	n_{22}	n_{2+}
Total	n_{+1}	n_{+2}	n_{++}

Let D denote the disease status and E denote the exposure status. The probabilities of having the disease when a subject is exposed or not exposed to the risk factor are respectively,

$$\begin{aligned} p_{11} &= P(D = \text{Yes} \mid E = \text{Yes}), \\ p_{12} &= P(D = \text{Yes} \mid E = \text{No}). \end{aligned}$$

The relative risk of having the disease for a subject having exposed to the risk factor versus not having exposed is

$$RR(D = Yes) = \frac{p_{11}}{p_{12}}. \tag{1.1}$$

Whether the disease is associated with the exposure or not can be determined based on whether $RR = 1$ or not. The exposure increases (decreases) the disease risk if $RR > 1$ ($RR < 1$). A commonly used estimator of the relative risk based on the observed data in Table 1.1 is

$$\widehat{RR}(D = Yes) = \frac{n_{11}/n_{+1}}{n_{12}/n_{+2}}. \tag{1.2}$$

Estimation of the relative risk from the data is based on the assumption that the epidemiological study is a prospective study that randomly samples subjects from the study population and follows the subjects for a pre-defined period of time to obtain the disease status. When the disease incidence rate is low, the prospective study design is very expensive to conduct. In practice, many epidemiological studies of the association of a disease and a risk factor are carried out using the retrospective study design. The retrospective design samples subjects with or without the disease separately from the respective subpopulations and measures the samples' exposure history to the risk factor retrospectively. Data generated from the retrospective design can also be summarized in the same way as shown in Table 1.1. However, the relative risk of the disease cannot be estimated well by (1.2).

DOI: 10.1201/9781351049757-1

In other words, a large bias may occur in estimating the relative risk of the disease when (1.2) is applied to the retrospective design. This is because the retrospective design has the sampling probability of a subject in the population depending on the subject's disease status. A subject in the general population with the disease is often much more likely to be sampled than a subject without the disease.

Although data from the retrospective sampling design present a distorted picture of the probabilities of the disease given the exposure status, they do not distort the probabilities of exposure to the risk factor given the disease status. For example, for the sample of the diseased subpopulation,

$$\begin{aligned} q_{11} &= P(E = \text{Yes} \mid D = \text{Yes}), \\ q_{21} &= P(E = \text{No} \mid D = \text{Yes}), \end{aligned}$$

are the same as in the general population. The odds of exposure to the risk factor versus non-exposure for a subject in the sample from the diseased subpopulation,

$$\text{Odds}(D = \text{Yes}) = \frac{q_{11}}{q_{21}},$$

is the same as in the general population. Similarly, for the sample of the non-diseased subpopulation

$$\begin{aligned} q_{12} &= P(E = \text{Yes} \mid D = \text{No}), \\ q_{22} &= P(E = \text{No} \mid D = \text{No}), \end{aligned}$$

are the same as in the general population. The odds of exposure to the risk factor versus non-exposure for subjects in the sample from the non-diseased subpopulation,

$$\text{Odds}(D = \text{No}) = \frac{q_{12}}{q_{22}},$$

is also the same as in the general population. The odds ratio parameter, defined as the ratio of the two odds, i.e.,

$$OR = \frac{\text{Odds}(D = \text{Yes})}{\text{Odds}(D = \text{No})}, \tag{1.3}$$

is not affected by the retrospective sampling design. When the conditional probabilities are plugged in the odds ratio parameter, it can be expressed as

$$OR = \frac{q_{11}q_{22}}{q_{12}q_{21}}. \tag{1.4}$$

The conditional probabilities of the exposure status given the disease status link to the conditional probabilities of the disease status given the exposure status through the Bayesian formula,

$$q_{11} = \frac{P(D = \text{Yes} \mid E = \text{Yes})P(E = \text{Yes})}{P(D = \text{Yes})} = \frac{p_{11}P(E = \text{Yes})}{P(D = \text{Yes})}.$$

Applying this equality and similar equalities for other relevant probabilities, we can rewrite the odds ratio parameter as the ratio of the relative risks,

$$OR = \frac{p_{11}p_{22}}{p_{12}p_{21}} = \frac{RR(D = \text{Yes})}{RR(D = \text{No})}, \tag{1.5}$$

where the relative risk for no disease is

$$RR(D = \text{No}) = \frac{P(D = \text{No} \mid E = \text{Yes})}{P(D = \text{No} \mid E = \text{No})} = \frac{p_{21}}{p_{22}}.$$

This means the odds ratio parameter is always the same as the ratio of the relative risks which can be estimated from either a prospective study or a retrospective study.

In a typical retrospective study design, $P(D = \text{Yes} \mid E = \text{Yes})$ and $P(D = \text{Yes} \mid E = \text{No})$ are usually small, which implies that $P(D = \text{No} \mid E = \text{Yes}) \approx 1 \approx P(D = \text{No} \mid E = \text{No})$, i.e.,

$$RR(D = \text{No}) \approx 1.$$

This in turn implies that

$$OR \approx RR(D = \text{Yes}).$$

Since we can estimate the ratio of the relative risks, equivalently, the odds ratio parameter by

$$\widehat{OR} = \frac{n_{11}n_{22}}{n_{12}n_{21}},$$

using data from the retrospective study design, the relative risk parameter can be estimated approximately in a retrospective study when the disease prevalence rate is low in the general population.

Approximating the relative risk parameter is only one of the many utilities of the odds ratio parameter. When the association between the disease and the exposure is the main focus, the odds ratio parameter serves as a better measure of the association than the relative risk parameter as we argue in this book. This means that the odds ratio parameter is also useful in its own right even when a prospective study is conducted. In addition, the odds ratio parameter and its extension, the odds ratio function, are found to be useful in testing independence of random variables, in analyzing contingency tables, in handling biased sampling designs, and in modeling complex data. This book studies the theory and methods formulated around the concepts of the odds ratio parameter and the odds ratio function for measuring association and explores applications of the results to a variety of statistical problems in scientific research.

In the remainder of this chapter, we will examine the extensions of the odds ratio parameter to more complex data structure than the 2×2 table. Estimation approaches are also discussed along the line. The exposition will pave the way for generalizing the odds ratio parameter to the odds ratio function and helps understand the methods to be developed for estimation and inference on the odds ratio functions in subsequent chapters.

1.2 Odds ratio parameters in a $J \times K$ contingency table

The concept of using the odds ratio parameter in a 2×2 table to measure the association of the row and column random variables can be extended to contingency tables with arbitrary numbers of rows and columns. Let the column random variable and the row random variable of such a contingency table be respectively Y and X, where Y takes categorical values $1, \cdots, J$ and X takes categorical values $1, \cdots, K$. See Table 1.2.

Let the conditional probability of Y given X be

$$p_{jk} = P(Y = j \mid X = k),$$

TABLE 1.2
Counts in a $J \times K$ table.

	X				
Y	1	2	$\cdots$	K	Total
1	n_{11}	n_{12}	$\cdots$	n_{1K}	n_{1+}
2	n_{21}	n_{22}	$\cdots$	n_{2K}	n_{2+}
$\vdots$	$\vdots$	$\vdots$	$\vdots$	$\vdots$	$\vdots$
J	n_{J1}	n_{J2}	$\vdots$	n_{JK}	n_{J+}
Total	n_{+1}	n_{+2}	$\vdots$	n_{+K}	n_{++}

where $j = 1, \cdots, J; k = 1, \cdots, K$. Define the odds ratio parameters with respect to the reference point (J, K) as

$$\theta_{jk}(J, K) = \frac{p_{jk} p_{JK}}{p_{jK} p_{Jk}}, \tag{1.6}$$

for $j = 1, \cdots, J; k = 1, \cdots, K$. It can be easily seen from (1.6) that

$$\theta_{jK}(J, K) = \theta_{Jk}(J, K) = \theta_{JK}(J, K) = 1,$$

for $j = 1, \cdots, J; k = 1, \cdots, K$. As a result, the total number of non-trivial odds ratio parameters thus defined is $(J-1)(K-1)$. Other sets of odds ratio parameters can be defined using different reference points. For example, let $(1, 1)$ be the reference point. Define

$$\theta_{jk}(1, 1) = \frac{p_{jk} p_{11}}{p_{j1} p_{1k}},$$

for $j = 2, \cdots, J; k = 2, \cdots, K$. It can be seen that this set of odds ratio parameters are related to the set $\theta_{jk}(J, K)$ with $j = 1, \cdots, J-1; k = 1, \cdots, K-1$ through

$$\begin{aligned}\theta_{jk}(1, 1) &= \frac{p_{jk} p_{JK}}{p_{jK} p_{Jk}} \times \frac{p_{jK} p_{J1}}{p_{j1} p_{JK}} \times \frac{p_{1K} p_{Jk}}{p_{1k} p_{JK}} \times \frac{p_{JK} p_{11}}{p_{J1} p_{1K}} \\ &= \frac{\theta_{jk}(J, K)\theta_{11}(J, K)}{\theta_{j1}(J, K)\theta_{1k}(J, K)},\end{aligned} \tag{1.7}$$

for $j = 2, \cdots, J; k = 2, \cdots, K$. Because $\theta_{Jk}(J, K) = \theta_{jK}(J, K) = \theta_{JK}(J, K) = 1$ for $j = 1, \cdots, J-1; k = 1, \cdots, K-1$, simpler relationship can be derived from (1.7) for the following odds ratio parameters,

$$\begin{aligned}\theta_{JK}(1, 1) &= \theta_{11}(J, K), \\ \theta_{jK}(1, 1) &= \theta_{11}(J, K)/\theta_{j1}(J, K), \\ \theta_{Jk}(1, 1) &= \theta_{11}(J, K)/\theta_{1k}(J, K).\end{aligned}$$

Relationship (1.7) for the odds ratio parameters with respect to different reference points suggests that we only need to know a set of odds ratio parameters with a fixed reference point such as $\theta_{jk}(J, K)$ for $j = 1, \cdots, J-1; k = 1, \cdots, K-1$. All other odds ratio parameters formed in the contingency table can be expressed as functions of this set of odds ratio parameters. The maximum number of odds ratio parameters we need to know in a $J \times K$ contingency table is $(J-1)(K-1)$.

We have defined the odds ratio parameters in a $J \times K$ contingency table based on the conditional probabilities of the column random variable Y given the row random variable

X. As in a 2×2 contingency table for the disease and exposure association, the odds ratio parameters can also be defined using the conditional probabilities of the row random variable X given the column random variable Y. Let

$$q_{kj} = P(X = k \mid Y = j)$$

for $k = 1, \cdots, K; j = 1, \cdots, J$. Using the Bayesian formula for

$$q_{kj} = \frac{p_{jk} P(X = k)}{P(Y = j)},$$

it is easy to see that

$$\theta_{jk}(J, K) = \frac{q_{kj} q_{KJ}}{q_{kJ} q_{Kj}}, \tag{1.8}$$

for $k = 1, \cdots, K-1; j = 1, \cdots, J-1$. Since a sampling design that depends on the column random variable Y only does not change the conditional probability of X given Y. These odds ratio parameters are not affected by the sampling design that depends on the column random variable Y only. Likewise, the conditional probability of the column random variable Y given the row random variable X is not affected by the sampling design that depends on the row random variable X only. These odds ratio parameters are not affected by the sampling design that depends on the row random variable X only. Put together, the odds ratio parameters are invariant to sampling design either depends on the row random variable X alone or depends on the column random variable Y alone.

Another way to obtain the odds ratio parameters in a $J \times K$ contingency table is through the joint probability of the column random variable Y and the row random variable X. Let the joint probability be

$$h_{jk} = P(Y = j, X = k),$$

for $j = 1, \cdots, J$ and $k = 1, \cdots, K$. The joint probability can be expressed as the conditional probability times the marginal probability as

$$h_{jk} = p_{jk} P(X = k) = q_{kj} P(Y = j).$$

It is straightforward to see that

$$\theta_{jk}(J, K) = \frac{h_{jk} h_{JK}}{h_{jK} h_{Jk}}, \tag{1.9}$$

for $j = 1, \cdots, J-1; k = 1, \cdots, K-1$.

In summary, we have seen that all the odds ratio parameters in a $J \times K$ contingency table are determined by a set of odds ratio parameters that have a common reference point, such as $\{\theta_{jk}(J, K) \mid j = 1, \cdots, J-1; k = 1, \cdots, K-1\}$. This set of odds ratio parameters can be derived from the conditional probability $\{p_{jk} \mid j = 1, \cdots, J; k = 1, \cdots, K\}$ of the column random variable Y given the row random variable X, or the conditional probability $\{q_{kj} \mid k = 1, \cdots, K; j = 1, \cdots, J\}$ of the row random variable X given the column random variable Y, or from the joint probability $\{h_{jk} \mid j = 1, \cdots, J; k = 1, \cdots, K\}$ of the column random variable Y and the row random variable X. On the other hand, we demonstrate in the next section that these probability distributions can be represented in term of the set of odds ratio parameters plus some simpler distribution functions.

1.3 Odds ratio representations of conditional and joint distributions for a $J \times K$ contingency table

From the definition of the odds ratio parameters $\theta_{jk}(J,K)$ in (1.6), we can solve for p_{jk} as

$$p_{jk} = \frac{\theta_{jk}(J,K)p_{jK}p_{Jk}}{p_{JK}}. \tag{1.10}$$

By summing over j on both sides of the equation, we have

$$1 = \sum_{j=1}^{J} p_{jk} = \frac{p_{Jk}}{p_{JK}} \sum_{j=1}^{J} \theta_{jk}(J,K)p_{jK}.$$

This implies that

$$\frac{p_{JK}}{p_{Jk}} = \sum_{j=1}^{J} \theta_{jk}(J,K)p_{jK}.$$

Plugging this expression into the right-hand side of the equation (1.10), we obtain a representation of the conditional probability distribution $\{p_{jk} \mid j = 1,\cdots,J; k = 1,\cdots,K\}$ in terms of the set of odds ratio parameters $\{\theta_{jk}(J,K) \mid j = 1,\cdots,J-1; k = 1,\cdots,K-1\}$ and the conditional probability distribution at a fixed condition $\{p_{jK} \mid j = 1,\cdots,J\}$ as

$$p_{jk} = \frac{\theta_{jk}(J,K)p_{jK}}{\sum_{j=1}^{J} \theta_{jk}(J,K)p_{jK}}. \tag{1.11}$$

Applying the same operation to (1.8) and notice that

$$\frac{q_{KJ}}{q_{Kj}} = \sum_{k=1}^{K} \theta_{jk}(J,K)q_{kJ},$$

we can also represent the conditional probability distribution $\{q_{kj} \mid k = 1,\cdots,K; j = 1,\cdots,J\}$ in terms of the set of odds ratio parameters $\{\theta_{jk}(J,K) \mid j = 1,\cdots,J-1; k = 1,\cdots,K-1\}$ and the conditional probability distribution at a fixed condition $\{q_{kJ} \mid k = 1,\cdots,K\}$ as

$$q_{kj} = \frac{\theta_{jk}(J,K)q_{kJ}}{\sum_{k=1}^{K} \theta_{jk}(J,K)q_{kJ}}. \tag{1.12}$$

In addition, we can solve from (1.9) for h_{jk} as

$$h_{jk} = \frac{\theta_{jk}(J,K)h_{jK}h_{Jk}}{h_{JK}}. \tag{1.13}$$

By summing over j and k on both sides of the equation, we have

$$1 = \sum_{j=1}^{J}\sum_{k=1}^{K} h_{jk} = \frac{1}{h_{JK}} \sum_{j=1}^{J}\sum_{k=1}^{K} \theta_{jk}(J,K)h_{jK}h_{Jk}.$$

This implies that

$$h_{JK} = \sum_{j=1}^{J}\sum_{k=1}^{K} \theta_{jk}(J,K)h_{jK}h_{Jk}.$$

Plugging this expression into the right-hand side of the equation (1.13), we obtain a representation of the joint probability distribution $\{h_{jk} \mid j = 1, \cdots, J; k = 1, \cdots, K\}$ as

$$h_{jk} = \frac{\theta_{jk}(J,K) h_{jK} h_{Jk}}{\sum_{j=1}^{J} \sum_{k=1}^{K} \theta_{jk}(J,K) h_{jK} h_{Jk}}.$$

Since $h_{jK} = p_{jK} P(X = K)$ and $h_{Jk} = q_{kJ} P(Y = J)$. Replacing h_{jK} and h_{Jk} in both the numerator and the denominator on the right-hand side of the foregoing equation, we obtain a representation of the joint probability distribution $\{h_{jk} \mid j = 1, \cdots, J; k = 1, \cdots, K\}$ in terms of the set of odds ratio parameters $\{\theta_{jk}(J,K) \mid j = 1, \cdots, J-1; k = 1, \cdots, K-1\}$ and the first conditional probability distribution at a fixed condition $\{p_{jK} \mid j = 1, \cdots, J\}$ and the second conditional probability distribution at a fixed condition $\{q_{kJ} \mid k = 1, \cdots, K\}$ as

$$h_{jk} = \frac{\theta_{jk}(J,K) p_{jK} q_{kJ}}{\sum_{j=1}^{J} \sum_{k=1}^{K} \theta_{jk}(J,K) p_{jK} q_{kJ}}. \tag{1.14}$$

for$j = 1, \cdots, J; k = 1, \cdots, K$.

These representations for the conditional distributions and the joint distribution highlight the central role played by the odds ratio parameters in the prospective and respective conditional distributions, and the joint distribution. These representations can serve as modeling frameworks for estimating the odds ratio parameters and the relevant probability distributions. The fact that the same odds ratio parameters appear in the three different models suggests that the odds ratio parameters may be estimated based on the likelihoods derived from any one of the models with either the prospective sampling design or the retrospective sampling design.

1.4 Maximum likelihood estimators of odds ratio parameters in a $J \times K$ contingency table

To estimate the odds ratio parameters from the observed data in Table 1.2, likelihoods can be formed based on the representations given in the previous section. Depending on which representation is being used in forming the likelihoods, the likelihoods are respectively termed the prospective likelihood, the retrospective likelihood, or the joint likelihood when the representations (1.11), (1.12), and (1.14) are used. Since the odds ratio parameters are invariant to the retrospective sampling design, the likelihood estimators for the odds ratio parameters from maximizing any one of the likelihoods may be used whether the study design is prospective or retrospective.

The prospective likelihood under the observed data in Table 1.2 is

$$\prod_{j=1}^{J} \prod_{k=1}^{K} \left\{ \frac{\theta_{jk}(J,K) p_{jK}}{\sum_{j=1}^{J} \theta_{jk}(J,K) p_{jK}} \right\}^{n_{jk}}, \tag{1.15}$$

which can be rewritten as

$$\frac{\left\{ \prod_{j=1}^{J-1} \prod_{k=1}^{K-1} \theta_{jk}^{n_{jk}}(J,K) \right\} \prod_{j=1}^{J} p_{jK}^{n_{j+}}}{\prod_{k=1}^{K} \left\{ \sum_{j=1}^{J} \theta_{jk}(J,K) p_{jK} \right\}^{n_{+k}}}.$$

The log-likelihood for the set of parameters $\{\theta_{jk}(J,K), j=1,\cdots,J-1; k=1,\cdots,K-1\}$ and $\{p_{jK} \mid p_{jK} \geq 0, j=1,\cdots,J; \sum_{j=1}^{J} p_{jK}=1\}$ is

$$l = \sum_{j=1}^{J-1}\sum_{k=1}^{K-1} n_{jk}\log\theta_{jk}(J,K) + \sum_{j=1}^{J} n_{j+}\log p_{jK} - \sum_{k=1}^{K-1} n_{+k}\log\left\{p_{JK} + \sum_{j=1}^{J-1}\theta_{jk}(J,K)p_{jK}\right\}.$$

The Lagrange multiplier, $l_\lambda = l + \lambda\left(\sum_{j=1}^{J} p_{jK} - 1\right)$, can be used to relax the constraint $\sum_{j=1}^{J} p_{jK} = 1$. The score equations for the set of parameters are

$$\begin{aligned}
\frac{n_{jk}}{\theta_{jk}(J,K)} &= \frac{n_{+k}p_{jK}}{p_{JK} + \sum_{j=1}^{J-1}\theta_{jk}(J,K)p_{jK}}, && j=1,\cdots,J-1; k=1,\cdots,K-1,\\
\frac{n_{j+}}{p_{jK}} &= \sum_{k=1}^{K-1}\frac{n_{+k}\theta_{jk}(J,K)}{p_{JK} + \sum_{j=1}^{J-1}\theta_{jk}(J,K)p_{jK}} - \lambda, && j=1,\cdots,J,\\
1 &= \sum_{j=1}^{J} p_{jK}.
\end{aligned}$$

The first two equations can be respectively rewritten as

$$\begin{aligned}
\frac{n_{jk}}{n_{+k}} &= \frac{\theta_{jk}(J,K)p_{jK}}{p_{JK} + \sum_{j=1}^{J-1}\theta_{jk}(J,K)p_{jK}}, && j=1,\cdots,J-1; k=1,\cdots,K-1,\\
n_{j+} &= \sum_{k=1}^{K-1}\frac{n_{+k}\theta_{jk}(J,K)p_{jK}}{p_{JK} + \sum_{j=1}^{J-1}\theta_{jk}(J,K)p_{jK}} - \lambda p_{jK}, && j=1,\cdots,J.
\end{aligned}$$

Plug the first equation into the second one and sum over j to obtain $\lambda = -n_{+1}$. Furthermore, plug the first equation and $\lambda = n_{+K}$ back into the foregoing equation 2 to obtain

$$\hat{p}_{jK} = \frac{n_{jK}}{n_{+K}}, \qquad j=1,\cdots,J.$$

Note further that

$$\begin{aligned}
\frac{n_{Jk}}{n_{+k}} &= \frac{p_{JK}}{p_{JK} + \sum_{j=1}^{J-1}\theta_{jk}(J,K)p_{jK}}, && k=1,\cdots,K-1,\\
\frac{\theta_{jk}(J,K)}{\theta_{j_1k}(J,K)} &= \frac{n_{jk}n_{j_1K}}{n_{jK}n_{j_1k}}, && j,j_1=1,\cdots,J; k=1,\cdots,K-1,
\end{aligned}$$

Plug-in the maximum likelihood estimator for p_{jK}, it follows that

$$\hat{\theta}_{jk}(J,K) = \frac{n_{jk}n_{JK}}{n_{jK}n_{Jk}}, \qquad j=1,\cdots,J-1; k=1,\cdots,K-1.$$

The retrospective likelihood under the observed data in Table 1.2 is

$$\prod_{j=1}^{J}\prod_{k=1}^{K}\left\{\frac{\theta_{jk}(J,K)q_{kJ}}{\sum_{k=1}^{K}\theta_{jk}(J,K)q_{kJ}}\right\}^{n_{jk}}, \tag{1.16}$$

which can be rewritten as

$$\frac{\left\{\prod_{j=1}^{J-1}\prod_{k=1}^{K-1}\theta_{jk}^{n_{jk}}(J,K)\right\}\prod_{k=1}^{K} q_{kJ}^{n_{+k}}}{\prod_{j=1}^{J}\left\{\sum_{k=1}^{K}\theta_{jk}(J,K)q_{kJ}\right\}^{n_{j+}}}.$$

By similar derivation for obtaining the maximum prospective likelihood estimator for $\{\theta_{jk}(J,K), j=1,\cdots,J-1; k=1,\cdots,K-1\}$ and $\{p_{jK} \mid p_{jK} \geq 0, j=1,\cdots,J; \sum_{j=1}^{J} p_{jK} = 1\}$, it can be shown that the maximum retrospective likelihood estimator of $\{\theta_{jk}(J,K), j = 1,\cdots,J-1; k=1,\cdots,K-1\}$ and $\{q_{kJ} \mid q_{kJ} \geq 0, k=1,\cdots,K; \sum_{k=1}^{K} q_{kJ} = 1\}$ can be obtained as

$$\begin{aligned}\hat{q}_{kJ} &= \frac{n_{Jk}}{n_{J+}}, \qquad k=1,\cdots,K,\\ \hat{\theta}_{jk}(J,K) &= \frac{n_{jk}n_{JK}}{n_{jK}n_{Jk}}, \qquad j=1,\cdots,J-1; k=1,\cdots,K-1.\end{aligned}$$

The joint likelihood under the observed data in Table 1.2 is

$$\prod_{j=1}^{J}\prod_{k=1}^{K}\left\{\frac{\theta_{jk}(J,K)p_{jK}q_{kJ}}{\sum_{k=1}^{K}\sum_{j=1}^{J}\theta_{jk}(J,K)p_{jK}q_{kJ}}\right\}^{n_{jk}}, \tag{1.17}$$

which can be rewritten as

$$\frac{\left\{\prod_{j=1}^{J-1}\prod_{k=1}^{K-1}\theta_{jk}^{n_{jk}}(J,K)\right\}\prod_{k=1}^{K}q_{kJ}^{n_{+k}}\prod_{j=1}^{J}p_{jK}^{n_{j+}}}{\left\{\sum_{j=1}^{J}\sum_{k=1}^{K}\theta_{jk}(J,K)p_{jK}q_{kJ}\right\}^{n_{++}}}.$$

The corresponding log-likelihood for parameters $\{\theta_{jk}(J,K), j=1,\cdots,J-1; k=1,\cdots,K-1\}$, $\{p_{jK} \mid p_{jK} \geq 0, j = 1,\cdots,J; \sum_{j=1}^{J} p_{jK} = 1\}$, and$\{q_{kJ} \mid q_{kJ} \geq 0, k = 1,\cdots,K; \sum_{k=1}^{K} q_{kJ} = 1\}$ is

$$\begin{aligned} l &= \sum_{j=1}^{J-1}\sum_{k=1}^{K-1} n_{jk}\log\theta_{jk}(J,K) + \sum_{j=1}^{J} n_{j+}\log p_{jK} + \sum_{k=1}^{K} n_{+k}\log q_{kJ}\\ &\quad -n_{++}\log\left\{\sum_{j=1}^{J}\sum_{k=1}^{K}\theta_{jk}(J,K)p_{jK}q_{kJ}\right\}.\end{aligned}$$

To take care of the two constraints $\sum_{j=1}^{J} p_{jK} = 1$ and $\sum_{k=1}^{K} q_{kJ} = 1$, the Lagrange multipliers are used:

$$l_{\lambda,\mu} = l + \lambda\left(\sum_{j=1}^{J} p_{jK} - 1\right) + \mu\left(\sum_{k=1}^{K} q_{kJ} - 1\right).$$

The score equations for the set of parameters are

$$\begin{aligned}\frac{n_{jk}}{\theta_{jk}(J,K)} &= \frac{n_{++}p_{jK}q_{kJ}}{\sum_{j=1}^{J}\sum_{k=1}^{K}\theta_{jk}(J,K)p_{jK}q_{kJ}}, \qquad j=1,\cdots,J-1; k=1,\cdots,K-1,\\ \frac{n_{j+}}{p_{jK}} &= n_{++}\frac{\sum_{k=1}^{K}\theta_{jk}(J,K)q_{kJ}}{\sum_{j=1}^{J}\sum_{k=1}^{K}\theta_{jk}(J,K)p_{jK}q_{kJ}} - \lambda, \qquad j=1,\cdots,J,\\ 1 &= \sum_{j=1}^{J} p_{jK}\\ \frac{n_{+k}}{q_{kJ}} &= n_{++}\frac{\sum_{j=1}^{J}\theta_{jk}(J,K)p_{jK}}{\sum_{j=1}^{J}\sum_{k=1}^{K}\theta_{jk}(J,K)p_{jK}q_{kJ}} - \mu, \qquad k=1,\cdots,K,\\ 1 &= \sum_{k=1}^{K} q_{kJ}.\end{aligned}$$

Plug the first equations into the second and the fourth into the fifth to obtain $\lambda = \mu = 0$. The score equation can be reduced to

$$\begin{aligned}
\frac{n_{jk}}{n_{++}} &= \frac{\theta_{jk}(J,K)p_{jK}q_{kJ}}{\sum_{j=1}^{J}\sum_{k=1}^{K}\theta_{jk}(J,K)p_{jK}q_{kJ}}, & j=1,\cdots,J-1; k=1,\cdots,K-1,\\
\frac{n_{j+}}{n_{++}} &= \frac{\sum_{k=1}^{K}\theta_{jk}(J,K)p_{jK}q_{kJ}}{\sum_{j=1}^{J}\sum_{k=1}^{K}\theta_{jk}(J,K)p_{jK}q_{kJ}}, & j=1,\cdots,J,\\
\frac{n_{+k}}{n_{++}} &= \frac{\sum_{j=1}^{J}\theta_{jk}(J,K)p_{jK}q_{kJ}}{\sum_{j=1}^{J}\sum_{k=1}^{K}\theta_{jk}(J,K)p_{jK}q_{kJ}}, & k=1,\cdots,K.
\end{aligned}$$

Comparing the first and the second, and the first and the third equations leads to

$$\frac{n_{jk}}{n_{++}} = \frac{\theta_{jk}(J,K)p_{jK}q_{kJ}}{\sum_{j=1}^{J}\sum_{k=1}^{K}\theta_{jk}(J,K)p_{jK}q_{kJ}}, \qquad j=1,\cdots,J; k=1,\cdots,K.$$

It can now be derived that the maximum likelihood estimators for parameters $\{p_{jK} \mid p_{jK} \geq 0, j=1,\cdots,J; \sum_{j=1}^{J} p_{jK} = 1\}$, $\{q_{kJ} \mid q_{kJ} \geq 0, k=1,\cdots,K; \sum_{k=1}^{K} q_{kJ} = 1\}$, and $\{\theta_{jk}(J,K), j=1,\cdots,J-1; k=1,\cdots,K-1\}$ are respectively,

$$\begin{aligned}
\hat{p}_{jK} &= \frac{n_{jK}}{n_{+K}}, & j=1,\cdots,J,\\
\hat{q}_{kJ} &= \frac{n_{Jk}}{n_{J+}}, & k=1,\cdots,K,\\
\hat{\theta}_{jk}(J,K) &= \frac{n_{jk}n_{JK}}{n_{jK}n_{Jk}}, & j=1,\cdots,J-1; k=1,\cdots,K-1.
\end{aligned}$$

The above derivations show that the maximum likelihood estimators for odds ratio parameters $\{\theta_{jk}(J,K), j=1,\cdots,J-1; k=1,\cdots,K-1\}$ from the prospective likelihood, the retrospective likelihood, or the joint likelihood are the same. Although the model for the $J \times K$ contingency table discussed in this section is special in that it is a saturated model, as will be seen in the later chapters, this is always the case even when the model for the odds ratio parameters is not saturated as long as the parameters for the conditional models p_{jK} and q_{kJ} are saturated and are variation independent of the odds ratio parameters.

For the $J \times K$ table, a direct use of the conditional probability models or the joint probability model for obtaining the maximum likelihood estimators of the odds ratio parameters is simpler. However, the way we model and estimate the odds ratio parameters provides a prototype for the generalization of models with explicit odds ratio parameters to problems with more complex structure. For example, when Y is an ordinal variable, we may model the odds ratio parameters parametrically as

$$\log \theta_{jk}(J,K) = \gamma_k(J-j),$$

for $k=1,\cdots,K-1; j=1,\cdots,J-1$ and $\gamma_K \equiv 0$. Estimation and inference may be proceeded using the likelihoods. In addition to handling various models for the odds ratio parameters, forming the likelihoods based on the odds ratio representations often enables us to ignore the difference between prospective and retrospective sampling designs.

1.5 Link to logit model and logistic regression

When we treat the column random variable Y as outcome and the row random variable X as covariate, the odds ratio representation of the conditional distribution of Y given X connects with the logit regression model. Let $Y = 1$ be the baseline category in a logit regression model. The odds ratio representation

$$p_{jk} = \frac{\theta_{jk}(J,K)p_{jK}}{\sum_{j=1}^{J}\theta_{jk}(J,K)p_{jK}}$$

can be expressed as

$$\log\frac{p_{jk}}{p_{Jk}} = \log\frac{p_{jK}}{p_{JK}} + \log\theta_{jk}(J,K).$$

Let

$$\begin{aligned}\beta_{j0} &= \log\frac{p_{jK}}{p_{JK}},\\ \beta_{jk} &= \log\theta_{jk}(J,K).\end{aligned}$$

The representation of the conditional distribution can be transformed into a baseline logit model

$$\log\frac{p_{jk}}{p_{Jk}} = \beta_{j0} + \beta_{jk},$$

for $j = 1, \cdots, J-1$. When $J = 2$, the model becomes a logistic regression model

$$\log\frac{p_{1k}}{p_{2k}} = \beta_0 + \beta_k,$$

where $\beta_0 = \beta_{10}$ and $\beta_k = \beta_{1k}, k = 1, \cdots, K-1$. It is well known the regression parameter β_k in a logistic regression model corresponds to the log-odds ratio parameter.

Both the logit model and the logistic regression model can have more than one covariates in the model. They can be fitted in a general framework of the odds ratio representation discussed in the next chapter where an odds function that generalizes the odds ratio parameter is introduced. We discuss the generalization of the odds ratio parameter from a different perspective in the remaining sections of this chapter.

1.6 Odds ratio parameters in stratified 2×2 tables

In studying the association of exposure to a risk factor and disease, epidemiologists often need to control possible confounders. In such a situation, the study population is usually divided into subpopulations defined by the values of the confounders. These subpopulations are also termed stratified subpopulations. Let Z denote the stratum indicator. Suppose that M strata are under consideration. They are labelled as $Z = 1, \cdots, M$ respectively. Within each stratum, a sample is drawn and the data are summarized in a 2×2 table. As a result, a series of M 2×2 tables are obtained. See Table 1.3 for the data from the stratum $Z = m$.

Let D denote the disease status and E denote the exposure status. The probabilities of having or not having disease for a subject in stratum m when exposed to the risk factor are respectively

$$\begin{aligned}p_{11m} &= P(D = \text{Yes} \mid E = \text{Yes}, Z = m),\\ p_{21m} &= P(D = \text{No} \mid E = \text{Yes}, Z = m).\end{aligned}$$

TABLE 1.3
Counts on disease and exposure status for stratum $Z = m$.

Disease	Exposure Yes	Exposure No	Total
Yes	n_{11m}	n_{12m}	n_{1+m}
No	n_{21m}	n_{22m}	n_{2+m}
Total	n_{+1m}	n_{+2m}	n_{++m}

The odds of having disease for a subject in stratum m when exposed to the risk is

$$\text{Odds}(E = \text{Yes} \mid Z = m) = \frac{p_{11m}}{p_{21m}}.$$

Similarly, the probabilities of having or not having disease for a subject in stratum m when not exposed to the risk are respectively

$$\begin{aligned} p_{12m} &= P(D = \text{Yes} \mid E = \text{No}, Z = m), \\ p_{22m} &= P(D = \text{No} \mid E = \text{No}, Z = m). \end{aligned}$$

The odds of having disease for a subject in stratum m when exposed to the risk is

$$\text{Odds}(E = \text{No} \mid Z = m) = \frac{p_{12m}}{p_{22m}}.$$

The odds ratio for stratum m is thus

$$OR(Z = m) = \frac{p_{11m}p_{22m}}{p_{12m}p_{21m}},$$

for $m = 1, \cdots, M$. We call $OR(Z = m)$ the conditional odds ratio with condition $Z = m$.

Suppose the reference point $(2, 2)$ is chosen. Denote the $OR(Z = m)$ by $\theta_{11}(2, 2)(m)$. Applying the odds ratio representation (1.11) to the conditional probabilities of disease given exposure in stratum m, we have

$$p_{jkm} = \frac{\theta_{jk}(2,2)(m)p_{j2m}}{\sum_{j=1}^{2} \theta_{jk}(2,2)(m)p_{j2m}}, \tag{1.18}$$

for $j = 1, 2; k = 1, 2$, and $m = 1, \cdots, M$. Similarly, let

$$\begin{aligned} q_{11m} &= P(E = \text{Yes} \mid D = \text{Yes}, Z = m), \\ q_{21m} &= P(E = \text{No} \mid D = \text{Yes}, Z = m), \\ q_{12m} &= P(E = \text{Yes} \mid D = \text{No}, Z = m), \\ q_{22m} &= P(E = \text{No} \mid D = \text{No}, Z = m). \end{aligned}$$

Applying the odds ratio representation (1.12) of the conditional probabilities of exposure given disease in stratum m, we have

$$q_{kjm} = \frac{\theta_{jk}(2,2)(m)q_{k2m}}{\sum_{k=1}^{2} \theta_{jk}(2,2)(m)q_{k2m}}, \tag{1.19}$$

for $k = 1, 2; j = 1, 2$, and $m = 1, \cdots, M$. Finally, let

$$\begin{aligned} h_{11m} &= P(D = \text{Yes}, E = \text{Yes} \mid Z = m), \\ h_{21m} &= P(D = \text{No}, E = \text{Yes} \mid Z = m), \\ h_{12m} &= P(D = \text{Yes}, E = \text{No} \mid Z = m), \\ h_{22m} &= P(D = \text{No}, E = \text{No} \mid Z = m). \end{aligned}$$

Applying the odds ratio representation (1.14) of the joint probabilities of disease and exposure in stratum m, we have

$$h_{kjm} = \frac{\theta_{jk}(2,2)(m) p_{j2m} q_{k2m}}{\sum_{j=1}^{2} \sum_{k=1}^{2} \theta_{jk}(2,2)(m) p_{j2m} q_{k2m}}, \tag{1.20}$$

for $k = 1, 2; j = 1, 2$, and $m = 1, \cdots, M$.

The odds ratio representations (1.18), (1.19), and (1.20) highlight the connections between the conditional probability distributions and the joint probability distribution through the conditional odds ratio parameters $\theta_{11}(2,2)(m), m = 1, \cdots, M$. The conditional odds ratio parameters can in principle be estimated by maximizing the prospective, retrospective, or the joint likelihoods as follow,

$$\begin{aligned} PL &= \prod_{m=1}^{M} \prod_{j=1}^{2} \prod_{k=1}^{2} \left\{ \frac{\theta_{jk}(2,2)(m) p_{j2m}}{\sum_{j=1}^{2} \theta_{jk}(2,2)(m) p_{j2m}} \right\}^{n_{jkm}}, \\ RL &= \prod_{m=1}^{M} \prod_{j=1}^{2} \prod_{k=1}^{2} \left\{ \frac{\theta_{jk}(2,2)(m) q_{k2m}}{\sum_{k=1}^{2} \theta_{jk}(2,2)(m) q_{k2m}} \right\}^{n_{jkm}}, \\ JL &= \prod_{m=1}^{M} \prod_{j=1}^{2} \prod_{k=1}^{2} \left\{ \frac{\theta_{jk}(2,2)(m) p_{j2m} q_{k2m}}{\sum_{j=1}^{2} \sum_{k=1}^{2} \theta_{jk}(2,2)(m) p_{j2m} q_{k2m}} \right\}^{n_{jkm}}. \end{aligned}$$

1.7 Common odds ratio parameter

In epidemiological studies, tables in the strata may have small counts, making the estimation of the odds ratio parameters from individual tables unreliable. It is often reasonable to assume the conditional odds ratio parameters between disease and exposure are the same after the confounders are controlled. That is, the odds ratio parameters are homogeneous across all the strata: $\theta_{11}(2,2)(1) = \cdots = \theta_{11}(2,2)(M)$. For simplicity, we use θ to denote the common odds ratio parameter in the following.

The likelihoods for the common odds ratio parameter reduce to

$$PL = \prod_{m=1}^{M} \frac{\theta^{n_{11m}} \prod_{j=1}^{2} p_{j2m}^{n_{j+m}}}{\{p_{22m} + \theta p_{12m}\}^{n_{++m}}}, \tag{1.21}$$

$$RL = \prod_{m=1}^{M} \frac{\theta^{n_{11m}} \prod_{k=1}^{2} q_{k2m}^{n_{+km}}}{\{q_{22m} + \theta q_{12m}\}^{n_{++m}}}, \tag{1.22}$$

$$JL = \prod_{m=1}^{M} \frac{\theta^{n_{11m}} \prod_{j=1}^{2} p_{j2m}^{n_{j+m}} \prod_{k=1}^{2} q_{k2m}^{n_{+km}}}{\{p_{22m} q_{22m} + p_{12m} q_{22m} + p_{22m} q_{12m} + \theta p_{12m} q_{12m}\}^{n_{++m}}}. \tag{1.23}$$

To estimate the common odds ratio parameter by maximizing any one of the likelihoods, we still need to deal with the nuisance distributions $p_{j2m}, j = 1, 2; m = 1, \cdots, M$ or $q_{k2m}, k = 1, 2; m = 1, \cdots, M$. Estimation of them is hampered by possible small cell counts in the tables. These nuisance parameters can be eliminated by conditioning on the order statistics. For a given stratum $Z = m$, conditioning on the order statistic of the column random variable Y or the order statistic of the row random variable X within a stratum corresponds to permuting the observed Y or X within a stratum respectively. Both lead to the same conditional distribution within a stratum, which can be used to form a conditional likelihood for estimating θ. Conditioning on the order statistic within a stratum is equivalent to conditioning on the observed margins of the 2×2 table from that stratum. The conditional distribution for the stratum $Z = m$ is

$$P(N_{jkm} = n_{jkm}, j = 1, 2; k = 1, 2 \mid N_{j+m} = n_{j+m}, j = 1, 2; N_{+km} = n_{+km}, k = 1, 2) \propto \frac{\theta^{n_{11m}}}{\sum_{\sigma} \theta^{n_{11m}(\sigma)}}, \tag{1.24}$$

where N_{jkm}, a random variable, denotes the cell count for the jth row and the kth column in stratum m, and $n_{11m}(\sigma)$ is the cell count for the cell in the first row and the first column of the table in stratum m after the permutation σ over the indices of all the individually observed disease outcome Y is applied.

To obtain a more explicit formula for the denominator of (1.24), note that the observed number of subjects having the disease in the sample in stratum m is n_{1+m} and the observed number of subjects having exposed to the risk in the sample in stratum m is n_{+1m}. The observed subjects having disease and also exposed to the risk in stratum m is n_{11m}. The number of different permutations that make the number of subjects having disease and exposed to the risk in stratum m exactly t is

$$\binom{n_{1+m}}{t}\binom{n_{2+m}}{n_{+1m} - t},$$

i.e., choose t from n_{1+m} diseased subjects in stratum m to have exposure to the risk and choose the rest $n_{+1m} - t$ exposed subjects from the n_{2+m} non-diseased subjects. The following diagram illustrates the combinatorial number.

$$\begin{array}{ccc} Y & \overbrace{1, \cdots, 1, 1, \cdots, 1,}^{n_{1+m}} & \overbrace{2, \cdots, 2, 2 \cdots, 2}^{n_{2+m}} \\ X & \underbrace{1, \cdots, 1,}_{t} 2, \cdots, 2, & \underbrace{1, \cdots, 1,}_{n_{+1m}-t} 2 \cdots, 2 \end{array}$$

The number t needs to satisfies $0 \le t \le n_{1+m}$, $t \le n_{+1m}$, and $n_{2+m} \ge n_{+1m} - t$. Put together, they restrict t in the range

$$\mathcal{T} = \{t \mid \max(0, n_{+1m} - n_{2+m}) \le t \le \min(n_{1+m}, n_{+1m})\}$$

The conditional likelihood for estimating the common odds ratio parameter can be written as

$$\prod_{m=1}^{M} \frac{\binom{n_{1+m}}{n_{11m}}\binom{n_{2+m}}{n_{+1m} - n_{11m}} \theta^{n_{11m}}}{\sum_{t \in \mathcal{T}} \binom{n_{1+m}}{t}\binom{n_{2+m}}{n_{+1m} - t} \theta^{t}}. \tag{1.25}$$

Estimation and inference on the common odds ratio parameter can be carried out by maximizing the conditional likelihood (1.25). Computationally, maximizing the conditional likelihood for θ may not be easily obtained when n_{1+m} and n_{2+m} are not small. Geyer and

Thompson (1992)[69], Diaconis and Sturmfels (1998)[55], and Mehta et al (2000)[124] provided Monte Carlo methods to approximate the likelihood. The Monte Carlo methods will be discussed in detail in Chapter 3.

In deriving the conditional likelihood (1.25), we assumed that the odds ratio parameters are homogeneous across different strata. The likelihood can also be used to estimate and make inference on the odds ratio parameters when they are not homogeneous, such as testing the hypothesis that the odds ratio parameters are homogeneous. To increase the power of such a test, we may model the odds ratio parameters across different strata. For example, if Z is an ordinal variable, one sensible model is

$$\log \theta(Z) = \gamma_0 + \gamma_1 Z.$$

Testing homogeneity of the odds ratio parameters under this model becomes testing the hypothesis $H_0 : \gamma_1 = 0$. Likelihood ratio test, likelihood score test, or Wald test may be performed.

1.8 Odds ratio representations for a $J \times K \times M$ contingency table

The results for a $2 \times 2 \times M$ table can be extended to a $J \times K \times M$ contingency table. Data in a $J \times K \times M$ contingency table can be presented in a series of $J \times K$ tables for each level of the third variable. See Table 1.4. Let the column random variable be Y, the row random variable be X, and the third variable be Z, where $Y = 1, \cdots, J$, $X = 1, \cdots, K$, and $Z = 1, \cdots, M$.

TABLE 1.4
Counts in the $J \times K$ table for $Z = m$.

	X				
Y	1	2	$\cdots$	K	Total
1	n_{11m}	n_{12m}	$\cdots$	n_{1Km}	n_{1+m}
2	n_{21m}	n_{22m}	$\cdots$	n_{2Km}	n_{2+m}
$\vdots$	$\vdots$	$\vdots$	$\vdots$	$\vdots$	$\vdots$
J	n_{J1m}	n_{J2m}	$\vdots$	n_{JKm}	n_{J+m}
Total	n_{+1m}	n_{+2m}	$\vdots$	n_{+Km}	n_{++m}

Let the cell probabilities be

$$g_{jkm} = P(Y = j, X = k, Z = m), \tag{1.26}$$

for $j = 1, \cdots, J; k = 1, \cdots, K; m = 1, \cdots, M$. Let the conditional probabilities and the joint conditional probabilities be respectively

$$\begin{aligned} p_{jkm} &= P(Y = j \mid X = k, Z = m), \\ q_{kjm} &= P(X = k \mid Y = j, Z = m), \\ h_{jkm} &= P(Y = j, X = k \mid Z = m). \end{aligned}$$

Let J, K be the reference categories for Y and X respectively. Given $Z = m$, define the conditional odds ratio parameter as

$$\theta_{jk}(J,K)(m) = \frac{p_{jkm}p_{JKm}}{p_{jKm}p_{Jkm}},$$

for $j = 1, \cdots, J-1; k = 1, \cdots, K-1$. For notational simplicity, we also use it when $j = J$ or $k = K$ with the understanding that the odds ratios are 1. As is seen in the previous section, these conditional odds ratio parameters can be equivalently defined using the conditional probabilities $q_{kjm}, k = 1, \cdots, K; j = 1, \cdots, M$, or the joint conditional probabilities $h_{jkm}, j = 1, \cdots, J; k = 1, \cdots, K$. By similar arguments used in section 1.3, we can obtain the odds ratio representations of these conditional probabilities as

$$p_{jkm} = \frac{\theta_{jk}(J,K)(m)p_{jKm}}{\sum_{j=1}^{J}\theta_{jk}(J,K)(m)p_{jKm}}, \tag{1.27}$$

$$q_{kjm} = \frac{\theta_{jk}(J,K)(m)q_{kJm}}{\sum_{k=1}^{K}\theta_{jk}(J,K)(m)q_{kJm}}, \tag{1.28}$$

$$h_{jkm} = \frac{\theta_{jk}(J,K)(m)p_{jKm}q_{kJm}}{\sum_{j=1}^{J}\sum_{k=1}^{K}\theta_{jk}(J,K)(m)p_{jKm}q_{kJm}}. \tag{1.29}$$

As was done in the previous section, the odds ratio representations (1.27), (1.28), and (1.29) can be respectively used to form the prospective, retrospective, and the joint likelihoods as follow.

$$PL = \prod_{m=1}^{M}\prod_{j=1}^{J}\prod_{k=1}^{K}\left\{\frac{\theta_{jk}(J,K)(m)p_{jKm}}{\sum_{j=1}^{J}\theta_{jk}(J,K)(m)p_{jKm}}\right\}^{n_{jkm}},$$

$$RL = \prod_{m=1}^{M}\prod_{j=1}^{J}\prod_{k=1}^{K}\left\{\frac{\theta_{jk}(J,K)(m)q_{kJm}}{\sum_{k=1}^{K}\theta_{jk}(J,K)(m)q_{kJm}}\right\}^{n_{jkm}},$$

$$JL = \prod_{m=1}^{M}\prod_{j=1}^{J}\prod_{k=1}^{K}\left\{\frac{\theta_{jk}(J,K)(m)p_{jKm}q_{kJm}}{\sum_{j=1}^{J}\sum_{k=1}^{K}\theta_{jk}(J,K)(m)p_{jKm}q_{kJm}}\right\}^{n_{jkm}}.$$

For a given stratum $Z = m$, to eliminate the nuisance distributions $\{p_{jKm}, j = 1, \cdots, J\}$ and/or $\{q_{kJm}, k = 1, \cdots, K\}$ from the likelihood, the order statistic for Y or the order statistic for X in the given stratum is conditioned on, which is equivalent to permuting the observed Y for all the subjects. This leads to the conditional probability of observing the data in stratum m given the order statistic of Y and the order statistic of X to be proportional to

$$\frac{\prod_{j=1}^{J-1}\prod_{k=1}^{K-1}\{\theta_{jk}(J,K)(m)\}^{n_{jkm}}}{\sum_{\sigma}\prod_{j=1}^{J-1}\prod_{k=1}^{K-1}\{\theta_{jk}(J,K)(m)\}^{n_{jkm}(\sigma)}},$$

where σ is a permutation of individual Y from all subjects in the sample of $Z = m$ stratum, $n_{jkm}(\sigma), j = 1, \cdots, J, k = 1, \cdots, K$ are the cell counts after the application of the permutation σ, and the summation in the denominator is over all possible such permutations.

For given margins of Y in a stratum, i.e., $n_{1+m}, \cdots, n_{J+m}$, the number of different permutations that do not change the marginal is $\prod_{j=1}^{J-1} n_{j+m}!$, where $n_{j+m}!$ is the factorial of n_{j+m}. This number is the same for both the numerator and the denominator, and is therefore cancelled in the conditional probability expression. For a given stratum, the total number of permutations that generate a particular table with cell counts $t_{jkm}, j = 1, \cdots, J; k = 1, \cdots, K$ in stratum m and with the same margins as the observed table, i.e., $t_{1+m} =$

$n_{1+m}, \cdots, t_{J+m} = n_{J+m}; t_{+1m} = n_{+1m}, \cdots, t_{+Km} = n_{+km}$, is $\prod_{j=1}^{J-1} n_{j+m}! \prod_{k=1}^{K-1} n_{+km}!$ times

$$\binom{n_{++m}}{t_{11m}, \cdots, t_{JKm}} = \frac{n_{++m}!}{\prod_{j=1}^{J} \prod_{k=1}^{K} t_{jkm}!}.$$

Let

$$\mathcal{T} = \{t = (t_{11m}, \cdots, t_{JKm}) \mid t_{j+m} = n_{j+m}, j = 1, \cdots, J; t_{+km} = n_{+km}, k = 1, \cdots, K\}.$$

The conditional likelihood is thus

$$\prod_{m=1}^{M} \frac{\binom{n_{++m}}{n_{11m}, \cdots, n_{JKm}} \prod_{j=1}^{J-1} \prod_{k=1}^{K-1} \{\theta_{jk}(J,K)(m)\}^{n_{jkm}}}{\sum_{t \in \mathcal{T}} \binom{n_{++m}}{t_{11m}, \cdots, t_{JKm}} \prod_{j=1}^{J-1} \prod_{k=1}^{K-1} \{\theta_{jk}(J,K)(m)\}^{n_{jkm}(\sigma)}}. \tag{1.30}$$

To reduce the computation burden in finding the combinatorial numbers, we may divide both the numerator and the denominator by

$$\binom{n_{++m}}{n_{1+m}, \cdots, n_{J+m}} = \frac{n_{++m}!}{\prod_{j=1}^{J} n_{j+m}!}.$$

The conditional likelihood reduces to

$$\prod_{m=1}^{M} \frac{\prod_{j=1}^{J} \binom{n_{j+m}}{n_{j1m}, \cdots, n_{jKm}} \prod_{j=1}^{J-1} \prod_{k=1}^{K-1} \{\theta_{jk}(J,K)(m)\}^{n_{jkm}}}{\sum_{t \in \mathcal{T}} \prod_{j=1}^{J} \binom{n_{j+m}}{t_{j1m}, \cdots, t_{jKm}} \prod_{j=1}^{J-1} \prod_{k=1}^{K-1} \{\theta_{jk}(J,K)(m)\}^{n_{jkm}(\sigma)}}. \tag{1.31}$$

The conditional likelihood (1.31) becomes (1.25) when $J = K = 2$. Alternatively, we may divide both the numerator and the denominator by

$$\binom{n_{++m}}{n_{+1m}, \cdots, n_{+Km}} = \frac{n_{++m}!}{\prod_{k=1}^{J} n_{+km}!}.$$

The conditional likelihood reduces to

$$\prod_{m=1}^{M} \frac{\prod_{k=1}^{K} \binom{n_{+km}}{n_{1km}, \cdots, n_{Jkm}} \prod_{j=1}^{J-1} \prod_{k=1}^{K-1} \{\theta_{jk}(J,K)(m)\}^{n_{jkm}}}{\sum_{t \in \mathcal{T}} \prod_{k=1}^{K} \binom{n_{+km}}{t_{1km}, \cdots, t_{Jkm}} \prod_{j=1}^{J-1} \prod_{k=1}^{K-1} \{\theta_{jk}(J,K)(m)\}^{n_{jkm}(\sigma)}}. \tag{1.32}$$

The conditional likelihood approach to estimating the odds ratio parameters may be computationally difficult to implement when the marginal counts are large. In this case, the prospective likelihood, the retrospective likelihood or the joint likelihood can be used to jointly maximize all the model parameters to obtain an estimator of the odds ratio parameters. We may have that the margins $n_{1+m}, \cdots, n_{J+m}$ and $n_{+1m}, \cdots, n_{+Km}$ may not be all large, but the margins $n_{1++}, \cdots, n_{J++}$ and $n_{+1+}, \cdots, n_{+K+}$ are large. In those cases, we can implement the maximum likelihood approach by further modeling the conditional distribution $\{p_{jKm}, j = 1, \cdots, J; m = 1, \cdots, M\}$ and $\{q_{Jkm}, k = 1, \cdots, K; m = 1, \cdots, M\}$. Define additional odds ratio parameters as

$$\begin{aligned} \omega_{jm}(J,M) &= \frac{p_{jKm} p_{JKM}}{p_{jKM} p_{JKm}}, \\ \delta_{km}(K,M) &= \frac{q_{kJm} q_{KJM}}{q_{kJM} q_{KJm}}. \end{aligned}$$

Apply the same idea of odds ratio representation of distrbutions, we can express

$$\begin{aligned} p_{jKm} &= \frac{\omega_{jm}(J,M)p_{jKM}}{\sum_{j=1}^{J}\omega_{jm}(J,M)p_{jKM}}, \\ q_{kJm} &= \frac{\delta_{km}(K,M)q_{kJM}}{\sum_{k=1}^{K}\delta_{km}(K,M)q_{kJM}}. \end{aligned}$$

plug-in the additional representations respectively to (1.27), (1.28), and (1.29), we obtain new representations for the conditional probabilities as

$$p_{jkm} = \frac{\theta_{jk}(J,K)(m)\omega_{jm}(J,M)p_{jKM}}{\sum_{j=1}^{J}\theta_{jk}(J,K)(m)\omega_{jm}(J,M)p_{jKM}}, \tag{1.33}$$

$$q_{kjm} = \frac{\theta_{jk}(J,K)(m)\delta_{km}(K,M)q_{kJM}}{\sum_{k=1}^{K}\theta_{jk}(J,K)(m)\delta_{km}(K,M)q_{kJM}}, \tag{1.34}$$

$$h_{jkm} = \frac{\theta_{jk}(J,K)(m)\delta_{km}(K,M)\omega_{jm}(J,M)p_{jKM}q_{kJM}}{\sum_{j=1}^{J}\sum_{k=1}^{K}\theta_{jk}(J,K)(m)\delta_{km}(K,M)\omega_{jm}(J,M)p_{jKM}q_{kJM}}. \tag{1.35}$$

The probabilities $\{p_{jKM}, j = 1, \cdots, J\}$ and $\{q_{kJM}, k = 1, \cdots, K\}$ may be well estimated when we model the odds ratio parameters $\{\theta_{jk}(J,K)(m), j = 1, \cdots, J-1; k = 1, \cdots, K-1; m = 1, \cdots, M\}$, $\{\omega_{jm}(J,M), j = 1, \cdots, J-1, m = 1, \cdots, M-1\}$, and $\{\delta_{km}(K,M), k = 1, \cdots, K-1; m = 1, \cdots, M-1\}$. For example, when Y, X, and Z are all ordinal variable, we may model the odds ratio parameters as

$$\begin{aligned} \log\theta_{jk}(J,K)(m) &= \gamma_0(j-J)(k-K) + \gamma_1 m(j-J)(k-K), \\ \log\omega_{jm}(J,M) &= \alpha(j-J)(m-M), \\ \log\delta_{km}(K,M) &= \beta(k-K)(m-M). \end{aligned}$$

Other models for the log-odds ratio parameters may be considered. The general rules for modeling odds ratio parameters will be systematically discussed in the next chapter.

1.9 Summary and discussion

In this Chapter, we have introduced the odds ratio representations for cell probabilities of contingency tables up to three-dimensions. Strategies for modeling the odds ratio parameters are also briefly discussed. One advantage of such parametrization of the probability model is that the same odds ratio parameter appears in the prospective likelihood, retrospective likelihood, and the joint likelihood. This feature suggests that the odds ratio parameter is invariant to the retrospective sampling design. As a result, we may estimate the odds ratio parameter using any one of the likelihoods without the need to pay particular attention to the retrospective sampling designs.

We have not discussed how to represent probability distributions in a contingency table of dimension higher than 3 although the representations in a three-dimensional contingency table shed some lights on what we can do for the higher dimensional contingency tables. Even for the 3-dimensional table, we still have something left untreated. Namely, we did not provide an odds ratio representation for the joint distribution g_{jkm} in Section 1.7. To deal with these more complex cases, we need to generalize the concept of odds ratio parameter to odds ratio function. This will be done in the next chapter. Once the concept is extended, we will be able to represent cell probabilities in any contingency table in terms of odds ratio functions plus certain simpler distributions. The odds ratio function extension also allows us to decompose the density of a continuous distribution into odds ratio functions plus conditional distributions at a fixed condition, which facilitates modeling odds ratio functions for estimation and inference.

Odds ratio parameter has been widely used to characterize disease risk in epidemiology and biostatistics. Definition and interpretation of odds ratio parameter are common sense for epidemiologists and other health researchers. It is hard to exhausts the long list of references on the estimation and inference of odds ratio parameters. One of the reasons for its popularity is its invariant to the case-control design (Cole, 1979)[45]. See Breslow (1996)[14] for a history on this topic. Important milestones in epidemiological and biostatistical development include Cornfield (1951)[46], Mantel and Haenszel (1959)[121], Prentice and Pyke (1979)[141]. Other important developments include Cochran (1954)[44], Breslow et al (1978)[15], Manski and Lerman (1977)[119], Cosslett(1981)[47], Davis (1985)[49], Liang (1984, 1985)[100, 101], Robins et al (1986a, 1986b)[150, 151], Philips and Holland (1987) [139], Scott and Wild (1986, 1991, 1997) [161, 162, 163], Carroll et al (1995)[21], Lawless et al (1999) [93], Liang and Qin (2000) [103] among others. A more comprehensive reference list can be found in Breslow (1996) [14] and in the reference therein.

1.10 Exercises

1. For the observed data in Table 1.2, the maximum likelihood estimator of the odds ratio parameter is
$$\hat{\theta} = \frac{n_{11}n_{22}}{n_{12}n_{21}}. \tag{1.36}$$
Use the likelihood based on the odds ratio representation to derive the variance estimate for the odds ratio parameter estimator. i.e., $\widehat{\mathrm{var}}(\hat{\theta}) = \hat{\theta}^2\hat{V}$ where
$$\hat{V} = \frac{1}{n_{11}} + \frac{1}{n_{12}} + \frac{1}{n_{21}} + \frac{1}{n_{22}}, \tag{1.37}$$
which is a consistent estimator of the variance of $\log\hat{\theta}$.

2. For a $J \times K$ table, the maximum likelihood estimator of the odds ratio parameter $\theta_{jk}(J,K)$ is
$$\hat{\theta}_{jk}(J,K) = \frac{n_{jk}n_{JK}}{n_{jK}n_{Jk}}.$$
Find the variance-covariance estimate for $\hat{\theta}_{jk}(J,K), j = 1,\cdots,J-1; k = 1,\cdots,K-1$.

3. For the following observed 2×2 table,
   (a) Find all tables with the fixed marginal counts as those in Table 1.5.
   (b) Estimate the odds ratio parameter by maximizing the likelihood conditional on fixed marginal counts.
   (c) Find a variance estimate of the above maximum likelihood estimator.
   (d) Test the hypothesis that there is no association between disease and exposure.

TABLE 1.5
Counts of disease and exposure.

Disease	Exposure Yes	Exposure No	Total
Yes	2	1	3
No	10	20	30
Total	12	21	33

4. In stratified 2×2 tables, if we assume the odds ratio parameters are homogeneous across all strata, Mantel-Haenszel approach (Mantel and Haenszel, 1959)[121] to estimating the common odds ratio parameter is

$$\hat{\theta} = \frac{\sum_{m=1}^{M} n_{11m} n_{22m} / n_{++m}}{\sum_{m=1}^{M} n_{12m} n_{21m} / n_{++m}}.$$

The variance estimate of the common odds ratio estimator is given in Robins et al (1986)[150].

For the following observed stratified 2×2 tables,

TABLE 1.6
Stratified disease and exposure counts.

Stratum	Disease	Exposure Yes	Exposure No
1	Yes	2	1
	No	5	7
2	Yes	2	2
	No	7	10
3	Yes	1	1
	No	3	5
4	Yes	2	0
	No	9	5

(a) Test the hypothesis that the common odds ratio is 1 by the Mantel-Haenszel approach.

(b) Find the common odds ratio estimation and its variance estimate by the Mantel-Haenszel approach and the Robin et al's approach.

(c) Test the hypothesis that the common odds ratio is 1 by the conditional likelihood approach.

(d) Estimate the common odds ratio by the conditional likelihood approach.

(e) Compare the results of the two approaches.

5. For the data in Table 1.6, assume the odds ratio parameters are the same. Find the odds ratio estimator and its variance estimate by maximizing the joint conditional likelihood (1.23).

2

Odds Ratio Function and Its Modeling

2.1 Odds ratio function

Let $a(y, x)$ be an arbitrary positive function on the domain $\mathcal{Y} \times \mathcal{X}$. Define the odds ratio function corresponding to the positive function relative to a reference point $(y_0, x_0) \in \mathcal{Y} \times \mathcal{X}$ as

$$\eta_a\{(y, y_0); (x, x_0)\} = \frac{a(y, x)a(y_0, x_0)}{a(y_0, x)a(y, x_0)}. \tag{2.1}$$

From the definition of the odds ratio function, it is straightforwardly seen that

$$\eta_a\{(y, y_0); (x, x_0)\} = [\eta_a\{(y_0, y); (x, x_0)\}]^{-1}, \tag{2.2}$$
$$\eta_a\{(y, y_0); (x, x_0)\} = [\eta_a\{(y, y_0); (x_0, x)\}]^{-1}. \tag{2.3}$$
$$\eta_a\{(y, y_0); (x, x_0)\} = \eta_a\{(y_0, y); (x_0, x)\}. \tag{2.4}$$

In the following, we may drop the subscript a when no confusion arises. Let (y_1, x_1) be another reference point. The odds ratio function with the reference point (y_1, x_1) relates to the odds ratio function with the reference point (y_0, x_0) in the following way.

$$\eta\{(y, y_1); (x, x_1)\} = \frac{\eta\{(y, y_0); (x, x_0)\}\eta\{(y_1, y_0); (x_1, x_0)\}}{\eta\{(y, y_0); (x_1, x_0)\}\eta\{(y_1, y_0); (x, x_0)\}}. \tag{2.5}$$

This is because

$$\begin{aligned} \frac{a(y, x)a(y_1, x_1)}{a(y_1, x)a(y, x_1)} &= \frac{a(y, x)a(y_0, x_0)}{a(y_0, x)a(y, x_0)} \frac{a(y, x_0)a(y_0, x_1)}{a(y, x_1)a(y_0, x_0)} \\ &\quad \frac{a(y_1, x_0)a(y_0, x)}{a(y_1, x)a(y_0, x_0)} \frac{a(y_1, x_1)a(y_0, x_0)}{a(y_1, x_0)a(y_0, x_1)}. \end{aligned}$$

It can also be verified from the definition (2.1) that

$$\frac{\eta\{(y, y_1); (x, x_1)\}}{\eta\{(y, y_1); (x_0, x_1)\}} = \frac{\eta\{(y, y_0); (x, x_0)\}}{\eta\{(y_1, y_0); (x, x_0)\}}. \tag{2.6}$$

It is therefore sufficient in many applications to consider odds ratio functions relative to one fixed reference point.

On the other hand, for a given bivariate positive function, if its odds ratio function with respect to a reference point is the function itself, the bivariate function is called an odds ratio function. For a fixed reference point, the odds ratio function has the following properties:

$$\eta\{(y, y_0); (x, x_0)\} > 0, \tag{2.7}$$
$$\eta\{(y_0, y_0); (x, x_0)\} = 1, \tag{2.8}$$
$$\eta\{(y, y_0); (x_0, x_0)\} = 1. \tag{2.9}$$

DOI: 10.1201/9781351049757-2

Any bivariate function having properties (2.7), (2.8), and (2.9) is an odds ratio function. This can be easily seen from that, if a bivariate function $a(y, x)$ satisfies (2.7), (2.8), and (2.9) for a given reference point (y_0, x_0), i.e.,

$$a(y, x_0) = a(y_0, x) = a(y_0, x_0) = 1,$$

then the odds ratio function of the positive function $a(y, x)$ is

$$\eta\{(y, y_0), (x, x_0)\} = \frac{a(y, x)a(y_0, x_0)}{a(y, x_0)a(y_0, x)} = a(y, x).$$

Note that the odds ratio function defined in this section can be extended to the case where either or both y and x are vectors instead of scalar variables. The properties of the odds ratio functions discussed in this section remain unchanged. This extension makes the odds ratio function more flexible to use as we see in further development.

2.2 Odds ratio decomposition of density functions

Let $p(y \mid x)$ denote the density of Y given X with respect to a known measure, usually Lebesgue measure if Y is a continuous random variable or count measure if Y is discrete. Suppose that the density satisfies the positivity condition, i.e., if the marginal density of Y, $g(y)$, is nonzero at y, then $p(y \mid x)$ is nonzero for all x values where the marginal density, $f(x)$, of X is nonzero. Let (y_0, x_0) be a fixed point in the sampling space. The odds ratio function of the conditional density with respect to a reference point (y_0, x_0) is

$$\eta_p\{(y, y_0); (x, x_0)\} = \frac{p(y \mid x)p(y_0 \mid x_0)}{p(y \mid x_0)p(y_0 \mid x)}. \tag{2.10}$$

The odds ratio function for a conditional density can be interpreted as follows. Given $\Delta > 0$, the odds of Y falls in the interval $[y - \Delta, y + \Delta]$ versus the interval $[y_0 - \Delta, y_0 + \Delta]$ when $X = x$ is approximately

$$\frac{P(Y \in [y - \Delta, y + \Delta] \mid x)}{P(Y \in [y_0 - \Delta, y_0 + \Delta] \mid x)} \approx \frac{p(y \mid x)}{p(y_0 \mid x)}.$$

The odds ratio function

$$\eta_p\{(y, y_0); (x, x_0)\} \approx \frac{P(Y \in [y - \Delta, y + \Delta] \mid x)}{P(Y \in [y_0 - \Delta, y_0 + \Delta] \mid x)} \Bigg/ \frac{P(Y \in [y - \Delta, y + \Delta] \mid x_0)}{P(Y \in [y_0 - \Delta, y_0 + \Delta] \mid x_0)}$$

is thus approximately the ratio of the odds of Y falls in the interval $[y-\Delta, y+\Delta]$ versus the interval $[y_0-\Delta, y_0+\Delta]$ when $X = x$ to the odds of Y falls in the interval $[y-\Delta, y+\Delta]$ versus the interval $[y_0 - \Delta, y_0 + \Delta]$ when $X = x_0$. Note that the qualification "approximately" can be removed when Y takes finite values and Δ is sufficiently small. For binary Y, this interpretation reduces to the commonly known interpretation for the odds ratio parameter in a logistic regression model when $x = x_0 + 1$ and $y = y_0 + 1$.

In the following examples, the odds ratio functions from the probability mass functions or the density functions of some commonly encountered statistical models are extracted.

Example 2.1.1: For the linear logistic regression model, the conditional probability mass function of Y given X is

$$p(y \mid x) = \frac{\exp\{y(\beta_0 + \beta_1 x)\}}{1 + \exp(\beta_0 + \beta_1 x)}. \tag{2.11}$$

The odds ratio function has the form

$$\eta\{(y, y_0); (x, x_0)\} = \exp\{\beta_1(y - y_0)(x - x_0)\}.$$

When $(y_0, x_0) = (0, 0)$, $\eta\{(y = 1, y_0 = 0); (x, x_0 = 0)\} = \exp(\beta_1 x)$. In particular,

$$\eta\{(y = 1, y_0 = 0); (x = 1, x_0 = 0)\} = \exp(\beta_1),$$

which is the odds ratio parameter in the logistic regression model.

Example 2.1.2: For the Poisson regression model, the probability mass function is

$$p(y \mid x) = \frac{\lambda^y(x)}{y!}\exp\{-\lambda(x)\}, \tag{2.12}$$

where $\log \lambda(x) = \beta_0 + \beta_1 x$. The odds ratio function with respect to the fixed point (y_0, x_0) is

$$\eta\{(y, y_0); (x, x_0)\} = \exp\{\beta_1(y - y_0)(x - x_0)\}.$$

Note that the odds ratio function has exactly the same functional form as in the logistic regression model discussed in Example 2.1.1.

Example 2.1.3: For the conditional normal regression model, the density function with respect to Lebesgue measure is

$$p(y \mid x) = \frac{1}{\sqrt{2\pi}\sigma}\exp\left\{-\frac{1}{2\sigma^2}(y - \beta_0 - \beta_1 x)^2\right\}, \tag{2.13}$$

The odds ratio function with respect to the reference point (y_0, x_0) is

$$\eta\{(y, y_0); (x, x_0)\} = \exp\left\{\frac{\beta_1}{\sigma^2}(y - y_0)(x - x_0)\right\}.$$

The functional form is just slightly different from the odds ratio function in Examples 2.1.1 and 2.1.2. The next example explains the reason for the similarity.

Example 2.1.4: In a general exponential model, given X, Y follows an exponential distribution with density

$$p(y \mid x) = \exp\left[\frac{1}{\phi}\{y\theta - b(\theta)\} + a(y)\right],$$

where $\theta = \theta(x, \gamma)$ is a function of x and unknown parameter γ, and ϕ is the dispersion parameter. The odds ratio function is

$$\eta\{(y, y_0); (x, x_0)\} = \exp\left[\frac{1}{\phi}(y - y_0)\{\theta(x, \gamma) - \theta(x_0, \gamma)\}\right].$$

For a generalized linear model, $\theta(x, \gamma) = h(x\gamma)$ where h is a known function. For the generalized linear model with the canonical link function, i.e., $\theta(x, \gamma) = x\gamma$, the odds ratio function reduces to

$$\eta\{(y, y_0); (x, x_0)\} = \exp\left\{\frac{\gamma}{\phi}(y - y_0)(x - x_0)\right\}.$$

Examples 2.1.1–2.1.3 are all special cases of the generalized linear model with the canonical link function.

Example 2.1.5: Let Y denote the survival time and X denote the covariate. The proportional hazards regression model has conditional density

$$p(y \mid x) = \lambda_0(y)e^{\beta x}\exp\left\{-\Lambda_0(y)e^{\beta x}\right\}, \tag{2.14}$$

where λ_0 and Λ_0 are respectively the baseline hazard density and the cumulative hazard function. The odds ratio function is

$$\eta\{(y, y_0); (x, x_0)\} = \exp\left[\{\Lambda_0(y) - \Lambda_0(y_0)\}(e^{\beta x} - e^{\beta x_0})\right]. \tag{2.15}$$

When the conditional density of Y given X is a positive function, the joint density denoted by $h(y, x)$ and the conditional density of X given Y denoted by $q(x \mid y)$ are both positive functions. There are odds ratio functions correspond to them respectively. For the joint density function, the corresponding odds ratio function

$$\begin{aligned}\eta_h\{(y, y_0); (x, x_0)\} &= \frac{h(y, x)h(y_0, x_0)}{h(y, x_0)h(y_0, x)} \\ &= \frac{p(y \mid x)f(x)p(y_0 \mid x_0)f(x_0)}{p(y \mid x_0)f(x_0)p(y_0 \mid x)f(x)} \\ &= \eta_p\{(y, y_0); (x, x_0)\}.\end{aligned}$$

For the conditional density of X given Y, the corresponding odds ratio function is

$$\begin{aligned}\eta_q\{(y, y_0); (x, x_0)\} &= \frac{q(x \mid y)q(x_0 \mid y_0)}{q(x \mid y_0)q(x_0 \mid y)} \\ &= \frac{q(x \mid y)g(y)q(x_0 \mid y_0)g(y_0)}{q(x \mid y_0)g(y_0)q(x_0 \mid y)g(y)} \\ &= \eta_p\{(y, y_0); (x, x_0)\}.\end{aligned}$$

These equalities suggest that the odds ratio functions extracted from the conditional densities and joint density are the same. In other words, the same odds ratio function is embeded in $p(y \mid x)$, $q(x \mid y)$, and $h(y, x)$. The following examples extract odds ratio functions from joint densities.

Example 2.1.6: In *Example 2.1.3*, if (Y, X) follows a joint normal distribution with means μ_y, μ_x and variance σ_y^2, σ_x^2 and correlation coefficient ρ, then

$$\beta_1 = \rho\frac{\sigma_y}{\sigma_x} \text{ and } \sigma^2 = (1 - \rho^2)\sigma_y^2.$$

The odds ratio function with reference point $(0, 0)$ can be rewritten as

$$\eta\{(y, y_0 = 0); (x, x_0 = 0)\} = \frac{\rho}{(1 - \rho^2)\sigma_x\sigma_y}yx.$$

For standardized Y, X, i.e., $\sigma_y = \sigma_x = 1$, the odds ratio parameter θ relates to the correlation coefficient through $\theta = \rho/(1 - \rho^2)$ or reversely $\rho = \theta/(\sqrt{\theta^2 + 0.25} + 0.5)$. Figure 2.1 displays the relationship for θ ranging from -10 to 10.

Example 2.1.7: For binary outcome Y, X given Y can be viewed as the two-sample problem. Assume X given $Y = 0$ follows $N(\mu_0, \sigma_0^2)$ and X given $Y = 1$ follows $N(\mu_1, \sigma_1^2)$. Then the odds ratio function

$$\begin{aligned}\eta\{(y = 1, y_0 = 0); (x, x_0)\} &= \exp\left\{\left(\frac{1}{2\sigma_0^2} - \frac{1}{2\sigma_1^2}\right)(x^2 - x_0^2) + \left(\frac{\mu_1}{\sigma_1^2} - \frac{\mu_0}{\sigma_0^2}\right)(x - x_0)\right\} \\ &= \exp\left[\left\{\left(\frac{1}{2\sigma_0^2} - \frac{1}{2\sigma_1^2}\right)(x + x_0) + \left(\frac{\mu_1}{\sigma_1^2} - \frac{\mu_0}{\sigma_0^2}\right)\right\}(x - x_0)\right]\end{aligned}$$

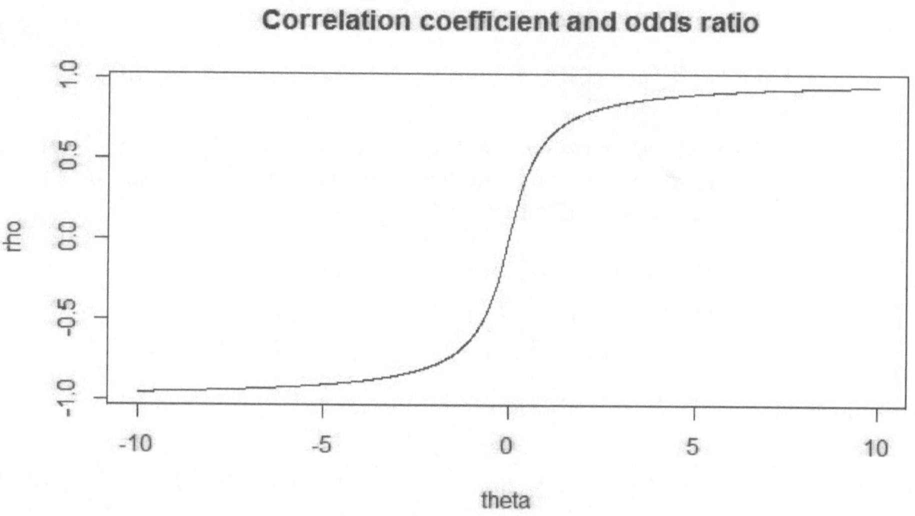

FIGURE 2.1
Relationship of correlation coefficient and odds ratio parameter in bivariate normal model.

which is quadratic in x. When the two samples have equal variance, the odds ratio function reduces to

$$\eta\{(y=1,y_0=0);(x,x_0)\} = \exp\left\{\frac{\mu_1-\mu_0}{\sigma_0^2}(x-x_0)\right\}.$$

Example 2.1.8. (Extension of Example 2.1.1): For the logistic regression model with multiple covariates, let $X=(X_1,\cdots,X_r)$. The conditional probability mass function for the distribution of Y given X is

$$p(y\mid x) = \frac{\exp\{y(\beta_0+\beta_1x_1+\cdots+\beta_rx_r)\}}{1+\exp(\beta_0+\beta_1x_1+\cdots+\beta_rx_r)}. \tag{2.16}$$

Let the reference point be (y_0,x_0) where $x_0=(x_{10},\cdots,x_{r0})$. The odds ratio function is

$$\eta\{(y,y_0);(x,x_0)\} = \exp\{\beta_1(y-y_0)(x_1-x_{10})+\cdots+\beta_r(y-y_0)(x_r-x_{r0})\}.$$

When $(y_0,x_0)=(0,0,\cdots,0)$, $\eta\{(y=1,y_0);(x,x_0)\}=\exp(\beta_1x_1+\cdots+\beta_rx_r)$. If $x^*=(x_{10},\cdots,x_{(j-1)0},x_j,x_{(j+1)0},\cdots,x_{r0})$, then the odds ratio function at such points of X becomes

$$\eta\{(y=1,y_0);(x^*,x_0)\} = \exp(\beta_jx_j).$$

If further $x_j=1$, the odds ratio function at that point becomes the odds ratio parameter associated with x_j in the logistic regression model.

Example 2.1.9. (Extension of Example 2.1.2) : For the Poisson regression model with multiple covariates $X=(X_1,\cdots,X_r)$, let

$$p(y\mid x_1,\cdots,x_r) = \frac{\lambda^y(x_1,\cdots,x_r)}{y!}\exp\{-\lambda(x_1,\cdots,x_r)\}, \tag{2.17}$$

where $\log \lambda(x_1, \cdots, x_r) = \beta_0 + \beta_1 x_1 + \cdots \beta_r x_r$. The odds ratio function for the reference point (y_0, x_0) where $x_0 = (x_{10}, \cdots, x_{r0})$ is

$$\eta\{(y, y_0); (x, x_0)\} = \exp\{\beta_1(y - y_0)(x_1 - x_{10}) + \cdots + \beta_r(y - y_0)(x_r - x_{r0})\}.$$

Example 2.1.10. (Extension of Example 2.1.3): Let $Y = (Y_1, \cdots, Y_s)^t$ given $X = (X_1, \cdots, X_r)^t$ be multivariate normal distributed with mean vector $\mu(X) = (\mu_1(X), \cdots, \mu_s(X))^t$ and variance covariance matrix Σ, where

$$\begin{aligned} \mu_1(x) &= \mu_{10} + \beta_{11}x_1 + \cdots + \beta_{1r}x_r, \\ \cdots & \quad \cdots, \\ \mu_s(x) &= \mu_{s0} + \beta_{s1}x_1 + \cdots + \beta_{sr}x_r. \end{aligned}$$

The conditional multivariate normal distribution has density

$$p(y \mid x) = \frac{1}{(2\pi)^{r/2}|\Sigma|^{1/2}} \exp\left\{-\frac{1}{2}(y - \mu_0 - Bx)^t \Sigma^{-1}(y - \mu_0 - Bx)\right\}, \tag{2.18}$$

where $\mu_0 = (\mu_{10}, \cdots, \mu_{r0})$ and

$$B = \begin{pmatrix} \beta_{11} & \beta_{12} & \cdots & \beta_{1r} \\ \beta_{21} & \beta_{22} & \cdots & \beta_{2r} \\ \vdots & \vdots & \vdots & \vdots \\ \beta_{s1} & \beta_{s2} & \cdots & \beta_{sr} \end{pmatrix}.$$

Let $y_0 = (y_{10}, \cdots, y_{s0})$ and $x_0 = (x_{10}, \cdots, x_{r0})$. The odds ratio function with respect to the reference point (y_0, x_0) is

$$\eta\{(y, y_0); (x, x_0)\} = \exp\{(y - y_0)^t \Sigma^{-1} B(x - x_0)\}.$$

When $s = 1$, the odds ratio function becomes the one from a multivariate linear regression, i.e.,

$$\eta\{(y, y_0); (x, x_0)\} = \exp\left\{\sum_{j=1}^{r} \frac{\beta_j}{\sigma^2}(y - y_0)(x_j - x_{j0})\right\},$$

where $\Sigma = (\sigma^2)$ and $\beta_j = \beta_{1j}, j = 1, \cdots, r$.

Example 2.1.11: Consider the $J \times K \times M$ contingency table discussed in Chapter 1, where the conditional odds ratio parameters were defined. For the conditional probability distribution

$$p_{jkm} = P(Y = j \mid X = k, Z = m), \qquad j = 1, \cdots, J; k = 1, \cdots, K; m = 1, \cdots, M.$$

Let the reference categories for Y, X, and Z be repectively J, K, and M. The odds ratio function can be defined as

$$\eta[(j, J); \{(k, m), (K, M)\}] = \frac{p_{jkm} p_{JKM}}{p_{jKM} p_{Jkm}},$$

for $j = 1, \cdots, J - 1; (k, m) = (1, 1), \cdots, (J, M - 1)$.

2.3 Odds ratio representation of densities

Since the odds ratio function $\eta(y;x)$ is shared by $p(y \mid x)$, $q(x \mid y)$, and $h(y,x)$, one interesting question is if we can explicitly express the odds ratio function in these densities. From the odds ratio function expression in (2.10), it can be seen that

$$p(y \mid x) = \eta\{(y,y_0);(x,x_0)\}p(y \mid x_0)\frac{p(y_0 \mid x)}{p(y_0 \mid x_0)}. \tag{2.19}$$

Since $\int p(y \mid x)dy = 1$, it follows that

$$\int \eta\{(y,y_0);(x,x_0)\}p(y \mid x_0)dy = \frac{p(y_0 \mid x_0)}{p(y_0 \mid x)}, \tag{2.20}$$

where we use dy to denote the known measure on which the density is obtained. This leads to an odds ratio representation of the conditional density as

$$p(y \mid x) = \frac{\eta\{(y,y_0);(x,x_0)\}p(y \mid x_0)}{\int \eta\{(y,y_0);(x,x_0)\}p(y \mid x_0)dy}. \tag{2.21}$$

A similar odds ratio representation for the conditional density $q(x|y)$ can also be obtained by a similar argument as

$$q(x \mid y) = \frac{\eta\{(y,y_0);(x,x_0)\}q(x \mid y_0)}{\int \eta\{(y,y_0);(x,x_0)\}q(x \mid y_0)dx}. \tag{2.22}$$

Furthermore, for the joint density,

$$h(y,x) = \frac{\eta\{(y,y_0);(x,x_0)\}h(y,x_0)h(y_0,x)}{h(y_0,x_0)}. \tag{2.23}$$

It follows from $\int\int h(y,x)dydx = 1$ that

$$h(y,x) = \frac{\eta\{(y,y_0);(x,x_0)\}h(y,x_0)h(y_0,x)}{\int\int \eta\{(y,y_0);(x,x_0)\}h(y,x_0)h(y_0,x)dydx}. \tag{2.24}$$

Renormalized $h(y,x_0)$ and $h(y_0,x)$, we can rewrite the odds ratio representation of a joint density into

$$h(y,x) = \frac{\eta\{(y,y_0);(x,x_0)\}p(y \mid x_0)q(x \mid y_0)}{\int\int \eta\{(y,y_0);(x,x_0)\}p(y \mid x_0)q(x \mid y_0)dydx}. \tag{2.25}$$

The odds ratio representations can be applied to the density functions in the previous examples as follow.

Example 2.2.1: For the simple logistic regression model, let $(y_0,x_0) = (0,0)$. The odds ratio representation is,

$$p(y \mid x) = \frac{\exp(\beta_1 yx)p(y \mid x_0 = 0)}{\sum_{y=0}^{1}\exp(\beta_1 yx)p(y \mid x_0 = 0)}, \tag{2.26}$$

where

$$p(y \mid x_0 = 0) = \frac{\exp(\beta_0 y)}{1+\exp(\beta_0)}.$$

Example 2.2.2: For the logistic regression model with multiple covariates, let $(y_0, x_0) = (0, \cdots, 0)$. The odds ratio representation is

$$p(y \mid x) = \frac{\exp\{\beta_1 y x_1 + \cdots + \beta_r y x_r)\} p(y \mid x_0 = 0)}{p(y = 0 \mid x = 0) + \exp(\beta_1 x_1 + \cdots + \beta_r x_r) p(y = 1 \mid x_0 = 0)}. \tag{2.27}$$

where

$$p(y \mid x_0 = 0) = \frac{\exp(\beta_0 y)}{1 + \exp(\beta_0)}.$$

Example 2.2.3: For the simple Poisson regression model, let $(y_0, x_0) = (0, 0)$. The odds ratio representation is

$$p(y \mid x) = \frac{\exp(\beta_1 y x) p(y \mid x_0 = 0)}{\sum_{y=0}^{\infty} \exp(\beta_1 y x) p(y \mid x_0 = 0)}, \tag{2.28}$$

where

$$p(y \mid x_0 = 0) = \frac{\exp(\beta_0 y)}{y!} \exp(-e^{\beta_0}).$$

Example 2.2.4: For the Poisson regression model with multiple covariates, let $(y_0, x_0) = (0, \cdots, 0)$. The odds ratio representation is

$$p(y \mid x) = \frac{\exp\{\beta_1 y x_1 + \cdots + \beta_r y x_r)\} p(y \mid x_0 = 0)}{\sum_{y=0}^{\infty} \exp(\beta_1 x_1 y + \cdots + \beta_r x_r y) p(y \mid x_0 = 0)}. \tag{2.29}$$

where

$$p(y \mid x_0 = 0) = \frac{\exp(\beta_0 y)}{y!} \exp(-e^{\beta_0}).$$

Example 2.2.5: For the simple normal regression model, let $(y_0, x_0) = (0, 0)$. The odds ratio representation is

$$p(y \mid x) = \frac{\exp\left\{\frac{\beta_1}{\sigma^2} y x\right\} p(y \mid x_0 = 0)}{\int \exp\left\{\frac{\beta_1}{\sigma^2} y x\right\} p(y \mid x_0 = 0) dy}, \tag{2.30}$$

where

$$p(y \mid x_0 = 0) = \frac{1}{\sqrt{2\pi}\sigma} \exp\left\{-\frac{1}{2\sigma^2}(y - \beta_0)^2\right\}.$$

Example 2.2.6: For the multiple outcome normal model with multivariate covariates, let $(y_0, x_0) = (0, \cdots, 0)$. The odds ratio representation is

$$p(y \mid x) = \frac{\exp(y^t \Sigma^{-1} B x) p(y \mid x_0 = 0)}{\int \cdots \int \exp(y^t \Sigma^{-1} B x) p(y \mid x_0 = 0) dy_1, \cdots dy_s}, \tag{2.31}$$

where

$$p(y \mid x_0 = 0) = \frac{1}{(2\pi)^{r/2} |\Sigma|^{1/2}} \exp\left\{-\frac{1}{2}(y - \mu_0)^t \Sigma^{-1} (y - \mu_0)\right\}, \tag{2.32}$$

When $s = 1$, the odds ratio representation reduces to

$$p(y \mid x) = \frac{\exp\left\{\frac{\beta_1}{\sigma^2} y x + \cdots + \frac{\beta_r}{\sigma^2} y x_r\right\} p(y \mid x_0 = 0)}{\int \exp\left\{\frac{\beta_1}{\sigma^2} y x + \cdots + \frac{\beta_r}{\sigma^2} y x_r\right\} p(y \mid x_0 = 0) dy},$$

where

$$p(y \mid x_0 = 0) = \frac{1}{\sqrt{2\pi}\sigma} \exp\left\{-\frac{1}{2\sigma^2}(y - \beta_0)^2\right\}.$$

Example 2.2.7: For the generalized linear model with canonical link function, let $(y_0, x_0) = (0, 0)$. Then,

$$p(y \mid x) = \frac{\exp\left(\frac{\gamma}{\phi} yx\right) p(y \mid x_0 = 0)}{\int \exp\left(\frac{\gamma}{\phi} yx\right) p(y \mid x_0 = 0) dy}, \tag{2.33}$$

where

$$p(y \mid x_0 = 0) = \frac{\exp\{a(y)\}}{\int \exp\{a(y)\} dy}.$$

Example 2.2.8: For the Cox regression model, let $(y_0, x_0) = (0, 0)$. Then,

$$p(y \mid x) = \frac{\exp\left\{-\Lambda_0(y)(e^{\beta x} - 1)\right\} p(y \mid x_0 = 0)}{\int \exp\left\{-\Lambda_0(y)(e^{\beta x} - 1)\right\} p(y \mid x_0 = 0) dy}, \tag{2.34}$$

where

$$p(y \mid x_0 = 0) = \lambda_0(y) \exp\left\{-\Lambda_0(y)\right\}. \tag{2.35}$$

Example 2.2.9: The odds ratio representation of the conditional cell probabilities in a three-way contingency table is

$$p_{jkm} = \frac{\eta[(j, J); \{(k, m), (K, M)\}] p_{jKM}}{\sum_{j=1}^{J} \eta[(j, J); \{(k, m), (K, M)\}] p_{jKM}}.$$

for $j = 1, \cdots, J; k = 1, \cdots, K; m = 1, \cdots, M$.

Example 2.2.10: The joint normal density can be represented as

$$h(y, x) = \frac{\eta\{(y, y_0 = 0); (x, x_0 = 0)\} p(y \mid x_0 = 0) q(x \mid y_0 = 0)}{\int\int \eta\{(y, y_0 = 0); (x, x_0 = 0)\} p(y \mid x_0 = 0) q(x \mid y_0 = 0) dy dx},$$

where

$$\begin{aligned} p(y \mid x_0 = 0) &= \frac{1}{\sigma_y \sqrt{2\pi(1 - \rho^2)}} \exp\left\{\frac{1}{2(1 - \rho^2)\sigma_y^2}\left(y - \mu_y - \rho\frac{\sigma_y}{\sigma_x}\mu_x\right)^2\right\}, \\ q(x \mid y_0 = 0) &= \frac{1}{\sigma_x \sqrt{2\pi(1 - \rho^2)}} \exp\left\{\frac{1}{2(1 - \rho^2)\sigma_x^2}\left(x - \mu_x - \rho\frac{\sigma_x}{\sigma_y}\mu_x\right)^2\right\}. \end{aligned}$$

Example 2.2.11: For the two-sample normal model in *Example 2.1.7*,

$$h(y, x) = \frac{\eta\{(y, y_0 = 0); (x, x_0)\} p(y \mid x_0) q(x \mid y_0 = 0)}{\int\int \eta\{(y, y_0 = 0); (x, x_0)\} p(y \mid x_0) q(x \mid y_0 = 0) dy dx},$$

where

$$\begin{aligned} p(y \mid x_0) &= \frac{\exp[y \log\{\pi/(1 - \pi)\} + yu(x_0)]}{1 + \exp[\log\{\pi/(1 - \pi)\} + u(x_0)]}, \\ q(x \mid y_0 = 0) &= \frac{1}{\sqrt{2\pi}\sigma_0} \exp\left\{\frac{1}{2\sigma_0^2}(x - \mu_0)^2\right\}. \end{aligned}$$

where $\pi = P(Y = 1)$ and $u(x) = (\sigma_0/\sigma_1)^2 \exp\{-(x-\mu_0)^2/(2\sigma_0^2) - (x-\mu_1)^2/(2\sigma_1^2)\}$.

Example 2.2.12: For the $J \times K \times M$ contingency table, the odds ratio representation for the joint cell probabilities is

$$g_{jkm} = \frac{\eta[(j,J);\{(k,m),(K,M)\}]p_{jKM}q_{kmJ}}{\sum_{m=1}^{M}\sum_{k=1}^{K}\sum_{j=1}^{J}\eta[(j,J);\{(k,m),(K,M)\}]p_{jKM}q_{kmJ}}.$$

2.4 Conditional odds ratio function

Let $b(y,x,z)$ be an arbitrary positive function on the domain $\mathcal{Y} \times \mathcal{X} \times \mathcal{Z}$. Define the conditional odds ratio function of y versus x conditional on z relative to a reference point $(y_0, x_0) \in \mathcal{Y} \times \mathcal{X}$ corresponding to the positive function b as

$$\eta_b\{(y,y_0);(x,x_0) \mid z\} = \frac{b(y,x,z)b(y_0,x_0,z)}{b(y_0,x,z)b(y,x_0,z)}. \tag{2.36}$$

Similar to the unconditional odds ratio function, it can be straightforwardly seen from the definition of the conditional odds ratio function that

$$\begin{aligned}
\eta_b\{(y,y_0);(x,x_0) \mid z\} &= [\eta_b\{(y_0,y);(x,x_0) \mid z\}]^{-1},\\
\eta_b\{(y,y_0);(x,x_0) \mid z\} &= [\eta_b\{(y,y_0);(x_0,x) \mid z\}]^{-1},\\
\eta_b\{(y,y_0);(x,x_0) \mid z\} &= \eta_b\{(y_0,y);(x_0,x) \mid z\}.
\end{aligned}$$

Let (y_1, x_1) be another reference point. Similar relationship for the conditional odds ratio functions with the reference point (y_1, x_1) and the conditional odds ratio functions with the reference point (y_0, x_0) holds, i.e.,

$$\eta_b\{(y,y_1);(x,x_1) \mid z\} = \frac{\eta_b\{(y,y_0);(x,x_0) \mid z\}\eta_b\{(y_1,y_0);(x_1,x_0) \mid z\}}{\eta_b\{(y,y_0);(x_1,x_0) \mid z\}\eta_b\{(y_1,y_0);(x,x_0) \mid z\}}. \tag{2.37}$$

Furthermore,

$$\frac{\eta_b\{(y,y_1);(x,x_1) \mid z\}}{\eta_b\{(y,y_1);(x_0,x_1) \mid z\}} = \frac{\eta_b\{(y,y_0);(x,x_0) \mid z\}}{\eta_b\{(y_1,y_0);(x,x_0) \mid z\}}. \tag{2.38}$$

It is therefore sufficient to consider a set of conditional odds ratio functions relative to one fixed reference point.

For a given tri-variate positive function, if its conditional odds ratio function with respect to a reference point is the function itself, the tri-variate function is called a conditional odds ratio function. For a fixed reference point, the conditional odds ratio function has the following properties:

$$\begin{aligned}
\eta_b\{(y,y_0);(x,x_0) \mid z\} &> 0,\\
\eta_b\{(y_0,y_0);(x,x_0) \mid z\} &= 1,\\
\eta_b\{(y,y_0);(x_0,x_0) \mid z\} &= 1.
\end{aligned}$$

Any tri-variate function having these properties is a conditional odds ratio function. This can be easily seen from that if a tri-variate function $b(y,x,z)$ satisfies these conditions for a given reference point (y_0, x_0), then the odds ratio function of the positive function $b(y,x,z)$ is

$$\eta_b\{(y,y_0),(x,x_0) \mid z\} = \frac{b(y,x,z)b(y_0,x_0,z)}{b(y,x_0,z)b(y_0,x,z)} = b(y,x,z).$$

Note that y, x and z in the foregoing discussion can all be vectors.

2.5 Odds ratio representation of a joint conditional density

Let the conditional density of Y given X and Z with respect to a given measure be $p(y \mid x, z)$. Suppose that the positivity condition is satisfied: if the marginally conditional density of Y given Z, i.e. $f(y \mid z) > 0$ and the marginally conditional density of X given Z, i.e. $g(x \mid z) > 0$, then $p(y \mid x, z) > 0$. The conditional odds ratio function of y versus x given z with reference point (y_0, x_0) for the positive function $p(y \mid x, z)$ is

$$\eta_{cp}\{(y, y_0); (x, x_0) \mid z\} = \frac{p(y \mid x, z)p(y_0 \mid x_0, z)}{p(y_0 \mid x, z)p(y \mid x_0, z)}. \tag{2.39}$$

The conditional odds ratio function for a conditional density can be interpreted similarly as for the unconditional odds ratio function of a density. Given $Z = z$ and $\Delta > 0$, the conditional odds ratio function

$$\eta_{cp}\{(y, y_0); (x, x_0) \mid z\} \approx \frac{P(Y \in [y - \Delta, y + \Delta]|x, z)}{P(Y \in [y_0 - \Delta, y_0 + \Delta]|x, z)} \Big/ \frac{P(Y \in [y - \Delta, y + \Delta]|x_0, z)}{P(Y \in [y_0 - \Delta, y_0 + \Delta]|x_0, z)}$$

is approximately the ratio of the odds of Y falls in the interval $[y - \Delta, y + \Delta]$ versus in the interval $[y_0 - \Delta, y_0 + \Delta]$ when $X = x$ to the odds of Y falls in the interval $[y - \Delta, y + \Delta]$ versus in the interval $[y_0 - \Delta, y_0 + \Delta]$ when $X = x_0$. The conditional odds ratio function for the conditional density of X given Y and Z, $q(x \mid y, z)$, is

$$\eta_{cq}\{(y, y_0); (x, x_0) \mid z\} = \frac{q(x \mid y, z)q(x_0 \mid y_0, z)}{q(x_0 \mid y, z)q(x \mid y_0, z)}, \tag{2.40}$$

The conditional odds ratio function can also be obtained from the joint conditional density of (Y, X) given Z, $(y, x \mid z)$, as

$$\eta_{ch}\{(y, y_0); (x, x_0) \mid z\} = \frac{h(y, x \mid z)h(y_0, x_0 \mid z)}{h(y_0, x \mid z)h(y, x_0 \mid z)}. \tag{2.41}$$

It is easy to see that

$$\eta_{cp}\{(y, y_0); (x, x_0) \mid z\} = \eta_{cq}\{(y, y_0); (x, x_0) \mid z\} = \eta_{ch}\{(y, y_0); (x, x_0) \mid z\}.$$

For simplicity, we will use $\eta_c\{(y, y_0); (x, x_0) \mid z\}$ to denote these odds ratio functions.

By similar arguments as in the unconditional odds ratio representation, an odds ratio representation of the conditional density $p(y \mid x, z)$ has the form

$$p(y \mid x, z) = \frac{\eta_c\{(y, y_0); (x, x_0) \mid z\}p(y \mid x_0, z)}{\int \eta_c\{(y, y_0); (x, x_0) \mid z\}p(y \mid x_0, z)dy}. \tag{2.42}$$

An odds ratio representation of the conditional density $q(x \mid y, z)$ can be obtained as

$$q(x \mid y, z) = \frac{\eta_c\{(y, y_0); (x, x_0) \mid z\}q(x \mid y_0, z)}{\int \eta_c\{(y, y_0); (x, x_0) \mid z\}q(x \mid y_0, z)dx}. \tag{2.43}$$

An odds ratio representation for the joint conditional density, $h(y, x \mid z)$, can be obtained as

$$h(y, x \mid z) = \frac{\eta_c\{(y, y_0); (x, x_0) \mid z\}p(y \mid, x_0, z)q(x \mid y_0, z)}{\int \int \eta_c\{(y, y_0); (x, x_0) \mid z\}p(y \mid x_0, z)q(x \mid y_0, z)dydx}. \tag{2.44}$$

Those representations indicate that the conditional odds ratio function is shared by the three conditional densities.

Example 2.3.1: Consider the partial linear regression model

$$Y = \beta_0 + \beta_1 X + h(Z) + \epsilon,$$

where $\epsilon \sim N(0, \sigma^2)$. The conditional density of Y given X and Z is

$$p(y \mid x, z) = \frac{1}{2\sigma^2} \exp\left[-\frac{1}{2\sigma^2}\{y - \beta_0 - \beta_1 x - h(z)\}^2\right].$$

The conditional odds ratio function with reference point $(y_0 = 0, x_0)$ is

$$\eta_c\{(y, y_0); (x, x_0) \mid z\} = \exp\left\{\frac{\beta_1}{\sigma^2} y(x - x_0)\right\}.$$

Taking $x_0 = 0$, the odds ratio representation of the conditional density is

$$p(y \mid x, z) = \frac{\eta_c\{(y, y_0); (x, x_0) \mid z\} p(y \mid x_0, z)}{\int \eta_c\{(y, y_0); (x, x_0) \mid z\} p(y \mid x_0, z) dy},$$

where

$$p(y \mid x_0 = 0, z) = \frac{1}{2\sigma^2} \exp\left[-\frac{1}{2\sigma^2}\{y - \beta_0 - h(z)\}^2\right].$$

Example 2.3.2: For the generalized partial linear model with binary outcome, let

$$P(Y = 1 \mid X, Z) = \frac{\exp\{\beta_0 + \beta_1 X + h(Z)\}}{1 + \exp\{\beta_0 + \beta_1 X + h(Z)\}},$$

The conditional odds ratio function with reference point $(y_0 = 0, x_0)$ is

$$\eta_c\{(y, y_0 = 0); (x, x_0) \mid z\} = \exp\{\beta_1 y(x - x_0)\}.$$

Taking $x_0 = 0$, the odds ratio representation of the conditional density is

$$p(y \mid x, z) = \frac{\eta_c\{(y, y_0); (x, x_0) \mid z\} p(y \mid x_0, z)}{\int \eta_c\{(y, y_0); (x, x_0) \mid z\} p(y \mid x_0, z) dy},$$

where

$$p(y \mid x_0 = 0, z) = \frac{\exp\{\beta_0 + h(Z)\}}{1 + \exp\{\beta_0 + h(Z)\}},$$

2.6 Odds ratio representation of a complex joint density

The framework for obtaining odds ratio representation of a density or a probability mass function can be further extended to multivariate densities as follows. Consider the case where variables in Y are divided into G groups as $Y_1, \cdots, Y_G$. Each Y_j is a vector of $s_j \geq 1$ dimension for $j = 1, \cdots, G$. For simplicity and without loss of generality, assume that no X or Z is involved. Let $h(y_1, \cdots, y_G)$ be the density of $(Y_1, \cdots, Y_G)$ under an appropriate product of count measures and Lebesgue measures. Assume the positivity condition holds, i.e., $p(y_1, \cdots, y_G) > 0$ if $f_j(y_j) > 0$ for $j = 1, \cdots, G$, where $f_j(y_j)$ denotes the marginal density of Y_j. To simplify the notation in this section, we fix the reference point at $y_0 = (y_{10}, \cdots, y_{G0})$ and suppress the reference point in the expression for the odds ratio functions whenever possible.

Note first that the unconditional odds ratio representation of the joint density for the two groups y_1 versus $y_{-1} = (y_2, \cdots, y_G)$ is

$$h(y_1, y_{-1}) = \frac{\eta_1(y_1; y_{-1})p(y_1 \mid y_{-10})h(y_{-1} \mid y_{10})}{\int \cdots \int \eta_1(y_1; y_{-1})p(y_1 \mid y_{-10})h(y_{-1} \mid y_{10}) \prod_{k=1}^{G} dy_k}. \tag{2.45}$$

Next, the conditional odds ratio representation for the density $h(y_{-1} \mid y_{10})$ for the two groups y_2 versus $y_{-(1,2)} = (y_3, \cdots, y_G)$ is

$$h(y_{-1} \mid y_{10}) = \frac{\eta_2(y_2; y_{-(1,2)} \mid y_{10})p(y_2 \mid y_{-20})h(y_{-(1,2)} \mid y_{10}, y_{20})}{\int \cdots \int \eta_2(y_2; y_{-(1,2)} \mid y_{10})p(y_2 \mid y_{-20})h(y_{-(1,2)} \mid y_{10}, y_{20}) dy_2 \cdots dy_G}. \tag{2.46}$$

It follows from plugging (2.46) into (2.45) that the joint density $f(y_1, \cdots, y_G)$ can be expressed as

$$\frac{\eta_1(y_1; y_{-1})\eta_2(y_2; y_{-(1,2)} \mid y_{10}) \prod_{k=1}^{2} p(y_k \mid y_{-k0})h(y_{-(1,2)} \mid y_{10}, y_{20})}{\int \cdots \int \eta_1(y_1; y_{-1})\eta_2(y_2; y_{-(1,2)} \mid y_{10}) \prod_{k=1}^{2} p(y_k \mid y_{-k0})h(y_{-(1,2)} \mid y_{10}, y_{20}) \prod_{k=1}^{G} dy_k}. \tag{2.47}$$

By repeatedly applying the odds ratio representation for $h(y_{-(1,2)} \mid y_{10}, y_{20})$, the joint density $h(y_1, \cdots, y_G)$ can eventually be represented in a form with odds ratio functions and conditional densities at fixed conditions as

$$h(y_1, \cdots, y_G) = \frac{\eta(y_1; \cdots; y_G) \prod_{k=1}^{G} p(y_k \mid y_{-k0})}{\int \cdots \int \eta(y_1; \cdots; y_G) \prod_{k=1}^{G} p(y_k \mid y_{-k0}) dy_k}, \tag{2.48}$$

where

$$\eta(y_1; \cdots; y_G) = \prod_{k=1}^{G-1} \eta_k\{y_k; (y_G, \cdots, y_{(k+1)}) \mid y_{(k-1)0}, \cdots, y_{10}\}.$$

Any conditional odds ratio function can be uniquely determined by the product of the odds ratio functions as follows.

$$\eta_c(y_i; y_j \mid y_{-(i,j)}) = \frac{\eta(y_1; \cdots; y_i; \cdots; y_j; \cdots; y_G)\eta(y_1; \cdots; y_{i0}; \cdots; y_{j0}; \cdots; y_G)}{\eta(y_1; \cdots; y_{i0}; \cdots; y_j; \cdots; y_G)\eta(y_1; \cdots; y_i; \cdots; y_{j0}; \cdots; y_G)},$$

for $1 \le i < j \le G$, where y_{-A} denotes $(y_j, j \notin A)$. Since

$$\eta_k\{y_k; (y_G, \cdots, y_{(k+1)}) \mid y_{(k-1)0}, \cdots, y_{10}\} = \eta_k\{y_k; (y_G, \cdots, y_{(k+1)}, y_{(k-1)0}, \cdots, y_{10})\},$$

the product of the odds ratio functions can be rewritten as

$$\eta(y_1; \cdots; y_G) = \prod_{k=1}^{G-1} \eta_k\{y_k; (y_G, \cdots, y_{(k+1)}, y_{(k-1)0}, \cdots, y_{10})\}.$$

Another form of the odds ratio representation is obtained by replacing the odds ratio functions in (2.48) by the foregoing expression.

The following examples demonstrate the odds ratio functions contained in various joint density functions and their odds ratio representations.

Example 2.4.1: For the joint normal model with mean μ and variance Σ, when $s_j = 1$ for $j = 1, \cdots, G$, the joint density is

$$f(y_1, \cdots, y_G) = \frac{1}{(2\pi)^{G/2}|\Sigma|^{1/2}} \exp\left\{\frac{1}{2}(Y-\mu)^t\Omega(Y-\mu)\right\},$$

where $\Omega = (w_{ij}) = -\Sigma^{-1}$. The odds ratio function

$$\eta_k\{y_k; (y_G, \cdots, y_{(k+1)}) \mid y_{(k-1)0}, \cdots, y_{10}\} = \exp\left\{\sum_{j=k+1}^{G} \omega_{kj}(y_k - y_{k0})(y_j - y_{j0})\right\}$$

The conditional odds ratio function becomes

$$\eta_c(y_i; y_j \mid y_{-(i,j)}) = \exp\{\omega_{ij}(y_i - y_{i0})(y_j - y_{j0})\}.$$

The conditional independence of Y_i and Y_j given $y_{-(i,j)}$ reduces to $w_{ij} = 0$. The odds ratio representation of the joint normal density is

$$f(y_1, \cdots, y_G) = \frac{\eta(y_1; \cdots; y_G)\prod_{k=1}^{G} f_k(y_k \mid y_{-k0})}{\int \cdots \int \eta(y_1; \cdots; y_G)\prod_{k=1}^{G} f_k(y_k \mid y_{-k0})dy_k},$$

where

$$f_k(y_k \mid y_{-k0}) = \frac{1}{\sqrt{2\pi|w_{kk}|^{-1}}} \exp\left\{\frac{1}{2}(y_k - \mu_{k0})^t w_{kk}^{-1}(y_k - \mu_{k0})\right\},$$

and

$$\mu_{k0} = \mu_k - \sum_{j \neq k} w_{kk}^{-1} w_{kj}(y_{j0} - \mu_j)$$

for $k = 1, \cdots, G$. Note that the odds ratio representation still holds when some or all Y_ks are vectors. In this case, w_{jk}, $j = 1, \cdots, n; k = 1, \cdots, n$ become the corresponding block matrices in the precision matrix.

Example 2.4.2: In the log-linear model for a general G-dimensional contingency table, the cell probability $h(y_1, \cdots, y_G) = P(Y_1 = y_1, \cdots, Y_G = y_G)$ is modeled as

$$h(y_1, \cdots, y_G) = \exp\left(\lambda_0 + \lambda_{y_1}^{Y_1} + \cdots + \lambda_{y_G}^{Y_G} + \cdots + \lambda_{y_1 \cdots y_G}^{Y_1 \cdots Y_G}\right),$$

where

$$\exp(-\lambda_0) = \sum_{y_1} \cdots \sum_{y_G} \exp\left(\lambda_{y_1}^{Y_1} + \cdots + \lambda_{y_G}^{Y_G} + \cdots + \lambda_{y_1 \cdots y_G}^{Y_1 \cdots Y_G}\right).$$

Let $(y_{10}, \cdots, y_{G0})$ be the fixed point in the sample space. The conditional probability of $Y_j = y_j$ given that $Y_{-j} = y_{-j0}$ is

$$p_j(y_j \mid y_{-j0}) = \frac{\exp\left(\lambda_{y_j}^{Y_j} + \lambda_{y_j + y_{-j0}}^{Y_j + Y_{-j0}}\right)}{\sum_{y_j} \exp\left(\lambda_{y_j}^{Y_j} + \lambda_{y_j + y_{-j0}}^{Y_j + Y_{-j0}}\right)},$$

where $\lambda_{y_j + y_{-j0}}^{Y_j + Y_{-j0}}$ denotes all the λ terms with $Y_j = y_j$ in and at least one additional $k \neq j$ in and $Y_k = y_{k0}$ for any $k \neq j$ that is in. The odds ratio function

$$\eta(y_1; \cdots; y_G) = \exp\left(\lambda_{y_1 y_2}^{Y_1 Y_2} + \cdots + \lambda_{y_1 \cdots y_G}^{Y_1 \cdots Y_G} - \sum_{j=1}^{G} \lambda_{y_j + y_{-j0}}^{Y_j + Y_{-j0}}\right\}.$$

The odds ratio representation of the joint probabilities is

$$h(y_1, \cdots, y_G) = \frac{\eta(y_1; \cdots; y_G) \prod_{j=1}^{G} p_j(y_j \mid y_{-j0})}{\sum_{y_1, \cdots, y_G} \eta(y_1; \cdots; y_G) \prod_{j=1}^{G} p_j(y_j \mid y_{-j0})} \tag{2.49}$$

As a special case, the cell probabilities of the $J \times K \times M$ table discussed in section 1.7, $h_{jkm}, j = 1, \cdots, J; k = 1, \cdots, K; m = 1, \cdots, M$, can be represented by the odds ratio functions and the conditional probabilities at a fixed condition.

The conditional odds ratio function becomes

$$\begin{aligned}\eta_c(y_i; y_j \mid y_{-(i,j)}) &= \exp\left(\lambda_{y_i y_j}^{Y_i Y_j} - \lambda_{y_{i0} y_j}^{Y_i Y_j} - \lambda_{y_i y_{j0}}^{Y_i Y_j} + \lambda_{y_{i0} y_{j0}}^{Y_i Y_j}\right) \\ &\quad \times \exp\left(\lambda_{y_i y_j + y_{-(ij)}}^{Y_i Y_j + Y_{-(ij)}} - \lambda_{y_{i0} y_j + y_{-(ij)}}^{Y_i Y_j + Y_{-(ij)}} - \lambda_{y_i y_{j0} + y_{-(ij)}}^{Y_i Y_j + Y_{-(ij)}} + \lambda_{y_{i0} y_{j0} + y_{-(ij)}}^{Y_i Y_j + Y_{-(ij)}}\right),\end{aligned}$$

where $\lambda_{y_i y_j + y_{-(ij)}}^{Y_i Y_j + Y_{-(ij)}}$ denotes all the λ terms with $Y_i Y_j$ in and at least one additional Y_k in with $k \neq i$ nor j.

2.7 Relationship between conditional and unconditional odds ratio functions

The conditional odds ratio functions connect to the unconditional odds ratio functions in the following ways.

$$\begin{aligned}\eta[(y, y_0); \{(x, z), (x_0, z_0)\}] &= \frac{b(y, x, z) b(y_0, x_0, z_0)}{b(y_0, x, z) b(y, x_0, z_0)} \\ &= \frac{b(y, x, z) b(y_0, x_0, z)}{b(y_0, x, z) p(y, x_0, z)} \frac{b(y, x_0, z) b(y_0, x_0, z_0)}{b(y_0, x_0, z) b(y, x_0, z_0)} \\ &= \eta\{(y, y_0); (x, x_0) \mid z\} \eta\{(y, y_0); (z, z_0) \mid x_0\}. \end{aligned} \tag{2.50}$$

That is,

$$\frac{\eta[(y, y_0); \{(x, z), (x_0, z_0)\}]}{\eta\{(y, y_0); (x, x_0) \mid z\}} = \eta\{(y, y_0); (z, z_0) \mid x_0\}.$$

Note that $\eta\{(y, y_0); (z, z_0) \mid x_0\} = \eta[(y, y_0); \{(x_0, z), (x_0, z_0)\}]$. The conditional odds ratio can be expressed in terms of unconditional odds ratio as

$$\eta\{(y, y_0); (x, x_0) \mid z\} = \frac{\eta[(y, y_0); \{(x, z), (x_0, z_0)\}]}{\eta[(y, y_0); \{(x_0, z), (x_0, z_0)\}]}. \tag{2.51}$$

Since x and z are in symmetric positions, it also holds that

$$\eta[(y, y_0); \{(x, z), (x_0, z_0)\}] = \eta\{(y, y_0); (z, z_0) \mid x) \eta\{(y, y_0); (x, x_0) \mid z_0\}, \tag{2.52}$$

or that

$$\frac{\eta[(y, y_0); \{(x, z), (x_0, z_0)\}]}{\eta\{(y, y_0); (z, z_0) \mid x)} = \eta\{(y, y_0); (x, x_0) \mid z_0\},$$

and that

$$\eta\{(y, y_0); (z, z_0) \mid x\} = \frac{\eta[(y, y_0); \{(x, z), (x_0, z_0)\}]}{\eta[(y, y_0); \{(x, z_0), (x_0, z_0)\}]}. \tag{2.53}$$

The relationship can be extended to the case with multiple groups of outcomes where

$$\eta(y_1;\cdots;y_G) = \prod_{k=1}^{G-1} \eta_k\{y_k;(y_{10},\cdots,y_{(k-1)0},y_{k+1},\cdots,y_G)\}$$

By the relationship of the unconditional odds ratio function to the conditional odds ratio function, for any k, it follows

$$\begin{aligned}\eta_k\{y_k;(y_{10},\cdots,y_{(k-1)0},y_{k+1},\cdots,y_G)\} \\ &= \eta_k\{y_k;(y_{10},\cdots,y_{(k-1)0},y_{k+1},\cdots,y_{G-1}) \mid y_G\} \\ &\quad \times\eta\{y_k;y_G \mid y_{10},\cdots,y_{(k-1)0},y_{(k+1)0},\cdots,y_{(G-1)0}\}.\end{aligned}$$

Plugging the expression into the multi-group odds ratio functions to obtain

$$\begin{aligned}\eta(y_1;\cdots;y_G) &= \prod_{k=1}^{G-1} \eta_k\{y_k;(y_{10},\cdots,y_{(k-1)0},y_{k+1},\cdots,y_{G-1}) \mid y_G\} \\ &\quad \times \prod_{k=1}^{G-1} \eta\{y_k;y_G \mid y_{10},\cdots,y_{(k-1)0},y_{(k+1)0},\cdots,y_{(G-1)0}\} \\ &= \eta(y_1;\cdots;y_{G-1} \mid y_G) \prod_{k=1}^{G-1} \eta\{y_k;y_G \mid y_{10},\cdots,y_{(k-1)0},y_{(k+1)0},\cdots,y_{(G-1)0}\}.\end{aligned}$$

This means

$$\frac{\eta(y_1;\cdots;y_G)}{\eta(y_1;\cdots;y_{G-1} \mid y_G)} = \prod_{k=1}^{G-1} \eta\{y_k;y_G \mid y_{10},\cdots,y_{(k-1)0},y_{(k+1)0},\cdots,y_{(G-1)0}\}. \tag{2.54}$$

This formula is useful in dealing with conditional models. For example, a commonly encountered situation is two outcomes with a vector of covariates where $G = 3$ and $y_G = x$ denoting the covariates. Then

$$\eta(y_1;y_2;x) = \eta(y_1;y_2 \mid x)\eta(y_1;x \mid y_{20})\eta(y_2;x \mid y_{10}).$$

These are odds ratio functions in the joint density for (Y_1, Y_2, X) or the joint conditional density $(Y_1, Y_2 \mid X)$. For the conditional density $(Y_1 \mid Y_2, X)$, the odds ratio functions are

$$\eta(y_1;y_2 \mid x)\eta(y_1;x \mid y_{20}),$$

and the odds ratio functions corresponding to the conditional density for $(Y_2 \mid Y_1, X)$ are

$$\eta(y_1;y_2 \mid x)\eta(y_2;x \mid y_{10}).$$

If we have three outcomes and a vector of covariates, then

$$\eta(y_1;y_2;y_3;x) = \eta(y_1;y_2;y_3 \mid x)\eta(y_1;x \mid y_{20},y_{30})\eta(y_2;x \mid y_{10},y_{30})\eta(y_3;x \mid y_{10},y_{20}).$$

These odds ratio functions are for densities of either (Y_1, Y_2, Y_3, X) or $(Y_1, Y_2, Y_3 \mid X)$. For the density $(Y_1, Y_2 \mid Y_3, X)$, the odds ratio functions are

$$\frac{\eta(y_1;y_2;y_3;x)}{\eta(y_3;x \mid y_{10},y_{20})} = \eta(y_1;y_2;y_3 \mid x)\eta(y_1;x \mid y_{20},y_{30})\eta(y_2;x \mid y_{10},y_{30}).$$

Similarly, the odds ratio functions for $(Y_1, Y_3 \mid Y_2, X)$ and $(Y_2, Y_3 \mid Y_1, X)$ are respectively

$$\frac{\eta(y_1; y_2; y_3; x)}{\eta(y_2; x \mid y_{10}, y_{30})} = \eta(y_1; y_2; y_3 \mid x)\eta(y_1; x \mid y_{20}, y_{30})\eta(y_3; x \mid y_{10}, y_{20})$$

and

$$\frac{\eta(y_1; y_2; y_3; x)}{\eta(y_1; x \mid y_{20}, y_{30})} = \eta(y_1; y_2; y_3 \mid x)\eta(y_2; x \mid y_{10}, y_{30})\eta(y_3; x \mid y_{10}, y_{20}).$$

For the conditional density $(Y_1 \mid Y_2, Y_3, X)$, the odds ratio function is

$$\frac{\eta(y_1; y_2; y_3; x)}{\eta(y_2; y_3; x \mid y_{10})} = \eta\{y_1; (y_2, y_3) \mid x\}\eta(y_1; x \mid y_{20}, y_{30}).$$

Similarly, odds ratio functions for the conditional densities $(Y_2 \mid Y_1, Y_3, X)$ and $(Y_3 \mid Y_1, Y_2, X)$ can be obtained respectively as

$$\frac{\eta(y_1; y_2; y_3; x)}{\eta(y_1; y_3; x \mid y_{10})} = \eta\{y_2; (y_1, y_3) \mid x\}\eta(y_2; x \mid y_{10}, y_{30}).$$

and

$$\frac{\eta(y_1; y_2; y_3; x)}{\eta(y_1; y_2; x \mid y_{10})} = \eta\{y_3; (y_1, y_2) \mid x\}\eta(y_3; x \mid y_{10}, y_{20}).$$

Odds ratio functions for any conditional density can be obtained in the same way.

If we apply the same formula again to $\eta(y_1; \cdots; y_{G-1} \mid y_G)$ in (2.54), we can obtain

$$\frac{\eta(y_1; \cdots; y_{G-1} \mid y_G)}{\eta(y_1; \cdots; y_{G-2} \mid y_{G-1}, y_G)} = \prod_{k=1}^{G-2} \eta\{y_k : y_{G-1} \mid y_{10}, \cdots, y_{(k-1)0}, y_{(k+1)0}, \cdots, y_{(G-2)0}, y_G\},$$

$$\cdots \quad \cdots$$

$$\frac{\eta(y_1; y_2; y_3 \mid y_4 \cdots, y_G)}{\eta(y_1; y_2 \mid y_3, \cdots, y_G)} = \prod_{k=1}^{2} \eta\{y_k; y_3 \mid y_{(\{1,2\}-\{k\})0}, y_4, \cdots, y_G\}.$$

Combining (2.54) with the foregoing equalities, the joint odds ratio functions can be expressed in terms of the conditional odds ratio functions as

$$\eta(y_1; \cdots; y_G) = \prod_{j=2}^{G} \prod_{k=1}^{j-1} \eta\{y_k : y_j \mid y_{10}, \cdots, y_{(k-1)0}, y_{(k+1)0}, \cdots, y_{(j-1)0}, y_{(j+1)}, \cdots, y_G\}. \tag{2.55}$$

By plugging (2.55) into (2.48), we obtain another odds ratio representation in terms of pairwise conditional odds ratio functions and conditional densities at fixed conditions. Variations of the pairwise conditional odds ratio representation can be obtained by changing the order of the variables.

2.8 Hierarchical odds ratio representation for a joint density

The relationship between the conditional odds ratio function and the unconditional odds ratio functions can be used to obtain a representation of a joint density in terms of the hierarchical odds ratio functions. To see that, consider first the simple case with three variables (y_1, y_2, y_3). Applying (2.54), we have

$$\frac{\eta(y_1; y_2; y_3)}{\eta(y_1; y_2 \mid y_3)} = \eta(y_1; y_3 \mid y_{20})\eta(y_2; y_3 \mid y_{10}).$$

This implies

$$\eta(y_1; y_2; y_3) = \frac{\eta(y_1; y_2 \mid y_3)}{\eta(y_1; y_2 \mid y_{30})}\eta(y_1; y_2 \mid y_{30})\eta(y_1; y_3 \mid y_{20})\eta(y_2; y_3 \mid y_{10}).$$

Using a different condition, we can also obtain

$$\frac{\eta(y_1; y_2; y_3)}{\eta(y_1; y_3 \mid y_2)} = \eta(y_1; y_2 \mid y_{30})\eta(y_2; y_3 \mid y_{10}),$$

and thus

$$\eta(y_1; y_2; y_3) = \frac{\eta(y_1; y_3 \mid y_2)}{\eta(y_1; y_2 \mid y_{20})}\eta(y_1; y_2 \mid y_{30})\eta(y_1; y_3 \mid y_{20})\eta(y_2; y_3 \mid y_{10}).$$

Furthermore, from

$$\frac{\eta(y_1; y_2; y_3)}{\eta(y_2; y_3 \mid y_1)} = \eta(y_2; y_3 \mid y_{10})\eta(y_1; y_3 \mid y_{20}),$$

it follows that

$$\eta(y_1; y_2; y_3) = \frac{\eta(y_2; y_3 \mid y_1)}{\eta(y_2; y_3 \mid y_{10})}\eta(y_1; y_2 \mid y_{30})\eta(y_1; y_3 \mid y_{20})\eta(y_2; y_3 \mid y_{10}).$$

Hence,

$$\frac{\eta(y_1; y_2 \mid y_3)}{\eta(y_1; y_2 \mid y_{30})} = \frac{\eta(y_1; y_3 \mid y_2)}{\eta(y_1; y_3 \mid y_{20})} = \frac{\eta(y_2; y_3 \mid y_1)}{\eta(y_2; y_3 \mid y_{10})}. \tag{2.56}$$

Define the ratio of the odds ratios as $\eta(y_1 : y_2 : y_3)$ which represents the higher order association. The odds ratio functions can now be represented as

$$\eta(y_1; y_2; y_3) = \eta(y_1 : y_2 : y_3)\eta(y_1; y_2 \mid y_{30})\eta(y_1; y_3 \mid y_{20})\eta(y_2; y_3 \mid y_{10}).$$

The hierarchical representation can also be carried out for the joint density with G groups of variables. Note first that

$$\begin{aligned}
\eta(y_1; \cdots; y_G) &= \eta(y_1; \cdots; y_{G-1} \mid y_G) \prod_{k=1}^{G-1} \eta\{y_k; y_G \mid y_{-(k,G)0}\} \\
&= \frac{\eta(y_1; \cdots; y_{G-1} \mid y_G)}{\prod_{j \neq G} \prod_{k \neq j} \eta(y_k; y_j \mid y_{-(k,j)0})} \prod_{j=1}^{G} \prod_{k \neq j} \eta\{y_k; y_j \mid y_{-(k,j)0}\} \\
&= \frac{\eta(y_1; \cdots; y_{s-1}; y_{s+1}; \cdots; y_G \mid y_s)}{\prod_{j \neq s} \prod_{k \neq j} \eta(y_k; y_j \mid y_{-(k,j)0})} \prod_{j=1}^{G} \prod_{k \neq j} \eta(y_k; y_j \mid y_{-(k,j)0}),
\end{aligned}$$

for $s = 1, \cdots, G-1$. It can be seen that

$$\frac{\eta(y_1; \cdots; y_{G-1} \mid y_G)}{\prod_{j \neq G} \prod_{k \neq j} \eta(y_k; y_j \mid y_{-(k,j)0})} = \frac{\eta(y_1; \cdots; y_{s-1}; y_{s+1}; \cdots; y_G \mid y_s)}{\prod_{j \neq s} \prod_{k \neq j} \eta(y_k; y_j \mid y_{-(k,j)0})},$$

for $s = 1, \cdots, G-1$. The term

$$\xi_1(y_1 : \cdots : y_G) = \prod_{j=1}^{G} \prod_{k \neq j} \eta(y_k; y_j \mid y_{-(k,j)0})$$

is of the first order. Note next that

$$\begin{aligned}\eta(y_1; \cdots; y_{G-1} \mid y_G) &= \eta(y_1; \cdots; y_{G-2} \mid y_{G-1}, y_G) \prod_{k=1}^{G-2} \eta\{y_k; y_{G-1} \mid y_{-(k,G-1,G)0)0}, y_G\} \\ &= \frac{\eta(y_1; \cdots; y_{G-2} \mid y_{G-1}, y_G)}{\prod_{j \neq G-1,G} \prod_{k \neq j} \eta\{y_k; y_j \mid y_{-(j,k,G)0}, y_G\}} \prod_{j \neq G} \prod_{k \neq j} \eta\{y_k; y_j \mid y_{-(j,k,G)0}, y_G\} \\ &= \frac{\eta(y_1; \cdots; y_{t-1}; y_{t+1}; \cdots; y_{G-1} \mid y_t, y_G)}{\prod_{j \neq t,G} \prod_{k \neq j} \eta\{y_k; y_j \mid y_{-(j,k,G)0}, y_G\}} \prod_{j \neq G} \prod_{k \neq j} \eta\{y_k; y_j \mid y_{-(j,k,G)0}, y_G\}.\end{aligned}$$

It follows that

$$\frac{\eta(y_1; \cdots; y_{G-2} \mid y_{G-1}, y_G)}{\prod_{j \neq G-1,G} \prod_{k \neq j} \eta\{y_k; y_j \mid y_{-(j,k,G)0}, y_G\}} = \frac{\eta(y_1; \cdots; y_{t-1}; y_{t+1}; \cdots; y_{G-1} \mid y_t, y_G)}{\prod_{j \neq t,G} \prod_{k \neq j} \eta\{y_k; y_j \mid y_{-(j,k,G)0}, y_G\}},$$

for $t = 1, \cdots, G-2$. This implies that

$$\xi_2(y_1 : \cdots : y_{G-1} \mid y_G) = \prod_{j \neq G} \prod_{k \neq j} \frac{\eta(y_k; y_j \mid y_{-(j,k)0}, y_G)}{\eta(y_k; y_j \mid y_{-(k,j)0})} = \prod_{j \neq s} \prod_{k \neq j} \frac{\eta(y_k; y_j \mid y_{-(j,k)0}, y_s)}{\eta(y_k; y_j \mid y_{-(k,j)0})},$$

which is of the second order. Continue the operation to obtain higher order terms. Lastly,

$$\begin{aligned}\eta(y_1; y_2; y_3 \mid y_4 \cdots, y_G) &= \eta(y_1; y_2 \mid y_3, \cdots, y_G) \prod_{k=1}^{2} \eta\{y_k; y_3 \mid y_{(\{1,2\}-\{k\})0}, y_4, \cdots, y_G\} \\ &= \frac{\eta(y_1; y_2 \mid y_3, \cdots, y_G)}{\prod_{j=1}^{2} \prod_{k \neq j} \eta\{y_k; y_j \mid y_{-(\{1,2,3\}-\{j,k\})0}, y_4, \cdots, y_G\}} \\ &\quad \prod_{j=1}^{3} \prod_{k \neq j} \eta\{y_k; y_j \mid y_{-(\{1,2,3\}-\{j,k\})0}, y_4, \cdots, y_G\}.\end{aligned}$$

The term

$$\xi_{G-1}(y_1 : y_2 \mid y_3, \cdots, y_G) = \frac{\eta(y_1; y_2 \mid y_3, \cdots, y_G)}{\prod_{j=1}^{2} \prod_{k \neq j} \eta\{y_k; y_j \mid y_{-(\{1,2,3\}-\{j,k\})0}, y_4, \cdots, y_G\}}$$

is the highest order term, and

$$\xi_{G-2}(y_1 : y_2 : y_3 \mid y_4, \cdots, y_G) = \prod_{j=1}^{3} \prod_{k \neq j} \frac{\eta\{y_k; y_j \mid y_{-(\{1,2,3\}-\{j,k\})0}, y_4, \cdots, y_G\}}{\eta\{y_k; y_j \mid y_{-(\{1,2,3\}-\{j,k\})0}, y_5, \cdots, y_G\}}$$

is the next to the highest term. The hierarchical representation is

$$\eta(y_1; \cdots; y_G) = \prod_{k=1}^{G-1} \xi_k(y_1 : \cdots : y_{G-k+1} \mid y_{G-k+2}, \cdots, y_G).$$

The hierarchical representation of the odds ratio functions of a joint density function acquired symmetry in the expression. This can be useful in theoretical proofs, see for example, Malinsky et al (2021)[118]. However, in terms of model specification, constraints that may not be easy to keep track of are introduced.

2.9 Odds ratio functions embedding in a density function

A bivariate function is an odds ratio function with respect to a reference point (y_0, x_0) if and only if

$$\begin{aligned} \eta\{(y,y_0);(x,x_0)\} &> 0, \\ \eta\{(y_0,y_0);(x,x_0)\} &= 1, \\ \eta\{(y,y_0);(x_0,x_0)\} &= 1. \end{aligned}$$

For a given odds ratio function, it is interesting to know if a density function can be found to generate the given odds ratio function. A sufficient condition for a given odds ratio function to be generated by a density function is that we can find some density functions $p(y)$ on $\mathcal{Y}$ and $q(x)$ on $\mathcal{X}$ such that

$$\int\int \eta(\{y,y_0);(x,x_0)\}p(y)q(x)d\mu_y d\mu_x < \infty, \tag{2.57}$$

for some known measures μ_y on $\mathcal{Y}$ and μ_x on $\mathcal{X}$. When (2.57) holds, the following density function

$$h(y,x) = \frac{\eta(\{y,y_0);(x,x_0)\}p(y)q(x)}{\int \eta(\{y,y_0);(x,x_0)\}p(y)q(x)d\mu_y d\mu_x}$$

has the given odds ratio function embedded in it.

For a given odds ratio function, such densities $p(y)$ and $q(x)$ may not always exist such that (2.57) is satisfied. For example, consider the following bivariate function defined on $[0,\infty) \times [0,\infty)$ such that

$$\zeta(y,x) = \begin{cases} \frac{1}{|1-xy|} & \text{if } xy \neq 1, \\ 1 & \text{if } xy = 1. \end{cases}$$

It can be seen that conditions (2.7)-(2.9) are satisfied with reference point $(0,0)$, i.e., $\zeta(y,x) > 0$ and

$$\zeta(y,0) = 1 = \zeta(0,x),$$

for all $(x,y) \in [0,\infty) \times [0,\infty)$. This means that $\zeta(y,x)$ is an odds ratio function. Since, for any given density function $q(x)$,

$$\int \frac{q(x)}{|1-xy|}dx = \int_0^{1/y} \frac{q(x)}{|1-xy|}dx + \int_{1/y}^{\infty} \frac{q(x)}{|1-xy|}dx.$$

Either of the integral is $+\infty$ for almost all $y \in (0,\infty)$. Therefore, we cannot find non-trivial density functions $p(y)$ and $q(x)$ such that

$$\int\int \zeta(y,x)p(y)q(x)dxdy < \infty.$$

This means $\zeta(y,x)$ cannot be embedded into a density function.

For an odds ratio function $\eta\{(y,y_0);(x,x_0)\}$, if there exist two functions $\xi(x)$ and $\delta(y)$ such that

$$\eta\{(y,y_0);(x,x_0)\} \leq \xi(x)\delta(y), \tag{2.58}$$

then $\eta(y;x)$ can be embedded into a density function. For the previously given odds ratio function $\zeta(y;x)$, it can be seen that, for any $x \in (0,\infty)$ and any large number A, we can find a y value such that $\zeta(y;x) > A$. This means $\zeta(y;x)$ does not satisfy (2.58).

Theorem 2.1. Suppose that $\eta\{(y,y_0);(x,x_0)\}$ is a measurable bivariate function defined on the measurable space $(\mathcal{Y}\times\mathcal{X}, B, \mu_y\times\mu_x)$ satisfying (2.7)-(2.9). If there exist two measurable functions $\xi(x)$ and $\delta(y)$ such that (2.58) holds for all x, y, then there exist positive probability density functions $p(y) > 0, y \in \mathcal{Y}$ with respect to the σ-finite measure μ_y, and $q(x) > 0, x \in \mathcal{X}$ with respect to the σ-finite measure μ_x, such that

$$\int\int \eta\{(y,y_0);(x,x_0)\}p(y)q(x)d\mu_y d\mu_x < +\infty.$$

It follows from this that the odds ratio function $\eta\{(y,y_0);(x,x_0)\}$ can be embedded into the density

$$h(y,x) = \frac{\eta\{(y,y_0);(x,x_0)\}p(y)q(x)}{\int\int \eta\{(y,y_0);(x,x_0)\}p(y)q(x)d\mu_y d\mu_x}.$$

Proof: Under (2.58), there exist two densities $p(y)$ and $q(x)$ such that

$$\int\int \eta\{(y,y_0);(x,x_0)\}p(y)q(x)d\mu_y d\mu_x < \infty.$$

The density $p(y)$ can be found in the following way. Let $\phi(y)$ be the density function such that $\phi(y) \geq 0$ and $\int \phi(y)d\mu_y = 1$. Let $A_m = \{y \mid m-1 < \delta(y) \leq m\}$. Define the density as

$$p(y) = \frac{\sum_{m=1}^{\infty} m^{-1}2^{-m}1_{\{y\in A_m\}}\phi(y)}{\sum_{m=1}^{\infty} m^{-1}2^{-m}\Phi(A_m)},$$

where $\Phi(A_m) = \int_{A_m}\phi(y)d\mu_y$. It follows that

$$\int \delta(y)p(y)d\mu_y \leq \frac{\sum_{m=1}^{\infty} 2^{-m}\Phi(A_m)}{\sum_{m=1}^{\infty} m^{-1}2^{-m}\Phi(A_m)} < \infty.$$

Similarly, we can find a density $q(x)$ such that $\int \xi(x)q(x)d\mu_x < \infty$.

Condition (2.58) is satisfied for frequently encountered odds ratio functions. First, if there are k pairs of such functions such that

$$\eta(y;x) \leq \delta_1(y)\xi_1(x) + \cdots + \delta_k(y)\xi_k(x),$$

(2.58) is satisfied. This is because

$$\begin{aligned} \eta(y;x) &\leq |\delta_1(y)\xi_1(x) + \cdots + \delta_k(y)\xi_k(x)| \\ &\leq \{|\delta_1(y)| + \cdots + |\delta_k(y)|\}\{|\xi_1(x)| + \cdots + |\xi_k(x)|\}. \end{aligned}$$

Furthermore, if we have

$$\log\eta(y;x) \leq \delta_1(y)\xi_1(x),$$

then condition (2.58) is satisfied because

$$\eta(y;x) \leq \exp\left\{\frac{1}{2}\delta_1^2(y)\right\}\exp\left\{\frac{1}{2}\xi_1^2(x)\right\}.$$

In addition, if there are k pairs of such functions such that

$$\log\eta(y;x) \leq \delta_1(y)\xi_1(x) + \cdots + \delta_k(y)\xi_k(x),$$

(2.58) is also satisfied because

$$\log \eta(y;x) \;\leq\; |\log \eta(y;x)| \leq \{|\delta_1(y)| + \cdots + |\delta_k(y)|\}\{|\xi_1(x)| + \cdots + |\xi_k(x)|\}.$$

The combination of these basic inequalities can cover fairly complicated functions.

Condition (2.58), however, is not necessary for an odds ratio function to be embedded into a density. For example, consider the following joint density defined on $\mathcal{X} \times \mathcal{Y} = [0,2] \times [0,2]$,

$$f(y,x) = \begin{cases} \frac{cy}{\sqrt{|1-xy|}} & \text{if } xy \neq 1, \\ 1 & \text{if } xy = 1. \end{cases}$$

where $c = (9\sqrt{3} - 15)/4$. This is a density function because

$$\begin{aligned} \int_0^2 \int_0^2 f(y,x)dxdy &= \int_0^{1/2} \int_0^2 \frac{cy}{\sqrt{|1-xy|}} dxdy \\ &+ \int_{1/2}^2 \left\{ \int_0^{1/y} \frac{cy}{\sqrt{|1-xy|}} dx + \int_{1/y}^2 \frac{cy}{\sqrt{|1-xy|}} dx \right\} dy \\ &= \frac{c}{3} + 3c + 2\sqrt{3}c = \frac{10 + 6\sqrt{3}}{3} c. \end{aligned}$$

The odds ratio function with reference point $(0,0)$,

$$\eta(y,x) = \begin{cases} \frac{1}{\sqrt{|1-xy|}} & \text{if } xy \neq 1, \\ 1 & \text{if } xy = 1, \end{cases}$$

is embedded in the density function. On the other hand, for any $x \in [1/2, 2]$ and any large number A, there exists a y value such that $f(x,y) > A$. There do not exist functions $\xi(x)$ on $x \in [1/2, 2]$ and $\delta(y)$ on $y \in [1/2, 2]$ such that

$$\eta(y;x) \leq \xi(x)\delta(y),$$

for $(y,x) \in [1/2,2] \times [1/2,2]$. This means (2.58) is not satisfied.

In general, Theorem 2.1 can be extended to including multivariate cases. Suppose that $\eta_j\{y_j; (y_{10}, \cdots, y_{(j-1)0}, y_{j+1}, \cdots, y_G)\}, j = 1, \cdots, G-1$ are a set of odds ratio functions with reference point $(y_{10}, \cdots, y_{G0})$. Let

$$\eta(y_1; \cdots; y_G) = \prod_{j=1}^{G-1} \eta_j\{y_j; (y_{10}, \cdots, y_{(j-1)0}, y_{j+1}, \cdots, y_G)\}.$$

If there exist density functions $p_j(y_j), j = 1, \cdots, G$ such that

$$\int \cdots \int \eta(y_1; \cdots; y_G) \prod_{j=1}^{G} \{p_j(y_j)dy_j\} < \infty, \tag{2.59}$$

then $\eta(y_1; \cdots; y_G)$ can be embedded into the joint density

$$h(y_1, \cdots, y_G) = \frac{\eta(y_1; \cdots; y_G) \prod_{j=1}^{G} p_j(y_j)}{\int \cdots \int \eta(y_1; \cdots; y_G) \prod_{j=1}^{G} \{p_j(y_j)dy_j\}}. \tag{2.60}$$

Corollary 2.1. Suppose that

$$\eta(y_1; \cdots; y_G) = \prod_{j=1}^{G-1} \eta_j\{y_j; (y_{10}, \cdots, y_{(j-1)0}, y_{j+1}, \cdots, y_G)\}.$$

where $\eta_j\{y_j; (y_{10}, \cdots, y_{(j-1)0}, y_{j+1}, \cdots, y_G)\}, j = 1, \cdots, G-1$ are measurable odds ratio functions with reference point $(y_{10}, \cdots, y_{G0})$. If there exist measurable functions $\xi_j(y_j), j = 1, \cdots, G$ such that

$$\eta(y_1; \cdots; y_G) \leq \prod_{j=1}^{G} \xi_j(y_j).$$

Then there exists density functions $p_j(y_j), j = 1, \cdots, G$ such that

$$\int \cdots \int \eta(y_1; \cdots; y_G) \prod_{j=1}^{G} \{p_j(y_j) dy_j\} < \infty.$$

It follows from this that $\eta(y_1; \cdots; y_G)$ is embedded in the joint density (2.60).

2.10 Modeling odd ratio functions in a family of densities

Modeling odds ratio function may start from an arbitrary positive bivariate function of (y, x) denote by $a(y, x)$. Let

$$\eta\{(y, y_0); (x, x_0)\} = \frac{a(y, x)a(y_0, x_0)}{a(y, x_0)a(y_0, x)}. \tag{2.61}$$

By taking logarithm on both side, it follows that

$$\log \eta\{(y, y_0); (x, x_0)\} = \log a(y, x) - \log a(y, x_0) - \log a(y_0, x) + \log a(y_0, x_0). \tag{2.62}$$

Since the only restriction required for a is a positive function, there is no restriction on the form of the bivariate function $\log a(y, x)$. In other words, for arbitrary bivariate function $\xi(y, x)$,

$$b(y, x) = \xi(y, x) - \xi(y_0, x) - \xi(y, x_0) + \xi(y_0, x_0), \tag{2.63}$$

is the logarithm of an odds ratio function or $\eta\{(y, y_0); (x, x_0)\} = \exp\{b(y, x)\}$ is an odds ratio function. If an odds ratio function satisfies the integrable condition (2.57) for some densities $p(y)$ and $q(x)$, the odds ratio function can be embedded into a density. For many odds ratio functions, (2.58) is satisfied. For example, if $\xi(y, x) = h_1(y) + h_2(x) + \gamma h_3(y) h_4(x)$, the logarithm of the pre-odds ratio function generated by $\xi(y, x)$ has the form

$$\log \eta\{(y, y_0); (x, x_0)\} = \gamma\{h_3(y) - h_3(y_0)\}\{h_4((x) - h_4(x_0)\}.$$

Condition (2.58) is satisfied and therefore the odds ratio function can be embedded into a density. More specifically, if we choose a linear function for $\xi(y, x) = \beta_0 + \beta_1 y + \beta_2 x$, then $\xi(y, x) = 0$. This corresponds to independence of randoma variables Y and X. The simplest nontrivial odds ratio function can be obtained when $\xi(y, x) = \gamma y x$. In this case, the logarithm of odds ratio function generated by ξ is

$$\begin{aligned} b(y, x) &= \gamma y x - \gamma y_0 x - \gamma y x_0 + \gamma y_0 x_0 \\ &= \gamma (y - y_0)(x - x_0). \end{aligned} \tag{2.64}$$

This log-odds ratio function is embedded the generalized linear model with canonical link function, which includes the linear logistic regression, the linear Poisson regression, and the normal linear regression as special cases. For a more general function $\xi(y,x)=\gamma y e^{\beta x}$,

$$\begin{aligned} b(y,x) &= \gamma y e^{\beta x} - \gamma y_0 e^{\beta x} - \gamma y e^{\beta x_0} + \gamma y_0 e^{\beta x_0} \\ &= \gamma(y-y_0)(e^{\beta x}-e^{\beta x_0}). \end{aligned} \tag{2.65}$$

This odds ratio function is embedded in the proportional hazard regression with exponential baseline survival distribution. More general odds ratio function can be obtained when more complex $\xi(y,x)$ is specified. For example, $\xi(y,x)=\Lambda(y)e^{\beta x}$ yields the odds ratio function (2.15) for the Cox regression model when Λ is a monotonic increasing function. Use this approach, other less familiar odds ratio function can be generated. For example, for a quadratic function $\xi(y,x)=\gamma y^2 x^2$,

$$\begin{aligned} b(y,x) &= \gamma y^2x^2 - \gamma y_0^2x^2 - \gamma y^2 x_0^2 + \gamma y_0^2 x_0^2 \\ &= \gamma(y^2-y_0^2)(x^2-x_0^2) \qquad (2.66) \\ &= \gamma(y+y_0)(x+x_0)(y-y_0)(x-x_0) \qquad (2.67) \end{aligned}$$

Conditional odds ratio function can be obtained through the unconditional odds ratio function using stratification arguments. For example, conditional odds ratio function can be obtained by setting the γ parameter to be a function of z. For the log-bilinear odds ratio function, a conditional odds ratio function is

$$\log \eta\{(y,y_0);(x,x_0)\mid z\} = \gamma(z)(y-y_0)(x-x_0).$$

$\gamma(z)$ may be further modeled by a linear function as

$$\gamma(z)=\gamma_0+\gamma_1 z.$$

More generally, a conditional log-odds ratio function may be specified in the form

$$\log \eta\{(y,y_0);(x,x_0)\mid z\} = \gamma(x,y,z)(y-y_0)(x-x_0).$$

By specifying different functional forms for $\gamma(x,y,z)$, different models for the conditional odds ratio can be obtained.

For the joint odds ratio functions, a frequently employed model is the bilinear odds ratio model, i.e.,

$$\log \eta(y_1;\cdots;y_G) = \sum_{j=2}^{G}\sum_{k=1}^{j-1}\theta_{jk}(y_j-y_{j0})(y_k-y_{k0}).$$

A morel complex model specifies parametric models for each pairwise conditional odds ratio function in the joint odds ratio model. The pairwise conditional odds ratio models are

$$\begin{aligned} &\log \eta\{y_k : y_j \mid y_{10},\cdots,y_{(k-1)0},y_{(k+1)0},\cdots,y_{(j-1)0},y_{(j+1)},\cdots,y_G\} \\ &= \gamma_{kj}\{y_{(j+1)},\cdots,y_G\}(y_j-y_{j0})(y_k-y_{k0}), \end{aligned}$$

where $\gamma_{kj}\{y_{(j+1)},\cdots,y_G\}$ may be further modeled by

$$\gamma_{kj}\{y_{(j+1)},\cdots,y_G\} = \gamma_{kj0}+\gamma_{kj(j+1)}y_{j+1}+\cdots+\gamma_{kjG}y_G.$$

for $k=1,\cdots,j-1; j=2,\cdots,G$.

Once the odds ratio function is modeled, the odds ratio model may be embedded into a density function by specifying the conditional density functions at a fixed condition. These conditional densities at a fixed condition need to satisfy (2.57). When the ranges of Y and X are finite and the odds ratio function is bounded, requirement (2.57) is always satisfied. In general, requirement (2.57) may impose restrictions on the models $p(y)$ and $q(x)$. For example, let $(y_0, x_0) = (0, 0)$,

$$\eta(y; x) = \exp(\gamma y x),$$

and

$$\begin{aligned} \frac{dP}{dy} &= \frac{1}{\sqrt{2\pi}\sigma_1} \exp\left\{-\frac{1}{2\sigma_1^2}(y-\mu_1)^2\right\}, \\ \frac{dQ}{dx} &= \frac{1}{\sqrt{2\pi}\sigma_2} \exp\left\{-\frac{1}{2\sigma_2^2}(x-\mu_2)^2\right\}. \end{aligned}$$

It then follows that the integrable condition (2.57) is equivalent to the requirement that

$$\sigma_1^2\sigma_2^2\gamma^2 < 1.$$

This requirement corresponds to the squared correlation coefficient $\rho^2 < 1$ in a bivariate normal model. This can be seen by noticing that in a bivariate normal model, $\sigma_1^2 = (1-\rho^2)\sigma_y^2$, $\sigma_2^2 = (1-\rho^2)\sigma_x^2$, and $\gamma = \rho/\{(1-\rho^2)\sigma_y\sigma_x\}$, which means

$$\rho^2 = \sigma_1^2\sigma_2^2\gamma^2.$$

Further along this line, when either y or x or both are vectors, the linear odds ratio function is

$$\begin{aligned} \log \eta\{(y, y_0); (x, x_0)\} &= \sum_{j=1}^{s}\sum_{k=1}^{r} \gamma_{jk}(y_j - y_{j0})(x_k - x_{k0}) \\ &= (y-y_0)'\Sigma_{12}(x-x_0). \end{aligned}$$

If the densities of $P(y)$ and $Q(x)$ are modeled respectively by

$$\begin{aligned} \frac{dP}{dy} &= \frac{1}{\sqrt{2\pi|\Sigma_{11}|}} \exp\left\{-\frac{1}{2}(y-\mu_1)'\Sigma_{11}^{-1}(y-\mu_1)\right\}, \\ \frac{dQ}{dx} &= \frac{1}{\sqrt{2\pi|\Sigma_{22}|}} \exp\left\{-\frac{1}{2}(x-\mu_2)'\Sigma_{22}^{-1}(x-\mu_2)\right\}. \end{aligned}$$

Note that

$$-\frac{1}{2}(y-\mu_1)'\Sigma_{11}^{-1}(y-\mu_1) - \frac{1}{2}(x-\mu_2)'\Sigma_{22}^{-1}(x-\mu_2) + (y-y_0)'\Sigma_{12}(x-x_0)$$

$$= Const - \frac{1}{2}(y-\mu_y, x-\mu_x)\begin{pmatrix} \Sigma_{11}^{-1} & -\Sigma_{12} \\ -\Sigma_{12}' & \Sigma_{22}^{-1} \end{pmatrix}\begin{pmatrix} y-\mu_y \\ x-\mu_x \end{pmatrix},$$

where

$$\begin{pmatrix} \mu_y \\ \mu_x \end{pmatrix} = \begin{pmatrix} \Sigma_{11}^{-1} & -\Sigma_{12} \\ -\Sigma_{12}' & \Sigma_{22}^{-1} \end{pmatrix}^{-1}\begin{pmatrix} \Sigma_{11}^{-1}\mu_1 - \Sigma_{12}x_0 \\ \Sigma_{11}^{-1}\mu_2 - \Sigma_{12}y_0 \end{pmatrix}.$$

Requirement (2.57) is true if and only if

$$\begin{pmatrix} \Sigma_{11}^{-1} & -\Sigma_{12} \\ -\Sigma_{12}' & \Sigma_{22}^{-1} \end{pmatrix} > 0,$$

That is, the components in the odds ratio representation were derived from a true joint normal density.

2.11 Semiparametric odds ratio model

When a model of the odds ratio function can be embedded into densities to form a family of densities, we obtain an odds ratio model. Most existing models may be reformulated in this framework. On the other hand, we may model the odds ratio functions parametrically, and leave the conditional distributions at a fixed condition unspecified. We terms such a model semiparametric odds ratio model.

When the variables are divided into two groups, the semiparametric model appears as

$$p(y,x)dydx = \frac{\eta(y;x \mid \theta)dP(y)dQ(x)}{\int\int \eta(y;x \mid \theta)dP(y)dQ(x)}. \tag{2.68}$$

where θ is an unknown parameter and P and Q are unknown distribution functions satisfying

$$\int\int \eta(y;x \mid \theta)dP(y)dQ(x) < \infty,$$

for all $\theta \in \Theta$, the parameter space for θ.

When the variables are divided into G groups, the semiparametric odds ratio model appears as

$$\frac{\prod_{k=1}^{G-1} \eta_k\{y_k;(y_G,\cdots,y_{(k+1)} \mid y_{(k-1)0},\cdots,y_{10},\theta\} \prod_{k=1}^{G} dF_k(y_k \mid y_{-k0})}{\int\cdots\int \prod_{k=1}^{G-1} \eta_k\{y_k;(y_G,\cdots,y_{(k+1)})\mid y_{(k-1)0},\cdots,y_{10},\theta\} \prod_{k=1}^{G} dF_k(y_k \mid y_{-k0})}, \tag{2.69}$$

where θ is the unknown parameter and $F_k, k=1,\cdots,G$ are unspecified distribution functions satisfying

$$\int\cdots\int \prod_{k=1}^{G-1} \eta_k\{y_k;(y_G,\cdots,y_{(k+1)})\mid y_{(k-1)0},\cdots,y_{10},\theta\} \prod_{k=1}^{G} dF_k(y_k \mid y_{-k0}) < \infty, \tag{2.70}$$

for all $\theta \in \Theta$. Formally, the semiparametric odds ratio model is defined as follows. Let $(\mathcal{Y},\mathcal{B},\mu)$ be a measurable space, where $\mathcal{Y} = \mathcal{Y}_1 \times \cdots \times \mathcal{Y}_G$. Let $\eta(y_1,\cdots,y_G \mid \theta)$ be a measurable function on $(\mathcal{Y},\mathcal{B},\mu)$ for $\theta \in \Theta$ satisfying

$$\eta(y_1;\cdots,y_p \mid \theta) = \prod_{k=1}^{G-1} \eta_k\{y_k;(y_{k+1},\cdots,y_G) \mid (y_{10},\cdots,y_{(k-1)0}),\theta\}$$

and $\eta_k\{y_k;(y_{k+1},\cdots,y_G) \mid (y_{10},\cdots,y_{(k-1)0}), k=1,\cdots,G-1$ are odds ratio functions with reference point $(y_{10},\cdots,y_{G0})$. That is, $\eta_k\{y_k;(y_{k+1},\cdots,y_G) \mid (y_{10},\cdots,y_{(k-1)0}) > 0$ and

$$\eta_k\{y_{k0};(y_{k+1},\cdots,y_G)|(y_{10},\cdots,y_{(k-1)0}) = \eta_k\{y_k;(y_{(k+1)0},\cdots,y_{G0})|(y_{10},\cdots,y_{(k-1)0}) = 1.$$

Let

$$\mathcal{F} = \left\{F=(F_1,\cdots,F_G) \ll \mu \mid \int\cdots\int \eta(y_1;\cdots;y_G \mid \theta)\prod_{k=1}^{G} dF_k(y_k) < \infty, \forall\theta \in \Theta\right\}.$$

The semiparametric odds ratio model for $(Y_1,\cdots,Y_G)$ is the family of distributions with densities with respect to μ as

$$p(y_1,\cdots,y_G \mid \theta) = \frac{\eta(y_1;\cdots;y_G \mid \theta)\prod_{k=1}^{G} dF_k(y_k)}{\int\cdots\int \eta(y_1;\cdots;y_G \mid \theta)\prod_{k=1}^{G} dF_k(y_k)},$$

for $(\theta, F) \in \Theta \times \mathcal{F}$. If additional restrictions are imposed on the set of distribution $\mathcal{F}$, the model is called restricted semiparametric odds ratio model. The family of conditional densities derived from the semiparametric odds ratio model is termed conditional semiparametric odds ratio model, e.g., $\{p(y_1, \cdots, y_j \mid y_{j+1}, \cdots, y_G, \theta, F), (\theta, F) \in \Theta \times \mathcal{F}\}$.

When restricted to $dF(y) > 0$ for all $y \in \mathcal{Y}$, an equivalent reparametrization of the baseline distribution in the semiparametric odds ratio model is more convenient to use. Let P_k be the marginal distribution for Y_k. Let

$$\mathcal{H}_k = \left\{ H_k(y_k) \mid \exp\{H_k(y_k)\} = \frac{dF_k(y \mid y_{-k0})}{dP_k}, P_k \ll F_k \ll P_k \right\},$$

where $F_k \ll P_k$ means F_k is absolutely continuous with respect to P_k. Then $dF_k(y_k \mid y_{-k0})$ in (2.69) can be replaced by $\exp\{H_k(y_k)\}dP_k(y_k)$. The integrable requirement (2.70) becomes

$$\int \cdots \int \eta(y_1; \cdots ; y_G \mid \theta) \prod_{k=1}^{G} e^{H_k(y_k)} dP_k(y_k) < \infty, \tag{2.71}$$

Example 2.7.1. The log-odds ratio functions in Examples 2.2.1, 2.2.2, 2.2.3, 2.2.4, 2.2.5, and 2.2.6 when $s = 1$ all have the same form as

$$\log \eta\{(y, y_0); (x, x_0) \mid \theta\} = \sum_{j=1}^{r} \theta_j (y - y_0)(x_j - x_{j0}).$$

The differences are in the conditional density at a fixed condition. The semiparametric odds ratio model

$$p(y \mid x) = \frac{\eta\{(y, y_0); (x, x_0) \mid \theta\} p(y \mid x_0)}{\int \eta\{(y, y_0); (x, x_0) \mid \theta\} p(y \mid x_0) dy}$$

that leaves the conditional density at a fixed condition unspecified generalizes and unifies all the parametric models. Another semiparametric odds ratio model that uses the reverse conditioning is

$$q(x \mid y) = \frac{\eta\{(y, y_0); (x, x_0) \mid \theta\} q(x \mid y_0)}{\int \eta\{(y, y_0); (x, x_0) \mid \theta\} q(x \mid y_0) dx},$$

where $q(x \mid y_0)$ is the density of an unspecified distribution of X given $Y = y_0$. The joint semiparametric odds ratio model is

$$h(y, x) = \frac{\eta\{(y, y_0); (x, x_0) \mid \theta\} p(y \mid x_0) q(x \mid y_0)}{\int \int \eta\{(y, y_0); (x, x_0) \mid \theta\} p(y \mid x_0) q(x \mid y_0) dy dx}.$$

This covers the models in examples 2.2.10 and 2.2.11.

Example 2.7.2. The log-odds ratio functions conditional on Z in Examples 2.3.1 and 2.3.2 can be extended to

$$\log \eta\{(y, y_0); (x, x_0) \mid z, \theta\} = \sum_{j=1}^{r} \theta_j(z)(y - y_0)(x_j - x_{j0}).$$

The semiparametric odds ratio model covers the two examples is

$$p(y \mid x, z) = \frac{\eta\{(y, y_0); (x, x_0) \mid z, \theta\} p(y \mid x_0, z)}{\int \eta\{(y, y_0); (x, x_0 \mid z, \theta\} p(y \mid x_0, z) dx},$$

where $p(y \mid x_0, z)$ is the density of an unspecified conditional distribution. A counterpart semiparametric odds ratio model covers the two examples is

$$q(x \mid y, z) = \frac{\eta\{(y, y_0); (x, x_0) \mid z, \theta\}q(x \mid y_0, z)}{\int \eta\{(y, y_0); (x, x_0 \mid z, \theta\}q(x \mid y_0, z)dx},$$

where $p(y \mid x_0, z)$ is the density of an unspecified conditional distribution. The joint semiparametric odds ratio model is

$$h(y, x \mid z) = \frac{\eta\{(y, y_0); (x, x_0) \mid z, \theta\}\}p(y \mid x_0, z)q(x \mid y_0, z)}{\int\int \eta\{(y, y_0); (x, x_0) \mid z, \theta\}p(y \mid x_0, z)q(x \mid y_0, z)dydx},$$

where neither $p(y \mid x_0, z)$ nor $q(x \mid y_0, z)$ is specified.

Example 2.7.3. For the joint normal model with mean μ and variance Σ, the odds ratio functions given in Example 2.4.1 are

$$\eta_k\{y_k; (y_G, \cdots, y_{(k+1)}) \mid y_{(k-1)0}, \cdots, y_{10}\} = \exp\left\{\sum_{j=k+1}^{G} \omega_{kj}(y_k - y_{k0})(y_j - y_{j0})\right\},$$

where $\Omega = (w_{ij}) = -\Sigma^{-1}$. A semiparametric odds ratio model extended the joint normal model is

$$\frac{\eta(y_1; \cdots; y_r \mid \omega_{jk}, j > k) \prod_{i=1}^{r} f(y_i \mid y_{-i0})}{\int \cdots \int \eta(y_1; \cdots; y_r \mid \omega_{jk}, j > k) \prod_{i=1}^{r}\{f(y_i \mid y_{-i0})dy_i\}},$$

where $f(y_i \mid y_{-i0}), i = 1, \cdots, r$ are unspecified.

2.12 Extension to relax the positivity condition for the odds ratio representation

In obtaining the odds ratio representation of a joint density, we require the population density satisfying the positivity condition, i.e., if all the marginal densities are positive at a given point, the joint density is also positive. This requirement may be relaxed in the following way once a reference point in the odds ratio representation is selected. Consider first the case with two groups of variables (Y_1, Y_2) having a joint density $p(y_1, y_2)$ with respect to the product measure $\mu_1 \times \mu_2$.. Let the reference point be selected as (y_{10}, y_{20}) with $p(y_{10}, y_{20}) > 0$. Let

$$\begin{aligned} \mathcal{Y}_1 &= \{y_1 \mid p(y_1, y_{20}) > 0\}, \\ \mathcal{Y}_2 &= \{y_2 \mid p(y_{10}, y_2) > 0\}. \end{aligned}$$

Instead of the positivity condition, we assume that

$$\mathcal{Y}_1 \times \mathcal{Y}_2 \supset \{(y_1, y_2) \mid p(y_1, y_2) > 0\}.$$

For any $(y_1, y_2) \in \mathcal{Y}_1 \times \mathcal{Y}_2$, define the odds ratio function with respect to the reference point (y_{10}, y_{20}) as

$$\eta(y_1; y_2) = \begin{cases} \frac{p(y_1, y_2)p(y_{10}, y_{20})}{p(y_1, y_{20})p(y_{10}, y_2)} & \text{if } p(y_1, y_2) > 0, \\ 0 & \text{if } p(y_1, y_2) = 0. \end{cases}$$

This η is well defined on $\mathcal{Y}_1 \times \mathcal{Y}_2$ because if $p(y_1, y_2) > 0$, then $p(y_1, y_{20}) > 0$ and $p(y_{10}, y_2) > 0$ by the assumption. When $p(y_1, y_2) > 0$, we have

$$p(y_1, y_2) = \frac{\eta(y_1; y_2) p(y_{10}, y_2) p(y_1, y_{20})}{p(y_{10}, y_{20})}.$$

Since $\int \int p(y_1, y_2) d\mu_1(y_1) d\mu_2(y_2) = 1$, it follows that

$$p(y_{10}, y_{20}) = \int \int \eta(y_1; y_2) p(y_{10}, y_2) p(y_1, y_{20}) d\mu_1(y_1) d\mu_2(y_2) > 0.$$

The odds ratio representation holds as

$$p(y_1, y_2) = \frac{\eta(y_1; y_2) p(y_{10}, y_2) p(y_1, y_{20})}{\int \int \eta(y_1; y_2) p(y_{10}, y_2) p(y_1, y_{20}) d\mu_1(y_1) d\mu_2(y_2)},$$

or as

$$p(y_1, y_2) = \frac{\eta(y_1; y_2) p(y_1 \mid y_{20}) p(y_2 \mid y_{10})}{\int \int \eta(y_1; y_2) p(y_1 \mid y_{20}) p(y_2 \mid y_{10}) d\mu_1(y_1) d\mu_2(y_2)}.$$

From this representation, it is straightforward to obtain the conditional odds ratio representations respectively as

$$p(y_1 \mid y_2) = \frac{\eta(y_1; y_2) p(y_1 \mid y_{20})}{\int \eta(y_1; y_2) p(y_1 \mid y_{20}) d\mu_1(y_1)}.$$

and

$$p(y_2 \mid y_1) = \frac{\eta(y_1; y_2) p(y_2 \mid y_{10})}{\int \eta(y_1; y_2) p(y_2 \mid y_{10}) d\mu_2(y_2)}.$$

Consider now the general case with G groups of variables $(Y_1, \cdots, Y_G)$ having density $p(y_1, \cdots, y_G)$ with respect to the product measure $\prod_{k=1}^G \mu_k$. Let $(y_{10}, \cdots, y_{G0})$ be a reference point such that $p(y_{10}, \cdots, y_{G0}) > 0$. Let

$$\mathcal{Y}_k = \{y_k \mid p(y_{10}, \cdots, y_{(k-1)0}, y_k, y_{(k+1)0}, \cdots, y_G) > 0\},$$

for $k = 1, \cdots, G$. Instead of positivity condition, we assume that

$$\mathcal{Y}_1 \times \cdots \times \mathcal{Y}_G \supset \{(y_1, \cdots, y_G) \mid p(y_1, \cdots, y_G) > 0\}.$$

In addition, we assume that $p(y_1, \cdots, y_G) > 0$ implies $p(z_1, \cdots, z_G) > 0$ where for any k, $z_k = y_k$ or $z_k = y_{k0}$.

Define odds ratio function for y_1 versus $(y_2, \cdots, y_G)$ with respect to the reference point $(y_{10}, \cdots, y_{G0})$ as

$$\eta\{y_1; (y_2, \cdots, y_G)\} = \begin{cases} \frac{p(y_1, y_2, \cdots, y_G) p(y_{10}, y_{20}, \cdots, y_{G0})}{p(y_1, y_{20}, \cdots, y_{G0}) p(y_{10}, y_2 \cdots, y_G)} & \text{if } p(y_1, y_2, \cdots, y_G) > 0, \\ 0 & \text{if } p(y_1, y_2, \cdots, y_G) = 0. \end{cases}$$

This odds ratio function is well defined because $p(y_1, \cdots, y_G) > 0$ implies $p(y_{10}, y_2, \cdots, y_G) > 0$. By similar arguments as in the case of two groups,

$$p(y_1, \cdots, y_G) = \frac{\eta\{y_1; (y_2, \cdots, y_G)\} p(y_{10}, y_2, \cdots, y_G) p(y_1, y_{20}, \cdots, y_{G0})}{\int \cdots \int \eta\{y_1; (y_2, \cdots, y_G)\} p(y_{10}, y_2, \cdots, y_G) p(y_1, y_{20}, \cdots, y_{G0}) \prod_{k=1}^G d\mu_k(y_k)}.$$

Similarly, define the odds ratio function for (y_{10}, y_2) versus $(y_3, \cdots, y_G)$ as

$$\eta\{(y_{10}, y_2); (y_3, \cdots, y_G)\} = \begin{cases} \dfrac{p(y_{10}, y_2, \cdots, y_G)p(y_{10}, y_{20}, \cdots, y_{G0})}{p(y_{10}, y_2, y_{30} \cdots, y_{G0})p(y_{10}, y_{20}, y_3 \cdots, y_G)} & \text{if } p(y_{10}, y_2, \cdots, y_G) > 0, \\ 0 & \text{if } p(y_{10}, y_2, \cdots, y_G) = 0. \end{cases}$$

Again, because $p(y_{10}, y_2, \cdots, y_G) > 0$ implies $p(y_{10}, y_{20}, y_3 \cdots, y_G) > 0$, the odds ratio function $\eta\{(y_{10}, y_2); (y_3, \cdots, y_G)\}$ is well-defined and the odds ratio representation for $p(y_{10}, y_2, \cdots, y_G)$ also holds. Continue this operation until $p(y_{10}, \cdots, y_{(G-2)0}, y_{(G-1)}, y_G)$. By the same arguments, define

$$\eta\{(y_{10}, \cdots, y_{(G-2)0}, y_{(G-1)}); y_G\} =$$

$$\begin{cases} \dfrac{p(y_{10}, \cdots, y_{(G-2)0}, y_{(G-1)}, y_G)p(y_{10}, \cdots, y_{G0})}{p(y_{10}, \cdots, y_{(G-2)0}, y_{(G-1)}, y_{G0})p(y_{10}, \cdots, y_{(G-1)0}, y_G)} & \text{if } p(y_{10}, \cdots, y_{(G-2)0}, y_{(G-1)}, y_G) > 0, \\ 0 & \text{if } p(y_{10}, \cdots, y_{(G-2)0}, y_{(G-1)}, y_G) = 0. \end{cases}$$

This odds ratio function is well-defined because both $p(y_{10}, \cdots, y_{(G-2)0}, y_{(G-1)}, y_{G0}) > 0$ and $p(y_{10}, \cdots, y_{(G-1)0}, y_G) > 0$ when $p(y_1, \cdots, y_G) > 0$. It follows that

$$p(y_{10}, \cdots, y_{(G-2)0}, y_{(G-1)}, y_G) =$$

$$\frac{\eta\{(y_{10}, \cdots, y_{(G-2)0}, y_{(G-1)}); y_G\}p_0(y_{(G-1)})p_0(y_G)}{\int\int \eta\{(y_{10}, \cdots, y_{(G-2)0}, y_{(G-1)}); y_G\}p_0(y_{(G-1)})p_0(y_G)d\mu_{(G-1)}(y_{(G-1)})d\mu_G(y_G)},$$

where $p_0(y_k) = p(y_{10}, \cdots, y_{(k-1)0}, y_k, y_{(k+1)0} \cdots, y_{G0})$, for $k = 1, \cdots, G$. Combining the odds ratio representations to obtain

$$p(y_1, \cdots, y_G) = \frac{\eta(y_1; \cdots; y_G)\prod_{k=1}^G p_0(y_k)}{\int \cdots \int \eta(y_1; \cdots; y_G)\prod_{k=1}^G p_0(y_k)d\mu_k(y_k)},$$

when $p(y_1, \cdots, y_G) > 0$, where

$$\begin{aligned} \eta(y_1; \cdots; y_G) &= \prod_{k=1}^{G-1} \eta\{(y_{10}, \cdots, y_{(k-1)0}, y_k); (y_{(k+1)}, \cdots, y_G)\} \\ &= \prod_{k=1}^{G-1} \eta\{y_k; (y_{(k+1)}; \cdots; y_G) \mid y_{10}, \cdots, y_{(k-1)0})\}. \end{aligned}$$

Other representations for conditional densities can be obtained from the joint odds ratio representation.

The extension allows us to study certain applications of the odds ratio model with more generality and flexibility, such as in the application to the truncated data where the sampling probability can be 0 in certain region of the full data. To keep the development simple in the future, we still assume the positivity assumption holds in general and stated explicitly if the extended assumption instead of the positivity condition holds.

2.13 Literature on odds ratio functions and relevant statistical models

A variety of special cases of the semiparametric odds ratio model were proposed in the literature. The simplest is perhaps the density ratio model (Qin, 1998)[142] which assumes

$$\log \frac{q(x \mid Y=1)}{q(x \mid Y=0)} = b_0 + b_1 x.$$

where $q(x \mid y)$ denote the conditional density X given Y, and Y takes only two values 0 or 1. This model is simply the logistic regression model if Y given X is of primary interest. The density ratio model concentrates on the study of the conditional density of X instead and has been applied to different problems often along with empirical likelihood (Qin and Zhang, 1997; Qin, 1998; Qin and Zhang, 2003; Fokianos, 2004; Cheng and Chu, 2004)[143, 142, 144, 62, 43]. The exponential tilting model (Tan, 2009)[175] extends the density ratio model to including more than two densities, where one density compares with another in the form

$$q(x \mid y_1) = q(x \mid y_0)\exp\{b_0(y_1, y_0) + b_1(y_1, y_0)x\},$$

where

$$\int q(x \mid y_0)\exp\{b_1(y_1, y_0)x\}\,dx = \exp\{-b_0(y_1, y_0)\}.$$

The exponential tilting model can be equivalently written as

$$q(x \mid y_1) = \frac{q(x \mid y_0)\exp\{b_1(y_1, y_0)x\}}{\int q(x \mid y_0)\exp\{b_1(y_1, y_0)x\}dx}, \tag{2.72}$$

where $\int q(x \mid y_0)\exp\{b_1(y_1, y_0)x\}dy < \infty$. When $b_1(y_1, y_0) = \beta_1(y_1 - y_0)$, model (2.72) becomes the proportional likelihood model (Luo and Tsai, 2012)[116]. It is not difficult to see that all these models are special cases of the semiparametric odds ratio model with a reference point (x_0, y_0),

$$q(x \mid y) = \frac{\eta(x; y)q(x \mid y_0)}{\int \eta(x; y)q(x \mid y_0)dx}, \tag{2.73}$$

where the odds ratio function $\eta(x; y)$ satisfies $\int \eta(x; y)q(x \mid y_0)dx < \infty$ for all y.

Satten and Carroll (2000) [159] introduced the odds ratio parameters with a reference point in the analysis of contingency tables. See also Satten and Kupper (1993)[158]. Odds ratio function for continuous outcomes appeared in Liang and Qin (2000)[103] where it was called "generalized odds ratio". Chen (2003)[26] applied such odds ratio functions for studying the biased sampling problem. Chen (2004)[27] gave a representation of a conditional density in terms of the odds ratio function and an unspecified baseline density and applied the representation to modeling the covariate distribution in a missing covariate problem in regression analysis. Osius (2005; 2009)[131, 132] studied the issue of uniqueness of the odds ratio function for a bivariate distribution and semiparametric inference of the odds ratio parameter. Chen (2007)[28] proposed a general representation for a joint density for two random variables conditional on a third one in the discussion on measuring association between random variables. Chen (2010)[29] extended the odds ratio representation to multivariate outcomes and discussed its use for checking model compatibility for conditionally specified models. Chen (2011)[30] focused on parameter identifiability with the odds ratio representation of a density function and discussed its application to outcome-dependent biased sample.

Besag(1974)[7] provided a different representation of the joint distribution under the positivity condition. That representation is often termed Brook expansion (Brook, 1964; Besag and Kooperberg, 1995; Besag, 1996)[16, 10, 8]. Such models are usually termed auto-regression models. A disadvantage of the representation is that the model identifiability is often vague and it is less flexible than models based on the odds ratio representations.

Other related literatures include the discussion of the general biased sampling problems, such as Vardi (1985)[191] on length-biased sampling, Gilbert et al (1999)[71] and Gilbert (2000)[70] on the general biased sampling problem and statistical estimation and inference. Rabinowitz (1997)[146] on efficient inference based on the case-control design, Scott and Wild (1997)[163] on generalized case-control sample with some known supplemental information, and Lawless, Kalbfleisch, and Wild (1999)[93] on stratified outcome-dependent sampling. Rathouz and Gao (2009)[147] and Huang and Rathouz (2012)[81] on the restricted proportional likelihood ratio model which has ties with the traditional generalized linear model.

2.14 Exercises

1. The correlation coefficient between two random variables is sometimes used to measure the association between two random variables. However, zero correlation between two random variables does not imply no association between two random variables. Let ϵ_1, ϵ_2 be independent standard normally distributed random variables. Let

$$\begin{aligned} X &= \epsilon_1^2, \\ Y &= \epsilon_1 + \epsilon_2. \end{aligned}$$

   (a) Show that X and Y are uncorrelated.

   (b) Show that the density for the joint distribution of (Y, X) is

$$h(y,x) = \frac{1_{\{x \geq 0\}}}{4\pi\sqrt{x}} e^{-\frac{y^2}{2}-x}\left(e^{y\sqrt{x}} + e^{-y\sqrt{x}}\right).$$

   (c) Find the odds ratio function between X and Y with respect to the reference point $(0, 0)$.

2. The odds ratio function can be regarded as a measure with local property, while the correlation coefficient can be regarded as a measure with global property. Let X and Y be two random variables with finite second moments. Let $a < b$ be two constants in the range of X and $c < d$ be two constants in the range of Y. Let

$$\begin{aligned} Z &= Y 1_{\{a \leq Y \leq b\}}, \\ W &= X 1_{\{c \leq X \leq d\}}. \end{aligned}$$

   Let $\eta_1(y; x)$ be the odds ratio function for Y and X and $\eta_2(z, w)$ be the odds ratio function for Z and W. Show that

$$\eta_1(z, w) = \eta_2(z, w)$$

   for $a \leq z \leq b$ and $c \leq w \leq d$. But in general, $\text{corr}(Z, W) \neq \text{corr}(X, Y)$.

3. Show that the correlation coefficient is invariant to linear transformation of individual variables, i.e.

$$\begin{aligned} Z &= a + bY, \\ W &= c + dX. \end{aligned}$$

Then

$$\text{corr}(Z, W) = \text{corr}(X, Y).$$

Let $\eta_1(y; x)$ be the odds ratio function for Y and X and $\eta_2(z, w)$ be the odds ratio function for Z and W. In general,

$$\eta_1(z, w) \neq \eta_2(z, w).$$

4. A Gaussian copula function is defined on$[0, 1]^r$ with

$$c(u_1, \cdots, u_r) = \Phi_R\{\Phi^{-1}(u_1), \cdots, \Phi^{-1}(u_r)\}$$

where Φ is the standard normal distribution function and Φ_R is a joint normal distribution function with mean 0 and variance-covariance matrix R and R is a correlation matrix. Let F_j be the marginal distribution function of Y_j for $j = 1, \cdots, r$ and the distribution function of $\{F_1(Y_1), \cdots, F_r(Y_r)\}$ is the Gaussian copula.

Find the odds ratio functions for the random variables $(Y_1, \cdots, Y_r)$ with reference point $(0, \cdots, 0)$.

5. For the case of three random variables: X, Y, and Z, write down a semiparametric odds ratio model that has the conditional models for X given $Y = y_0, Z = z_0$, Y given $X = x_0, Z = z_0$, and Z given $X = x_0, Y = y_0$, as unspecified, and the odds ratio functions are all in log-bilinear forms.

3

Estimation and Inference on Semiparametric Odds Ratio Model

3.1 An introduction to likelihood-based approaches

Consider first the simple semiparametric odds ratio model

$$f(y \mid x) = \frac{\eta(y; x \mid \theta) f(y \mid x_0)}{\int \eta(y; x \mid \theta) f(y \mid x_0) d\mu(y)},$$

where $\eta(y; x \mid \theta)$ is a known function with unknown parameter $\theta \in \Theta$, and $f(y \mid x_0)$ is an unspecified density function subject to

$$\int \eta(y; x \mid \theta) f(y \mid x_0) d\mu(y) < +\infty,$$

for all $\theta \in \Theta$. Let the observed data be n independent identically distributed copies: $(Y_i, X_i), i = 1, \cdots, n$. The likelihood for parameters θ and the distribution function $F(y \mid x_0) = \int_{-\infty}^{y} f(t \mid x_0) d\mu(t)$ is

$$\prod_{i=1}^{n} \frac{\eta(Y_i; X_i \mid \theta) dF(Y_i \mid x_0)}{\int \eta(y; X_i \mid \theta) dF(y \mid x_0)}. \tag{3.1}$$

Two different types of likelihood-based approaches to estimating θ are introduced in this chapter. The first is based on the permutation likelihoods that eliminate the functional parameters $F(\cdot \mid x_0)$ through conditioning. Because the permutation likelihood is challenging to maximize. Two computational feasible approaches are employed. One constructs a pseudo-likelihood based on small subsets of subjects to reduce the computation complexity. The other computes the permutation likelihood through Monte Carlo approximation. Either the pseudo-likelihood or the approximate permutation likelihood is maximized to obtain estimators for θ. Once an estimator of the parameter θ is obtained, the functional parameters can then be estimated by plugging the θ estimator in the appropriate estimating equations for the functional parameter. The second type of approaches is based on the semiparametric likelihood that restricts $F(\cdot \mid x_0)$ to be a probability measure with finite support. The semiparametric likelihood is then jointly maximized with respect to (θ, F) to obtain the parameter estimator.

The likelihood-based approaches to the simple odds ratio model can be extended to the joint odds ratio model for complex data, where

$$f(y_1, \cdots, y_G) = \frac{\eta(y_1; \cdots; y_G \mid \theta) \prod_{k=1}^{G} f_k(y_k \mid y_{-k0})}{\int \cdots \int \eta(y_1; \cdots; y_G \mid \theta) \prod_{k=1}^{G} f_k(y_k \mid y_{-k0}) d\mu_k(y_k)}.$$

Let $F_k(y_k \mid y_{-k0}) = \int_{-\infty}^{y_k} f_k(t \mid y_{-k0}) d\mu_k(t)$ for $k = 1, \cdots, G$. Let the observed data be

DOI: 10.1201/9781351049757-3

$(Y_{i1}, \cdots, Y_{iG}), i = 1, \cdots, n$. The likelihood for the observed data is

$$\prod_{i=1}^{n} \frac{\eta(Y_{i1}; \cdots; Y_{iG} \mid \theta) \prod_{k=1}^{G} dF_k(Y_{ik} \mid y_{-k0})}{\int \cdots \int \eta(y_1; \cdots; y_G \mid \theta) \prod_{k=1}^{G} dF_k(y_k \mid y_{-k0})}. \tag{3.2}$$

Similar to the simple case, the permutation likelihood approach first eliminates all the nuisance parameter $F_k, k = 1, \cdots, G$ by conditioning on the order statistics and then maximize the permutation likelihoods to obtain estimators for θ. The maximum semiparametric likelihood approach restricts F_k to be a probability measure with finite support for $k = 1, \cdots, G$, and jointly maximizes the likelihood to obtain an estimator for $(\theta, F_k, k = 1, \cdots, G)$.

3.2 Pseudo-likelihood approaches

3.2.1 Pairwise and group-wise pseudo-likelihood approaches

Conditional on the order statistics of $Y_1, \cdots, Y_n$, likelihood (3.1) reduces to

$$\begin{aligned} & P(Y_1, \cdots, Y_n \mid Y_{(1)}, \cdots, Y_{(n)}, X_1, \cdots, X_n) \\ &= \frac{\prod_{i=1}^{n} f(Y_i \mid X_i)}{\sum_{\sigma \in \mathcal{P}_n} \prod_{i=1}^{n} f(Y_{\sigma(i)} \mid X_i)} \qquad (3.3) \\ &= \frac{\prod_{i=1}^{n} \eta(Y_i; X_i \mid \theta)}{\sum_{\sigma \in \mathcal{P}_n} \prod_{i=1}^{n} \eta(Y_{\sigma(i)}; X_i \mid \theta)}, \qquad (3.4) \end{aligned}$$

where $(Y_{(1)}, \cdots, Y_{(n)})$ is the order statistic of $(Y_1, \cdots, Y_n)$ and $\mathcal{P}_n$ is the collection of all permutations on the indices $\{1, \cdots, n\}$. The permutation likelihood was originally proposed in Kalbfleisch (1978)[86] and was mainly used for hypothesis testing due to the computation challenges involved in estimation. Let the observed Y takes K distinctive values with respective frequencies $n_k, k = 1, \cdots, K$. The total number of different items in the summation of the denominator is the combinatorial number

$$\binom{n}{n_1, \cdots, n_K} = \frac{n!}{\prod_{k=1}^{K} n_k!},$$

where $n! = n(n-1)\cdots 1$, the factorial of n. This number can be huge even for a binary outcome Y. Computationally, enumerating all possible permutations is often very difficult to impossible unless n is very small. One idea to circumvent the difficulty is to consider the permutation likelihood for a subset of the whole sample, and then combine the likelihoods from all possible distinctive subsets to form a pseudo-likelihood to be maximized to obtain the parameter estimator. This section examines methods based on this idea.

The pairwise pseudo-likelihood approach considers the conditional likelihood (3.4) for two subjects at a time and combines all such pairwise likelihoods to form a composite pseudo-likelihood. Specifically, if subjects i and j are taken as a sample, the conditional likelihood is

$$P(Y_i, Y_j \mid Y_{(i)}, Y_{(j)}, X_i, X_j) = \frac{\eta(Y_i; X_i \mid \theta)\eta(Y_j; X_j \mid \theta)}{\eta(Y_i; X_i \mid \theta)\eta(Y_j; X_j \mid \theta) + \eta(Y_j; X_i \mid \theta)\eta(Y_i; X_j \mid \theta)}.$$

A composite likelihood is constructed by combining all possible pairwise likelihoods as

$$\prod_{i=1}^{n} \prod_{j>i} \frac{\eta(Y_i; X_i \mid \theta)\eta(Y_j; X_j \mid \theta)}{\eta(Y_i; X_i \mid \theta)\eta(Y_j; X_j \mid \theta) + \eta(Y_j; X_i \mid \theta)\eta(Y_i; X_j \mid \theta)}.$$

This approach was originally proposed in Liang and Qin (2000)[103]. The composite likelihood is not a true likelihood, and is therefore termed pseudo-likelihood in Liang and Qin (2000)[103]. The pseudo-likelihood can be further rewritten as

$$\prod_{i=1}^{n}\prod_{j>i}\frac{\exp\{\xi(Y_i,Y_j;X_i,X_j\mid\theta)\}}{1+\exp\{\xi(Y_i,Y_j;X_i,X_j\mid\theta)\}}, \tag{3.5}$$

where

$$\xi(Y_i,Y_j;X_i,X_j\mid\theta)=\log\eta(Y_i;X_i\mid\theta)+\log\eta(Y_j;X_j\mid\theta)-\log\eta(Y_j;X_i\mid\theta)-\log\eta(Y_i;X_j\mid\theta).$$

Liang and Qin (2000)[103] demonstrated that the maximum pseudo-likelihood approach yields a consistent estimator for θ.

Computationally, the pairwise pseudo-likelihood approach is attractive when n is not too large. Consider for example the bilinear log-odds ratio model

$$\log\eta(y;x\mid\theta)=\theta(y-y_0)(x-x_0).$$

Under this model, the pseudo-likelihood reduces to

$$\prod_{i=1}^{n}\prod_{j>i}\frac{\exp\{\theta(Y_i-Y_j)(X_i-X_j)\}}{1+\exp\{\theta(Y_i-Y_j)(X_i-X_j)\}}. \tag{3.6}$$

The likelihood maximization can be carried out by first creating pseudo covariate data $(Y_i-Y_j)(X_i-X_j)$ and pseudo-outcome data 1 for all $j>i$, then applying the linear logistic regression model without intercept to obtain the estimator for θ. The variance for the maximum pseudo-likelihood estimator for θ, however, cannot be obtained as the inverse of the minus second derivative of the log-pseudo-likelihood evaluated at the maximum point. Denote the pseudo-likelihood estimator of θ, i.e., the maximizer of (3.6), by $\tilde\theta$. The variance for the pseudo-likelihood estimator $\tilde\theta$ can be obtained as follows. The score equation for estimating θ is

$$\sum_{i=1}^{n}\sum_{j>i}\frac{(Y_i-Y_j)(X_i-X_j)}{1+\exp\left\{\tilde\theta(Y_i-Y_j)(X_i-X_j)\right\}}=0.$$

By Taylor expansion around θ_0,

$$\sum_{i=1}^{n}\sum_{j>i}(Y_i-Y_j)(X_i-X_j)(1-\pi_{ij})$$
$$-\sum_{i=1}^{n}\sum_{j>i}(Y_i-Y_j)^2(X_i-X_j)^{\otimes 2}\pi_{ij}(1-\pi_{ij})\left\{(\tilde\theta-\theta_0)+o_p(|\tilde\theta-\theta_0|)\right\}=0,$$

where $v^{\otimes 2}=vv'$for a vector v, and

$$\pi_{ij}=\frac{\exp\{\theta_0(Y_i-Y_j)(X_i-X_j)\}}{1+\exp\{\theta_0(Y_i-Y_j)(X_i-X_j)\}}.$$

This leads to

$$\begin{aligned}\sqrt{n}(\tilde\theta-\theta_0) &= \left[\frac{2}{n(n-1)}\sum_{i=1}^{n}\sum_{j>i}(Y_i-Y_j)^2(X_i-X_j)^{\otimes 2}\pi_{ij}(1-\pi_{ij})\{1+o_p(1)\}\right]^{-1}\\ &\quad \frac{2}{\sqrt{n}(n-1)}\sum_{i=1}^{n}\sum_{j>i}(Y_i-Y_j)(X_i-X_j)(1-\pi_{ij}).\end{aligned} \tag{3.7}$$

To understand the large sample behavior of the pairwise pseudo-likelihood estimator, note that both

$$\frac{1}{n(n-1)}\sum_{i=1}^{n}\sum_{j>i}(Y_i - Y_j)^2(X_i - X_j)^{\otimes 2}\pi_{ij}(1-\pi_{ij})$$

and

$$\frac{2}{n(n-1)}\sum_{i=1}^{n}\sum_{j>i}(Y_i - Y_j)(X_i - X_j)(1-\pi_{ij})$$

are U-statistics. The asymptotic properties of the estimator can be obtained by applying the asymptotic theory for the U-statistics, which will be discussed in the next section. For the general pairwise pseudo-likelihood (3.5), similar arguments can be used in studying the asymptotic behavior of the θ estimator.

The foregoing approach of maximizing the pairwise pseudo-likelihood, though simple, yields an inefficient estimator for θ. To increase the efficiency, permutation likelihoods of more than two subjects at a time may be considered. Such likelihoods are termed groupwise likelihood in this book. For example, for a group of three subjects with indices $\{i, j, k\}$, the likelihood is

$$\frac{\eta(Y_i; X_i \mid \theta)\eta(Y_j; X_j \mid \theta)\eta(Y_k; X_k \mid \theta)}{\sum_{\sigma\in\mathcal{P}_3}\eta(Y_{\sigma(i)}; X_i \mid \theta)\eta(Y_{\sigma(j)}; X_j \mid \theta)\eta(Y_{\sigma(k)}; X_k \mid \theta)},$$

where $\mathcal{P}_3$ is the collection of all permutations for $\{i, j, k\}$. There are a total of 6 permutations in $\mathcal{P}_3$. They transforms (i, j, k) respectively into

$$(i, j, k), (i, k, j), (j, i, k), (k, j, i), (j, k, i), (k, i, j).$$

The pseudo-likelihood thus formed is

$$\prod_{i=1}^{n}\prod_{j>i}\prod_{k>j}\frac{\eta(Y_i; X_i \mid \theta)\eta(Y_j; X_j \mid \theta)\eta(Y_k; X_k \mid \theta)}{\sum_{\sigma\in\mathcal{P}_3}\eta(Y_{\sigma(i)}; X_i \mid \theta)\eta(Y_{\sigma(j)}; X_j \mid \theta)\eta(Y_{\sigma(k)}; X_k \mid \theta)}.$$

Under the bilinear odds ratio model, the pseudo-likelihood reduces to

$$\prod_{i=1}^{n}\prod_{j>i}\prod_{k>j}\left[\sum_{\sigma\in\mathcal{P}_3}\exp\left\{\theta(y_{\sigma(i)} - y_i)x_i + \theta(y_{\sigma(j)} - y_j)x_j + \theta(y_{\sigma(k)} - y_k)x_k\right\}\right]^{-1}.$$

The score equation is

$$\sum_{1\le i<j<k\le n}\frac{\sum_{\sigma\in\mathcal{P}_3} z_{ijk,\sigma}\exp(\theta z_{ijk,\sigma})}{\sum_{\sigma\in\mathcal{P}_3}\exp(\theta z_{ijk,\sigma})} = 0,$$

where

$$z_{ijk,\sigma} = (y_{\sigma(i)} - y_i)x_i + (y_{\sigma(j)} - y_j)x_j + (y_{\sigma(k)} - y_k)x_k$$

The pseudo-likelihood estimator $\hat{\theta}$ satisfies

$$\begin{aligned}\sqrt{n}(\hat{\theta} - \theta_0) &= \left[\frac{1}{C(3,n)}\sum_{i<j<k}\frac{\sum_{\sigma\in\mathcal{P}_3} z_{ijk,\sigma}^{\otimes 2}\exp(\theta_0 z_{ijk,\sigma})}{\sum_{\sigma\in\mathcal{P}_3}\exp(\theta_0 z_{ijk,\sigma})}\{1+o_p(1)\}\right]^{-1} \\ &\quad \frac{\sqrt{n}}{C(3,n)}\sum_{i<j<k}\frac{\sum_{\sigma\in\mathcal{P}_3} z_{ijk,\sigma}\exp(\theta_0 z_{ijk,\sigma})}{\sum_{\sigma\in\mathcal{P}_3}\exp(\theta_0 z_{ijk,\sigma})},\end{aligned} \tag{3.8}$$

where $C(3, n)$ denotes the combinatorial number of choosing 3 out of n. Both terms on the right-hand side are U-statistics. Again, the maximum groupwise pseudo-likelihood estimator can also be shown to be asymptotically normally distributed using the asymptotic theory on U-statistic outlined in the next section.

3.2.2 Asymptotic theory for U-statistics

For independent identically distributed random vectors $Z_1, \cdots, Z_n$ and a fixed number $m < n$, a general U-statistic is defined as

$$U_n = \frac{1}{C(m,n)} \sum_{1 \le i_1 < \cdots < i_m \le n} h(Z_{i_1}, \cdots, Z_{i_m}), \tag{3.9}$$

where h is called the kernel function satisfying

$$h(Z_1, \cdots, Z_m) = h(Z_{j_1}, \cdots, Z_{j_m})$$

for any $(j_1, \cdots, j_m)$ that is a permutation of indices $1, \cdots, m$, and $C(m,n)$ denotes the combinatorial number of choosing m from n. When $E|h(Z_1, \cdots, Z_m)| < \infty$, Hoeffding (1961)[79] showed that the strong law of large number holds, i.e.,

$$\frac{1}{C(m,n)} \sum_{1 \le i_1 < \cdots < i_m \le n} h(Z_{i_1}, \cdots, Z_{i_m}) \to E\{h(Z_1, \cdots, Z_m)\},$$

almost sure. The asymptotic normality of the U-statistic was obtained in Hoeffding (1948)[77]. Assume further $E\{h(Z_1, \cdots, Z_m)\} = 0$. For $k \le m$, define

$$h_k(Z_1, \cdots, Z_k) = E\left\{h(Z_1, \cdots, Z_m) \mid Z_1, \cdots, Z_k\right\}.$$

The variance of the U-statistic is

$$\begin{aligned} Var(U_n) &= \frac{1}{C^2(m,n)} \sum_{1 \le i_1 < \cdots < i_m \le n} \sum_{1 \le j_1 < \cdots < j_m \le n} E\left\{h(Z_{i_1}, \cdots, Z_{i_m}) h(Z_{j_1}, \cdots, Z_{j_m})\right\} \\ &= \frac{1}{C^2(m,n)} \sum_{k=\max(1,2m-n)}^{m} C(m,n)C(k,m)C(m-k,n-m)E\left\{h_k^2(Z_1, \cdots, Z_k)\right\} \\ &= \frac{1}{C(m,n)} \sum_{k=\max(1,2m-n)}^{m} C(k,m)C(m-k,n-m)\sigma_k^2, \end{aligned}$$

where, for $k = 1, \cdots, m$,

$$\sigma_k^2 = Var\left\{h_k(Z_1, \cdots, Z_k)\right\}.$$

For fixed k and m such that $1 \le k \le m$,

$$\frac{C(k,m)C(m-k,n-m)}{C(m,n)} = \frac{k!C^2(k,m)}{n^k} + o\left(\frac{1}{n^k}\right),$$

when $n \to \infty$. For a fixed m, if $\sigma_m^2 < \infty$ and $\sigma_1^2 > 0$, then

$$Var(U_n) = \frac{m^2}{n}\sigma_1^2 + o\left(\frac{1}{n}\right).$$

In this case,

$$Var\left\{\sqrt{n}U_n - \frac{1}{\sqrt{n}} \sum_{i=1}^{n} h_1(Z_i)\right\} \to 0$$

as $n \to \infty$, which implies

$$\sqrt{n}U_n - \frac{1}{\sqrt{n}} \sum_{i=1}^{n} h_1(Z_i) \to 0$$

in probability. Furthermore, because

$$\frac{1}{\sqrt{n}}\sum_{i=1}^{n} h_1(Z_i) \to N(0, m^2\sigma_1^2),$$

It follows that

$$\sqrt{n}U_n \longrightarrow N\left(0, m^2\sigma_1^2\right),$$

in distribution by the Slutsky lemma. More detailed theoretical analysis of U-statistics can be found in Hoeffding (1948)[77] and Lee (1990)[94].

3.2.3 Asymptotic distributions of the pseudo-likelihood estimators

To apply the asymptotic results on the U-statistics to the pairwise pseudo-likelihood score, note that $Z_i = (Y_i, X_i), i = 1, \cdots, n$ and

$$h(Z_i, Z_j) = (Y_i - Y_j)(X_i - X_j)(1 - \pi_{ij}).$$

It follows from the central limit theorem for the U-statistics that

$$\frac{2}{n(n-1)}\sum_{i=1}^{n}\sum_{j>i}(Y_i - Y_j)(X_i - X_j)(1 - \pi_{ij}) \longrightarrow N(0, A),$$

where

$$A = 4Var\left[E\left\{(Y_1 - Y_2)(X_1 - X_2)(1 - \pi_{12}) \mid X_1, Y_1\right\}\right].$$

Furthermore, it follows from the law of large number for U-statistics that

$$\frac{2}{n(n-1)}\sum_{i=1}^{n}\sum_{j>i}(Y_i - Y_j)^2(X_i - X_j)^{\otimes 2}\pi_{ij}(1 - \pi_{ij})\left\{1 + o_p(1)\right\} \longrightarrow B, a.s.,$$

where

$$B = E\left\{(Y_1 - Y_2)^2(X_1 - X_2)^{\otimes 2}\pi_{12}(1 - \pi_{12})\right\}.$$

We can conclude by combining the two results that

$$\sqrt{n}(\tilde{\theta} - \theta_0) \longrightarrow N(0, \Omega),$$

where $\Omega = B^{-1}AB^{-1}$. A consistent estimate for Ω can be obtained by $\hat{B}^{-1}\hat{A}\hat{B}^{-1}$, where

$$\hat{A} = \frac{4}{n-1}\sum_{i=1}^{n}\left\{h_1(Y_i, X_i) - \frac{1}{n}\sum_{j=1}^{n} h_1(Y_j, X_j)\right\}^{\otimes 2},$$

$$\hat{B} = \frac{1}{n(n-1)}\sum_{i=1}^{n}\sum_{j\neq i}(Y_i - Y_j)^2(X_i - X_j)^{\otimes 2}\pi_{ij}(1 - \pi_{ij}),$$

where

$$h_1(Y_1, X_1) = E\left\{(Y_1 - Y_2)(X_1 - X_2)(1 - \pi_{12}) \mid X_1, Y_1\right\},$$

which can be estimated by

$$\hat{h}_1(y, x) = \frac{1}{n}\sum_{i=1}^{n}\frac{(y - Y_i)(x - X_i)}{1 + \exp\left\{\hat{\theta}(y - Y_i)(x - X_i)\right\}}.$$

A direct estimate given in Chan (2013)[22] is

$$\hat{A}^* = \frac{4}{n-1}\sum_{i=1}^{n}\left[\frac{1}{n-1}\sum_{j=1}^{n}\frac{(Y_j - Y_i)(X_j - X_i)}{1+\exp\left\{\hat{\theta}(Y_j - Y_i)(X_j - X_i)\right\}}\right]^{\otimes 2}.$$

Similarly, by applying the asymptotic results of the U-statistics to the groupwise pseudo-likelihood score with group size 3, we can conclude from (3.8) that

$$\sqrt{n}(\tilde{\theta} - \theta_0) \longrightarrow N(0, \Omega),$$

where $\Omega = B^{-1}AB^{-1}$ and

$$\begin{aligned} A &= 9Var\left[E\left\{\frac{\sum_{\sigma\in\mathcal{P}_3} z_{123,\sigma}\exp(\theta_0 z_{123,\sigma})}{\sum_{\sigma\in\mathcal{P}_3}\exp(\theta_0 z_{123,\sigma})} \mid X_1, Y_1\right\}\right], \\ B &= E\left\{\frac{\sum_{\sigma\in\mathcal{P}_3} z_{123,\sigma}^{\otimes 2}\exp(\theta_0 z_{123,\sigma})}{\sum_{\sigma\in\mathcal{P}_3}\exp(\theta_0 z_{123,\sigma})}\right\}. \end{aligned}$$

A consistent estimate for Ω can be obtained by $\hat{A}^{-1}\hat{B}\hat{A}^{-1}$, where

$$\begin{aligned} \hat{A} &= \frac{9}{n-1}\sum_{i=1}^{n}\left\{h_1(Y_i, X_i) - \frac{1}{n}\sum_{j=1}^{n} h_1(Y_j, X_j)\right\}^{\otimes 2}, \\ \hat{B} &= \frac{1}{C(3,n)}\sum_{i<j<k}\frac{\sum_{\sigma\in\mathcal{P}_3} z_{ijk,\sigma}^{\otimes 2}\exp(\theta_0 z_{ijk,\sigma})}{\sum_{\sigma\in\mathcal{P}_3}\exp(\theta_0 z_{ijk,\sigma})}, \end{aligned}$$

where

$$h_1(Y_1, X_1) = E\left\{\frac{\sum_{\sigma\in\mathcal{P}_3} z_{123,\sigma}\exp(\theta_0 z_{123,\sigma})}{\sum_{\sigma\in\mathcal{P}_3}\exp(\theta_0 z_{123,\sigma})} \mid X_1, Y_1\right\},$$

which can be estimated by

$$\hat{h}_1(Y_1, X_1) = \frac{2}{n(n-1)}\sum_{j<k}\frac{\sum_{\sigma\in\mathcal{P}_3} z_{1jk,\sigma}\exp\left(\hat{\theta} z_{1jk,\sigma}\right)}{\sum_{\sigma\in\mathcal{P}_3}\exp\left(\hat{\theta} z_{1jk,\sigma}\right)}.$$

3.3 Permutation likelihood approach

Although the groupwise pseudo-likelihood estimator with group size more than 2 is more efficient than that of the pairwise pseudo-likelihood estimator, the efficiency increase is at the cost of a rapid increase in computation complexity and the groupwise likelihoods estimators are themselves inefficient. In addition, the computation burden of the groupwise likelihood approach increases exponentially in the sample size because of the combinatorial nature of the items in the estimating equations. To resolve these problems, we consider the approximation to the original permutation likelihood by the Monte Carlo approach in this section.

The permutation likelihood (3.4) is also impractical to maximize directly because of a huge number of permutations need to be evaluated. One way to circumvent the need to evaluate the huge number of permutations is to take a simple random sample of the permutations and use the sample version to approximate the summation in the permutation likelihood. Such an idea has been widely used in hypothesis testing where a simple random sample of the permutations is sufficient (Pesarin and Salmaso, 2010)[136]. However, for parameter estimation, such an approach fails to deliver the results one had hoped. The Monte Carlo approach to be introduced in this section aims to overcoming the difficulty by using the idea of adaptive importance sampling weights. An approximation approach using the Metropolis sampling algorithm is proposed to find the maximum permutation likelihood estimator of (3.4). Before discussing the Monte Carlo approach that works, we examine some plausible approaches that does not work well for the estimation problem in practice.

3.3.1 Approximations using simple Monte Carlo or asymptotics

Let $L(\theta)$ denote the likelihood (3.4) and $M(\theta) = L^{-1}(\theta)$. It follows that

$$M(\theta) = \sum_{\sigma \in \mathcal{P}_n} \exp\left\{\sum_{i=1}^{n} \log \frac{\eta(Y_{\sigma(i)}; X_i \mid \theta)}{\eta(Y_i; X_i \mid \theta)}\right\}. \tag{3.10}$$

Maximizing the permutation likelihood (3.4) is thus equivalent to minimizing the objective function $M(\theta)$. Let

$$S_n(\Gamma \mid \theta) = \sum_{i=1}^{n} a\{i, \Gamma(i) \mid \theta\},$$

where Γ is a random permutation of $\{1, \cdots, n\}$ and

$$a\{i, \Gamma(i) \mid \theta\} = \log \frac{\eta(Y_{\Gamma(i)}; X_i \mid \theta)}{\eta(Y_i; X_i \mid \theta)}.$$

$M(\theta)$ can be rewritten in terms of the expectation with respect to a probability distribution over the permutations as

$$M(\theta) = n! E\left[\exp\left\{S_n(\Gamma \mid \theta)\right\}\right],$$

where

$$P(\Gamma = \sigma \mid Y, X) = \frac{1}{n!},$$

and $(Y, X) = (Y_1, X_1, \cdots, Y_n, X_n)$. One simple approach that appears attractive is to sample permutations following the uniform distribution and then approximate $M(\theta)$ by the sample average,

$$M(\theta) \approx \frac{n!}{N} \sum_{k=1}^{N} \exp\left\{S_n(\sigma_k \mid \theta)\right\},$$

where $\sigma_1 \cdots, \sigma_N$ are permutations following the uniform distribution. This simulation approach is often used in hypothesis testing. It does not work well in parameter estimation, especially when $S_n(\sigma \mid \theta)$ may take very large values for some permutations. This is because the approximation may require a practically unattainable hugh sample size in order to achieve reasonable approximation accuracy for estimation when $\theta \neq 0$.

Under some weak conditions, Hoeffding (1951)[78] showed that

$$\frac{S_n(\Gamma \mid \theta) - E\left\{S_n(\Gamma \mid \theta) \mid Y, X\right\}}{\sqrt{Var\left\{S_n(\Gamma \mid \theta) \mid Y, X\right\}}} \longrightarrow N(0, 1),$$

or equivalently $S_n(\Gamma \mid \theta)$ is approximately normal with mean $E\{S_n(\Gamma \mid \theta) \mid Y, X\}$ and variance $Var\{S_n(\Gamma \mid \theta) \mid Y, X\}$, where

$$\begin{aligned} E\{S_n(\Gamma \mid \theta) \mid Y, X\} &= \frac{1}{n}\sum_{i=1}^{n}\sum_{j=1}^{n} a(i,j \mid \theta), \\ Var\{S_n(\Gamma \mid \theta) \mid Y, X\} &= \frac{1}{n-1}\sum_{i=1}^{n}\sum_{j=1}^{N} d^2(i,j \mid \theta), \end{aligned}$$

and

$$d(i,j) = a(i,j \mid \theta) - \frac{1}{n}\sum_{k=1}^{n} a(k,j \mid \theta) - \frac{1}{n}\sum_{l=1}^{n} a(i,l \mid \theta) + \frac{1}{n^2}\sum_{k=1}^{n}\sum_{l=1}^{n} a(k,l \mid \theta).$$

This motivates the use of the approximate distribution for $S_n(\Gamma \mid \theta)$ in obtaining $M(\theta)$. Such an approximation leads to

$$\begin{aligned} M(\theta) &\approx n! \int \exp\left[E\{S_n(\Gamma \mid \theta) \mid Y, X\} + \sqrt{Var\{S_n(\Gamma \mid \theta) \mid Y, X\}} z\right] \frac{1}{\sqrt{2\pi}} \exp(-z^2/2) dz \\ &= n! \exp\left[E\{S_n(\Gamma \mid \theta) \mid Y, X\} + Var\{S_n(\Gamma \mid \theta) \mid Y, X\}/2\right]. \end{aligned}$$

However, minimizing the last expression with respect to θ does not in general yield a consistent estimator of θ.

To see that, consider the simple example where the data were generated from a normal linear regression model,

$$Y_i = \beta_0 + \beta_1 X_i + \epsilon_i,$$

for $i = 1, \cdots, n$, and $\epsilon_1, \cdots, \epsilon_n$ are i.i.d. $N(0, \tau^2)$. If we analyzing the data using the semiparametric odds ratio model, the correct model should be

$$p(y \mid x) = \frac{\eta(y; x \mid \theta) dF(y \mid x_0 = 0)}{\int \eta(y; x \mid \theta) dF(y \mid x_0 = 0)},$$

where $\theta = \beta/\tau^2$, and

$$\log \eta(y; x \mid \theta) = \theta yx,$$

for the reference point is $(x_0, y_0) = (0, 0)$. In this case,

$$\begin{aligned} E\{S_n(\Gamma \mid \theta) \mid Y, X\} &= \theta\left(\sum_{i=1}^{n} Y_i X_i - n\bar{Y}\bar{X}\right), \\ Var\{S_n(\Gamma \mid \theta) \mid Y, X\} &= \frac{\theta^2}{n-1}\sum_{i=1}^{n}\sum_{j=1}^{N}(Y_i - \bar{Y})^2 (X_j - \bar{X})^2. \end{aligned}$$

When plugging in these expressions into the approximation, it leads to

$$M(\theta) \approx n! \exp\left(\theta n \hat{\tau}_{xy} + \theta^2 (n-1)\hat{\tau}_y^2 \hat{\tau}_x^2/2\right),$$

where $\hat{\tau}_{xy}$, $\hat{\tau}_y^2$, and $\hat{\tau}_x^2$ are respectively the empirical covariance between Y and X, the empirical variances of Y and X. The minimizer can be obtained as

$$\tilde{\theta} = \frac{n\hat{\tau}_{xy}}{(n-1)\hat{\tau}_x^2 \hat{\tau}_y^2}.$$

When $n \to \infty$, in probability or almost sure,

$$\tilde{\theta} \to \frac{\rho}{\tau_x \tau_y} = (1 - \rho^2)\theta,$$

where ρ is the correlation coefficient of Y and X. This means that the minimizer of the approximate objective function, i.e., $\tilde{\theta}$, is not a consistent estimator for the odds ratio parameter θ.

3.3.2 Adaptive Monte Carlo approximation to permutation likelihood

To obtain better approximations to $M(\theta)$ so that the minimizer of the approximate objective function leads to a consistent estimator of θ, consider a non-uniform distribution for Γ,

$$P(\Gamma = \delta \mid Y, X, \theta_b) = \frac{\prod_{i=1}^n \eta(Y_{\delta(i)}; X_i \mid \theta_b)}{\sum_{\sigma \in \mathcal{P}_n} \prod_{i=1}^n \eta(Y_{\sigma(i)}; X_i \mid \theta_b)}. \tag{3.11}$$

This distribution can be rewritten as

$$\begin{aligned} P(\Gamma = \delta \mid Y, X, \theta_b) &= \frac{\prod_{i=1}^n \eta(Y_{\delta(i)}; X_i \mid \theta_b)}{\prod_{i=1}^n \eta(Y_i; X_i \mid \theta_b)} \frac{\prod_{i=1}^n \eta(Y_i; X_i \mid \theta_b)}{\sum_{\sigma \in \mathcal{P}_n} \prod_{i=1}^n \eta(Y_{\sigma(i)}; X_i \mid \theta_b)} \\ &= \frac{\exp\{S_n(\delta \mid \theta_b)\}}{M(\theta_b)}. \end{aligned}$$

Using this distribution, $M(\theta)$ in (3.10) can be rewritten as

$$\begin{aligned} M(\theta) &= \sum_{\sigma \in \mathcal{P}_n} \exp\{S_n(\sigma \mid \theta)\} P(\sigma \mid Y, X, \theta_b) \frac{M(\theta_b)}{\exp\{S_n(\sigma \mid \theta_b)\}} \\ &= M(\theta_b) \sum_{\sigma \in \mathcal{P}_n} \exp\{S_n(\sigma \mid \theta) - S_n(\sigma \mid \theta_b)\} P(\sigma \mid Y, X, \theta_b) \end{aligned}$$

It follows that

$$\frac{M(\theta)}{M(\theta_b)} = E_{\theta_b}\left[\exp\{S_n(\Gamma \mid \theta) - S_n(\Gamma \mid \theta_b)\}\right], \tag{3.12}$$

where E_{θ_b} denotes the expectation with regard to (3.11). Note further that,

$$\begin{aligned} \log \frac{M(\theta)}{M(\theta_b)} &= \log E_{\theta_b}\left[\exp\{S_n(\Gamma \mid \theta) - S_n(\Gamma \mid \theta_b)\}\right] \\ &\geq E_{\theta_b}\{S_n(\Gamma \mid \theta) - S_n(\Gamma \mid \theta_b)\}, \end{aligned}$$

where the last inequality is due to the Jensen's inequality. Similar arguments for the EM algorithm cannot be applied here because we need to minimize the objective function, not maximize it.

To compute the expectation on the right-hand side of (3.12), consider Taylor expansion of the exponent

$$S_n(\Gamma \mid \theta) - S_n(\Gamma \mid \theta_b) \approx \sqrt{n}(\theta - \theta_b)^t \frac{1}{\sqrt{n}} \frac{\partial}{\partial \theta} S_n(\Gamma \mid \theta_b) + \frac{1}{2}(\theta - \theta_b)^t \frac{\partial^2}{\partial \theta^2} S_n(\Gamma \mid \theta_b)(\theta - \theta_b)$$

Note that, in general,

$$\begin{aligned} \frac{1}{\sqrt{n}} \frac{\partial}{\partial \theta} S_n(\Gamma \mid \theta_b) &\to N\{\mu_n(\theta_b), \Sigma_n(\theta_b)\}, \\ \frac{1}{n} \frac{\partial^2}{\partial \theta^2} S_n(\Gamma \mid \theta_b) &\to \Sigma_n(\theta_b), \end{aligned}$$

where $X_n \to N(\mu_n, \Sigma_n)$ means, in distribution,

$$\Sigma_n^{-1/2}(X_n - \mu_n) \to N(0, I).$$

and $X_n \to \Omega_n$ means $\alpha^t(X_n - \Omega_n)\alpha \to 0$ in probability for all α such that $||\alpha||_2 = 1$. In addition, we may approximate the permutation distribution by

$$P(\Gamma \mid Y, X, \theta_b) \approx P\left(\Gamma \mid \frac{1}{\sqrt{n}} S_n(\Gamma \mid \theta_b) = x, \theta_b\right) \phi\{x \mid \mu_n(\theta_b), \Sigma_n(\theta_b)\},$$

where $\phi\{x \mid \mu_n(\theta_b), \Sigma_n(\theta_b)\}$ is the normal density with mean $\mu_n(\theta_b)$ and variance matrix $\Sigma_n(\theta_b)$. Applying the approximations in computing the expectation in (3.12), we obtain the approximation

$$\frac{M(\theta)}{M(\theta_b)} \approx \exp\left\{\sqrt{n}(\theta - \theta_b)^t \mu_n(\theta_b) + \frac{1}{2}\sqrt{n}(\theta - \theta_b)^t\{\Sigma_n(\theta_b) + \Omega_n(\theta_b)\}\sqrt{n}(\theta - \theta_b)\right\}.$$

Minimizing the approximation on the right-hand side of the expression to obtain

$$\sqrt{n}(\theta - \theta_b) = -\{\Sigma_n(\theta_b) + \Omega_n(\theta_b)\}^{-1}\mu_n(\theta_b).$$

For a given θ_b, the foregoing equation can be used to update θ estimate. At the updated θ, denote by θ_{new},

$$\frac{M(\theta_{new})}{M(\theta_b)} \approx \exp\left\{-\frac{1}{2}\left[\mu_n(\theta_b)\right\}^t \{\Sigma_n(\theta_b) + \Omega_n(\theta_b)\}^{-1}\mu_n(\theta_b)\right\}.$$

If $\Sigma_n(\theta_b) + \Omega_n(\theta_b)$ is positive definite, it follows that

$$\frac{M(\theta_{new})}{M(\theta_b)} \lesssim 1,$$

which implies that $M(\theta_{new}) \lesssim M(\theta_b)$. The updating steps may be repeated until convergence is achieved.

One remaining question is how to compute $\mu_n(\theta_b)$, $\Sigma_n(\theta_b)$, and $\Omega_n(\theta_b)$. From the derivation, we know that

$$\begin{aligned}
\mu_n(\theta_b) &= E_{\theta_b}\left\{\frac{1}{\sqrt{n}}\frac{\partial}{\partial\theta}S_n(\Gamma \mid \theta_b)\right\}, \\
\Sigma_n(\theta_b) &= V_{\theta_b}\left\{\frac{1}{\sqrt{n}}\frac{\partial}{\partial\theta}S_n(\Gamma \mid \theta_b)\right\}, \\
\Omega_n(\theta_b) &= E_{\theta_b}\left\{\frac{1}{n}\frac{\partial^2}{\partial\theta^2}S_n(\Gamma \mid \theta_b)\right\}.
\end{aligned}$$

Computation of the means and variance directly using the permutation distribution (3.11) is impractical due to the large number of permutations need to be evaluated. A Monte Carlo sample of the permutations may be used to approximate these quantities. Direct sampling from (3.11) is also difficult, but a Markov chain Monte Carlo sampling approach can be applied to this problem. A Metropolis sampling algorithm is implemented in the following section to obtain a sample of permutations that follow the distribution in (3.11).

3.3.3 Metropolis algorithm for sampling permutations for estimation

The basic idea of a Metropolis-Hastings sampling approach (Metropolis et al, 1953; Hastings, 1970) is as follows. To sample from a known distribution function with density $p(y)$, start from an initial value Y_0. Generate Y_1^* from the candidate distribution $q(y \mid Y_0)$ that depends on the current value Y_0. Usually, it is easy to generate Y_1^* from $q(y \mid Y_0)$ for any given Y_0. Y_1^* is accepted, i.e., the new sample $Y_1 = Y_1^*$, with probability

$$\pi = \min\left\{1, \frac{q(Y_0 \mid Y_1^*)p(Y_1^*)}{q(Y_1^* \mid Y_0)p(Y_0)}\right\},$$

and Y_1^* is rejected, i.e., $Y_1 = Y_0$, with probability $1-\pi$. Repeat this step with initial value Y_0 replaced by Y_1 in generating Y_2. Generate Y_3 etc. in the same fashion. The distribution of Y_N in the generated Markov chain converges to $p(y)$ when $N \to \infty$ if the generated Markov chain is irreducible and aperiodic (Liu, 2001)[111]. When q is symmetric, i.e., $q(y \mid y^*) = q(y^* \mid y)$ for any y, y^* in the domain, the acceptance probability reduces to

$$\pi = \min\left\{1, \frac{p(Y_1^*)}{p(Y_0)}\right\}.$$

The resulting sampling algorithm is Metropolis algorithm. More theoretical analysis of the Markov chain sampling approaches can be found in Peskun (1973)[137] and the convergence rate of the Markov chain Monte Carlo sampling approach can be found in Rosenthal (1995)[155]. One important property of the Markov chain Monte Carlo sampling approach is that $p(y)$ may be known only up to a normalizing constant. This allows us to circumvent the need to compute the normalizing constant which can be difficult to do, such as the denominator in the permutation distribution (3.11).

In generating a Markov chain with limiting distribution (3.11), we choose a random walk over permutations with a given maximum cycle length m as the candidate distribution. More specifically, randomly choose m indices without replacement from $\{1, \cdots, n\}$. Denote the indices by $I_m = \{i_1, \cdots, i_m\}$. Let σ_{I_m} be the permutation such that $\sigma_{I_m}(i_1, \cdots, i_m) = (i_2, \cdots, i_m, i_1)$ and keep all other indices unchanged. Suppose that the current permutation is σ_k. Let $\sigma_{k+1}^* = \sigma_{I_m}(\sigma_k)$ and

$$\pi = \min\left\{1, \frac{\prod_{i=1}^n \eta(Y_{\sigma_{k+1}^*(i)}; X_i \mid \theta_b)}{\prod_{i=1}^n \eta(Y_{\sigma_k(i)}; X_i \mid \theta_b)}\right\}.$$

Accept σ_{k+1}^*, i.e., $\sigma_{k+1} = \sigma_{k+1}^*$, with probability π. Reject σ_{k+1}^*, i.e., $\sigma_{k+1} = \sigma_k$, with probability $1 - \pi$. This Markov chain converges to the target distribution in (3.11).

To generate a sample of size N for estimating $\mu_n(\theta_b)$, $\Sigma_n(\theta_b)$, and $\Omega_n(\theta_b)$, sampling the Markov chain generated above after a burn-in period and making the consecutive samples far apart such that their correlation is close to 0. Denote such a sample by $\sigma^{(1)}, \cdots, \sigma^{(N)}$. With such a sample, we can approximate $\mu_n(\theta_b)$, $\Sigma_n(\theta_b)$, and $\Omega_n(\theta_b)$ respectively by

$$\begin{aligned}
\hat{\mu}_n(\theta_b) &= \frac{1}{N}\sum_{k=1}^N \frac{1}{\sqrt{n}}\frac{\partial}{\partial\theta} S_n(\sigma^{(k)} \mid \theta_b), \\
\hat{\Sigma}_n(\theta_b) &= \frac{1}{N-1}\sum_{k=1}^N \left\{\frac{1}{\sqrt{n}}\frac{\partial}{\partial\theta} S_n(\sigma^{(k)} \mid \theta_b) - \hat{\mu}_n(\theta)\right\}^2, \\
\hat{\Omega}_n(\theta_b) &= \frac{1}{N}\sum_{k=1}^N \frac{1}{n}\frac{\partial^2}{\partial\theta^2} S_n(\sigma^{(k)} \mid \theta_b).
\end{aligned}$$

The update of θ_b can be completed by

$$\theta_b^{new} = \theta_b - \left\{\hat{\Sigma}_n(\theta_b) + \hat{\Omega}_n(\theta_b)\right\}^{-1} \hat{\mu}_n(\theta_b)/\sqrt{n}.$$

Denote the convergent point by $\hat{\theta}_N$. It can be shown that when N is sufficiently large, $\hat{\theta}_N$ is approximately equivalent to the maximum permutation likelihood estimator. The variance of the maximum permutation likelihood estimator can be approximated by

$$\left\{\hat{\Sigma}_n(\hat{\theta}_N) + \hat{\Omega}_n(\hat{\theta}_N)\right\}^{-1} /n.$$

For finite sample sizes in simulation studies, the estimator has good performance. In general, the permutation likelihood estimator is less biased and has smaller variance than the pairwise pseudo-likelihood estimator.

3.3.4 Permutation likelihood for the joint model

The Monte Carlo simulation approach to maximizing the permutation likelihood for a single outcome can be extended to the general case with multiple outcomes. The permutation likelihood for multivariate outcomes can be derived from (3.2) as

$$L(\theta) = \frac{\prod_{i=1}^n \eta(Y_{i1}; \cdots ; Y_{iG} \mid \theta)}{\sum_{\sigma \in \mathcal{P}_n(G)} \prod_{i=1}^n \eta(Y_{\sigma_1(i)1}; \cdots ; Y_{\sigma_G(i)G} \mid \theta)},$$

where $\sigma = (\sigma_1, \cdots, \sigma_G)$ with σ_k being a permutation of indices $\{1, \cdots, n\}$ for $k = 1, \cdots, G$ and $\mathcal{P}_n(G)$ is the collection of all vectors of permutations of the form σ. Because the permutation is invertible, it can be seen that

$$\begin{aligned}
&\sum_{\sigma \in \mathcal{P}_n(G)} \prod_{i=1}^n \eta(Y_{\sigma_1(i)1}; \cdots ; Y_{\sigma_G(i)G} \mid \theta) \\
&\quad = n! \sum_{\sigma \in \mathcal{P}_n(G)} \prod_{i=1}^n \eta(Y_{\sigma_1\{\sigma_G^{-1}(i)\}1}; \cdots ; Y_{\sigma_G\{\sigma_G^{-1}(i)\}(G-1)}; Y_{iG} \mid \theta)
\end{aligned}$$

the permutation likelihood can be simplified to

$$L(\theta) = \frac{\prod_{i=1}^n \eta(Y_{i1}; \cdots ; Y_{iG} \mid \theta)}{n! \sum_{\sigma \in \mathcal{P}_n(G-1)} \prod_{i=1}^n \eta(Y_{\sigma_1(i)1}; \cdots ; Y_{\sigma_{G-1}(i)(G-1)}; Y_{iG} \mid \theta)}, \tag{3.13}$$

Maximizing the likelihood is equivalent to minimizing the following objective function

$$M(\theta) = \sum_{\sigma \in \mathcal{P}_n(G-1)} \exp\left\{\sum_{i=1}^n \log \frac{\eta(Y_{\sigma_1(i)1}; \cdots ; Y_{\sigma_{G-1}(i)(G-1)}; Y_{iG} \mid \theta)}{\eta(Y_{i1}; \cdots ; Y_{iG} \mid \theta)}\right\}. \tag{3.14}$$

Let $\Gamma = (\Gamma_1, \cdots, \Gamma_{G-1})$ be a random vector taking values in $\mathcal{P}_n(G-1)$. Then

$$M(\theta) = (n!)^{G-1} E\left[\exp\left\{S_n(\Gamma \mid \theta)\right\}\right],$$

where

$$S_n(\Gamma \mid \theta) = \sum_{i=1}^n \log \frac{\eta(Y_{\Gamma_1(i)1}; \cdots ; Y_{\Gamma_{G-1}(i)(G-1)}; Y_{iG} \mid \theta)}{\eta(Y_{i1}; \cdots ; Y_{iG} \mid \theta)}$$

and the expectation is taken under the distribution

$$P(\Gamma = \sigma \mid Y) = \frac{1}{(n!)^{G-1}},$$

with Y being the observed data matrix. Similar to the univariate case, the sample of permutations based on the uniform distribution does not in general yield a good approximation to the permutation likelihood.

To obtain a better approximation to $M(\theta)$, consider sampling a vector of permutations from the following distribution,

$$P(\Gamma = \delta \mid Y, \theta) = \frac{\prod_{i=1}^{n} \eta(Y_{\delta_1(i)1}; \cdots ; Y_{\delta_{G-1}(i)(G-1)}; Y_{iG} \mid \theta)}{\sum_{\sigma \in \mathcal{P}_n(G-1)} \prod_{i=1}^{n} \eta(Y_{\sigma_1(i)1}; \cdots ; Y_{\sigma_{G-1}(i)(G-1)}; Y_{iG} \mid \theta)}. \tag{3.15}$$

The distribution can be rewritten as

$$P(\Gamma = \delta \mid Y, \theta) = \frac{\exp\{S_n(\delta \mid \theta)\}}{M(\theta)}.$$

For a fixed θ_b, this leads to

$$\frac{M(\theta)}{M(\theta_b)} = E\left[\exp\{S_n(\Gamma \mid \theta) - S_n(\Gamma \mid \theta_b)\}\right],$$

where E_{θ_b} is the expectation taken under the distribution (3.15). By exactly the same arguments as in the case of a univariate outcome,

$$\frac{M(\theta)}{M(\theta_b)} \approx \exp\left\{\sqrt{n}(\theta - \theta_b)^t \mu_n(\theta_b) + \frac{1}{2}\sqrt{n}(\theta - \theta_b)^t \{\Sigma_n(\theta_b) + \Omega_n(\theta_b)\}\sqrt{n}(\theta - \theta_b)\right\},$$

where

$$\begin{aligned}
\mu_n(\theta_b) &= E_{\theta_b}\left\{\frac{1}{\sqrt{n}}\frac{\partial}{\partial\theta} S_n(\Gamma \mid \theta_b)\right\}, \\
\Sigma_n(\theta_b) &= V_{\theta_b}\left\{\frac{1}{\sqrt{n}}\frac{\partial}{\partial\theta} S_n(\Gamma \mid \theta_b)\right\}, \\
\Omega_n(\theta_b) &= E_{\theta_b}\left\{\frac{1}{n}\frac{\partial^2}{\partial\theta^2} S_n(\Gamma \mid \theta_b)\right\}.
\end{aligned}$$

Again, at the updated θ, denote by θ_{new},

$$\frac{M(\theta_{new})}{M(\theta_b)} \approx \exp\left\{-\frac{1}{2}\left[\mu_n(\theta_b)\right\}^t \{\Sigma_n(\theta_b) + \Omega_n(\theta_b)\}^{-1} \mu_n(\theta_b)\right\}.$$

If $\Sigma_n(\theta_b) + \Omega_n(\theta_b)$ is positive definite, it follows that

$$M(\theta_{new}) \lesssim M(\theta_b),$$

which means the update decreases the objective function $M(\theta)$.

To generate a sample of permutations for estimating $\mu_n(\theta_b)$, $\Sigma_n(\theta_b)$, and $\Omega_n(\theta_b)$, Metropolis sampling algorithm can be used. Starting from an initial vector of permutation $(\sigma_{10}, \cdots, \sigma_{(G-1)0})$ and the first variable index 1, randomly choose m_1 indices without replacement from $\{1, \cdots, n\}$, where m_1 is the maximum cycle to be chosen for variable 1.

The score equation for estimating h_1 is

$$h_{k1} = -\log \sum_{i=1}^{n} \frac{\sum_{k_2=1}^{K_2} \cdots \sum_{k_G=1}^{K_G} \eta(y_{k1}; \cdots ; y_{k_G G}; X_i \mid \theta) \exp\left(\sum_{s=2}^{G} h_{k_s s}\right) \prod_{s=2}^{G} n_{k_s s}}{\sum_{k_1=1}^{K_1} \cdots \sum_{k_G=1}^{K_G} \eta(y_{k_1 1}; \cdots ; y_{k_G g}; X_i \mid \theta) \exp\left(\sum_{s=1}^{G} h_{k_s s}\right) \prod_{s=1}^{G} n_{k_s s}}.$$

Likewise, an iterative algorithm may be used to update h_1. When all other parameters are fixed, the algorithm computes

$$\begin{aligned} h_{k1}^{*(t)} &= -\log \sum_{i=1}^{n} \frac{\sum_{k_2=1}^{K_2} \cdots \sum_{k_G=1}^{K_G} \eta(y_{k1}; \cdots ; y_{k_G G}; X_i \mid \theta) \exp\left(\sum_{s=2}^{G} h_{k_s s}^{(t-1)}\right) \prod_{s=2}^{G} n_{k_s s}}{\sum_{k_1=1}^{K_1} \cdots \sum_{k_G=1}^{K_G} \eta(y_{k_1 1}; \cdots , y_{k_G g}; X_i \mid \theta) \exp\left(\sum_{s=1}^{G} h_{k_s s}^{(t-1)}\right) \prod_{s=1}^{G} n_{k_s s}}, \\ h_{k1}^{(t)} &= h_{k1}^{*(t)} - \log \sum_{k=1}^{K_1} \exp\{h_{k1}^{*(t)}\}, k = 1, \cdots , K_1. \end{aligned}$$

Similarly, it can be shown by Jensen's inequality that

$$l\{\theta, h_1^{(t)}, h_2^{(t-1)}, \cdots , h_G^{(t-1)}\} \geq l\{\theta, h_1^{(t-1)}, \cdots , h_G^{(t-1)}\}.$$

The other baseline functions can be updated similarly. Updating θ can be done similarly to the case with two groups.

Although the maximum likelihood approach can in principle be applied to multivariate outcomes with arbitrary groups. In practice, such an approach may encounter difficulties in computation implementation because of the challenge in computing the normalizing constants, i.e., the denominators with G-fold of summations, each of which has $\prod_{j=1}^{G} K_j$ terms. The total number of terms increase exponentially as the number of groups increase. For example, if we have $G = 5$ and the sample size is 100 and there is no tied observations, the total number of items to sum over is 100^5. Several compromised approaches may be applied to partially address the computation problem. One approach is to categorize the continuous data before fitting a semiparametric odds ratio model. This may be done by rounding the numeric values. For example, if the number of categories is reduced to 10, the total number of items to sum over is 10^5 which is more manageable. Another approach is to perform multiple analyses, each with one group of outcomes. The parameter estimators from the latter analyses may be less efficient when compared with the joint analysis. Still another approach is to use the Monte Carlo approximation to the integrals in the likelihood maximization. This method will be discussed in detail in Chapter 8 when data are incompletely observed due to missing or censoring.

3.4.5 Large sample behavior of the maximum likelihood estimator

The maximum semiparametric likelihood estimator for the parameters (θ, F) in the semiparametric odds ratio model can be shown to be consistent and asymptotically efficient under some regularity conditions. The variance of the asymptotically normally distributed estimator for θ can be estimated by the inverse of the profile information matrix for θ, i.e., the minus second derivative of the profile likelihood with respect to θ. See for example, Osius (2009)[132], Luo and Tsai (2012)[116], and Chen et al (2015)[35].

Both the result on the convergence of the algorithm to the maximum likelihood estimator in Chen (2015)[31] and the asymptotic results on the maximum likelihood estimator can also be extended to the multivariate cases. Chen et al (2015)[35] showed that the maximum likelihood estimator is consistent and asymptotically normally distributed. The variance of the maximum likelihood estimator can be obtained by the inverse of the information matrix of the profile likelihood for θ. The proofs of these results along with other theoretical justifications are collected in Chapter 10 of this book.

3.5 Comparison of different likelihood approaches

The pseudo-likelihood approaches are not efficient for the semiparametric odds ratio model. But they, especially the pairwise pseudo-likelihood approach, can be computationally more efficient. In practice, they may be used directly or as a way to obtain the initial values for the more efficient but computationally slower approaches.

The permutation likelihood approach is in essence a conditional likelihood approach. It eliminates the nuisance baseline functions from the conditional likelihood. It can be the most efficient approach to estimating the odds ratio parameter. This follows the similar argument of Liang (1983, 1984)[99, 100] for parametric models. For semiparametric models, it can be argued similarly (Bickel et al, 1993)[11]. The exact verification remains to be done.

The maximum semiparametric likelihood estimator is the most efficient approach for estimating both the odds ratio parameter and the baseline functions. Due to the computation complexity, it is only feasible for very few number of groups of variables. For estimating the odds ratio parameter only, the maximum permutation likelihood estimator is likely to be equivalent to the maximum semiparametric likleihood estimator. The maximum permutation likelihood estimator is expected to have better small sample performance.

3.6 The R package `SPORM` for semiparametric odds ratio model

We have so far introduced three likelihood-based estimation and inference approaches to the semiparametric odds ratio model. They are respectively: the maximum pairwise pseudo-likelihood approach, the maximum permutation likelihood approach, and the maximum semiparametric likelihood approach. These methods are relatively new and are not typically incorporated into the major statistical software packages. To make the methods available for use in data analysis, we have programmed the methods in FORTRAN and wrapped them in an R package named `SPORM` (SemiParametric Odds Ratio Model). There are three R functions corresponding respectively to the three approaches for fitting the semiparametric odds ratio model. Specifically,

1. The R-function `pwlkh` fits the model using the maximum pairwise pseudo-likelihood approach. The function call is as follows.

   ```
   pwlkh(y, x, niter, eps, vect).
   ```

   The output of the function gives the parameter estimation and the estimated variance-covariance matrix. In the function call, the argument `y` in the function specifies the outcome data which is a matrix composed of the observed values of a vector of multiple outcome variables. The argument `x` specifies the covariate data which is a matrix composed of the observed values of a vector of multiple covariates. `niter` specifies the maximum number of iterations if convergence is not achieved. `eps` specifies the convergence criterion. `vect` specifies the form of vectorization of the odds ratio parameters used. One way (`vect`="col") is the column-wise vectorization which is also the default choice. The other way (`vect`="row") is the row-wise vectorization. The vectorization method only takes effect when there are multiple outcome variables and the way to vectorize only affects the listing order of the estimated parameters and their variance estimates. Default values are set for `niter, eps, vect`.

2. The R-function `splkh` fits the model using the maximum semiparametric likelihood approach. The function call is as follows.

```
splkh(y, x, niter, eps, vect).
```

The form of output from this function is the same as that of `pwlkh`. The arguments of the function have also the same meanings as in the function `pwlkh`. Default values are set for `niter, eps, vect`.

3. The R-function `pmlkh` fits the model using the maximum permutation likelihood approach. The function call is as follows.

```
pmlkh(data, group, niter, nlag, nburnin, nsamp, nintv, maxcyc).
```

The form of output from this function is similar to the other two functions. In the function call, the argument `data` specifies the name of the data matrix. The argument `group` is a vector whose components specify the size of grouped variables of the data matrix. For example, if the data matrix has, say 10 variables and we divide them into 4 groups with sizes $2, 1, 4, 3$, then `group` is a vector of length 4 with `group[1]=2, group[2]=1, group[3]=4, group[4]=3`. `group` is used to specify a multivariate odds ratio model as given by (2.69). The argument `niter` specifies the maximum number of iterations used in finding the maximum permutation likelihood estimator. `nlag` specifies the number of iterations used to assess the convergence of the algorithm for finding the parameter estimator. The parameter estimate output is the average of the last `nlag` estimates. `nburnin` specifies the number of iterations discarded in the Markov chain Monte Carlo sampling algorithm before a Monte Carlo sample is taken for estimating the mean and variance matrix in the simulation algorithm. `nsamp` specifies the size of the sample taken for mean and variance approximation in each iteration step. `nintv` specifies the interval in taking a sample from the Monte Carlo Markov chain for the Monte Carlo approximation. `maxcyc` specifies the length of the permutation cycle of the candidate permutation in Metropolis sampling of the permutations. Parameter vectorization in the permutation approach is always column-wise within each group of variables and row-wise for at the block level. Default values are set for `niter, nlag, nburnin, nsamp, nintv, maxcyc`. As a result, only `data, group` are necessary for the function call.

Currently, the choice of odds ratio function in the package is limited to bilinear odds ratio function. Within this type of models, if some of the odds ratio parameters are known to be 0, i.e., the model is structurally restricted, a set of alternative functions (`pwlkhfix, splkhfix, pmlkhfix`) within the packages can be used to fit such models. The ways to call these functions are as follow.

```
pwlkhfix(y, x, fixstruct, niter, eps, vect),
splkhfix(y, x, fixstruct, niter, eps, vect),
pmlkhfix(data, group, fixstruct, niter, nlag,
         nburnin, nsamp, nintv, maxcyc).
```

In each case, a new variable `fixstruct` is added to specify which coefficients in the bilinear odds ratio model are set to 0. The rest of the arguments have the same meanings in the respective functions **pwlkh, splkh, pmlkh**. Naturally, the previously introduced functions (**pwlkh, splkh, pmlkh**) are special cases of the functions (**pwlkhfix, splkhfix, pmlkhfix**). The former is easy to call as no restriction is put on the coefficients. The latter generates the same results if the struct matrix does not put any restrictions on the odds ratio parameters. In addition to the input difference, the output of the latter also include the maximum log-likelihoods which can be useful for hypothesis testing and model selection.

3.7 Simulation study using SPORM

In this section, we use the R package **SPORM** to conduct simulation studies to evaluate the performance of the likelihood approaches. Since the package can be applied to univariate outcome, multivariate outcomes as a group, as well as multi-groups of multivariate outcomes in the case of the permutation likelihood approach, the following evaluations are done for different scenarios. The models for data generation are linear, binary, or Poisson regression model with or without biased sampling. The model for data analysis is always the semiparametric odds ratio model.

3.7.1 Univariate outcome

The univariate outcome is generated by one of the following models: The linear regression model,

$$Y = \beta_0 + \beta_1 X_1 + \beta_2 X_2 + \beta_3 X_3 + \sigma\epsilon,$$

where X_1, X_2, X_3 are independent uniformly distributed on $[-1, 1]$, and ϵ follows a standard normal distribution and $\sigma > 0$; The binary regression model,

$$P(Y = 1 \mid X_1, X_2, X_3) = \frac{\exp(\beta_0 + \beta_1 X_1 + \beta_2 X_2 + \beta_3 X_3)}{1 + \exp(\beta_0 + \beta_1 X_1 + \beta_2 X_2 + \beta_3 X_3)},$$

where X_1, X_2, X_3 are generated in the same way as in the normal-model case;

The Poisson regression model,

$$P(Y = k \mid X_1, X_2, X_3) = \frac{\lambda^k(X_1, X_2, X_3)}{k!} \exp\{-\lambda(X_1, X_2, X_3)\},$$

and

$$\log \lambda(X_1, X_2, X_3) = \beta_0 + \beta_1 X_1 + \beta_2 X_2 + \beta_3 X_3,$$

where X_1, X_2, X_3 are generated similarly to previous models. We model the data by the semiparametric odds ratio model in all the cases,

$$p(y \mid x_1, x_2, x_3) = \frac{\exp(\theta_1 y x_1 + \theta_2 y x_2 + \theta_3 y x_3) p(y)}{\int \exp(\theta_1 y x_1 + \theta_2 y x_2 + \theta_3 y x_3) p(y) dy}. \tag{3.24}$$

In simulating the data, we chose $\theta = (\beta_1, \beta_2, \beta_3)/\sigma^2$, $\beta_0 = 0$, and $\sigma^2 = 1$. Sample sizes of 100 and 400 are simulated. In the permutation likelihood approach, we take a sample of **nsamp** $= 2,000$ permutations with a sampling interval of **sintv** $= 200$ steps after **nburn** $= 40,000$ burn-in steps. The final estimate is the average of the last 20 iterations with **niter=50** and **nstart** $= 30$. The number of replicates is set to 500.

Table 3.1 lists the simulation results. The simulation results show that the permutation likelihood (PM) estimator has better performance than both the semiparametric likelihood

TABLE 3.1
Simulation results for univariate outcome.

Model	Method	$\theta_1=1$	$\theta_2=1$	$\theta_3=-2$	$\theta_1=1$	$\theta_2=1$	$\theta_3=-2$
		Sample size=100			Sample size=400		
Normal	PW	1.064	1.0446	-2.114	1.017	1.009	-2.031
		0.062	0.063	0.142	0.015	0.014	0.032
		0.049	0.048	0.105	0.017	0.015	0.026
	PM	1.051	1.031	-2.090	1.010	1.004	-2.022
		0.054	0.055	0.129	0.013	0.012	0.029
		0.057	0.056	0.128	0.013	0.013	0.029
	SP	1.081	1.061	-2.145	1.017	1.011	-2.035
		0.058	0.059	0.139	0.013	0.013	0.030
		0.059	0.058	0.132	0.013	0.013	0.029
Poisson	PW	1.063	1.070	-2.124	1.019	1.012	-2.033
		0.082	0.078	0.198	0.016	0.017	0.039
		0.058	0.075	0.159	0.014	0.013	0.037
	PM	1.053	1.054	-2.095	1.014	1.012	-2.028
		0.066	0.061	0.152	0.012	0.012	0.033
		0.052	0.051	0.145	0.011	0.011	0.032
	SP	1.103	1.103	-2.189	1.028	1.026	-2.055
		0.074	0.068	0.187	0.013	0.013	0.034
		0.054	0.054	0.152	0.011	0.011	0.032
Binary	PW	1.037	1.068	-2.156	1.020	1.021	-2.038
		0.210	0.203	0.283	0.049	0.046	0.056
		0.195	0.159	0.184	0.050	0.044	0.056
	PM	1.025	1.053	-2.128	1.019	1.019	-2.034
		0.202	0.193	0.274	0.048	0.046	0.055
		0.208	0.210	0.259	0.048	0.048	0.058
	SP	1.037	1.066	-2.153	1.022	1.022	-2.034
		0.207	0.198	0.283	0.048	0.046	0.055
		0.210	0.210	0.262	0.048	0.048	0.058

PW: pairwise pseudo-likelihood; PM: permutation likelihood; SP: semiparametric likelihood. The first entry under each method is the parameter estimates, the second is the empirical variances of the estimates, and the third is the average of the estimated variances.

(SP) estimator and the pairwise likelihood (PW) estimator. Since the SP estimator is semiparametric efficient in large sample, the SP estimator is asymptotically as good as the PM estimator, if not better. This can be partially seen from the case with a sample size 400. The SP estimator and the PM estimator are very close to each other. But for the sample size of 100, the PM estimator is better in terms of both bias and the variation. The variance estimate generally approximates the empirical variance well. For small sample sizes, the PM variance estimates still perform well. However, the PW and the SP variance estimates can be of the target. In particular, the variance estimate for the PW estimator may have poor performance at times.

In terms of computation time, the PM approach is a little slower in computation than the SP approach for the sample size of 100. When the sample size increases to 400, the PM approach becomes faster than the SP approach because the computation time for the PM approach is much less affected by the sample size. The computation saving is expected to be much more prominent when dealing with multivariate continuous outcomes and/or larger sample sizes. For discrete outcomes, the SP approach is faster than the PM approach. The computation time for the PW approach can increase very fast as the sample size increases.

For different data generation models, the relative merits of different approaches remain similar.

3.7.2 Multivariate outcomes as a group

In this section, we simulate data with two outcomes and two covariates. The covariates X_1, X_2 are independent uniformly distributed on $[-1, 1]$. The multivariate outcome is generated by one of the following models: The linear regression model has

$$\begin{aligned} Y_1 &= \beta_{10} + \beta_{11}X_1 + \beta_{12}X_2 + \sigma_1\epsilon_1, \\ Y_2 &= \beta_{20} + \beta_{21}X_1 - \beta_{22}X_2 + \sigma_2\epsilon_2, \end{aligned}$$

where ϵ_1, ϵ_2 follow independent standard normal distributions; The binary regression model has

$$\begin{aligned} P(Y_1 = 1 \mid X_1, X_2) &= \frac{\exp(\beta_{10} + \beta_{11}X_1 + \beta_{12}X_2)}{1 + \exp(\beta_{10} + \beta_{11}X_1 + \beta_{12}X_2)}, \\ P(Y_2 = 1 \mid X_1, X_2) &= \frac{\exp(\beta_{20} + \beta_{21}X_1 + \beta_{22}X_2)}{1 + \exp(\beta_{20} + \beta_{21}X_1 + \beta_{22}X_2)}, \end{aligned}$$

where X_1, X_2 are generated in the same way as in the case of normal model;

The Poisson regression model has

$$\begin{aligned} P(Y_1 = k \mid X_1, X_2) &= \frac{\lambda_1^k(X_1, X_2)}{k!} \exp\{-\lambda_1(X_1, X_2)\}, \\ P(Y_2 = k \mid X_1, X_2) &= \frac{\lambda_2^k(X_1, X_2)}{k!} \exp\{-\lambda_2(X_1, X_2)\} \end{aligned}$$

and

$$\begin{aligned} \log \lambda_1(X_1, X_2) &= \beta_{10} + \beta_{11}X_1 + \beta_{12}X_2, \\ \log \lambda_2(X_1, X_2) &= \beta_{20} + \beta_{21}X_1 + \beta_{22}X_2, \end{aligned}$$

where X_1, X_2 are generated similarly to previous models. We model the data by the semiparametric odds ratio model in all the cases,

$$p(y_1, y_2 \mid x_1, x_2) = \frac{\exp(\theta_{11}y_1x_1 + \theta_{12}y_1x_2 + \theta_{21}y_2x_1 + \theta_{22}y_2x_2)p(y_1, y_2)}{\int\int \exp(\theta_{11}y_1x_1 + \theta_{12}y_1x_2 + \theta_{21}y_2x_1 + \theta_{22}y_2x_2)p(y_1, y_2)dy_1dy_2} \quad (3.25)$$

In the simulation, we choose $\theta = (\beta_{11}/\sigma_1^2, \beta_{12}/\sigma_1^2, \beta_{21}/\sigma_2^2, \beta_{22}/\sigma_2^2) = (1, -1, 1, 1)$, $\beta_{10} = \beta_{20} = 0$, and $\sigma_1^2 = \sigma_2^2 = 1$. Sample sizes of 100 and 400 are simulated respectively. In the permutation likelihood approach, we take a sample of `nsamp` $= 2,000$ permutations with a sampling interval of `sintv` $= 200$ steps after `nburnin` $= 40,000$ burn-in steps. The final estimate is the average of the last 20 iterations with `niter=50` and `nstart` $= 30$. The number of replicates is set to 500.

The simulation results are listed in Table 3.2. The results demonstrate the same patterns as seen in the case of univariate outcome. Basically, both the permutation likelihood approach and the semiparametric likelihood approach are more efficient than the pairwise pseudo-likelihood approach. The variance estimates approximate the empirical variance well, especially, the variance estimate of the permutation likelihood approach. The permutation likelihood approach appears to have a better performance than the semiparametric likelihood approach, at least, for the simulated finite samples.

TABLE 3.2
Simulation results for multiple outcomes as one group. The abbreviations in the table are the same as in Table 3.1.

Model	Method	$\theta_{11}=1$	$\theta_{12}=-1$	$\theta_{21}=1$	$\theta_{21}=1$	$\theta_{11}=1$	$\theta_{12}=-1$	$\theta_{21}=1$	$\theta_{22}=1$
		Sample size=100				Sample size=400			
Normal	PW	1.078	-1.040	1.053	1.067	1.013	-1.010	1.011	1.012
		0.076	0.082	0.076	0.079	0.018	0.018	0.018	0.018
		0.064	0.053	0.075	0.078	0.023	0.013	0.018	0.014
	PM	1.064	-1.028	1.035	1.042	1.011	-1.005	1.009	1.007
		0.066	0.072	0.071	0.071	0.016	0.017	0.016	0.016
		0.070	0.069	0.069	0.069	0.016	0.016	0.016	0.016
	SP	1.086	-1.050	1.057	1.064	1.016	-1.010	1.014	1.012
		0.070	0.076	0.069	0.073	0.016	0.017	0.016	0.016
		0.072	0.071	0.071	0.071	0.016	0.016	0.016	0.016
Poisson	PW	1.047	1.063	1.056	1.062	1.013	-1.004	1.011	1.011
		0.075	0.089	0.079	0.082	0.019	0.019	0.017	0..019
		0.086	0.064	0.080	0.081	0.016	0.021	0.015	0.017
	PM	1.027	-1.045	1.035	1.058	1.008	-0.998	1.011	1.008
		0.059	0.071	0.065	0.071	0.016	0.017	0.015	0.016
		0.065	0.069	0.065	0.069	0.015	0.016	0.015	0.016
	SP	1.138	-1.165	1.062	1.091	1.030	-1.022	1.018	1.015
		0.122	0.151	0.071	0.082	0.017	0.018	0.014	0.015
		0.075	0.080	0.068	0.072	0.015	0.016	0.015	0.016
Binary	PW	1.049	-1.045	1.118	1.053	1.0199	-1.016	1.015	1.032
		0.188	0.184	0.171	0.181	0.043	0.047	0.040	0.046
		0.152	0.187	0.148	0.178	0.049	0.048	0.044	0.046
	PM	1.037	-1.033	1.101	1.041	1.015	-1.012	1.010	1.027
		0.172	0.176	0.165	0.171	0.041	0.046	0.038	0.045
		0.178	0.175	0.180	0.177	0.041	0.041	0.041	0.041
	SP	1.049	-1.045	1.101	1.053	1.019	-1.014	1.013	1.030
		0.176	0.181	0.169	0.175	0.042	0.046	0.039	0.045
		0.179	0.177	0.181	0.178	0.041	0.041	0.041	0.041

3.7.3 Multivariate outcomes

The semiparametric odds ratio model is a flexible tool for modeling multivariate data. In comparison to the joint normal model conventionally used for the multivariate data, the semiparametric odds ratio model can handle mixed continuous and discrete data types. It is also flexible in that the model allows the variables to be arbitrarily grouped into blocks as needed in practice. For the analysis under such a flexible model, the pairwise pseudo-likelihood approach no longer works well. The semiparametric likelihood approach can in principle be applied. However, the computation burden of the method can quickly render the method impractical. The permutation likelihood approach appears to be the only viable approach currently available. The function `pmlkh` in the SPORM package does provide the computational ability to handle such models. This subsection is therefore concentrated on evaluating the performance of the permutation likelihood approach in such a setting using the function `pmlkh` in SPORM.

We first simulate a multivariate normal distribution with nine variables to examine the performance of the permutation likelihood approach. The mean of the normal distribution is set to 0 and the variance-covariance matrix is generated by first multiplying the transpose of a random matrix with the random matrix itself and then selecting the resulting matrix with elements in reasonable ranges. For the selected parameter values, data sets with sample sizes 100 and 400 are generated.

In the data analysis, the variables are grouped into 3 groups: (y_1, y_2), (y_3, y_4, y_5), and (y_6, y_7, y_8, y_9). the corresponding semiparametric odds ratio model is

$$\frac{\eta_1\{(y_1,y_2);(y_3,\cdots,y_9)\}\eta_2\{(y_3,y_4,y_5);(y_6\cdots,y_9)|(y_{10},y_{20})\}dF_1(y_1,y_2)dF_2(y_3,y_4,y_5)}{\int\cdots\int\eta_1\{(y_1,y_2);(y_3,\cdots,y_9)\}\eta_2\{(y_3,y_4,y_5);(y_6\cdots,y_9)|(y_{10},y_{20})\}dF_1(y_1,y_2)dF_2(y_3,y_4,y_5)}.$$

The model for the odds ratio functions are respectively

$$\begin{aligned}
\log\eta_1\{(y_1,y_2);(y_3,\cdots,y_9)\} &= \theta_{13}y_1y_3+\theta_{23}y_2y_3+\theta_{14}y_1y_4+\theta_{24}y_2y_4\\
&\quad+\theta_{15}y_1y_5+\theta_{25}y_2y_5+\theta_{16}y_1y_6+\theta_{26}y_2y_6+\theta_{17}y_1y_7+\theta_{27}y_2y_7\\
&\quad+\theta_{18}y_1y_8+\theta_{25}y_2y_8+\theta_{19}y_1y_9+\theta_{29}y_2y_9,\\
\log\eta_2\{(y_3,y_4,y_5);(y_6\cdots,y_9)\mid(y_{10},y_{20})\} &= \theta_{36}y_3y_6+\theta_{46}y_4y_6+\theta_{56}y_5y_6+\theta_{37}y_3y_7\\
&\quad+\theta_{47}y_4y_7+\theta_{57}y_5y_7+\theta_{38}y_3y_8+\theta_{48}y_4y_8+\theta_{58}y_5y_8\\
&\quad+\theta_{39}y_3y_9+\theta_{49}y_4y_9+\theta_{59}y_5y_9.
\end{aligned}$$

Use the `SPORM` package for the analysis of this model, we have **group**$= c(2,3,4)$, and the function call as

```
pmlkh(dat,group, niter=50, nstart=30, nburnin=50000, nsamp=5000, nint=500,
                        maxcyc=5),
```

`dat` is the observed data matrix for variables $(y_1, \cdots, y_9)$.

Table 3.3 lists the simulation results which are based on 500 replicates. It can be seen from the table that the bias of the maximum permutation likelihood estimator is in general small. The biases for the sample size 400 are substantially smaller than those of the sample size 100. The variance estimates in general track the empirical variance very well. Inference based on the asymptotic normal with the estimate variances are expected to behave well.

3.8 Data analysis using SPORM

We demonstrate the use of the package `SPORM` for analysis of multivariate data using the lipoprotein data (`lipid.Rdata`) in the package. The objective of lipoprotein study is to understand which single nucleotide polymorphisms (SNPs) affect the high-density lipoprotein level in human blood serum. The study employed an extreme-value sampling design. It samples cases and controls with HDL lower than 30 percentile or greater than 90 percentile in respective age and gender groups. In addition, the low-density lipoprotein (LDL), the total cholesterol (TC), and the triglycerides (TR) were also measured. This study genotyped 625 cases and 606 controls. More details of the study design can be found in Edmondson et al (2011) [58].

Chen et al (2015) [35] analyzed the `lipid` data for identifying SNPs associated with the HDL. The basic odds ratio model employed is as follows.

$$\begin{aligned}
\log\eta\{(HDL,LDL,TR,TC);SNP\mid age,sex,casestatus\} &= \theta_1 SNP*HDL\\
&\quad+\psi_{11}SNP*LDL+\psi_{12}SNP*TR+\psi_{13}SNP*TC, \qquad (3.26)
\end{aligned}$$

By examining SNPs one at a time, 12 SNPs as listed in Table 3.4 were identified to have significant association with the HDL. Because different lipoprotein measurements can be

TABLE 3.3

Permutation likelihood estimator for data with 3 groups and varying numbers of variables in the goups.

		Normal, SS=100 *			Normal, SS=400 **		
Param.	Truth	Est.	Emp.v	Est.v	Est.	Emp.v	Est.v
θ_{13}	2.405	2.633	0.290	0.289	2.438	0.059	0.063
θ_{23}	0.657	0.706	0.055	0.054	0.674	0.011	0.012
θ_{14}	0.455	0.492	0.109	0.110	0.467	0.023	0.024
θ_{24}	0.068	0.067	0.025	0.024	0.078	0.005	0.053
θ_{15}	-5.457	-6.003	0.888	0.836	-5.524	0.171	0.182
θ_{25}	-2.171	-2.379	0.172	0.162	-2.200	0.035	0.036
θ_{16}	-3.483	-3.833	0.388	0.355	-3.516	0.072	0.077
θ_{26}	-1.226	-1.343	0.069	0.064	-1.237	0.013	0.014
θ_{17}	3.726	4.110	0.658	0.628	3.751	0.128	0.135
θ_{27}	1.705	1.882	0.144	0.139	1.711	0.029	0.030
θ_{18}	4.676	5.129	0.733	0.707	4.722	0.146	0.155
θ_{28}	1.928	2.119	0.149	0.144	1.947	0.031	0.032
θ_{19}	-1.154	-1.274	0.070	0.070	-1.167	0.013	0.015
θ_{29}	-0.223	-0.239	0.013	0.012	-0.227	0.002	0.003
θ_{36}	1.145	1.264	0.152	0.140	1.171	0.030	0.030
θ_{46}	-0.071	-0.074	0.063	0.062	-0.058	0.013	0.013
θ_{56}	-3.917	-4.300	0.489	0.452	-3.958	0.093	0.099
θ_{37}	0.769	0.834	0.330	0.302	0.725	0.061	0.065
θ_{47}	1.572	1.732	0.191	0.184	1.548	0.036	0.040
θ_{57}	4.748	5.231	0.868	0.876	4.799	0.182	0.192
θ_{38}	-0.818	-0.901	0.303	0.276	-0.862	0.059	0.061
θ_{48}	0.461	0.502	0.147	0.139	0.438	0.028	0.030
θ_{58}	5.631	6.170	0.960	0.957	5.689	0.199	0.211
θ_{39}	1.081	1.189	0.049	0.050	1.081	0.009	0.011
θ_{49}	0.432	0.469	0.022	0.020	0.430	0.004	0.004
θ_{59}	-1.073	-1.184	0.083	0.083	-1.082	0.016	0.018

SS: Sample size; Param.: Name of the parameters for the variable blocks (1,2), (3,4,5), (6,7,8,9); Est.: Permutation likelihood estimates; Emp.v: Empirical variance estimate; Est.v: Average of the estimated variance. Results are based on 500 replicates. * There are 46 cases removed from the calculation due to unconvergence. ** There are 6 cases removed due to unconvergence.

highly correlated and the SNPs themselves can also be correlated, a joint analysis can yield more information on the nature of the associations between SNPs and lipoproteins. The covariates in the `lipid` data include `BMI, age, sex, smoke status,` and `case status.` A variation of the basic model that includes all the 12 SNPs in the model, termed basic joint model, is used for the joint analysis. Two sets of joint analyses are performed based the basic joint model: One with adjustment of the additional covariates body mass index (`BMI`) and `smoke status`, the other without.

The analysis that includes additional covariate adjustment is performed using the R function

```
pmlkh(as.matrix(lipid),group).
```

The one without additional covariates are done by removing the corresponding variables from data object `lipid`, and then apply the function `pmlkh` similarly. Table 3.4 shows the results of the analyses. From Table 3.4, it can be seen that both analyses yield similar

TABLE 3.4
Odds ratio estimators for SNPs and lipoprotein association.

SNPs	HDL	LDL	TG	TC
		Without adjusting smoke and BMI		
rs3764261	-1.23 (0.54)*	-1.04 (0.61)	-0.46 (0.22)*	1.48(0.85)
rs1800775	-0.02 (0.61)	0.12 (0.69)	0.05 (0.25)	-0.16(0.97)
rs711752	0.90 (1.58)	1.16 (1.78)	0.24 (0.63)	-1.54(2.49)
rs7203984	-0.83 (1.61)	-0.40 (1.82)	0.01 (0.65)	0.55(2.55)
rs11508026	-0.81 (2.16)	-0.61 (2.41)	-0.28 (0.85)	0.63(3.38)
rs12720922	0.99 (1.49)	0.51 (1.66)	0.21 (0.59)	-0.88(2.33)
rs9939224	-0.15 (0.72)	-0.06 (0.81)	-0.10 (0.29)	0.16(1.14)
rs1532624	1.11 (1.63)	0.60 (1.81)	0.46 (0.64)	-0.75(2.54)
rs11076175	-2.07 (2.83)	-2.13 (3.15)	-1.18 (1.15)	2.89(4.35)
rs7499892	2.01 (2.56)	1.84 (2.85)	0.96 (1.05)	-2.46(3.92)
rs11076176	-0.20 (2.35)	-0.53 (2.68)	-0.08 (0.95)	0.89(3.75)
rs289714	1.11 (2.37)	1.26 (2.70)	0.38 (0.96)	-1.85(3.78)
		With adjusting smoke and BMI		
rs3764261	-1.26(0.55)*	-1.03(0.62)	-0.46 (0.22)*	1.50(0.87)
rs1800775	-0.02(0.62)	0.13(0.69)	0.05 (0.24)	-0.16(0.97)
rs711752	0.84(1.59)	1.09(1.80)	0.25 (0.64)	-1.49(2.52)
rs7203984	-0.97(1.58)	-0.47 (1.78)	0.00 (0.63)	0.67(2.49)
rs11508026	-0.78(2.16)	-0.56 (2.41)	-0.29 (0.85)	0.61(3.38)
rs12720922	1.10(1.46)	0.59 (1.63)	0.21 (0.58)	-0.98(2.28)
rs9939224	-0.12(0.72)	-0.06 (0.81)	-0.11 (0.29)	0.15(1.14)
rs1532624	1.17(1.64)	0.61 (1.82)	0.46 (0.64)	-0.80(2.55)
rs11076175	-1.81(2.75)	-2.06 (3.08)	-1.21 (1.12)	2.78(4.24)
rs7499892	1.76(2.50)	1.73 (2.79)	0.99(1.02)	-2.35(3.83)
rs11076176	-0.31(2.45)	-0.59 (2.79)	-0.08(0.99)	1.00(3.91)
rs289714	1.25(2.47)	1.33 (2.81)	0.38(1.00)	-1.98(3.94)

results. Most of the significant associations obtained in the univariate analyses are no longer significant in the joint analyses. An exception is SNP rs3764261 which remains significantly associated with all the measured lipoproteins.

To explain the difference between the univariate analyses and the joint analyses, two possible reasons may be considered. One might be that the previously observed associations are marginal, and no direct association exist once other variables are adjusted. The other could be the limited sample size does not have enough power to detect potential associations.

3.9 Exercises

1. For the U-statistic, when the kernel function has mean 0 and $\sigma_m^2 < \infty$, show that

$$\sqrt{n}U_n \longrightarrow N(0, m^2\sigma_1^2),$$

in distribution by showing the following statements are true.

(a) $\sqrt{n}T_n \to N(0, m^2\sigma_1^2)$ in distribution, where

$$T_n = \sum_{i=1}^{n} E\{U_n \mid X_i\},$$

(b) $nE(U_n - T_n)^2 \to 0$.

(c) $\sqrt{n}U_n \to N(0, m^2\sigma_1^2)$ in distribution because $\sqrt{n}(U_n - T_n) \to 0$ in probability.

2. Let $Y_1, \cdots, Y_n$ be iid random variables. The sample variance

$$\hat{\sigma}^2 = \frac{1}{n-1}\sum_{i=1}^{n}(Y_i - \bar{Y})^2$$

can be expressed as a U-statistic.

(a) Find h function for the U-statistic.

(b) Use the asymptotic theory for the U-statistic, derive an asymptotic variance for the estimator.

3. In multiple linear regression model, we have

$$\begin{aligned} Y_1 &= \beta_{10} + \beta_{11}X_1 + \cdots + \beta_{1p}X_p + \epsilon_1, \\ &\vdots, \\ Y_q &= \beta_{q0} + \beta_{q1}X_1 + \cdots + \beta_{qp}X_p + \epsilon_q, \end{aligned}$$

Suppose that $(\epsilon_1, \cdots, \epsilon_q) \sim N(0, \Sigma)$.

(a) Write down the conditional density for $Y = (Y_1, \cdots, Y_q)^t$ given $X = (X_1, \cdots, X_p)^t$.

(b) Find the odds ratio decomposition of the conditional density.

(c) What are the odds ratio parameters in the odds ratio representation?

(d) Suppose that the diagonal elements of Σ are all 1. If we consider one regression alone, for example, the first one, do the β parameters in the regression differ from the odds ratio parameters in the odds ratio decomposition of the density for Y given X?

(e) How to interpret the odds ratio parameters in the odds ratio model?

4. Using the lipid data to answer the following questions.

(a) Compare the analysis based on the basic odds ratio model (3.26) and the following model

$$\log \eta\{\text{HDL;SNP} \mid \text{age, sex, case-status}\} = \theta_1\text{SNP*HDL}, \qquad (3.27)$$

what is the difference in the interpretation of the odds ratio parameter θ_1?

(b) Perform an analysis of the data using model (3.27) and compare the results with those from model (3.26). Is there any major difference?

(c) How do you specify a model such that the (conditional) association between `HDL` and `LDL` is also estimated?

5. Can the permutation likelihood be expressed in terms of the U-statistic? Can the asymptotic theory for the U-statistic be applied to the permutation likelihood?

4

Estimation and Inference on Conditional Odds Ratio Function

4.1 A general formulation of the problem

In many applications, it is of interest to examine the association of the outcome and the exposure conditional on a set of potential confounders. The inference on the association can be formulated as inference on whether the odds ratio function for the joint density of Y_1 and Y_2 conditional on X is a constant 1. More specifically, the joint density conditional on covariate X has the following odds ratio representation

$$p(y_1, y_2 \mid x) = \frac{\eta(y_1; y_2 \mid x)p(y_1 \mid y_{20}, x)p(y_2 \mid y_{10}, x)}{\int\int \eta(y_1; y_2 \mid x)p(y_1 \mid y_{20}, x)p(y_2 \mid y_{10}, x)dy_1 dy_2}. \tag{4.1}$$

The estimation problem is to find a good estimator of the conditional odds ratio function assuming no knowledge of $p(y_1 \mid y_{20}, x)$ and $p(y_2 \mid y_{10}, x)$ other than that from the observed data. The inference is on the parameters in the conditional odds ratio function $\eta(y_1; y_2 \mid x)$ that measures the association of Y_1 and Y_2 conditional on X. Suppose that $\eta(y_1; y_2 \mid x)$ is modeled parametrically by θ as $\eta(y_1; y_2 \mid x, \theta)$ for $\theta \in \Theta$. In Chapter 3, parametric solutions to this problem have been introduced. One approach is to model $\eta(y_1; x \mid y_{20})$ parametrically and maximizes the semiparametric conditional likelihood based on

$$p(y_1 \mid y_2, x) = \frac{\eta(y_1; y_2 \mid x)\eta(y_1; x \mid y_{20})p(y_1 \mid y_{20}, x_0)}{\int \eta(y_1; y_2 \mid x)\eta(y_1; x \mid y_{20})p(y_1 \mid y_{20}, x_0)dy_1}.$$

Another approach is to model $\eta(y_2; x \mid y_{10})$ parametrically and maximize the semiparametric conditional likelihood based on

$$p(y_2 \mid y_1, x) = \frac{\eta(y_1; y_2 \mid x)\eta(y_2; x \mid y_{10})p(y_2 \mid y_{10}, x_0)}{\int \eta(y_1; y_2 \mid x)\eta(y_2; x \mid y_{10})p(y_2 \mid y_{10}, x_0)dy_2}.$$

Alternatively, we may choose to model both $\eta(y_1; x \mid y_{20})$ and $\eta(y_2; x \mid y_{10})$ parametrically, and maximize the semiparametric conditional likelihood based on (4.1). These likelihood approaches are respectively termed prospective, retrospective, and joint likelihood approaches. One problem with these semiparametric modeling approaches is that the validity of the approaches relies on the correctness of the models for $\eta(y_1; x \mid y_{20})$ and/or $\eta(y_2; x \mid y_{10})$. When the potential confounders X is of moderate to high dimension, misspecification of the models for $\eta(y_1; x \mid y_{20})$ and/or $\eta(y_2; x \mid y_{10})$ is of serious concerns because it may lead to bias in estimating the parameters in $\eta(y_1; y_2 \mid x)$, which are of primary interest.

To make the estimation of parameters in $\eta(y_1; y_2 \mid x)$ robust to model misspecification, it is desirable to keep both $p(y_1 \mid y_{20}, x)$ and $p(y_2 \mid y_{10}, x)$ nonparametric. However, a reasonably efficient solution satisfies such a requirement may not always achievable in practice. In this chapter, we examine approaches to making robust inference on the parameters in the conditional odds ratio function.

DOI: 10.1201/9781351049757-4

4.2 Permutation approach for stratified sample

4.2.1 Permutation likelihood approach

When the observed sample forms clusters based on the observed values of X, the approach of permutation within each cluster can be used to eliminate the baseline distribution functions. Let the obseved data be organized as $(Y_{kj1}, Y_{kj2}, X_{kj}), j = 1, \cdots, n_k; k = 1, \cdots, m$, where $X_{kj} = X_k$, i.e., all the subjects within a cluster or stratum have the same X value. For a given stratum $X = X_k$, the joint density of the observed data $(Y_{kj1}, Y_{kj2}, X_k), j = 1, \cdots, n_k$ is

$$\prod_{j=1}^{n_k} \frac{\eta(Y_{kj1}; Y_{kj2} \mid X_k)p(Y_{kj1} \mid y_{20}, X_k)p(Y_{kj2} \mid y_{10}, X_k)}{\int\int \eta(y_1; y_2 \mid X_k)p(y_1 \mid y_{20}, X_k)p(y_2 \mid y_{10}, X_k)dy_1dy_2}. \tag{4.2}$$

Following the same argument as in obtaining the permutation likelihood in Chapter 3, the permutation likelihood derived from (4.2) by conditioning on order statistics of Y_1 and Y_2 within each stratum defined by X_k can be written as

$$\frac{\prod_{j=1}^{n_k} \eta(Y_{kj1}; Y_{kj2} \mid X_k, \theta)}{\sum_{\sigma \in \mathcal{P}_{k2}} \prod_{j=1}^{n_k} \eta(Y_{i\sigma_1(j)1}; Y_{k\sigma_2(j)2} \mid X_k, \theta)},$$

where $\sigma = (\sigma_1, \sigma_2)$ with $\sigma_k, k = 1, 2$ are permutations of indices $\{1, \cdots, n_k\}$, and $\mathcal{P}_{k2}$ is the collection of all such permutations. Because permutations form a group, the permutation likelihood can be simplified to

$$\frac{\prod_{j=1}^{n_k} \eta(Y_{kj1}; Y_{kj2} \mid X_k, \theta)}{n_k! \sum_{\sigma \in \mathcal{P}_k} \prod_{j=1}^{n_k} \eta(Y_{k\sigma(j)1}; Y_{kj2} \mid X_k, \theta)},$$

where $\mathcal{P}_k$ is the collection of all permutations of $\{1, \cdots, n_k\}$. Apart from a constant, the permutation likelihood for the observed data is a product of the likelihood from individual strata as

$$\prod_{k=1}^{m} \frac{\prod_{j=1}^{n_k} \eta(Y_{kj1}; Y_{kj2} \mid X_k, \theta)}{\sum_{\sigma \in \mathcal{P}_k} \prod_{j=1}^{n_k} \eta(Y_{k\sigma(j)1}; Y_{kj2} \mid X_k, \theta)}. \tag{4.3}$$

The likelihood can be rewritten as

$$\prod_{k=1}^{m} \left[\sum_{\sigma \in \mathcal{P}_k} \exp\left\{ \sum_{j=1}^{n_k} \log \frac{\eta(Y_{k\sigma(j)1}; Y_{kj2} \mid X_k, \theta)}{\eta(Y_{kj1}; Y_{kj2} \mid X_k, \theta)} \right\} \right]^{-1}. \tag{4.4}$$

Maximizing likelihood (4.4) in the conventional setting where the number of subjects in each stratum is small, such as in matched case-control study with 1 : 1 matching or 1 : 2 matching, can be done by enumerating all possible permutations within each stratum.

When the size of the sample in some or all of the strata are large such that enumerating all the permutations within a stratum becomes very time-consuming, the Metropolis sampling method discussed in Chapter 3 can be used. Let

$$M_k(\theta) = \sum_{\sigma \in \mathcal{P}_k} \exp\left\{ \sum_{j=1}^{n_k} \log \frac{\eta(Y_{k\sigma(j)1}; Y_{kj2} \mid X_k, \theta)}{\eta(Y_{kj1}; Y_{kj2} \mid X_k, \theta)} \right\},$$

for $k = 1, \cdots, m$. Let $Y_{k1} = (Y_{kj1}, j = 1, \cdots, n_k)$ and $Y_{k2} = (Y_{kj2}, j = 1, \cdots, n_k)$. Let Γ_k

be a random element taking values in $\mathcal{P}_k$. Let

$$P(\Gamma_k = \sigma \mid Y_{k1}, Y_{k2}, X_k, \theta) = \frac{\prod_{j=1}^{n_k} \eta(Y_{kj1}; Y_{kj2} \mid X_k, \theta)}{\sum_{\sigma \in \mathcal{P}_k} \prod_{j=1}^{n_k} \eta(Y_{k\sigma(j)1}; Y_{kj2} \mid X_k, \theta)}. \tag{4.5}$$

It follows by the same argument as in Chapter 3 that

$$\frac{M_k(\theta)}{M_k(\theta_b)} = E_{\mathcal{P}_k(\theta_b)} \left[\exp\left\{S_k(\Gamma_k \mid \theta) - S_k(\Gamma_k \mid \theta_b)\right\}\right],$$

where $E_{\mathcal{P}_k(\theta_b)}$ denotes expectation taken under distribution (4.5), and

$$S_k(\Gamma_k \mid \theta) = \sum_{j=1}^{n_k} \log \frac{\eta(Y_{k\Gamma_k(j)1}; Y_{kj2} \mid X_k, \theta)}{\eta(Y_{kj1}; Y_{kj2} \mid X_k, \theta)}.$$

Let

$$M(\theta) = \prod_{k=1}^{m} M_k(\theta).$$

It follows that

$$\frac{M(\theta)}{M(\theta_b)} = \prod_{k=1}^{m} \frac{M_k(\theta)}{M_k(\theta_b)},$$

where, for large n_k,

$$\frac{M(\theta)}{M(\theta_b)} \approx \exp\left[\sqrt{N}(\theta - \theta_b)^t \mu(\theta_b) + \frac{1}{2}\sqrt{N}(\theta - \theta_b)^t \left\{\Sigma(\theta_b) + \Omega(\theta_b)\right\} \sqrt{N}(\theta - \theta_b)\right],$$

$N = \sum_{k=1}^{m} n_k$, and

$$\begin{aligned}
\mu(\theta_b) &= \frac{1}{\sqrt{N}} \sum_{k=1}^{m} E_{\mathcal{P}_k(\theta_b)} \left\{\frac{\partial}{\partial\theta} S_k(\Gamma_k \mid \theta_b)\right\}, \\
\Sigma(\theta_b) &= \frac{1}{N} \sum_{k=1}^{m} V_{\mathcal{P}_k(\theta_b)} \left\{\frac{\partial}{\partial\theta} S_k(\Gamma_k \mid \theta_b)\right\}, \\
\Omega(\theta_b) &= \frac{1}{N} \sum_{k=1}^{m} E_{\mathcal{P}_k(\theta_b)} \left\{\frac{\partial^2}{\partial\theta^2} S_k(\Gamma_k \mid \theta_b)\right\}.
\end{aligned}$$

θ_b can be updated by

$$\theta_b^{new} = \theta_b - \frac{1}{\sqrt{N}} \left\{\Sigma(\theta_b) + \Omega(\theta_b)\right\}^{-1} \mu(\theta_b).$$

At convergence, θ_b is equivalent to the maximum permutation likelihood estimator. The variance of the estimator can be estimated by

$$\left\{\Sigma(\theta_b) + \Omega(\theta_b)\right\}^{-1}.$$

Computation of $\mu(\theta_b)$, $\Sigma(\theta_b)$ and $\Omega(\theta_b)$ can be done through computing

$$\begin{aligned}
\mu_k(\theta_b) &= E_{\mathcal{P}_k(\theta_b)} \left\{\frac{\partial}{\partial\theta} S_k(\Gamma_k \mid \theta_b)\right\}, \\
\Sigma_k(\theta_b) &= V_{\mathcal{P}_k(\theta_b)} \left\{\frac{\partial}{\partial\theta} S_k(\Gamma_k \mid \theta_b)\right\}, \\
\Omega_k(\theta_b) &= E_{\mathcal{P}_k(\theta_b)} \left\{\frac{\partial^2}{\partial\theta^2} S_k(\Gamma_k \mid \theta_b)\right\}.
\end{aligned}$$

If n_k is relative small, enumeration of all permutation can be used. For medium to large n_k, Metropolis sampling algorithm can be used to draw a sample for approximating the means and variance.

4.2.2 Application to common odds ratio estimation in stratified 2×2 tables

As an illustration of the use of the Monte Carlo sampling approach to estimating the odds ratio parameter, consider the estimation of the common odds ratio parameter in stratified 2×2 tables as given in Table 1.3. Mantel-Haenszel common odds ratio estimator for the stratified tables is

$$\hat{\gamma}_{MH} = \frac{\sum_{k=1}^{m} n_{11k} n_{22k}/n_{++k}}{\sum_{k=1}^{m} n_{12k} n_{21k}/n_{++k}}.$$

This estimator is not efficient. The conditional likelihood

$$\prod_{k=1}^{m} \frac{C(n_{11k}, n_{1+k}) C(n_{+1k} - n_{11k}, n_{2+k}) \gamma^{n_{11k}}}{\sum_{t \in \mathcal{T}_k} C(t, n_{1+k}) C(n_{+1k} - t, n_{2+k}) \gamma^t}, \tag{4.6}$$

may be maximized to obtain a more efficient estimator of the common odds ratio parameter, where $\mathcal{T}_k$ is the collection of table compatible with the margins of the observed kth table. See Section 1.7 for more details. Maximizing the likelihood can be challenging when a large number of compatible tables exist. Notice that this conditional likelihood is equivalent to the permutation likelihood which can be rewritten as

$$\prod_{k=1}^{m} \left(\sum_{\sigma \in \mathcal{P}_k} \exp\left[\{t(\sigma) - n_{11k}\}\theta\right] \right)^{-1},$$

where $t(\sigma)$ is the cell counts in the first row and the first column of the kth table, $\sigma \in \mathcal{P}_k$ is the permutation of $\{1, \cdots, n_{++k}\}$, and $\theta = \log \gamma$. The Monte Carlo simulation approach can be used to find the maximum likelihood estimator by sampling permutations for a given parameter value θ_b. When permutations generating the same table are collected together, the permutation likelihood can be simplified to

$$\prod_{k=1}^{m} \left(\sum_{t \in \mathcal{T}_k} C(t, n_{1+k}) C(n_{+1k} - t, n_{2+k}) \exp\left[\{t - n_{11k}\}\theta\right] \right)^{-1}.$$

A more efficient approach directly samples t from the distribution

$$p(t \mid \theta_b) = \frac{C(t, n_{1+k}) C(n_{+1k} - t, n_{2+k}) \exp(t\theta_b)}{\sum_{u \in \mathcal{T}_k} C(u, n_{1+k}) C(n_{+1k} - u, n_{2+k}) \exp(u\theta_b)},$$

where t as the first cell count is compatible with the observed margins of the kth table. See Diaconis and Sturmfels (1998). Let $T_{1k}, \cdots, T_{N_k k}$ be the sample drawn at θ_b. The likelihood can be approximated by

$$\prod_{k=1}^{m} M_k(\theta_b) \prod_{k=1}^{m} \frac{1}{N_k} \sum_{i=1}^{N_k} \exp\{(T_i - n_{11k})(\theta - \theta_b)\}.$$

The log-likelihood to be maximized is

$$l(\theta) - l(\theta_b) \approx \sum_{k=1}^{m} \log\left[\sum_{i=1}^{N_k} \exp\{(T_i - n_{11k})(\theta - \theta_b)\}\right].$$

Two approaches to the likelihood maximization may be taken. One uses the normal approximation by computing

$$\hat{\mu}_k(\theta_b) = \frac{1}{N_k}\sum_{i=1}^{N_k}\frac{T_i - n_{11k}}{\sqrt{n_{++k}}},$$
$$\hat{\Sigma}_k(\theta_b) = \frac{1}{N_k - 1}\sum_{i=1}^{N_k}\left\{\frac{T_i - n_{11k}}{\sqrt{n_{++k}}} - \hat{\mu}_k(\theta_b)\right\}^2,$$

and updating θ_b by

$$\theta_b^{new} = \theta_b - \left\{\sum_{k=1}^{m} n_k \hat{\Sigma}_k(\theta_b)\right\}^{-1}\sum_{k=1}^{m}\sqrt{n_k}\hat{\mu}_k(\theta_b).$$

Another approach is to directly maximize by Newton-Raphson algorithm. The first and second derivatives are respectively

$$D_1(\theta) = \sum_{k=1}^{m}\frac{\sum_{i=1}^{N_k}(T_i - n_{11k})\exp\{(T_i - n_{11k})(\theta-\theta_b)\}}{\sum_{i=1}^{N_k}(T_i - n_{11k})\exp\{(T_i - n_{11k})(\theta-\theta_b)\}},$$
$$D_2(\theta) = \sum_{k=1}^{m}\frac{\sum_{i=1}^{N_k}(T_i - n_{11k})^2\exp\{(T_i - n_{11k})(\theta-\theta_b)\}}{\sum_{i=1}^{N_k}(T_i - n_{11k})\exp\{(T_i - n_{11k})(\theta-\theta_b)\}}$$
$$-\sum_{k=1}^{m}\left[\frac{\sum_{i=1}^{N_k}(T_i - n_{11k})\exp\{(T_i - n_{11k})(\theta-\theta_b)\}}{\sum_{i=1}^{N_k}(T_i - n_{11k})\exp\{(T_i - n_{11k})(\theta-\theta_b)\}}\right]^2.$$

The parameter θ is updated by

$$\theta^{new} = \theta^{old} - D_2^{-1}(\theta^{old})D_1(\theta^{old}).$$

When $\theta^{old} = \theta_b$, the update step is the same as the first approach. This means the first updating approach is the same as the one-step Newton-Raphson approach. On the other hand, keep updating θ without updating the Monte Carlo sample for estimating the sum in the denominators in the likelihood may produce estimate with large bias just as a random sample with $\theta_b = 0$ yields. In this respect, the one-step update by normal approximation is sufficient. At convergence of the iteration between estimation and simulation, let $\hat{\theta}$ denote the final estimate. The variance of the maximum permutation likelihood estimator can be estimated by

$$\left\{\sum_{k=1}^{m} n_k\hat{\Sigma}_k(\hat{\theta})\right\}^{-1}.$$

For data summarized in $2\times 2\times m$ contingency tables, sampling permutations in the kth table can be made more efficient by excluding permutations that keep the counts in the contingency table unchanged, i.e., $\sum_{i=1}^{n} Y_{\sigma(i)1k}Y_{i2k}$ unchanged. If only two indices of $Y_{i1}, i = 1,\cdots,n$ are switched and $\sum_{i=1}^{n} Y_{\sigma(i)1k}Y_{i2k}$ is changed, then it can only be increased by 1 or decreased by 1. Sampling compatible tables following distribution

$$p(n_{11k}, n_{12k}, n_{21k}, n_{22k}, \theta_b) = \frac{C(n_{11k}, n_{1+k})C(n_{+1k} - n_{11k}, n_{2+k})\exp(n_{11k}\theta_b)}{\sum_{t\in\mathcal{T}_k} C(t, n_{1+k})C(n_{+1k} - t, n_{2+k})\exp(t\theta_b)},$$

can be carried out using Metropolis sampling algorithm. This sampling approach can be described as follows.

1. Let the current cell counts be $(n_{11k}, n_{12k}, n_{21k}, n_{22k})$. Randomly choose changes $(\delta_{11k}, \delta_{12k}, \delta_{21k}, \delta_{22k})$ where either $\delta_{11} = \delta_{22} = 1$ and $\delta_{12} = \delta_{21} = -1$ or $\delta_{11} = \delta_{22} = -1$ and $\delta_{12} = \delta_{21} = 1$ and $n_{ijk} + \delta_{ij} \geq 0$ for $i = 1, 2, j = 1, 2$.
2. Update the table to $(n_{11k} + \delta_{11}, n_{12k} + \delta_{12}, n_{21k} + \delta_{21}, n_{22k} + \delta_{22})$ with probability

$$\pi = \min\left\{1, \frac{p(n_{11} + \delta_{11k}, n_{12k} + \delta_{12}, n_{21k} + \delta_{21}, n_{22k} + \delta_{22}, \theta_b)}{p(n_{11k}, n_{12k}, n_{21k}, n_{22k}, \theta_b)}\right\}.$$

   Keep the table unchanged with probability $1 - \pi$.
3. Repeat the two steps

This algorithm is a special case of that proposed by Diaconis and Sturmfels (1998) [55].

4.3 Semiparametric efficient score for estimating conditional odds ratio function

4.3.1 Characterization of the nuisance score space and its orthogonal complement

The clustering structure for the conditional odds ratio model exists only in special circumstances, such as the matched case-control sampling design, or the stratified sampling design. When the clustering structure is not present in the data, one approach is to create artificial matching by categorizing the confounding variables and forming clusters based on the categories defined by the confounding variables. The permutation likelihood approach can then be applied to the artificially clustered data. The advantages of this approach includes simple and robust estimation without dealing with the nuisance distributions. The disadvantages include potential bias of imperfect matching and efficiency loss due to matching. In addition, clustering by categorizing covariates does not work when the dimension of the covariates is high. To further study the parameter estimation problem, we start from examining the semiparametric efficient estimation of the conditional odds ratio function under model (4.1) in the following.

For estimating the conditional odds ratio function, both conditional baseline functions $p(y_1 \mid y_{20}, x)$ and $p(y_2 \mid y_{10}, x)$ are nuisance parameters. The likelihood scores for estimating the nuisance parameters are

$$\begin{aligned} \dot{l}_1(h_1) &= h_1(Y_1, X) - E_0\left\{h_1(Y_1, X) \mid X\right\}, \\ \dot{l}_2(h_2) &= h_2(Y_2, X) - E_0\left\{h_2(Y_2, X) \mid X\right\}, \end{aligned}$$

for all $h_1(Y_1, X), h_2(Y_2, X) \in L^2(P_0)$, where P_0 is the probability law generating the data and E_0 is the expectation under P_0. The nuisance score space for $p(y_1 \mid y_{20}, x)$ is

$$H_1 = \{h_1(y_1, x) \mid E_0\{h_1(Y_1, X) \mid X\} = 0, E_0\{h_1^2(Y_1, X)\} < \infty\}.$$

The nuisance score space for $p(y_2 \mid y_{20}, x)$ is

$$H_2 = \{h_2(y_1, x) \mid E_0\{h_2(Y_2, X) \mid X\} = 0, E_0\{h_2^2(Y_2, X)\} < \infty\}.$$

The score space for estimating the conditional odds ratio function $\eta(y_1; y_2 \mid x)$ is the orthogonal complement of the nuisance score space $H_1 + H_2$, i.e., $(H_1 + H_2)^{\perp}$. Any $t(Y_1, Y_2, X) \in (H_1 + H_2)^{\perp}$ satisfies $E_0\{t(Y_1, Y_2, X) \mid X\} = 0$ and

$$\begin{aligned} E_0\left\{t(Y_1, Y_2, X) h_1(Y_1, X)\right\} &= 0, \text{ for all } h_1 \in H_1, \\ E_0\left\{t(Y_1, Y_2, X) h_2(Y_2, X)\right\} &= 0, \text{ for all } h_2 \in H_2. \end{aligned}$$

These equations can be equivalently rewritten as

$$E_0\left\{t(Y_1, Y_2, X) \mid Y_1, X\right\} = 0, \tag{4.7}$$

$$E_0\left\{t(Y_1, Y_2, X) \mid Y_2, X\right\} = 0. \tag{4.8}$$

It follows that the complement of the nuisance score space is

$$\begin{aligned} (H_1 + H_2)^{\perp} = \{t(Y_1, Y_2, X) \in L^2(P_0) \mid & E_0\{t(Y_1, Y_2, X) \mid Y_1, X\} \\ &= E_0\{t(Y_1, Y_2, X) \mid Y_2, X\} = 0\}. \end{aligned}$$

4.3.2 The semiparametric efficient score

Let the conditional odds ratio function be modeled by parameter θ as $\eta(y_1; y_2 \mid X, \theta)$. The likelihood score for estimating θ is

$$\dot{l}_\theta(a) = a^t \frac{\partial}{\partial \theta} \log \eta(Y_1; Y_2 \mid X, \theta) - E_0\left\{a^t \frac{\partial}{\partial \theta} \log \eta(Y_1; Y_2 \mid X, \theta) \mid X\right\}$$

for bounded conformable vector a. The efficient score for estimating θ is the projection in $L^2(P_0)$ space of the likelihood score for estimating θ to the orthogonal complement of the nuisance score space, i.e.,

$$S^{eff}(\theta)(a) = \text{Proj}\left\{\dot{l}_\theta(a) \mid (H_1 + H_2)^{\perp}\right\}.$$

This projection has the form

$$S^{eff}(\theta)(a) = \dot{l}_\theta(a) - \dot{l}_1(h_1^*) - \dot{l}_2(h_2^*),$$

for some $h_1^* \in H_1$ and $h_2^* \in H_2$. The projection can be found by solving the following equations:

$$\begin{aligned} E_0\left[\left\{\dot{l}_\theta(a) - \dot{l}_1(h_1^*) - \dot{l}_2(h_2^*)\right\} \dot{l}_1(h_1)\right] &= 0, \\ E_0\left[\left\{\dot{l}_\theta(a) - \dot{l}_1(h_1^*) - \dot{l}_2(h_2^*)\right\} \dot{l}_2(h_2)\right] &= 0, \end{aligned}$$

for all $h_1 \in H_1$ and $h_2 \in H_2$. These equations can be equivalently expressed as

$$\begin{aligned} \dot{l}_1(h_1^*) &= E_0\left\{\dot{l}_\theta(a) - \dot{l}_2(h_2^*) \mid Y_1, X\right\}, \\ \dot{l}_2(h_2^*) &= E_0\left\{\dot{l}_\theta(a) - \dot{l}_1(h_1^*) \mid Y_2, X\right\}. \end{aligned}$$

When plugged these expressions in the efficient score for estimating θ, the efficient score appears as

$$\dot{l}_\theta(a) - E_0\left\{\dot{l}_\theta(a) \mid Y_1, X\right\} - E_0\left\{\dot{l}_\theta(a) \mid Y_2, X\right\} + E_0\left\{\dot{l}_1(h_1^*) \mid Y_2, X\right\} + E_0\left\{\dot{l}_2(h_2^*) \mid Y_1, X\right\}.$$

By repeating applying the expressions for $\dot{l}_1(h_1^*)$ and $\dot{l}_2(h_2^*)$, we can obtain the efficient score for estimating θ in the conditional semiparametric odds ratio model in terms of alternating conditional expectations.

$$\dot{l}_\theta(a) - E_0\left\{\dot{l}_\theta(a) \mid Y_1, X\right\} - E_0\left\{\dot{l}_\theta(a) \mid Y_2, X\right\} + \sum_{k=1}^{\infty}(-1)^{k-1}\left\{E_0^{(Y_2,X)}E_0^{(Y_1,X)}\right\}^k\left\{\dot{l}_\theta(a)\right\}$$
$$+\sum_{k=1}^{\infty}(-1)^{k-1}\left\{E_0^{(Y_2,X)}E_0^{(Y_1,X)}\right\}^k\left\{\dot{l}_\theta(a)\right\},$$

where $\left\{E_0^{(Y_2,X)}E_0^{(Y_1,X)}\right\}^k$ and $\left\{E_0^{(Y_2,X)}E_0^{(Y_1,X)}\right\}^k$ are respectively k-fold of the following conditional expectations,

$$\begin{aligned} E_0^{(Y_2,X)}E_0^{(Y_1,X)}(\cdot) &= E_0\{E_0(\cdot \mid Y_1, X) \mid Y_2, X\}, \\ E_0^{(Y_1,X)}E_0^{(Y_2,X)}(\cdot) &= E_0\{E_0(\cdot \mid Y_2, X) \mid Y_1, X\}. \end{aligned}$$

The efficient score for estimating θ derived above is in general very complex. When Y_1 and Y_2 are independent conditional on X, the efficient score reduces to

$$\dot{l}_\theta(a) - E_0\left\{\dot{l}_\theta(a) \mid Y_1, X\right\} - E_0\left\{\dot{l}_\theta(a) \mid Y_2, X\right\},$$

which can be explicitly written as

$$\begin{aligned} S_0^{eff} &= a^t\frac{\partial}{\partial\theta}\log\eta(Y_1;Y_2 \mid X,\theta) - E_0\left\{a^t\frac{\partial}{\partial\theta}\log\eta(Y_1;Y_2 \mid X,\theta) \mid Y_1, X\right\} \\ &\quad -E_0\left\{a^t\frac{\partial}{\partial\theta}\log\eta(Y_1;Y_2 \mid X,\theta) \mid Y_2, X\right\} + E_0\left\{a^t\frac{\partial}{\partial\theta}\log\eta(Y_1;Y_2 \mid X,\theta) \mid X\right\}. \end{aligned}$$

Although this efficient score may not be very helpful for estimating θ, it is useful for testing conditional independence of Y_1 and Y_2 given X. More detailed discussion on testing conditional independence can be found in Chapter 6.

4.3.3 Locally efficient estimator

The semiparametric efficient scores are often not readily applicable to estimating the conditional odds ratio parameter or the conditional odds ratio function because the efficient score itself may depends on the unknown nuisance parameters. One way to resolve this problem is to first obtain a consistent estimator of the nuisance parameters and then use it in the efficient score calculation. However, this is often unattainable because the nuisance model is of high-dimension and may not be easily estimated due to the curse of dimensionality. A compromise in practice is to model the nuisance distribution parametrically or semiparametrically in the efficient score computation. If all the model are correctly specified, the efficient score is obtained and the estimator is called locally efficient estimator. If some of the models are not correctly specified, the estimator may still be unbiased. Such a property is termed the doubly robust property which will be discussed more generally in the next section for the odds ratio function estimation.

To obtain a consistent estimator of the nuisance parameters, $p(y_1 \mid y_2 = y_{20}, x)$ and $p(y_2 \mid y_{10}, x)$ can be modeled and estimated in the following way. First, apply the odds ratio decomposition of a conditional density

$$p(y_1 \mid y_{20}, x) = \frac{\eta_1(y_1; x)p(y_1 \mid y_{20}, x_0)}{\int \eta_1(y_1; x)p(y_1 \mid y_{20}, x_0)dy_1}.$$

and model $\eta_1(y_1; x)$ parametrically by $\eta_1(y_1; x \mid \alpha)$. Then estimate the nuisance parameter $(\alpha, p(y_1 \mid y_{20}, x_0))$ by maximizing the conditional likelihood based on

$$p(y_1 \mid y_2, x) = \frac{\eta(y_1; y_2 \mid x, \theta)\eta_1(y_1; x \mid \alpha)p(y_1 \mid y_{20}, x_0)}{\int \eta(y_1; y_2 \mid x, \theta)\eta_1(y_1; x \mid \alpha)p(y_1 \mid y_{20}, x_0)dy_1}. \tag{4.9}$$

Denote the estimator by $(\hat{\alpha}, \hat{p}(y_1 \mid y_{20}, x_0))$. Likewise, $p(y_2 \mid y_{10}, x)$ can be decomposed as

$$p(y_2 \mid y_{10}, x) = \frac{\eta_2(y_2; x)p(y_2 \mid y_{10}, x_0)}{\int \eta_2(y_2; x)p(y_2 \mid y_{10}, x_0)dy_2}.$$

and model $\eta_2(y_2; x)$ parametrically by $\eta_2(y_2; x \mid \beta)$. Estimate the nuisance parameter $(\beta, p(y_2 \mid y_{10}, x_0))$ by maximizing the conditional likelihood based on

$$p(y_2 \mid y_1, x) = \frac{\eta(y_1; y_2 \mid x, \theta)\eta_2(y_2; x \mid \beta)p(y_2 \mid y_{10}, x_0)}{\int \eta(y_1; y_2 \mid x, \theta)\eta_2(y_2; x \mid \beta)p(y_2 \mid y_{10}, x_0)dy_2}. \tag{4.10}$$

Denote the estimator by $(\hat{\beta}, \hat{p}(y_2 \mid y_{10}, x_0))$.

The conditional expectations in the semiparametric efficient score can now be computed based on the joint conditional density

$$p(y_1, y_2 \mid x) = \frac{\eta(y_1; y_2 \mid x, \theta)\eta_1(y_1; x \mid \alpha)\eta_2(y_2; x \mid \beta)p(y_1 \mid y_{20}, x_0)p(y_2 \mid y_{10}, x_0)}{\int \eta(y_1; y_2 \mid x, \theta)\eta_1(y_1; x \mid \alpha)\eta_2(y_2; x \mid \beta)p(y_1 \mid y_{20}, x_0)p(y_2 \mid y_{10}, x_0)dy_1 dy_2}, \tag{4.11}$$

with nuisance parameters $\{\alpha, p(y_1 \mid y_{20}, x_0), \beta, p(y_2 \mid y_{10}, x_0)\}$ replaced by $\{\hat{\alpha}, \hat{p}(y_1 \mid y_{20}, x_0), \hat{\beta}, \hat{p}(y_2 \mid y_{10}, x_0)\}$. Denote the latter by $\hat{p}(y_1, y_2 \mid x)$. For example, in computing the efficient score for the case Y_1 is a categorical variable taking finite categories, both Cov_0 and Var_0 in the efficient score expression can be computed by $\hat{p}(y_1, y_2 \mid x)$. In addition, the probability function in the b vector can also be computed by this distribution. The parameter θ in $\hat{p}(y_1, y_2 \mid x)$ should be updated after each iteration in solving the efficient score equation for θ.

When both models for $\eta_1(y_1; x)$ and $\eta_2(y_2; x)$ are correctly specified, α and β can be consistently estimated. In this case, the estimator from solving the efficient score is the most efficient. Since the efficiency of the estimator depends on the correctness of the semiparametric models for $p(y_1 \mid y_{20}, x)$ and $p(y_2 \mid y_{10}, x)$, the estimator is termed locally efficient estimator. When one model is correctly specified and the other model is misspecified, the efficiency of the estimator may not hold anymore. But the estimator is still asymptotically unbiased. The estimator is termed doubly robust. The doubly robust property can hold for a more general class of estimating scores, not necessarily those computed from the semiparametric efficient score. This class of estimating scores will be discussed in later sections of this chapter.

4.4 The special case with one categorical outcome

An important special case where the efficient score can be reduced is when one outcome is categorical with finite categories. In this case, the efficient score has a closed-form expression. Let Y_1 be the categorical outcome taking values $0, 1, \cdots, K$. The nuisance function $p(y_1 \mid y_{20}, x)$ can be written as $p(y_1 = k \mid y_{20}, x), k = 0, 1, \cdots, K$. From model (4.1), the nuisance score space for the parameter $p(y_1 \mid y_{20}, x)$ is

$$H_1 = \bar{\mathcal{L}}\left\{1_{\{Y_1 = k\}} - P_0(y_1 = k \mid X = x), k = 1, \cdots, K\right\}, \tag{4.12}$$

i.e., the closure of the linear combination of the K scores. Note that the score for $k = 0$ is excluded from H_1 because

$$1_{\{Y_1=0\}} - P_0(Y_1 = 0 \mid X = x) = -\sum_{k=1}^{K} \left\{1_{\{Y_1=k\}} - P_0(y_1 = k \mid X = x)\right\}.$$

Computation of the projection of an estimating score to the complement of the nuisance score space can be simplified in this case because the projection to a finite dimensional space is relatively easy to compute.

Note first that

$$\overline{H_1 + H_2} = \text{Proj}(H_1 \mid H_2^{\perp}) \oplus \bar{H}_2,$$

where $H_2^{\perp}$ denotes the orthogonal complement of H_2 in the closed linear space of $\overline{H_1 + H_2}$. It follows that

$$\text{Proj}(h \mid \overline{H_1 + H_2}) = \text{Proj}\{h \mid \text{Proj}(H_1 \mid H_2^{\perp})\} + \text{Proj}(h \mid \bar{H}_2).$$

In the special case where $\bar{H}_1 \cap \bar{H}_2 = \emptyset$,

$$\text{Proj}(h \mid \bar{H}_1 \oplus \bar{H}_2) = \text{Proj}(h \mid \bar{H}_1) + \text{Proj}(h \mid \bar{H}_2).$$

In general,

$$\text{Proj}(1_{\{Y_1=k\}} - P_0(Y_1 = k \mid X) \mid H_2^{\perp}) = 1_{\{Y_1=k\}} - P_0(Y_1 = k \mid Y_2, X).$$

If H_1 is a finite dimensional nuisance score space as given in (4.12),

$$\text{Proj}(H_1 \mid H_2^{\perp})\} = \bar{\mathcal{L}}\left\{1_{\{Y_1=k\}} - P_0(Y_1 = k \mid Y_2, X), k = 1, \cdots, K\right\}.$$

This is because if $H_1 = \bar{\mathcal{L}}(a_1, \cdots, a_K)$, H_2 is a closed linear space, and $\text{proj}(a_j \mid H_2) = b_j$ for $k = 1, \cdots, K$, then

$$\text{Proj}(H_1 \mid \bar{H}_2) = \bar{\mathcal{L}}(b_1, \cdots, b_K).$$

It follows further that

$$\text{Proj}\{h \mid \text{Proj}(H_1 \mid H_2^{\perp})\} = \text{Cov}_0(h, b)\left\{\text{Var}_0(b)\right\}^{-} b.$$

where B^- denotes the generalized inverse of B and

$$b = \left(1_{\{Y_1=1\}} - P_0(Y_1 = 1 \mid Y_2, X), \cdots, 1_{\{Y_1=K\}} - P_0(Y_1 = K \mid Y_2, X)\right)^t.$$

The generalized inverse can be avoided when $E_0\{\text{Var}_0(b \mid X)\}$ is of full rank. Bickel et al (1993, page 445)[11] gives some weak conditions that guarantee the existence of the inverse. The projection to the complement of the nuisance score space can be expressed as

$$\text{Proj}(h \mid \overline{H_1 + H_2}) = E_0(h \mid Y_2, X) + \text{Cov}_0(h, b)\left\{\text{Var}_0(b)\right\}^{-1} b.$$

Furthermore,

$$\text{Proj}\left\{h \mid (H_1 + H_2)^{\perp}\right\} = h - E_0(h \mid Y_2, X) - \text{Cov}_0(h, b)\left\{\text{Var}_0(b)\right\}^{-1} b.$$

For the score function in estimating the odds ratio parameter θ,

$$\begin{aligned} h(Y_1, Y_2, X) &= \dot{l}_\theta(a) \\ &= a^t \frac{\partial}{\partial \theta} \log \eta(Y_1; Y_2 \mid X, \theta) - E\left\{a^t \frac{\partial}{\partial \theta} \log \eta(Y_1; Y_2 \mid X, \theta) \mid X\right\}. \end{aligned}$$

The efficient score

$$
\begin{aligned}
S^{eff}(\theta)(a) &= \text{Proj}\left\{\dot{l}_\theta(a) \mid (H_1 + H_2)^\perp\right\} \\
&= \dot{l}_\theta(a) - \text{Proj}\left\{\dot{l}_\theta(a) \mid \overline{H_1 + H_2}\right\} \\
&= \dot{l}_\theta(a) - E_0\{\dot{l}_\theta(a) \mid Y_2, X\} - \text{Cov}_0\{\dot{l}_\theta(a), b\}\left\{\text{Var}_0(b)\right\}^{-1} b.
\end{aligned}
$$

The variance for the efficient score is

$$
\begin{aligned}
\text{Var}_0\left\{S^{eff}(\theta)(a)\right\} &= E_0\left[\text{Var}_0\left\{\dot{l}_\theta(a) \mid Y_2, X\right\}\right] \\
&\quad -\text{Cov}_0\{\dot{l}_\theta(a), b\}\left\{\text{Var}_0(b)\right\}^{-1}\text{Cov}_0\{\dot{l}_\theta(a), b\}.
\end{aligned}
$$

The variance bound for estimating θ under the semiparametric odds ratio model (4.1) is

$$
\left[E_0\left\{\text{Var}_0(\dot{l}_\theta \mid Y_2, X)\right\} - \text{Cov}_0(\dot{l}_\theta, b)\left\{\text{Var}_0(b)\right\}^{-1}\text{Cov}_0(\dot{l}_\theta, b)\right]^{-1}.
$$

Because of the form of b, it can be seen that

$$
\begin{aligned}
\text{Var}_0(b) &= E_0\left\{\text{Var}_0(b \mid Y_2, X)\right\}, \\
\text{Cov}_0(\dot{l}_\theta, b) &= E_0\left\{\text{Cov}_0(\dot{l}_\theta, b \mid Y_2, X)\right\}.
\end{aligned}
$$

As in the general case, the efficient score, though simplified, is still not directly available for estimating θ because it depends on the unknown true distribution through E_0, Var_0, and Cov_0. To find a locally efficient estimator for θ, consider the models for $p(y_1 \mid y_{20}, x)$ and $p(y_2 \mid y_{10}, x)$ respectively. As outlined in the previous section on computing the locally efficient estimator, let $\eta_1(y_1; x \mid \alpha)$ be the odds ratio function for density $p(y_1 \mid y_{20}, x)$ and $\eta_2(y_2; x \mid \beta)$ for density $p(y_2 \mid y_{10}, x)$. We jointly estimate $(\theta, \alpha, p(y_1 \mid y_{20}, x_0))$ from the conditional likelihood based on $p(y_1 \mid y_2, x)$ and denote the maximum likelihood estimator by $(\hat{\theta}_1, \hat{\alpha}, \hat{p}(y_1 \mid y_{20}, x_0))$. Similarly, we jointly estimate $(\theta, \beta, p(y_2 \mid y_{10}, x_0)$ from the conditional likelihood based on $p(y_2 \mid y_1, x)$ and denote the estimator by $(\hat{\theta}_2, \hat{\beta}, \hat{p}(y_2 \mid y_{10}, x_0))$. The probability distribution used to compute the probability, the expectation, and variance-covariance matrices has parameter $(\theta, \hat{\alpha}, p(y_1 \mid y_{20}, x_0), \hat{\beta}, \hat{p}(y_2 \mid y_{10}, x))$, where the value of θ is to be determined. As a result, θ is the only unknown parameter in the efficient score. The locally efficient estimator is obtained by solving the resulting estimation equations for θ. This estimator has some robust properties.

When the model for $p(y_1 \mid y_{20}, x)$, or equivalently $\eta_1(y_1; x \mid \alpha)$, is specified correctly, the model for Y_1 given Y_2, X will be correct. In this case, it is easily seen that $S^{eff}(\theta)$ has mean 0 even if the model for $p(y_2 \mid y_{10}, x)$, or equivalently $\eta_2(y_2; x \mid \beta)$, is misspecified. On the other hand, if the model for η_2 is correctly specified and the model for η_1 is misspecified, the estimator may be biased. More discussion on robustness of estimating equations for θ is discussed in the next section.

4.5 Doubly robust estimation of conditional odds ratio function

For estimating the conditional odds ratio function, we have demonstrated that the efficient score can be complex. Furthermore, the efficient score for estimating the odds ratio function may itself depend on the odds ratio function. As a result, only locally efficient score may be available to use for the estimation. In this section, we examine the way to robustly estimate the conditional odds ratio function.

4.5.1 An alternative characterization of the orthogonal complement of the nuisance score space

We have given a representation of the orthogonal complement of the nuisance score space as

$$(H_1 + H_2)^{\perp} = \{t(Y_1, Y_2, X) \in L^2(P) \mid E\{t(Y_1, Y_2, X) \mid Y_1, X\} = E\{t(Y_1, Y_2, X) \mid Y_2, X\} = 0\}.$$

This characterization is implicit and is less amenable to constructing estimating equations. Tchetgen Tchetgen et al (2010)[179] provided an alternative characterization which is more conveniently expressed in terms of estimating scores. Let the true conditional joint density for Y_1, Y_2 given X be

$$p(y_1, y_2 \mid x) = \frac{\eta(y_1; y_2 \mid x)p(y_1 \mid y_{20}, x)p(y_2 \mid y_{10}, x)}{\int \eta(y_1; y_2 \mid x)p(y_1 \mid y_{20}, x)p(y_2 \mid y_{10}, x)d\mu(y_1, y_2)}$$

Let $p^*(y_1 \mid y_{20}, x)$ be the conditional density of Y_1 given $Y_2 = y_{20}$ and X, which may or may not be the same as $p(y_1 \mid y_{20}, x)$. Similarly, define $p^*(y_2 \mid y_{10}, x)$. Define a new conditional joint density for Y_1, Y_2 given X as

$$p^*(y_1, y_2 \mid x) = p^*(y_1 \mid y_{20}, x)p^*(y_2 \mid y_{10}, x),$$

where

$$0 < \frac{p^*(y_1, y_2 \mid x)}{p(y_1, y_2 \mid x)} < +\infty.$$

Under P^*, Y_1 and Y_2 are independent given X. The orthogonal complement of the nuisance space under P^* can be characterized as the collection of $t(Y_1, Y_2, X)$ such that

$$\begin{aligned} t(Y_1, Y_2, X) &= s(Y_1; Y_2, X) - E^*\{s(Y_1; Y_2, X) \mid Y_1, X\} \\ &\quad - E^*\{s(Y_1; Y_2, X) \mid Y_2, X\} + E^*\{s(Y_1; Y_2, X) \mid X\}, \end{aligned}$$

for all possible $s(Y_1, Y_2, X) \in L^2(P^*)$, where E^* is the expectation taken under P^*. The orthogonal complement of the nuisance space under $p(y_1, y_2 \mid x)$ can be characterized as the collection of $t(Y_1, Y_2, X)$ such that

$$\begin{aligned} t(Y_1, Y_2, X) &= \frac{p^*(Y_1, Y_2 \mid X)}{p(Y_1, Y_2 \mid X)} \Big[s(Y_1; Y_2, X) - E^*\{s(Y_1; Y_2, X) \mid Y_1, X\} \\ &\quad - E^*\{s(Y_1; Y_2, X) \mid Y_2, X\} + E^*\{s(Y_1; Y_2, X) \mid X\} \Big], \end{aligned}$$

for all possible $s(Y_1, Y_2, X) \in L^2(P^*)$. It can be seen that the foregoing $t(Y_1, Y_2, X)$ satisfies conditions (4.9) and (4.10), i.e.,

$$E\{t(Y_1, Y_2, X) | Y_1, X\} = -E^*[E^*\{s(Y_1; Y_2, X) \mid Y_2, X\} Y_1, X\} + E^*\{s(Y_1; Y_2, X) | X\} = 0,$$
$$E\{t(Y_1, Y_2, X) | Y_2, X\} = -E^*[E^*\{s(Y_1; Y_2, X) \mid Y_1, X\} Y_2, X\} + E^*\{s(Y_1; Y_2, X) | X\} = 0.$$

Furthermore, for any $t(Y_1, Y_2, X)$ satisfying conditions (4.9) and (4.10), let

$$d(Y_1, Y_2, X) = t(Y_1, Y_2, X)\frac{p(Y_1, Y_2 \mid X)}{p^*(Y_1, Y_2 \mid X)}.$$

It follows that

$$\begin{aligned} E^*\{d(Y_1, Y_2, X) \mid Y_1, X\} &= E\{t(Y_1, Y_2, X) \mid Y_1, X\} = 0, \\ E^*\{d(Y_1, Y_2, X) \mid Y_2, X\} &= E\{t(Y_1, Y_2, X) \mid Y_2, X\} = 0, \end{aligned}$$

which further implies

$$E^* \{d(Y_1, Y_2, X) \mid X\} = 0.$$

It follows that

$$\begin{aligned} t(Y_1, Y_2, X) &= \frac{p^*(Y_1, Y_2 \mid X)}{p(Y_1, Y_2 \mid X)} \Big[d(Y_1; Y_2, X) - E^* \{d(Y_1; Y_2, X) \mid Y_1, X\} \\ &\quad - E^* \{d(Y_1; Y_2, X) \mid Y_2, X\} + E^* \{d(Y_1; Y_2, X) \mid X\} \Big]. \end{aligned}$$

4.5.2 The doubly robust property

For any given estimating score $s(Y_1, Y_2, X) \in L^2(P^*)$ for θ, the corresponding $t(Y_1; Y_2, X)$ given in the previous subsection may depend on the nuisance parameters $p(y_1 \mid y_{20}, x)$ and $p(y_2 \mid y_{10}, x)$ in addition to $\eta(y_1; y_2 \mid x)$. However, when either $p(y_1 \mid y_{20}, x)$ or $p(y_2 \mid y_{10}, x)$ but not necessarily both are correct, the estimating score $t(Y_1, Y_2, X)$ is still unbiased. This can be seen as follows.

When the correct $p(y_1 \mid y_{20}, x)$ and an incorrect $p(y_2 \mid y_{10}, x)$, denote by $p_w(y_2 \mid y_{10}, x)$, are used for the estimating score computation, then $p(y_1, y_2 \mid x)$ in the score expression is taken as

$$p_w(y_1, y_2 \mid x) = \frac{\eta(y_1; y_2 \mid x) p(y_1 \mid y_{20}, x) p_w(y_2 \mid y_{10}, x)}{\int\int \eta(y_1; y_2 \mid x) p(y_1 \mid y_{20}, x) p_w(y_2 \mid y_{10}, x) dy_1 dy_2}.$$

It then follows that

$$\begin{aligned} t(Y_1, Y_2, X) &= \frac{p^*(Y_1, Y_2 \mid X)}{p_w(Y_1, Y_2 \mid X)} \Big(s(Y_1; Y_2, X) - E^* \{s(Y_1; Y_2, X) \mid Y_1, X\} \\ &\quad - E^* \{s(Y_1; Y_2, X) \mid Y_2, X\} + E^* \{s(Y_1; Y_2, X) \mid X\} \Big). \end{aligned}$$

Note that

$$\frac{p(y_1, y_2 \mid x)}{p_w(y_1, y_2 \mid x)} = \frac{p(y_2 \mid y_{10}, x)}{p_w(y_2 \mid y_{10}, x)} G(x),$$

where

$$G(x) = \frac{\int\int \eta(y_1; y_2 \mid x) p(y_1 \mid y_{20}, x) p_w(y_2 \mid y_{10}, x) dy_1 dy_2}{\int\int \eta(y_1; y_2 \mid x) p(y_1 \mid y_{20}, x) p(y_2 \mid y_{10}, x) dy_1 dy_2}$$

The limit of the sample average of the foregoing estimating score is

$$\begin{aligned} E\{t(Y_1, Y_2, X) \mid X\} &= G(X) E^* \left\{ \frac{p(Y_2 \mid y_{10}, X)}{p_w(Y_2 \mid y_{10}, X)} s(Y_1; Y_2, X) \mid X \right\} \\ &\quad - G(X) E^* \left[\frac{p(Y_2 \mid y_{10}, X)}{p_w(Y_2 \mid y_{10}, X)} E^* \{s(Y_1; Y_2, X) \mid Y_1, X\} \mid X \right] \\ &\quad - G(X) E^* \left[\frac{p(Y_2 \mid y_{10}, X)}{p_w(Y_2 \mid y_{10}, X)} E^* \{s(Y_1; Y_2, X) \mid Y_2, X\} \mid X \right] \\ &\quad + G(X) E^* \left[\frac{p(Y_2 \mid y_{10}, X)}{p_w(Y_2 \mid y_{10}, X)} E^* \{s(Y_1; Y_2, X) \mid X\} \mid X \right] \\ &= 0, \end{aligned}$$

because

$$E^* [E^* \{s(Y_1; Y_2, X) \mid Y_1, X\} \mid Y_2, X] = E^* \{s(Y_1; Y_2, X) \mid X\}.$$

Likewise, when an incorrect $p(y_1 \mid y_{20}, x)$, denote by $p_w(y_1 \mid y_{20}, x)$, and the correct $p(y_2 \mid y_{10}, x)$ are used in the estimating score computation, $p(y_1, y_2 \mid x)$ in the score expression is taken as

$$p_w(y_1, y_2 \mid x) = \frac{\eta(y_1; y_2 \mid x) p_w(y_1 \mid y_{20}, x) p(y_2 \mid y_{10}, x)}{\int\int \eta(y_1; y_2 \mid x) p_w(y_1 \mid y_{20}, x) p(y_2 \mid y_{10}, x) dy_1 dy_2}.$$

It then follows that

$$\begin{aligned} t(Y_1, Y_2, X) &= \frac{p^*(Y_1, Y_2 \mid X)}{p_w(Y_1, Y_2 \mid X)} \Big(s(Y_1; Y_2, X) - E^* \{ s(Y_1; Y_2, X) \mid Y_1, X \} \\ &\quad - E^* \{ s(Y_1; Y_2, X) \mid Y_2, X \} + E^* \{ s(Y_1; Y_2, X) \mid X \} \Big). \end{aligned}$$

Note that

$$\frac{p(y_1, y_2 \mid x)}{p_w(y_1, y_2 \mid x)} = \frac{p(y_1 \mid y_{20}, x)}{p_w(y_1 \mid y_{20}, x)} H(x),$$

where

$$H(x) = \frac{\int\int \eta(y_1; y_2 \mid x) p_w(y_1 \mid y_{20}, x) p(y_2 \mid y_{10}, x) dy_1 dy_2}{\int\int \eta(y_1; y_2 \mid x) p(y_1 \mid y_{20}, x) p(y_2 \mid y_{10}, x) dy_1 dy_2}$$

The limit of the sample average of the foregoing estimating score is

$$\begin{aligned} E\{t(Y_1, Y_2, X) \mid X\} &= H(X) E^* \left\{ \frac{p(Y_1 \mid y_{20}, X)}{p_w(Y_1 \mid y_{20}, X)} s(Y_1; Y_2, X) \mid X \right\} \\ &\quad - H(X) E^* \left[\frac{p(Y_1 \mid y_{20}, X)}{p_w(Y_1 \mid y_{20}, X)} E^* \{ s(Y_1; Y_2, X) \mid Y_1, X \} \mid X \right] \\ &\quad - H(X) E^* \left[\frac{p(Y_1 \mid y_{20}, X)}{p_w(Y_1 \mid y_{20}, X)} E^* \{ s(Y_1; Y_2, X) \mid Y_2, X \} \mid X \right] \\ &\quad + H(X) E^* \left[\frac{p(Y_1 \mid y_{20}, X)}{p_w(Y_1 \mid y_{20}, X)} E^* \{ s(Y_1; Y_2, X) \mid X \} \mid X \right] \\ &= 0, \end{aligned}$$

because

$$E^* [E^* \{ s(Y_1; Y_2, X) \mid Y_2, X \} \mid Y_1, X] = E^* \{ s(Y_1; Y_2, X) \mid X \}.$$

In practice, we may take

$$p^*(y_1, y_2 \mid x) = p(y_1 \mid y_{20}, x) p_w(y_2 \mid y_{10}, x),$$

when $p(y_2 \mid y_{10}, x)$ is misspecified, or

$$p^*(y_1, y_2 \mid x) = p_w(y_1 \mid y_{20}, x) p(y_2 \mid y_{10}, x),$$

when $p(y_2 \mid y_{10}, x)$ is misspecified. In these cases, the doubly robust estimating score can be simplified to

$$\begin{aligned} t(Y_1, Y_2, X) &= \frac{C(X)}{\eta(Y_1; Y_2 \mid X)} \Big[s(Y_1; Y_2, X) - E^* \{ s(Y_1; Y_2, X) \mid Y_1, X \} \\ &\quad - E^* \{ s(Y_1; Y_2, X) \mid Y_2, X \} + E^* \{ s(Y_1; Y_2, X) \mid X \} \Big], \end{aligned}$$

where $C(X)$ can be either $G(X)$ or $H(X)$. In fact, the doubly robust property of the estimating score holds for any choice of square-integrable $C(X)$.

4.5.3 A more specific case

As an illustrative example, consider the simple conditional odds ratio model,

$$\log \eta(y_1; y_2 \mid x) = \theta(y_1 - y_{10})(y_2 - y_{20}).$$

Let

$$p^*(y_1, y_2 \mid x) = p_1^*(y_1)p_2^*(y_1),$$

where $p_k^*(y_k), k = 1, 2$ may be taken as the marginal empirical probability mass functions. For $p(y_1 \mid y_{20}, x)$ and $p_2(y_2 \mid y_{10}, x)$ in $p(y_1, y_2 \mid x)$, they may be further specified respectively by the semiparametric odds ratio models as

$$p(y_1 \mid y_{20}, x) = \frac{\exp\{\alpha(y_1 - y_{10})(x - x_0)\} p_1(y_1 \mid y_{20}, x_0)}{\int \exp\{\alpha(y_1 - y_{10})(x - x_0)\} p_1(y_1 \mid y_{20}, x_0) dy_1},$$

and

$$p(y_2 \mid y_{10}, x) = \frac{\exp\{\beta(y_2 - y_{20})(x - x_0)\} p_2(y_2 \mid y_{10}, x_0)}{\int \exp\{\beta(y_2 - y_{20})(x - x_0)\} p_2(y_2 \mid y_{10}, x_0) dy_2}.$$

The method first estimate (θ, α, p_1) from the conditional likelihood based on

$$p(y_1 \mid y_2, x) = \frac{\exp\{\theta(y_1 - y_{10})(y_2 - Y_{20}) + \alpha(y_1 - y_{10})(x - x_0)\} p_1(y_1 \mid y_{20}, x_0)}{\int \exp\{\theta(y_1 - y_{10})(y_2 - Y_{20}) + \alpha(y_1 - y_{10})(x - x_0)\} p_1(y_1 \mid y_{20}, x_0) dy_1},$$

and estimate (θ, β, p_2) from the conditional likelihood based on

$$p(y_2 \mid y_1, x) = \frac{\exp\{\theta(y_1 - y_{10})(y_2 - Y_{20}) + \beta(y_2 - y_{20})(x - x_0)\} p_2(y_2 \mid y_{10}, x_0)}{\int \exp\{\theta(y_1 - y_{10})(y_2 - Y_{20}) + \beta(y_2 - y_{20})(x - x_0)\} p_2(y_2 \mid y_{10}, x_0) dy_2}.$$

Both the maximum semiparametric conditional likelihood and the maximum permutation likelihood with the weighted approach to estimating the baseline functions can be used for the computation. Denote the estimates of $(\alpha, p_1, \beta, p_2)$ by $(\hat{\alpha}, \hat{p}_1, \hat{\beta}, \hat{p}_2)$. The doubly robust estimating score for θ is

$$\begin{aligned} t(Y_1, Y_2, X \mid \theta) &= \frac{\hat{p}_1^*(Y_1)\hat{p}_2^*(Y_2)c(\theta, \hat{\alpha}, \hat{p}_1, \hat{\beta}, \hat{p}_2)}{\eta(Y_1; Y_2 \mid X, \theta)p_1(Y_1 \mid y_{20}, X, \hat{\alpha}, \hat{p}_1)p_2(Y_2 \mid y_{10}, X, \hat{\beta}, \hat{p}_2)} \\ &\quad \Big(s(Y_1; Y_2, X \mid \theta) - E^*\{s(Y_1; Y_2, X \mid \theta) \mid Y_1, X\} \\ &\quad - E^*\{s(Y_1; Y_2, X \mid \theta) \mid Y_2, X\} + E^*\{s(Y_1; Y_2, X \mid \theta) \mid X\} \Big), \end{aligned}$$

where $c(\theta, \hat{\alpha}, \hat{p}_1, \hat{\beta}, \hat{p}_2)$ is the normalizing constant for the joint density. The doubly robust estimator of θ solves the equation

$$\sum_{i=1}^{n} t(Y_{i1}, Y_{i2}, X_i \mid \theta) = 0.$$

Alternatively, we may take

$$p^*(y_1, y_2 \mid x) = p_1(Y_1 \mid y_{20}, X, \hat{\alpha}, \hat{p}_1)p_2(Y_2 \mid y_{10}, X, \hat{\beta}, \hat{p}_2).$$

In this case,

$$\begin{aligned} t(Y_1, Y_2, X \mid \theta) = \frac{c(\theta, \hat{\alpha}, \hat{p}_1, \hat{\beta}, \hat{p}_2)}{\eta(Y_1; Y_2 \mid X, \theta)} \Big(& s(Y_1; Y_2, X \mid \theta) - E^*\{s(Y_1; Y_2, X \mid \theta) \mid Y_1, X, \hat{\alpha}, \hat{p}_1\} \\ & - E^*\Big\{s(Y_1; Y_2, X \mid \theta) \mid Y_2, X, \hat{\beta}, \hat{p}_2\Big\} + E^*\Big\{s(Y_1; Y_2, X \mid \theta) \mid X, \hat{\alpha}, \hat{p}_1, \hat{\beta}, \hat{p}_2\Big\}\Big), \end{aligned}$$

One choice of the estimating score may be the log-likelihood score for θ, i.e.,

$$s(Y_1; Y_2, X \mid \theta) = (Y_1 - y_{10})(Y_2 - y_{20}).$$

4.6 Robust estimation of conditional odds ratio function in multiple outcome models

The results on the doubly robust estimating scores for bivariate outcomes can be extended to the case with multiple outcomes. Consider the semiparametric conditional odds ratio model for multivariate outcomes,

$$p(y_1, \cdots, y_G \mid x) = \frac{\eta(y_1; \cdots; y_G \mid x, \theta) \prod_{k=1}^G dF_k(y_k \mid x)}{\int \cdots \int \eta(y_1; \cdots; y_G \mid x, \theta) \prod_{k=1}^G dF_k(y_k \mid x)}.$$

The nuisance score space for estimating θ is composed of the closure of the linear combinations of $H_1, \cdots, H_G$, i.e.,

$$H = \bar{\mathcal{L}}(H_1 + \cdots + H_G),$$

where

$$H_k = \{h(Y_k, X) \mid E\{h(Y_k, X) \mid X\} = 0, h(y_k, X) \in L^2(P_0)\}$$

for $k = 1, \cdots, G$. The orthogonal complement of H is $\mathcal{L}^{\perp}(H_1 + \cdots + H_G)$. For any measurable function $s(Y_1, \cdots, Y_G, X) \in \mathcal{L}^{\perp}(H_1 + \cdots + H_G)$, we have

$$E\{s(Y_1, \cdots, Y_G, X)h(Y_k, X)\} = 0,$$

for all $h(Y_k, X) \in H_k$, $k = 1, \cdots, G$. Since $h(Y_k, X) \in H_k$ can be expressed as

$$h(Y_k, X) = r(Y_k, X) - E\{r(Y_k, X) \mid X\}$$

for some $r(Y_k, X) \in L^2(P)$, $E\{s(Y_1, \cdots, Y_G, X)h(Y_k, X)\} = 0$, is equivalent to

$$E\{s(Y_1, \cdots, Y_G, X) \mid Y_k, X\} - E\{s(Y_1, \cdots, Y_G, X) \mid X\} = 0.$$

For any score such that $E\{s(Y_1, \cdots, Y_G, X) \mid X\} = 0$, $s \in \mathcal{L}^{\perp}(H_1 + \cdots + H_G)$ if

$$E\{s(Y_1, \cdots, Y_G, X) \mid Y_k, X\} = 0,$$

for all $k = 1, \cdots, G$.

When $\eta(y_1; \cdots; y_G \mid x, \theta) \equiv 1$ for some θ, then

$$s(Y_1, \cdots, Y_G, X) - E\{s(Y_1, \cdots, Y_G, X) \mid X\} - \sum_{j=1}^{G} [E\{s(Y_1, \cdots, Y_G, X) \mid Y_k, X\} - E\{s(Y_1, \cdots, Y_G, X) \mid X\}],$$

is the projection of $s(Y_1, \cdots, Y_G, X)$ to $\mathcal{L}^{\perp}(H_1 + \cdots + H_G)$. Let

$$p^*(y_1, \cdots, y_G \mid x) = \prod_{j=1}^{G} p^*(y_k \mid x),$$

where

$$0 < \frac{p(y_1, \cdots, y_G \mid x)}{p^*(y_1, \cdots, y_G \mid x)} < \infty.$$

By similar arguments as in Tchetgen Tchetgen et al (2010), $\mathcal{L}^{\perp}(H_1 + \cdots + H_G)$ is characterized by

$$\begin{aligned} T(s, p^*) &= \frac{p^*(Y_1, \cdots, Y_G \mid X)}{p(Y_1, \cdots, Y_G \mid X)} \Big(s(Y_1, \cdots, Y_G, X) - E^* \{ s(Y_1, \cdots, Y_G, X) \mid X \} \\ &\quad - \sum_{j=1}^{G} [E^* \{ s(Y_1, \cdots, Y_G, X) \mid Y_k, X \} - E^* \{ s(Y_1, \cdots, Y_G, X) \mid X \}] \Big), \end{aligned}$$

for any $s(Y_1, \cdots, Y_G, X) \in L^2(P_0)$, where E^* is the expectation computed under $p^*(y_1, \cdots, y_G \mid x)$. When the model is misspecified, i.e., $p(y_1, \cdots, y_G \mid x)$ is misspecified, it means one of $F_k(Y_k \mid x)$ is misspecified since $\eta(y_1; \cdots, y_G \mid x)$ is always assumed to be correctly specified. Let the misspecified $F_k(Y_k \mid X)$ be $F_k^*(Y_k \mid X)$ and $dP^*(y_1, \cdots, y_G \mid x) = \prod_{k=1}^{G} dF_k^*(y_k \mid x)$, then

$$\begin{aligned} T(s, p^*) &= \frac{c(X)}{\eta(Y_1; \cdots; Y_G \mid X, \theta)} \Big(s(Y_1, \cdots, Y_G, X) - E^* \{ s(Y_1, \cdots, Y_G, X) \mid X \} \\ &\quad - \sum_{j=1}^{G} [E^* \{ s(Y_1, \cdots, Y_G, X) \mid Y_j, X \} - E^* \{ s(Y_1, \cdots, Y_G, X) \mid X \}] \Big), \end{aligned}$$

It follows that

$$\begin{aligned} E\{T(s, p^*) \mid X\} &= E\{s(Y_1, \cdots, Y_G, X) \mid X\} - E^*\{s(Y_1, \cdots, Y_G, X) \mid X\} \\ &\quad - \sum_{j=1}^{G} E\left[E^* \{ s(Y_1, \cdots, Y_G, X) \mid Y_j, X \} - E^* \{ s(Y_1, \cdots, Y_G, X) \mid X \} \mid X\right], \end{aligned}$$

When $F_k^*(y_k \mid x) \neq F_k(y_k \mid x)$ for a give k and $F_j^*(y_k \mid x) = F_j(y_k \mid x)$ for all $j \neq k$, then

$$E\left[E^* \{ s(Y_1, \cdots, Y_G, X) \mid Y_j, X \} \mid X\right] = E^* \{ s(Y_1, \cdots, Y_G, X) \mid X \},$$

for any $j \neq k$. Furthermore,

$$\begin{aligned} E\left[E^* \{ s(Y_1, \cdots, Y_G, X) \mid Y_k, X \} \mid X\right] &= E\left[E \{ s(Y_1, \cdots, Y_G, X) \mid Y_k, X \} \mid X\right] \\ &= E\{s(Y_1, \cdots, Y_G, X) \mid X\}. \end{aligned}$$

It follows that

$$E\{T(s, p^*) \mid X\} = 0,$$

which means that if one of the models for $\eta(y_k; x \mid y_{-k0}), k = 1, \cdots, G$ is misspecified, the estimating score is still unbiased.

In the foregoing odds ratio model, if we are interested in estimating parameter θ_1 in $\eta(y_1; y_2 \mid y_3, \cdots, y_G, x, \theta_1)$ only while $\eta(y_1; \cdots; y_G \mid x)$ is known to have correctly specified. From the relationship between conditional odds ratio functions in Chapter 2,

$$\eta\{y_1; (y_2, \cdots, y_G, x)\} = \eta\{y_1; (y_2, \cdots, y_G) \mid x\}\eta(y_1, x \mid y_{20}, \cdots, y_{G0}).$$

4.7 Summary and literature

In epidemiological studies, Mantel-Haenszel test (Mantel and Haenszel, 1959)[121] for stratified 2×2 tables and the common odds ratio estimator are often used for testing conditional independence. Extension to the testing approach beyond 2×2 tables, including continuous variables, is found in Mantel (1963)[120]. In addition, various tests for homogeneity of the odds ratio parameters across strata were proposed (Davis, 1985; Liang, 1985; 1987)[49, 101, 102]. All these can be recast in the semiparametric odds ratio modeling framework. The Mantel-Haenszel test for stratified tables and the Mantel test for continuous variables can both be derived as tests based on the semiparametric efficient scores (Chen, 2007).

Partial correlation coefficient is traditionally used for measuring conditional association. More often, measuring the strength of the association is formulated in regression models. Either the partial correlation or the regression coefficient is not invariant to the change of the marginal characteristics while an ideal association measure should be invariant to the change of the marginal characteristics. Conditional odds ratio function can be estimated using conditional likelihood approach (Breslow, et al, 1978; Liang, 1984; Chen et al, 2013)[15, 100, 36]. Because of the variation independence of the nuisance models and the the odds ratio function in the semiparametric odds ratio model, doubly robust estimation of the odds ratio function is possible (Chen, 2007; Tchetgen Tchetgen et al, 2010; Tchetgen Tchetgen and Rotnizky, 2011; Tchetgen Tchetgen, 2013; Tan, 2019; Zetterqvist, 2019)[28, 179, 181, 180, 176, 203].

4.8 Exercises

1. The Mantel-Haenszel estimator of the conditional odds ratio for a set of stratified 2×2 tables can be described as follows. Let the cell counts for a given stratum k be shown in Table 4.1 where $k = 1, \cdots, K$. Mantel-Haenszel common odds ratio

TABLE 4.1
Counts for stratum k.

	X		
Y	0	1	Total
0	n_{00k}	n_{01k}	n_{0+k}
1	n_{10k}	n_{11k}	n_{1+k}
Total	n_{+0k}	n_{+1k}	n_{++k}

estimator is

$$\hat{\theta} = \frac{\sum_{k=1}^{K} n_{00k} n_{11k} / n_{++k}}{\sum_{k=1}^{K} n_{10k} n_{01k} / n_{++k}}.$$

(a) Show that $\hat{\theta}$ is asymptotically unbiased if $n_{++k} \to \infty$ for all $k = 1, \cdots, K$, while K is fixed.

(b) Find the maximum likelihood estimator for θ.

(c) A generalization of the maximum likelihood estimation approach to the case where either Y or X or both can be any numeric variables assumes that

$$\log \eta(y; x \mid z) = \gamma y x,$$

with reference point $(y = 0, x = 0)$. Suppose that a sample of size n_k is taken from stratum $Z = z_k$, for $k = 1, \cdots, K$. Find the maximum likelihood estimator for γ.

(d) One advantage of the Mantel-Haenszel approach over the maximum likelihood approach is that the estimator is still consistent when the number of strata goes to ∞ while n_{++k} does not necessarily goes to ∞. Can the Mantel-Haenszel common odds estimator be generalized to the latter case?

(e) Find the permutation likelihood estimator of γ in the latter case.

2. For the special case where Y_1 takes finite discrete values $0, 1, \cdots, K$, the efficient score for estimating θ has a closed form

$$S^{eff}(\theta) = \dot{l}_\theta - E_0\{\dot{l}_\theta \mid Y_2, X\} - \text{Cov}_0\{\dot{l}_\theta, b\} \{\text{Var}_0(b)\}^{-1} b$$

where b is defined in section 4.4.

(a) Verify directly that the efficient score is orthogonal to both nuisance score space H_1 and H_2.

(b) Show that if the model for η_1 is correctly specified, the efficient score is unbiased.

(c) If the model for η_2 is correctly specified, is the efficient score still unbiased?

3. The partial linear model assumes that

$$E(Y \mid X, Z) = \beta_0 + \beta^t X + g(Z), \tag{4.13}$$

where $g(z)$ is an unknown function. If we further assume that the prediction error

$$\epsilon = Y - E(Y \mid X, Z)$$

follows $N(0, \sigma^2)$, where $\sigma^2 > 0$ is an unknown parameter.

(a) Show that the conditional odds ratio function $\eta(y; x \mid z)$ with reference point $(y = 0, x = 0)$ has the following form,

$$\log \eta(y; x \mid z) = \gamma^t x y, \tag{4.14}$$

where $\gamma = \beta/\sigma^2$.

(b) Find a doubly robust estimating score for estimating γ.

(c) Discuss the similarity and difference of the model that assume (4.13) only and the model that assumes (4.14) only.

4. When the outcome Y is binary, a generalized partial linear model assumes

$$E(Y \mid X, Z) = \psi\left\{\beta_0 + \beta^t X + g(Z)\right\}, \tag{4.15}$$

where ψ is a known function, e.g.,

$$\psi(u) = \frac{\exp(u)}{1 + \exp(u)}. \tag{4.16}$$

(a) Show that the conditional odds ratio function $\eta(y; x \mid z)$ with reference point $(y = 0, x = 0)$ is,

$$\log \eta(y; x \mid z) = \beta^t x y. \tag{4.17}$$

(b) Find a doubly robust estimator for β.

(c) Compare your doubly robust estimator with these proposed in Tan (2019).

(d) Discuss the similarity and difference of assuming model (4.15) under (4.16) and the model assume (4.17) alone.

5

Application to Biased Sampling Problems

5.1 The general biased sampling problem

To facilitate the study of the general biased sampling problem, let the groups of variables in the data be $Y_1, \cdots, Y_G$ and the sampling probability be

$$\pi(Y_1, \cdots, Y_G) = P(S = 1 \mid Y_1, \cdots, Y_G),$$

where $S = 1$ indicates a sample from the general population being selected. Assume that $\pi(y_1, \cdots, y_G) > 0$. Define the odds ratio function with reference point $(y_{10}, \cdots, y_{G0})$ for the biased sampling probability as

$$\eta_1^\pi\{y_1; (y_2, \cdots, y_G)\} = \frac{\pi(y_1, y_2, \cdots, y_G)\pi(y_{10}, y_{20}, \cdots, y_{G0})}{\pi(y_{10}, y_2, \cdots, y_G)\pi(y_1, y_{20}, \cdots, y_{G0})}.$$

It follows that

$$\pi(y_1, y_2, \cdots, y_G) \propto \eta_1^\pi\{y_1; (y_2, \cdots, y_G)\}\pi(y_1, y_{20}, \cdots, y_{G0})\pi(y_{10}, y_2, \cdots, y_G).$$

Applying the same idea to $\pi(y_{10}, y_2, \cdots, y_G)$ by defining an odds ratio function associated with $\pi(y_{10}, y_2, \cdots, y_G)$ as

$$\eta_2^\pi\{(y_{10}, y_2); (y_3, \cdots, y_G)\} = \frac{\pi(y_{10}, y_2, y_3, \cdots, y_G)\pi(y_{10}, y_{20}, y_{30}, \cdots, y_{G0})}{\pi(y_{10}, y_{20}, y_3 \cdots, y_G)\pi(y_{10}, y_2, y_{30} \cdots, y_{G0})}.$$

It follows that

$$\pi(y_{10}, y_2, y_3, \cdots, y_G) \propto \eta_2^\pi\{(y_{10}, y_2); (y_3, \cdots, y_G)\}\pi(y_{10}, y_2, y_{30} \cdots, y_{G0})\pi(y_{10}, y_{20}, y_3 \cdots, y_G).$$

Continuing this process until y_{G-1} is reached, that is,

$$\eta_{G-1}^\pi\{(y_{10}, \cdots, y_{(p-2)0}, y_{G-1}); y_G\} = \frac{\pi(y_{10}, \cdots, y_{(p-2)0}, y_{G-1}, y_G)\pi(y_{10}, y_{20}, y_{30}, \cdots, y_{G0})}{\pi(y_{10}, \cdots, y_{(p-1)0}, , y_G)\pi(y_{10}, \cdots, y_{(p-2)0}, y_{G-1}, y_{G0})}$$

and

$$\begin{aligned}\pi(y_{10}, \cdots, y_{(p-2)0}, y_{G-1}, y_G) \quad \propto \quad & \eta_{G-1}^\pi\{(y_{10}, \cdots, y_{(p-2)0}, y_{G-1}); y_G\} \\ & \pi(y_{10}, \cdots, y_{(p-1)0}, , y_G)\pi(y_{10}, \cdots, y_{(p-2)0}, y_{G-1}, y_{G0}).\end{aligned}$$

By combining the expressions, we obtain

$$\pi(y_1, \cdots, y_G) \propto \eta^\pi(y_1; \cdots; y_G) \prod_{k=1}^{G} \pi(y_{10} \cdots, y_{(k-1)0}, y_k, y_{(k+1)0} \cdots, y_{G0}), \tag{5.1}$$

DOI: 10.1201/9781351049757-5

where

$$\eta^{\pi}(y_1;\cdots;y_G) = \prod_{k=1}^{G-1} \eta_k^{\pi}\{(y_{10},\cdots,y_{(k-1)0},y_k);(y_{(k+1)},\cdots,y_G)\}.$$

Suppose the distribution of a simple random sample from the general population has a density accepting an odds ratio representation with the same reference point $(y_{10},\cdots,y_{G0})$ as

$$p(y_1,\cdots,y_G) = \frac{\eta(y_1;\cdots,y_G)\prod_{k=1}^G dF_k(y_k)}{\int\cdots\int \eta(y_1;\cdots,y_G)\prod_{k=1}^G dF_k(y_k)}.$$

Under the general biased sampling design, the density of the biased sample becomes

$$p(y_1,\cdots,y_G|S=1) = \frac{\pi(y_1,\cdots,y_G)p(y_1,\cdots,y_G)}{\int\cdots\int \pi(y_1,\cdots,y_G)p(y_1,\cdots,y_G)\prod_{k=1}^G d\mu_k(y_k)}.$$

Applying the decompositions for the biased sampling probability and for the density for the general population, we obtain

$$p(y_1,\cdots,y_G \mid S=1) = \frac{\eta^{\pi}(y_1;\cdots;y_G)\eta(y_1;\cdots,y_G)\prod_{k=1}^G dF_k(y_k \mid S=1)}{\int\cdots\int \eta^{\pi}(y_1;\cdots;y_G)\eta(y_1;\cdots,y_G)\prod_{k=1}^G dF_k(y_k \mid S=1)}, \quad (5.2)$$

where

$$dF_k(y_k \mid S=1) = \frac{\pi(y_{10},\cdots,y_{(k-1)0},y_k,y_{(k+1)0},\cdots,y_{G0})dF_k(y_k)}{\int \pi(y_{10},\cdots,y_{(k-1)0},y_k,y_{(k+1)0},\cdots,y_{G0})dF_k(y_k)},$$

for $k=1,\cdots,G$. It follows that the conditional odds ratio function for the distribution of the biased sample can be expressed as

$$\eta_c(y_i;y_j \mid y_{-(i,j)},S=1) = \eta_c^{\pi}(y_i;y_j \mid y_{-(i,j)})\eta_c(y_i;y_j \mid y_{-(i,j)}),$$

where

$$\eta_c^{\pi}(y_i;y_j \mid y_{-(i,j)}) = \frac{\eta^{\pi}(y_1;\cdots;y_i;\cdots;y_j;\cdots;y_G)\eta^{\pi}(y_1;\cdots;y_{i0};\cdots;y_{j0};\cdots;y_G)}{\eta^{\pi}(y_1;\cdots;y_{i0};\cdots;y_j;\cdots;y_G)\eta^{\pi}(y_1;\cdots;y_i;\cdots;y_{j0};\cdots;y_G)},$$

and $\eta_c(y_i;y_j \mid y_{-(i,j)})$ is defined similarly as given in Chapter 2. The general framework with the sampling problem (5.1) and the biased sampling density (5.2) forms the basis of the discussion in this chapter.

The requirement $\pi(y_1,\cdots,y_G)>0$ on the set

$$\{(y_1,\cdots,y_G) \mid p(y_1,\cdots,y_G)>0\}$$

can be relaxed. Assume the population density satisfies the positivity condition. We may allow the sampling probability to be 0 in some regions as long as the sampling probability satisfies the conditions $P(S=1 \mid y_{10},\cdots,y_{G0})>0$ along with additional conditions similar to that discussed in relaxing the positivity condition for odds ratio representation of a joint density in Chapter 2. Let

$$\mathcal{Y}_k = \{y_k \mid P(S=1 \mid y_{10},\cdots,y_{(k-1)0},y_k,y_{(k+1)0},\cdots,y_{G0})>0\},$$

for $k=1,\cdots,G$. We assume that

$$\mathcal{Y}_1\times\cdots\times\mathcal{Y}_G \supset \{(y_1,\cdots,y_G) \mid P(S=1 \mid y_1,\cdots,y_G)>0\},$$

and that $P(S=1 \mid y_1,\cdots,y_G)>0$ implies $P(S=1 \mid z_1,\cdots,z_G)>0$ where for any k, $z_k=y_k$ or $z_k=y_{k0}$. All the foregoing discussion is still valid under the relaxed assumption on the sampling probability. In particular, the extended sampling designs allow for $P(S=1 \mid y_1,\cdots,y_G)>0$ only in $(y_1,\cdots,y_G)\in\mathcal{X}_1\times\cdots\times\mathcal{X}_G$, where $\mathcal{X}_k\subset\mathcal{Y}_k$ for $k=1,\cdots,G$. Such sampling designs include truncation and extreme-value sampling designs.

5.2 Parameter identifiability for outcome-dependent sample

5.2.1 Identifiability of model components

One important special case of the general biased sampling framework has the sampling probability satisfying

$$\pi(y_1, \cdots, y_G) \propto \prod_{k=1}^{G} \pi_k(y_k). \tag{5.3}$$

This corresponds to the special case where $\eta^\pi(y_1; \cdots; y_G) \equiv 1$ in the odds ratio representation of the general sampling probability. This sampling method can be viewed as follows. Let $S_k = 1$ denote a sampling method that has sampling probability

$$P(S_k = 1 \mid Y_1, \cdots, Y_p) = \pi_k(Y_k),$$

i.e., the sampling probability depends on Y_k only. For $k = 1, \cdots, G$, let $S = 1$ if and only if $S_k = 1, k = 1, \cdots, G$. This means

$$\{S = 1\} = \cap_{k=1}^{G}\{S_k = 1\}.$$

Assume the sampling methods are applied to draw a sample sequentially and the sampling methods are independent of each other. It then follows that

$$P(S = 1 \mid Y_1, \cdots, Y_G) = \prod_{k=1}^{G} P(S_k = 1 \mid Y_1, \cdots, Y_G) = \prod_{k=1}^{G} \pi_k(Y_k).$$

Under this sampling approach, (5.2) reduces to

$$p(y_1, \cdots, y_G \mid S = 1) = \frac{\eta(y_1; \cdots, y_G)\prod_{k=1}^{G} dF_k(y_k \mid S = 1)}{\int \cdots \int \eta(y_1; \cdots, y_G)\prod_{k=1}^{G} dF_k(y_k \mid S = 1)}. \tag{5.4}$$

It can be seen that the odds ratio function $\eta(y_1; \cdots, y_G)$ is invariant to the sampling designs having sampling probability (5.3). This special case covers most of the commonly encountered biased sampling designs, such as the case-control sampling design, the extreme-value sampling design, the length-biased sampling design, and other outcome-dependent sampling designs.

Proposition 5.1. The odds ratio function $\eta(y_1; \cdots, y_G)$ and $dF_k(y_k \mid S = 1), k = 1, \cdots, G$ are identifiable from the biased sampling design (5.3).

Proof: The identifiable of $dF_k(y_k \mid S = 1), k = 1, \cdots, G$ can be straightforwardly obtained by setting all $y_j = y_{j0}$ for all indices except one, and use the fact $\int dF_k(y_k \mid S = 1) = 1$. For the identifiable of the odds ratio function, note that

$$\frac{\eta(y_1; \cdots, y_G)}{\eta^*(y_1; \cdots, y_G)} = \frac{\int \cdots \int \eta(y_1; \cdots, y_G)\prod_{k=1}^{G} dF_k(y_k \mid S = 1)}{\int \cdots \int \eta^*(y_1; \cdots, y_G)\prod_{k=1}^{G} dF_k(y_k \mid S = 1)}$$

implies $\eta(y_1; \cdots, y_G) = c\eta^*(y_1; \cdots, y_G)$ for a constant $c > 0$. Since both η and η^* are odds ratio functions, it is necessary $c = 1$.

The identifiability of the odds ratio function in proposition 5.1 is restricted to the domain

$$\mathcal{Y} = \{(y_1, \cdots, y_G) \mid \pi_k(y_k) > 0, k = 1, \cdots, G\}.$$

Beyond this domain, the odds ratio function is cut off by the biased sample. If the odds ratio function in $\mathcal{Y}^c$ cannot be deduced from the odds ratio function in $\mathcal{Y}$, the outside part of the odds ratio function is not identifiable. On the other hand, the odds ratio function can be regarded to have a local characteristic because truncations on any variables do not change the local identifiability of the odds ratio function.

Proposition 5.1 clarifies the identifiability of different components in the odds ratio representation of a biased sampling density. However, the identifiability of different components does not necessarily transform into the identifiability of the parameters in the population model. In practice, a large number of biased sampling designs have the sampling probability known qualitatively rather than quantitatively. That is, we may know the sampling probability depends on certain variables, but the exact functional form of the sampling probability is unknown. This allows us to have great flexibility in drawing the sample. However, parameter identifiability can be complex in such biased sampling problems. In certain cases, some of the parameters of the population model may not be identifiable while other parameters may be identifiable. In other cases, all the parameters may be identifiable. But the identifiability can be weak, meaning the identifiability can be lost under certain parameter configurations.

5.2.2 Parameter identifiability in case-control design

To illustrate different scenarios of parameter identifiability, consider a typical case-control design where we know the sampling probability depends on the case-control status, but the exact sampling probability is unknown. This sampling design can be fit into the special framework discussed in the previous section with $G = 2$. Let Y_1 denote the binary outcome and Y_2 be a vector denoting the covariates. The case-control sampling probability is

$$P(S = 1 \mid Y_1, Y_2) = \pi_1(Y_1).$$

This sampling probability corresponds to the special scenario (5.3) where $\pi_2(y_2) \equiv 1$. Let the odds ratio representation of the joint density of the distribution for (Y_1, Y_2) be

$$p(y_1; y_2) = \frac{\eta(y_1; y_2) dF_1(y_1 \mid y_{20}) dF_2(y_2 \mid y_{10})}{\int\int \eta(y_1; y_2) dF_1(y_1 \mid y_{20}) dF_2(y_2 \mid y_{10})}.$$

The case-control sample distribution has density

$$p(y_1, y_2 \mid S = 1) = \frac{\eta(y_1; y_2) dF_1(y_1 \mid y_{20}, S = 1) dF_2(y_2 \mid y_{10})}{\int\int \eta(y_1; y_2) dF_1(y_1 \mid y_{20}, S = 1) dF_2(y_2 \mid y_{10})}.$$

The conditional density for Y_1 given Y_2 under the case-control sample becomes

$$p(y_1 \mid y_2, S = 1) = \frac{\eta(y_1; y_2) dF_1(y_1 \mid y_{20}, S = 1)}{\int \eta(y_1; y_2) dF_1(y_1 \mid y_{20}, S = 1)}.$$

It can be seen from the expression that the odds ratio function of y_1 versus y_2 is invariant to the biased sampling design that depends only on the outcome Y_1. The odds ratio function and $F_1(y_1 \mid y_{20}, S = 1)$ are identifiable.

If the population model is a linear logistic regression, i.e.,

$$\log \eta(y_1; y_2) = \theta y_1 y_2,$$

θ is identifiable as long as $P(Y_1 Y_2 = 0) \neq 1$ because $\eta(y_1; y_2)$ is identifiable from the case-control design. The identifiable function has the form

$$dF_1(y_1 \mid S=1) = \frac{\pi_1(y_1)dF_1(y_1)}{\int \pi_1(y_1)dF_1(y_1 \mid y_2=0)} = \frac{\exp\left\{y_1\left(\log\frac{\pi_1(1)}{\pi_1(0)}+\beta_0\right)\right\}}{1+\exp\left(\log\frac{\pi_1(1)}{\pi_1(0)}+\beta_0\right)}.$$

Since $\pi_1(y_1)$ is unknown, β_0 is not identifiable. The non-identifiability of the intercept corresponds to the non-identifiable of $F_1(y_1 \mid y_{20})$. In summary, the parameters in the linear logistic regression model are divided into two parts, viz., intercept and the odds ratio parameters. The intercept is not identifiable and the odds ratio parameters are identifiable in general.

The landscape of parameter identifiability is different when a different binary regression model is used to model the population distribution. Consider, for example, the probit regression model, i.e.,

$$P(y_1=1 \mid y_2) = \Phi(\beta_0+\beta_1 y_2),$$

or in general

$$P(y_1 \mid y_2) = \{\Phi(\beta_0+\beta_1 y_2)\}^{y_1}\{1-\Phi(\beta_0+\beta_1 y_2)\}^{1-y_1}.$$

The odds ratio function with reference point $(y_{10}=0, y_{20}=0)$ is

$$\eta(y_1=1; y_2) = \frac{\Phi(\beta_0+\beta_1 y_2)\{1-\Phi(\beta_0)\}}{\{1-\Phi(\beta_0+\beta_1 y_2)\}\Phi(\beta_0)}$$

and the baseline function is

$$dF_1(y_1) = \Phi(\beta_0)^{y_1}\{1-\Phi(\beta_0)\}^{1-y_1}.$$

Under the case-control sampling design, the odds ratio function remains the same and is identifiable. The function

$$dF_1(y_1 \mid S=1) = \frac{\pi_1(y_1)dF_1(y_1)}{\int \pi_1(y_1)dF_1(y_1 \mid y_2=0)} = \frac{\exp\left\{y_1\left(\log\frac{\pi_1(1)}{\pi_1(0)}+\log\frac{\Phi(\beta_0)}{1-\Phi(\beta_0)}\right)\right\}}{1+\exp\left(\log\frac{\pi_1(1)}{\pi_1(0)}+\log\frac{\Phi(\beta_0)}{1-\Phi(\beta_0)}\right)},$$

is also identifiable. When $\beta_1 \neq 0$, β_0, and β_1 are identifiable from the odds ratio function if $Y_1=1$ and Y_2 takes at least two distinctive values with non-zero probabilities. Under this condition, the ratio of the sampling probabilities $\pi_1(1)/\pi_1(0)$ also becomes identifiable. When $\beta_1=0$, neither β_0 nor $\pi_1(1)/\pi_1(0)$ is identifiable. The identifiability of β_0 is termed weak identifiable because β_0 can become non-identifiable when $\beta_1=0$. Note also that when Y_2 takes at most two distinctive values, neither β_0 nor β_1 is identifiable from the odds ratio function or from the biased sampling distribution.

Even with the logistic regression model for the population, parameter identifiability can be complex with the case-control design when additional complication is involved. Consider the possible misclassification of case and control status in a case-control study. Let Y_1 denote the observed case-control status and U denote the true disease status. Assume that

$$\log\frac{P(U=1 \mid Y_2)}{P(U=0 \mid Y_2)} = \beta_0+\beta_1 Y_2,$$

and

$$P(Y_1 = 1 \mid U = 1, Y_2) = p_1 \text{ and } P(Y_1 = 0 \mid U = 0, Y_2) = p_0.$$

The conditional probability of Y_1 given Y_2 is

$$\begin{aligned} p(Y_1 = 1 \mid Y_2) &= p(Y_1 = 1 \mid U = 1, Y_2)p(U = 1 \mid Y_2) + p(Y_1 = 1 \mid U = 0, Y_2)p(U = 0 \mid Y_2) \\ &= \frac{p_0 + p_1 \exp(\beta_0 + \beta_1 Y_2)}{1 + \exp(\beta_0 + \beta_1 Y_2)}. \end{aligned}$$

The odds ratio function for $p(y_1 \mid y_2)$ with reference point $(y_{10} = 0, y_{20} = 0)$ is

$$\log \eta(y_1; y_2) = \log \frac{\{p_0 + p_1 \exp(\beta_0 + \beta_1 y_2)\}\{1 - p_0 + (1 - p_1)\exp(\beta_0)\}}{\{p_0 + p_1 \exp(\beta_0)\}\{1 - p_0 + (1 - p_1)\exp(\beta_0 + \beta_1 y_2)\}}.$$

For a case-control sampling design with Y_1 as the outcome, $\eta(y_1; y_2)$ is identifiable. The baseline function for the biased sample is

$$p(Y_1 = 1 \mid y_2 = 0, S = 1) = \frac{\frac{\pi_1(1)\{p_0 + p_1 \exp(\beta_0)\}}{\pi_1(0)\{1 - p_0 + (1 - p_1)\exp(\beta_0)\}}}{1 + \frac{\pi_1(1)\{p_0 + p_1 \exp(\beta_0)\}}{\pi_1(0)\{1 - p_0 + (1 - p_1)\exp(\beta_0)\}}}$$

which is also identifiable. When $\beta_1 = 0$, none of parameters (p_0, p_1, β_0) is identifiable. When $\beta_1 \neq 0$ and X takes at least 4 different values with nonzero probability, $(p_0, p_1, \beta_0, \beta_1)$ are identifiable. The identifiability of parameters (p_0, p_1, β_0) is weak because they become non-identifiable once $\beta_1 = 0$.

5.2.3 Parameter identifiability in outcome-dependent sampling design

Parameter identifiability for outcome-dependent sample with the outcome taking more than two distinctive values is similar to the case-control design. Let Y_1 denote the general outcome and Y_2 denote the covariate vector. Suppose that the population model has

$$p(y_1 \mid y_2) = \exp\left[\frac{1}{\phi}\{\theta y_1 - b(\theta)\} + a(y_1)\right],$$

where θ is a function of y_2. The odds ratio function has the form

$$\log \eta(y_1; y_2) = \frac{1}{\phi}\{\theta(y_2) - \theta(y_{20})\}(y_1 - y_{10}).$$

With the outcome-dependent sampling probability $P(S = 1 \mid Y_1, Y_2) = \pi_1(Y_1)$, the density for the outcome-dependent sampling distribution is

$$p(y_1, y_2 \mid S = 1) = \frac{\eta(y_1; y_2) dF_1(y_1 \mid y_{20}, S = 1) dF_2(y_2 \mid y_{10})}{\int\int \eta(y_1; y_2) dF_1(y_1 \mid y_{20}, S = 1) dF_2(y_2 \mid y_{10})}.$$

Parameter identifiability in the odds ratio function and the baseline function for the biased sample is illustrated in the following by specific examples.

Example 5.2.1. For an outcome variable with $K + 1$ categories, consider a baseline logit model with

$$\log \frac{P(Y_1 = k \mid Y_2)}{P(Y_1 = 0 \mid Y_2)} = \beta_{k0} + \beta_{k1} Y_2,$$

for $k = 1, \cdots, K$, the odds ratio function is

$$\log \eta(y_1; y_2) = (1_{\{y_1 = 1\}}, \cdots, 1_{\{y_1 = K\}})\theta y_2$$

where the base category is $y_1 = 0$ and $\theta = (\beta_{11}, \cdots, \beta_{K1})'$. Under the biased sampling design, θ is identifiable when $P(1_{\{Y_1=k\}}Y_2 = 0) \neq 1$ for $k = 1, \cdots, K$. The baseline function for the biased sample is

$$P(Y_1 = k \mid Y_2 = 0, S = 1) = \frac{\exp\left\{\beta_{k0} + \log\frac{\pi_1(k)}{\pi_1(0)}\right\}}{1 + \sum_{k=1}^{K}\exp\left\{\beta_{k0} + \log\frac{\pi_1(k)}{\pi_1(0)}\right\}},$$

for $k = 0, \cdots, K$. When $\pi_1(0), \cdots, \pi_1(K)$ are unknown, the intercepts $\beta_{k0}, k = 1, \cdots, K$ are non-identifiable.

Example 5.2.2. For an outcome variable taking countable number of values, consider a Poisson regression model,

$$P(Y_1 = k \mid Y_2) = \frac{\lambda^k(Y_2)}{k!}\exp\{-\lambda(Y_2)\},$$

where $\log\lambda(Y_2) = \theta_0 + \theta_1 Y_2$. The odds ratio function is

$$\log\eta(y_1, y_2) = \theta_1 y_1 y_2.$$

When $P(Y_1Y_2 = 0) \neq 1$, θ_1 is identifiable. The baseline function for the biased sample is

$$P(Y_1 = k \mid Y_2 = 0, S = 1) = \frac{\pi(k)\exp\left(k\theta_0 - e^{\theta_0}\right)/k!}{\sum_{k=0}^{\infty}\pi(k)\exp\left(k\theta_0 - e^{\theta_0}\right)/k!}.$$

If the sampling probabilities $\pi(k), k = 0, 1, \cdots$ are unknown, θ_0 is not identifiable.

Example 5.2.3. For an outcome variable taking continuous values, consider a normal linear regression model with density

$$p(y_1 \mid y_2) = \frac{1}{\sigma\sqrt{2\pi}}\exp\left\{-\frac{1}{2\sigma^2}(y_1 - \beta_0 - \beta_1 y_2)^2\right\}.$$

The odds ratio function is

$$\log\eta(y_1; y_2) = \frac{\beta_1}{\sigma^2}y_1 y_2.$$

The odds ratio parameter $\theta = \beta_1/\sigma^2$ is identifiable if $P(Y_1Y_2 \neq 0) > 0$. The baseline function for the biased sample is

$$P(y_1 \mid y_2 = 0, S = 1) = \frac{\pi_1(y_1)\exp\left\{-\frac{1}{2\sigma^2}(y_1 - \beta_0)^2\right\}}{\int\pi_1(y_1)\exp\left\{-\frac{1}{2\sigma^2}(y_1 - \beta_0)^2\right\}dy_1}.$$

For unknown $\pi_1(y_1)$, the parameters (β_0, σ^2) are not identifiable. Note that the regression coefficient β_1 is not identifiable in this case because σ^2 is not identifiable.

The foregoing examples have a common theme, i.e., they are generalized linear models with canonical link function. The parameters in the population model are divided into two groups. One group is identifiable and the other group is non-identifiable. For the generalized linear models with non-canonical link function, weak identifiability is commonly encountered as is seen in the case-control design with non-logistic regression model. Estimation of weakly identifiable parameters can be tricky because the loss of identifiability when $\beta = 0$ makes the estimation much more difficult.

5.2.4 Extreme-value sampling and length-biased sampling designs

Two special biased sampling designs, the extreme-value sampling design and the length-biased sampling design, are further discussed in this section. The extreme-value sampling design is similar to the case-control sampling design, but is often applied to continuous outcomes. The length-biased sampling design is often applied to positive continuous outcomes such as the survival times.

In the extreme-value sampling design, the sampling probability is large for large absolute values of Y_1. For small absolute values of Y_1, the sampling probability is typically small or even 0. A typical sampling probability has

$$P(S = 1 \mid Y_1, Y_2) = \begin{cases} 1 & \text{if } Y_1 \geq M, \\ 0 & \text{if } -M < Y_1 < M, \\ 1 & \text{if } Y_1 \leq -M. \end{cases}$$

More general designs may have sampling probabilities for either or both $Y_1 \geq M$ and $Y_1 \leq -M$ less than 1. When the population model is a generalized linear model with canonical link function, the regression parameters divided by the scale parameter are usually identifiable as shown in Examples 5.2.1, 5.2.2, 5.2.3. In this case, regression coefficients themselves may not be identifiable such as in the normal linear regression model. When a non-canonical link function is involved, the parameter identifiability is usually more involved. A case-by-case study of the parameter identifiability is usually required for the generalized linear model with non-canonical link functions.

In the length-biased sampling design, the sampling probability is assumed to be proportional to outcome value Y_1 which is a positive random variable, i.e.,

$$P(S = 1 \mid Y_1, Y_2) \propto Y_1.$$

Suppose that outcome Y_1 given covariate Y_2 follows a distribution whose density accepts an odds ratio decomposition as

$$f(y_1 \mid y_2) = \frac{\eta(y_1; y_2) f(y_1 \mid y_{20})}{\int \eta(y_1; y_2) f(y_1 \mid y_{20}) dy_1}.$$

The sampling distribution becomes

$$p(y_1 \mid y_2, S = 1) = \frac{\eta(y_1; y_2) p(y_1 \mid y_{20}, S = 1)}{\int \eta(y_1; y_2) p(y_1 \mid y_{20}, S = 1) dy_1},$$

where

$$p(y_1 \mid y_{20}, S = 1) = \frac{y_1 f(y_1 \mid y_{20})}{\int y_1 f(y_1 \mid y_{20}) dy_1},$$

and $\int y_1 f(y_1 \mid y_{20}) dy_1 < \infty$. Since the sampling probability is almost known, the parameters identifiable in the original model remains identifiable under the length-biased sampling design.

One remark to add to the length-biased sample is that in order for the sampling design to be practical, one needs to have

$$\int y_1 f(y_1 \mid y_{20}) dy < +\infty.$$

This may not always hold. For example, consider the population distribution with density

$$p(y_1 \mid y_{20}) = \frac{2}{\pi} \frac{\theta y_1}{1 + (\theta y_1)^2},$$

for $y_1 > 0$, where $\theta > 0$. The length-biased sampling design cannot be implemented in the original sample with such a distribution.

5.3 Outcome-dependent sampling designs with covariate matching

5.3.1 The general framework

Another special case which is more general than the special case discussed in the previous section has sampling probability

$$P(S=1 \mid Y_1, \cdots, Y_G) = \pi(Y_1, Y_{j+1}, \cdots, Y_G), \tag{5.5}$$

which can be represented as

$$\pi(y_1, y_{j+1}, \cdots, y_G) \propto \eta^\pi(y_1; y_{j+1}; \cdots; y_G) \prod_{k=1}^{G} \pi_k(y_k),$$

where $\pi_k(y_k) = 1$ for $k = 2, \cdots, j$ and $\eta^\pi(y_1; y_{j+1}; \cdots; y_G)$ does not depend on $y_2, \cdots, y_j$. This leads to the reduction of (5.2) to

$$p(y_1, \cdots, y_G \mid S=1) = \frac{\eta^\pi(y_1; y_{j+1}; \cdots; y_G)\eta(y_1; \cdots; y_G) \prod_{k=1}^{G} dF_k(y_k \mid S=1)}{\int \cdots \int \eta^\pi(y_1; y_{j+1}; \cdots, y_G)\eta(y_1; \cdots; y_G) \prod_{k=1}^{G} dF_k(y_k \mid S=1)}. \tag{5.6}$$

When $\pi(Y_1, Y_{j+1}, \cdots, Y_G)$ is unknown, $p(y_1, \cdots, y_G \mid S=1)$ can be further rewritten as

$$\frac{\eta(y_1; \cdots; y_j \mid y_{j+1}, \cdots, y_G) \prod_{k=1}^{j} dF_k(y_k \mid y_{j+1}, \cdots, y_G, S=1) dF(y_{j+1}, \cdots, y_G \mid S=1)}{\int \cdots \int \eta(y_1; \cdots, ; y_j \mid y_{j+1}, \cdots, y_G) \prod_{k=1}^{j} dF_k(y_k \mid y_{j+1}, \cdots, y_G, S=1) dF(y_{j+1}, \cdots, y_G \mid S=1)}, \tag{5.7}$$

where $F_k(y_k \mid y_{j+1}, \cdots, y_G, S=1) = F_k(y_k \mid y_{j+1}, \cdots, y_G)$ for $k = 2, \cdots, j$. This framework with the sampling probabilty (5.5) and the sampling density (5.6) covers matched case-control sampling design, matched extreme-value sampling design, and many other more complex sampling designs. Results on component identifiability similar to those given in Proposition 5.1 hold.

Proposition 5.2. When $\pi(y_1, y_{j+1}, \cdots, y_G)$ is unknown, $\eta(y_1; \cdots; y_j \mid y_{j+1}, \cdots, y_G)$, $F_k(y_k \mid y_{j+1}, \cdots, y_G, S=1)$ for $k = 2, \cdots, j$, and $dF(y_{j+1}, \cdots, y_G \mid S=1)$ are identifiable. When $\eta^\pi(y_1; y_{j+1}; \cdots; y_G)$ is known but $\pi_k(y_k), k = 1, j+1, \cdots, G$ are unknown, $\eta(y_1; \cdots; y_G)$, $F_k(y_k \mid S=1)$ for $k = 1, \cdots, G$ are identifiable.

Proof: Note first that

$$\eta(y_1; \cdots, y_G) = \prod_{k=1}^{G-1} \eta_k\{y_k; (y_{k+1} \cdots, y_G) \mid (y_{10}, \cdots, y_{k0})\}.$$

Note further that, for $k < j$,

$$\begin{aligned} &\eta_k\{y_k; (y_{k+1} \cdots, y_G) \mid (y_{10}, \cdots, y_{k0})\} \\ &\quad = \eta_k\{y_k; (y_{k+1} \cdots, y_j) \mid (y_{10}, \cdots, y_{k0}), (y_{j+1}, \cdots, y_G)\} \\ &\qquad \times \eta\{y_k; (y_{j+1}, \cdots, y_G) \mid (y_{10}, \cdots, y_{k0}), (y_{(k+1)0} \cdots, y_{j0}). \end{aligned}$$

It follows that

$$\begin{aligned}\eta(y_1;\cdots,y_G) &= \prod_{k=1}^{j-1}\eta_k\{y_k;(y_{k+1}\cdots,y_j)\mid(y_{10},\cdots,y_{k0}),(y_{j+1},\cdots,y_G)\}\\ &\quad\times\prod_{k=1}^{j}\eta\{y_k;(y_{j+1},\cdots,y_G)\mid(y_{10},\cdots,y_{k0}),(y_{(k+1)0}\cdots,y_{j0})\\ &\quad\times\prod_{k=j+1}^{G-1}\eta_k\{y_k;(y_{k+1}\cdots,y_G)\mid(y_{10},\cdots,y_{k0})\}.\end{aligned}$$

This implies that

$$\begin{aligned}&\eta(y_1;\cdots,y_G)\prod_{k=1}^{G}dF_k(y_k\mid y_{-k0})\\ &=\ \eta(y_1;\cdots;y_j\mid y_{j+1},\cdots,y_G)\prod_{k=1}^{j}dF_k\{y_k\mid(y_{10},\cdots,y_{(j-1)0}),(y_{j+1},\cdots,y_G)\}\\ &\quad dF\{(y_{j+1},\cdots,y_G\mid(y_{10},\cdots,y_{j0})\}.\end{aligned}$$

Note also that the odds ratio function for the sampling probability can be rewritten as

$$\begin{aligned}\eta^\pi(y_1;y_{j+1};\cdots;y_G) &= \eta^\pi\{y_1;(y_{j+1},\cdots,y_G)\mid y_{20},\cdots,y_{j0})\\ &\quad\times\prod_{k=j+1}^{G-1}\eta_k^\pi\{y_k;(y_{k+1}\cdots,y_G)\mid(y_{10},\cdots,y_{k0})\}.\end{aligned}$$

It follows by combining the equalities that

$$\begin{aligned}&\eta^\pi(y_1;y_{j+1};\cdots;y_G)\pi_1(y_1)\prod_{k=j+1}^{G}\pi_k(y_k)\eta(y_1;\cdots,y_G)\prod_{k=1}^{G}dF_k(y_k\mid y_{-k0})\\ &\propto\ \eta(y_1;\cdots;y_j\mid y_{j+1},\cdots,y_G)\prod_{k=1}^{j}dF_k\{y_k\mid(y_{10},\cdots,y_{(j-1)0}),(y_{j+1},\cdots,y_G),S=1)\}\\ &\quad dF\{(y_{j+1},\cdots,y_G\mid(y_{10},\cdots,y_{j0}),S=1\},\end{aligned}$$

where

$$dF_k\{y_k\mid(y_{10},\cdots,y_{(j-1)0}),(y_{j+1},\cdots,y_G,S=1)\}=dF_k\{y_k\mid(y_{10},\cdots,y_{(j-1)0}),(y_{j+1},\cdots,y_G)\}$$

for $k=2,\cdots,j$. This shows that (5.7) holds, which implies the first part of the Proposition. The second part of the Proposition follows from Proposition 5.1.

5.3.2 Matched case-control design

The matched case-control design is often used in studying disease risk in epidemiological research. The design can be described using three groups of variables: the outcome Y_1, the risk factor under study Y_2, and the confounding variable to be controlled Y_3. The matched case-control samples are obtained by drawing a conventional case-control sample from each stratum defined by the value of Y_3 and the case and control sampling probabilities may vary across different strata. As a result of this design, the matched case-control sample has sampling probability

$$P(S = 1 \mid Y_1, Y_2, Y_3) = \pi(Y_1, Y_3). \tag{5.8}$$

If $\pi(Y_1, Y_3)$ is unknown, the sampling probability can be expressed as

$$\begin{aligned} & p(y_1, y_2, y_3 \mid S = 1) \\ & = \frac{\eta(y_1; y_2 \mid y_3) dF_1(y_1 \mid y_{20}, y_3, S = 1) dF_2(y_2 \mid y_{10}, y_3) dF_3(y_3 \mid y_{10}, y_{20}, S = 1)}{\int \eta(y_1; y_2 \mid y_3) dF_1(y_1 \mid y_{20}, y_3, S = 1) dF_2(y_2 \mid y_{10}, y_3) dF_3(y_3 \mid y_{10}, y_{20}, S = 1)}. \end{aligned}$$

This implies that $\eta(y_1; y_2 \mid y_3)$, $F_1(y_1 \mid y_{20}, y_3, S = 1)$, $F_2(y_2 \mid y_{10}, y_3)$, and $F_3(y_3 \mid y_{10}, y_{20}, S = 1)$ are identifiable from the biased sampling design (5.8). The distribution function $F_3(y_3 \mid y_{10}, y_{20}, S = 1)$ is often not of interest and is variation independent of the rest of the components. It is then sufficient to consider

$$p(y_1, y_2 \mid y_3, S = 1) = \frac{\eta(y_1; y_2 \mid y_3) dF_1(y_1 \mid y_{20}, y_3, S = 1) dF_2(y_2 \mid y_{10}, y_3)}{\int \eta(y_1; y_2 \mid y_3) dF_1(y_1 \mid y_{20}, y_3, S = 1) dF_2(y_2 \mid y_{10}, y_3)}.$$

When the population model for the outcome Y_1 conditional on Y_2, Y_3 follows a partial linear logistic regression model, i.e.,

$$\log \frac{G(y_1 = 1 \mid y_2, y_3)}{G(y_1 = 0 \mid y_2, y_3)} = \beta_1 y_2 + h(y_3),$$

The conditional odds ratio function

$$\eta(y_1; y_2 \mid y_3) = \beta_1 (y_2 - y_{20})(y_3 - y_{30}).$$

This means β_1 is identifiable in general. Also identifiable is $F_2(y_2 \mid y_{10}, y_3)$. Other components of the population model are not identifiable.

If the population model is a probit model, i.e.,

$$p(Y_1 = 1 \mid Y_2, Y_3) = \Phi\{\beta_1 Y_2 + h(Y_3)\},$$

the conditional odds ratio function is

$$\log \eta(y_1; y_2 \mid y_3) = \log \frac{\Phi\{\beta_1 Y_2 + h(Y_3)\}[1 - \Phi\{h(Y_3)\}]}{[1 - \Phi\{\beta_1 Y_2 + h(Y_3)\}]\Phi\{h(Y_3)\}}.$$

where $y_{10} = 0, y_{20} = 0$. When $\beta_1 = 0$, $h(y_3)$ is non-identifiable from the conditional odds ratio function. When $\beta_1 \neq 0$, $h(y_3)$ can be potentially identifiable from the conditional odds ratio function.

5.3.3 Extreme-value sampling design with matching

The extreme-value sampling design with matching behaves in the same way as the matched case-control design except the outcome can take more than three categories or is continuous in nature. The same framework (5.6) and (5.7) can be applied. When the outcome given covariates follow a generalized linear model, i.e.,

$$p(y_1 \mid y_2, y_3) = \exp\left(\frac{1}{\phi}[\theta(y_2, y_3) y_1 - b\{\theta(y_2, y_3)\}] + a(\phi, y_1)\right),$$

the odds ratio function

$$\log \eta(y_1; y_2 \mid y_3) = \frac{1}{\phi}(y_1 - y_{10})\{\theta(y_2, y_3) - \theta(y_{20}, y_3)\}$$

is identifiable with the sampling design. When the canonical link function is used, i.e.,

$$\theta(y_2, y_3) = \beta y_2 + h(y_3),$$

the conditional odds ratio function simplifies to

$$\log \eta(y_1; y_2 \mid y_3) = \frac{\beta}{\phi}(y_1 - y_{10})(y_2 - y_{20}).$$

The parameter β/ϕ is identifiable from the conditional odds ratio function. The function $h(y_3)$ is not identifiable in general. When the link function is not canonical, for example,

$$\theta(y_2, y_3) = g\{\beta y_2 + h(y_3)\},$$

where $g(x) \neq x$, the odds ratio function becomes

$$\log \eta(y_1; y_2 \mid y_3) = \frac{1}{\phi}(y_1 - y_{10})\left[g\{\beta y_2 + h(y_3)\} - g\{\beta y_{20} + h(y_3)\}\right].$$

The function $h(y_3)$ may become identifiable when $\beta \neq 0$. When $\beta = 0$, $h(y_3)$ is not identifiable. The identifiability of $h(y_3)$ is thus weak.

5.4 Parameter estimation with biased sampling designs

Under the sampling design (5.3), the odds ratio function is invariant and identifiable. The baseline functions in the population model are not identifiable if the exact sampling probability is not known. The biased sampling density is a semiparametric odds ratio model irrespective of whether the original population model is in a form of parametric or semiparametric odds ratio model. Parameter estimation and inference under the biased sampling design can be carried out by the likelihood approaches discussed in Chapter 3. In particular, if the population model is a semiparametric odds ratio model, the model under the biased sampling design remains a semiparametric odds ratio model. As a result, the form of the model appears invariant to the biased sampling design although the baseline functions are different. That is, the population model is

$$p(y_1, \cdots, y_G) = \frac{\eta(y_1; \cdots; y_G)\prod_{k=1}^{G} dF_k(y_k \mid y_{-k0})}{\int \eta(y_1; \cdots; y_G)\prod_{k=1}^{G} dF_k(y_k \mid y_{-k0})},$$

where $F_k(y_k \mid y_{-k0}), k = 1, \cdots, G$ are unrestricted. The model for the biased sample is

$$p(y_1, \cdots, y_G \mid S = 1) = \frac{\eta(y_1; \cdots; y_G)\prod_{k=1}^{G} dF_k(y_k \mid y_{-k0}, S = 1)}{\int \eta(y_1; \cdots; y_G)\prod_{k=1}^{G} dF_k(y_k \mid y_{-k0}, S = 1)},$$

where $F_k(y_k \mid y_{-k0}, S = 1), k = 1, \cdots, G$ are unrestricted. These two models appear the same although $F_k(y_k \mid y_{-k0})$ may be different from $F_k(y_k \mid y_{-k0}, S = 1)$. There is no distinction in terms of the applicable estimation and inference approaches.

The likelihood based on the observed biased sampling data is

$$L(\theta, F_k^*, k = 1, \cdots, G) = \prod_{i=1}^{n} \frac{\eta(Y_{i1}; \cdots; Y_{ip})\prod_{k=1}^{G} dF_k^*(Y_{ik})}{\int \cdots \int \eta(y_1; \cdots; y_G)\prod_{k=1}^{G} dF_k^*(y_k)},$$

where $F_k^*(y_k) = F_k(y_k \mid y_{-k0}, S = 1)$. The analysis based on the joint likelihood with the outcome-dependent biased sampling designs is often termed the joint analysis. Alternative analyses that use conditioning to eliminate part of the joint likelihood to reduce complexity of the likelihood are often used. Such as in a case-control design, the prospective and retrospective likelihoods instead of the joint likelihood may be used. The prospective likelihood approach and the retrospective likelihood approach may yield the same estimator as the joint likelihood approach. But such equivalence is true only to a certain level of conditioning. Beyond that, they are no longer equivalent. To see this, consider the profile semiparametric likelihood that eliminates one of the baseline functions, $F_k^*(y_k)), k = 1, \cdots, G$. Suppose F_j^* is selected to be eliminated. The maximum likelihood estimator for F_j^*, while all other parameters are kept fixed, satisfies

$$\hat{F}_j^*(y_j) = \frac{\int \cdots \int \eta(y_1; \cdots ; y_G) \prod_{k=1}^G dF_k^*(y_k)}{n \int \cdots \int \eta(y_1; \cdots ; y_G) \prod_{k \neq j} dF_k^*(y_k)}.$$

The profile likelihood that eliminates F_j^* from the joint likelihood is

$$L_{pj}(\theta, F_k^*, k \neq j) = \prod_{i=1}^n \frac{\eta(Y_{i1}; \cdots ; Y_{ip}) \prod_{k=1, k\neq j}^G dF_k^*(Y_{ik})}{\int \cdots \int \eta(y_1; \cdots ; y_G) \prod_{k=1, k \neq j}^G dF_k^*(y_k)},$$

This likelihood is the same as the conditional likelihood based on Y_{-j} given Y_j under the biased sampling design. Denote the conditional likelihood by $L_{cj}(\theta, F_k^*, k \neq j)$. For the case-control design with $G = 2$, L_{p1} is the prospective likelihood and L_{p2} is the retrospective likelihood. As a result, the three likelihoods are equivalent in estimating the odds ratio parameter θ. This corresponds to the traditional use of the prospective likelihood approach which is efficient for the analysis of the case-control design.

For case-control design with $G \geq 3$, it corresponds to the modeling of the covariate distribution, such as the gene-environment independence. The three profile likelihoods at the first level of profiling are equal to three conditional likelihood respectively,

$$\begin{aligned} L_{p1}(\theta, F_2^*, F_3^*) &= L_{c1}(\theta, F_2^*, F_3^*), \\ L_{p2}(\theta, F_1^*, F_3^*) &= L_{c2}(\theta, F_1^*, F_3^*), \\ L_{p3}(\theta, F_1^*, F_2^*) &= L_{c3}(\theta, F_1^*, F_2^*). \end{aligned}$$

However, the profile likelihoods at the second level of profiling are not generally equal to the respective conditional likelihoods,

$$\begin{aligned} L_{p12}(\theta, F_3^*) &\neq L_{c12}(\theta, F_3^*), \\ L_{p13}(\theta, F_2^*) &\neq L_{c13}(\theta, F_2^*), \\ L_{p23}(\theta, F_1^*) &\neq L_{c23}(\theta, F_1^*). \end{aligned}$$

where L_{pjk} denotes the profile likelihood with both F_j^* and F_k^* eliminated, and L_{cjk} denotes the conditional likelihood of $Y_{-(jk)}$ conditional on (Y_j, Y_k). In this case, the retrospective likelihood usually refers to $L_{p1}(\theta, F_2^*, F_3^*)$ at the first level of profiling, while the prospective likelihood refers to $L_{c23}(\theta, F_1^*)$. These two likelihoods are not equivalent with respect to the efficiency in estimating the odds ratio parameter θ. Since the profiling likelihoods are equivalent to the joint likelihood in estimating the odds ratio parameter θ, $L_{p23}(\theta, F_1^*)$ is efficient in estimating θ. The prospective likelihood estimator of θ is usually inefficient in comparison.

Aside from the traditional prospective likelihood and the retrospective likelihood, it can be seen from the foregoing discussion that there are a number of other likelihoods that may be used for the odds ratio parameter estimation when the covariates are modeled in the analysis of the case-control design. For example, the retrospective likelihood $L_{p1}(\theta, F_2^*, F_3^*)$ has two equivalent likelihoods which are all equivalent to the joint likelihood estimator. In terms of computation, however, one may be favored over the others. Consider the case of studying gene-environment interactions by a case-control design. Genetic variables are often categorical in nature while the environment variables are often continuous in nature. If Y_2 contains genetic variables and Y_3 contains environment variables, then maximizing the likelihood $L_{p3}(\theta, F_1^*, F_2^*)$ to obtain the odds ratio parameter estimator is often favored because it yields the same estimator as maximizing the joint likelihood while the baseline functions to be estimated in the likelihood $L_{p3}(\theta, F_1^*, F_2^*)$ are of lower overall dimensions than those in the other equivalent likelihoods.

The foregoing discussion can also be applied to the matched case-control study design as well as other similar designs such as the matched extreme-values sampling design. As discussed in the previous sections, the odds ratio functions are also modified when a matched sampling design is applied. For any odds ratio function that is modified by the matched sampling design, the new odds ratio representation combines the relevant parts into a new group. At the new combined groups, the odds ratio functions are all identifiable. The joint likelihood is formed based on the new grouping. Similar likelihoods can be obtained. To demonstrate the idea, consider the following matched case-control study design with groups of variables Y_1, Y_2, Y_3, Y_4, where Y_1 denotes the outcome, Y_2 and Y_3 denote the covariates, and Y_4 denotes the group of matching variables. The sampling probability is

$$P(S = 1 \mid Y_1, Y_2, Y_3, Y_4) \propto \eta_\pi(Y_1; Y_4) \prod_{k=1}^{4} \pi_k(Y_k).$$

The sampling distribution has density

$$p(y_1, y_2, y_3, y_4 \mid S = 1) = \frac{\eta_\pi(y_1; y_4)\eta(y_1; y_2; y_3; y_4) \prod_{k=1}^4 dF_k^*(y_k)}{\int \cdots \int \eta_\pi(y_1; y_4)\eta(y_1; y_2; y_3; y_4) \prod_{k=1}^4 dF_k^*(y_k)}.$$

When $\eta_\pi(y_1; y_4)$ is unknown, we can group Y_1 and Y_4 into a new group. The odds ratio decomposition of the density for the new groups of random variables is

$$p\{(y_1, y_4), y_2, y_3 \mid S = 1) = \frac{\eta\{(y_1, y_4); y_2; y_3\} dF_{14}^*(y_1, y_4) dF_2^*(y_2) dF_3^*(y_3)}{\int \cdots \int \eta\{(y_1, y_4); y_2; y_3\} dF_{14}^*(y_1, y_4) dF_2^*(y_2) dF_3^*(y_3)},$$

where the distribution for the new combined group is

$$dF_{14}^*(y_1, y_4) = \frac{\eta_\pi(y_1; y_4)\eta(y_1; y_4 \mid y_{20}, y_{30}) dF^*(y_1) dF^*(y_4)}{\int \int \eta_\pi(y_1; y_4)\eta(y_1; y_4 \mid y_{20}, y_{30}) dF^*(y_1) dF^*(y_4)}.$$

Parameters in the odds ratio function $\eta\{(y_1, y_4); y_2; y_3\}$ is identifiable. Estimation of the parameters in the new model can be carried out using the likelihood approaches as discussed in the unmatched studies.

Identifiability of parameters under the biased sampling designs makes it possible to estimate the parameters using the likelihood approaches. However, identifiability alone does not guarantee the performance of the estimator based on maximizing the likelihoods. One thorny issue in biased sampling problems is the need to handle weakly identifiable parameters in estimation and inference when the parameters in the log-odds ratio function are not linear. This often happens under the generalized linear model with non-canonical link function or other complex models. Weakly identifiable parameters are often difficult to estimate due to the lack of information on such parameters in the observed data. Examples in the next section demonstrate such difficulties.

5.5 Analysis of misspecified odds ratio model

5.5.1 Misspecification in the semiparametric odds ratio model

In the semiparametric odds ratio model, the odds ratio function is modeled parametrically. The parametric model may be misspecified. There are also cases where the correct model is known but can be hard to estimate, while an approximate odds ratio model can be easier to estimate. See for example, the genetic applications discussed in the later sections of this chapter. In either case, a wrong model for the odds ratio function may be used for estimation. Since the baseline functions are nonparametric, no model misspecification of that part is involved.

More specifically, consider the univariate semiparametric odds ratio model. We know from the development in Chapter 2 that, when the positivity condition holds, the true model always accepts an odds ratio representation. Let the true density for the data be

$$p_0(y \mid x) = \frac{\eta_0(y;x)p_0(y \mid x_0)}{\int \eta_0(y;x)p_0(y \mid x_0)dy}, \tag{5.9}$$

and the parametric model specified be

$$p(y \mid x) = \frac{\eta(y;x \mid \theta)p(y \mid x_0)}{\int \eta(y;x \mid \theta)p(y \mid x_0)dy}, \tag{5.10}$$

where $\theta \in \Theta$. If there is no $\theta \in \Theta$ such that $\eta(y;x \mid \theta) = \eta_0(y;x)$ for all (y,x), the parametric odds ratio function is misspecified. For example,

$$\log \eta_0(y;x) = (y-y_0)\{(x-x_0)+(x-x_0)^2\},$$

may be misspecified as

$$\log \eta(y;x \mid \theta) = \theta(y-y_0)(x-x_0).$$

More complex misspecifications occur in the analysis of the biased sampling designs. Additional concrete examples can be found in §5.6.

For the misspecified semiparametric odds ratio model, we can use either the permutation likelihood approach or the semiparametric likelihood approach to estimating the parameter θ and $p(y \mid x_0)$. However, the variance of the estimator can no longer be consistently estimated by the inverse of the observed data information matrix. A sandwich type of variance estimates need to be implemented.

5.5.2 The permutation likelihood approach under misspecified models

The permutation likelihood score under the misspecified model is

$$\dot{l}(\theta) = \sum_{i=1}^{n} \frac{\partial}{\partial \theta} \log \eta(Y_{i1};\cdots;Y_{iG} \mid \theta) - E_\sigma \left\{ \sum_{i=1}^{n} \frac{\partial}{\partial \theta} \log \eta(Y_{\sigma_1(i)1};\cdots;Y_{\sigma_{G-1}(i)(G-1)};Y_{iG} \mid \theta) \right\},$$

where E_σ is computed under the distribution

$$P(\sigma \mid Y_{ord}) = \frac{\prod_{i=1}^{n} \eta(Y_{\sigma_1(i)1};\cdots;Y_{\sigma_{G-1}(i)(G-1)};Y_{iG} \mid \theta)}{\sum_{\tau \in \mathcal{P}_n(G-1)} \prod_{i=1}^{n} \eta(Y_{\tau_1(i)1};\cdots;Y_{\tau_{G-1}(i)(G-1)};Y_{iG} \mid \theta)}.$$

Assume that the maximum permutation likelihood estimator $\hat{\theta}$ satisfying $\dot{l}(\hat{\theta}) = 0$ converges to θ_0^*. The likelihood score can be expanded around θ_0^* as

$$0 = \dot{l}(\hat{\theta}) \approx \dot{l}(\theta_0^*) + \ddot{l}(\theta_0^*)(\hat{\theta} - \theta_0^*)$$

The maximum permutation likelihood estimator is

$$\sqrt{n}(\hat{\theta} - \theta_0^*) \approx \left\{ -\frac{1}{n}\ddot{l}(\theta_0^*) \right\}^{-1} \frac{1}{\sqrt{n}}\dot{l}(\theta_0^*).$$

The asymptotic variance of the maximum permutation likelihood estimator can be obtained by the sandwich formula as

$$\left[E\left\{ -\frac{1}{n}\ddot{l}(\theta_0^*) \right\} \right]^{-1} \mathrm{Var}\left\{ \frac{1}{\sqrt{n}}\dot{l}(\theta_0^*) \right\} \left[E\left\{ -\frac{1}{n}\ddot{l}(\theta_0^*) \right\} \right]^{-1}.$$

Because the information identity

$$\mathrm{E}\left\{ \frac{1}{\sqrt{n}}\dot{l}(\theta_0^*) \right\}^{\otimes 2} = E\left\{ -\frac{1}{n}\ddot{l}(\theta_0^*) \right\}$$

does not hold anymore under the misspecified model, the asymptotic variance does not reduce to

$$\left[E\left\{ -\frac{1}{n}\ddot{l}(\theta_0^*) \right\} \right]^{-1}.$$

While $E\{-\ddot{l}(\theta_0^*)/n\}$ can be consistently estimated by the observed data information matrix of the permutation likelihood under the misspecified odds ratio model, $\mathrm{Var}\{\dot{l}(\theta_0^*)/\sqrt{n}\}$ needs to be estimated separately when the odds ratio model is misspecified.

To find an estimate for the variance, note that

$$E_0\left\{ \sum_{i=1}^n \xi(i, \theta_0^*) \right\} = E_0\left\{ \sum_{i=1}^n \delta(i, \theta_0^*) \right\},$$

where

$$\begin{aligned} \xi(i,\theta) &= \frac{\partial}{\partial\theta} \log \eta(Y_{i1}; \cdots; Y_{iG} \mid \theta) \\ \delta(i,\theta) &= E_\sigma\left\{ \frac{\partial}{\partial\theta} \log \eta(Y_{\sigma_1(i)1}; \cdots; Y_{\sigma_{G-1}(i)(G-1)}; Y_{iG} \mid \theta) \right\}. \end{aligned}$$

It follows from $\dot{l}(\theta_0^*) = \sum_{i=1}^n \{\xi(i,\theta_0^*) - \delta(i,\theta_0^*)\}$ that

$$\begin{aligned} E_0\left\{ \frac{1}{\sqrt{n}}\dot{l}(\theta_0^*) \right\}^{\otimes 2} &= \frac{1}{n}E_0\left\{ \sum_{i=1}^n \xi(i,\theta_0^*) - E_0\left\{ \sum_{i=1}^n \xi(i,\theta_0^*) \right\} \right\}^{\otimes 2} \\ &-\frac{1}{n}E_0\left(\left[\sum_{i=1}^n \xi(i,\theta_0^*) - E_0\left\{ \sum_{i=1}^n \xi(i,\theta_0^*) \right\} \right] \left[\sum_{i=1}^n \delta(i,\theta_0^*) - E_0\left\{ \sum_{i=1}^n \delta(i,\theta_0^*) \right\} \right]^t \right) \\ &-\frac{1}{n}E_0\left(\left[\sum_{i=1}^n \delta(i,\theta_0^*) - E_0\left\{ \sum_{i=1}^n \delta(i,\theta_0^*) \right\} \right] \left[\sum_{i=1}^n \xi(i,\theta_0^*) - E_0\left\{ \sum_{i=1}^n \xi(i,\theta_0^*) \right\} \right]^t \right) \\ &+\frac{1}{n}E_0\left[\sum_{i=1}^n \delta(i,\theta_0^*) - E_0\left\{ \sum_{i=1}^n \delta(i,\theta_0^*) \right\} \right]^{\otimes 2}. \end{aligned}$$

The first variance on the right-hand side of the foregoing equation can be rewritten as

$$\frac{1}{n}\sum_{i=1}^{n} E_0\left[\xi(i,\theta_0^*) - E_0\left\{\xi(i,\theta_0^*)\right\}\right]^{\otimes 2}.$$

For the estimation of the rest terms, let $\tilde{h}(Y_{i_1 1},\cdots,Y_{i_G G})$ be an $L^2(P_0)$-function and

$$h(Y_{i_1 1},\cdots,Y_{i_G G}) = \tilde{h}(Y_{i_1 1},\cdots,Y_{i_G G}) - E_0\left\{\tilde{h}(Y_{i_1 1},\cdots,Y_{i_G G})\right\}.$$

It follows that $E_0\left\{h(Y_{i_1 1},\cdots,Y_{i_G G})\right\} = 0$, for all $(i_1,\cdots,i_G)$. Let $\sigma = (\sigma_1,\cdots,\sigma_{G-1})$. For a fixed j, let

$$\delta_h(j) \;=\; E_\sigma\left\{h(Y_{\sigma_1(j)1},\cdots,Y_{\sigma_{G-1}(j)(G-1)},Y_{jG})\right\}.$$

We then have $E_0\{\delta_h(j)\} = 0$. For any $L^2(P_0)$-function $u(Y_{i1},\cdots,Y_{iG})$,

$$\begin{aligned}
&E_0\left\{u(Y_{i1},\cdots,Y_{iG})\sum_{j=1}^{n}\delta_h(j)\right\}\\
&= \sum_{(i_1,\cdots,i_G)=i^{(0)}} E_0\left[u(Y_{i1},\cdots,Y_{iG})h(Y_{i_1 1},\cdots,Y_{i_G G})P\{\sigma(i_G)=(i_1,\cdots,i_{G-1})\}\right]\\
&\quad + \sum_{(i_1,\cdots,i_G)=i^{(1)}} E_0\left[u(Y_{i1},\cdots,Y_{iG})h(Y_{i_1 1},\cdots,Y_{i_G G})P\{\sigma(i_G)=(i_1,\cdots,i_{G-1})\}\right]\\
&\quad + \sum_{k=2}^{G}\sum_{(i_1,\cdots,i_G)=i^{(k)}} E_0\left[u(Y_{i1},\cdots,Y_{iG})h(Y_{i_1 1},\cdots,Y_{i_G G})P\{\sigma(i_G)=(i_1,\cdots,i_{G-1})\}\right].
\end{aligned}$$

where $(i_1,\cdots,i_G) = i^{(k)}$ denotes that, among the indices $i_1,\cdots,i_G$, exact k indices are equal to i. Note that

$$P\{\sigma(i_G)=(i_1,\cdots,i_{G-1})\} \le \frac{C}{n^{G-1}}.$$

It follows that

$$\sum_{k=2}^{G}\sum_{(i_1,\cdots,i_G)=i^{(k)}} E_0\left[u(Y_{i1},\cdots,Y_{iG})h(Y_{i_1 1},\cdots,Y_{i_G G})P\{\sigma(i_G)=(i_1,\cdots,i_{G-1})\}\right] = O(n^{-1}).$$

Let

$$g_{hs}(Y_{is}) = \sum_{i_s=i,(i_1,\cdots,i_G)=i^{(1)}} h(Y_{i_1 1},\cdots,Y_{i_G G})P\{\sigma(i_G)=(i_1,\cdots,i_{G-1})\}.$$

It follows that

$$\begin{aligned}
&\sum_{(i_1,\cdots,i_G)=i^{(1)}} E_0\left[u(Y_{i1},\cdots,Y_{iG})h(Y_{i_1 1},\cdots,Y_{i_G G})P\{\sigma(i_G)=(i_1,\cdots,i_{G-1})\}\right]\\
&\quad = E_0\left\{u(Y_{i1},\cdots,Y_{iG})\sum_{s=1}^{G} g_h(Y_{is})\right\}.
\end{aligned}$$

For $(i_1,\cdots,i_G) = i^{(0)}$, it can be derived that

$$\sum_{(i_1,\cdots,i_G)=i^{(0)}} E_0\left[u(Y_{i1},\cdots,Y_{iG})h(Y_{i_1 1},\cdots,Y_{i_G G})P\{\sigma(i_G)=(i_1,\cdots,i_{G-1})\}\right] \;=\; O(n^{-1}),$$

when $E_0\{u(Y_{i1},\cdots,Y_{iG})\}=0$. It follows by combining the results that

$$E_0\left\{u(Y_{i1},\cdots,Y_{iG})\sum_{i=1}^{n}\delta_h(i)\right\}=\sum_{s=1}^{G}E_0\left\{u(Y_{i1},\cdots,Y_{iG})g_{hs}(Y_{is})\right\}+O(n^{-1}).$$

Furthermore, for two $L^2(P_0)$-functions h_1,h_2, we have

$$\begin{aligned}&\frac{1}{n}E_0\left\{\sum_{i=1}^{n}\delta_{h_1}(i)\sum_{j=1}^{n}\delta_{h_2}(j)\right\}\\&=\frac{1}{n}\sum_{i_1,\cdots,i_G}\sum_{j_1,\cdots,j_G}E_0\Big[h_1(Y_{i_11},\cdots,Y_{i_GG})P\{\sigma(i_G)=(i_1,\cdots,i_{G-1})\}\\&\qquad\times h_2(Y_{j_11},\cdots,Y_{j_GG})P\{\sigma(j_G)=(j_1,\cdots,j_{G-1})\}\Big].\end{aligned}$$

For those terms where $(i_1,\cdots,i_G,j_1,\cdots,j_G)$ have at least two pairs of equal indices (including one common index in the two pairs), the sum of those terms is of $O(1/n)$. For the remaining terms, they either have no common index among $(i_1,\cdots,i_G,j_1,\cdots,j_G)$, or have one common index only. Among those having one common index, they are either among $(i_1,\cdots,i_G)$, or $(j_1,\cdots,j_G)$, or between $(i_1,\cdots,i_G)$ and $(j_1,\cdots,j_G)$. The sum of terms in the last case is

$$\frac{1}{n}\sum_{i=1}^{n}E_0\left\{\sum_{s=1}^{G}g_{h_1}(Y_{is})\sum_{t=1}^{G}g_{h_2}(Y_{it})\right\}.$$

For the rest terms, it can be shown that their sum is of the order $O(1/n)$ when $E\{\delta_{h_1}(i)\}=0$ and $E\{\delta_{h_2}(i)\}=0$, for all $i=1,\cdots,n$.

To apply the general result to estimate the rest terms in $E_0\left\{\frac{1}{\sqrt{n}}\dot{l}(\theta_0^*)\right\}^{\otimes 2}$, let

$$h_s(Y_1,\cdots,Y_G\mid\theta)=\frac{\partial}{\partial\theta_s}\log\eta(Y_1;\cdots;Y_G\mid\theta)-E_0\left\{\frac{\partial}{\partial\theta_s}\log\eta(Y_1;\cdots;Y_G\mid\theta)\right\},$$

for $s=1,\cdots,p$, where p is the dimension of parameter θ. Let

$$g^*_{h_sk}(Y_{ik},\theta)=\sum_{\{(i_1,\cdots,i_G)\mid i_k=i\}}h_s(Y_{i_11},\cdots,Y_{i_GG}\mid\theta)P\left\{\sigma(i_G)=(i_1,\cdots,i_{G-1})\mid\theta\right\}.$$

Let $h(Y_{i1},\cdots,Y_{iG}\mid\theta)=(h_1(Y_1,\cdots,Y_{iG}\mid\theta),\cdots,h_G(Y_{i1},\cdots,Y_{iG}\mid\theta))^t$, $g^*_{hk}(Y_{ik},\theta)=(g^*_{h_1}(Y_{ik},\theta),\cdots,g^*_{h_G}(Y_{ik},\theta))^t$. It follows that

$$\frac{1}{n}E_0\left(\left[\sum_{i=1}^{n}\xi(i,\theta_0^*)-E_0\left\{\sum_{i=1}^{n}\xi(i,\theta_0^*)\right\}\right]\left[\sum_{i=1}^{n}\delta(i,\theta_0^*)-E_0\left\{\sum_{i=1}^{n}\delta(i,\theta_0^*)\right\}\right]^t\right)$$

is asymptotically equivalent to

$$\frac{1}{n}\sum_{i=1}^{n}E_0\left(\left[\xi(i,\theta_0^*)-\frac{1}{n}\sum_{j=1}^{n}E_0\{\xi(j,\theta_0^*)\}\right]\left[\sum_{k=1}^{G}g^*_{hk}(Y_{ik},\theta_0^*)-\frac{1}{n}\sum_{s=1}^{n}E_0\left\{\sum_{k=1}^{G}g^*_{hk}(Y_{sk},\theta_0^*)\right\}\right]^t\right).$$

Furthermore,

$$\frac{1}{n}E_0\left(\left[\sum_{i=1}^{n}\delta(i,\theta_0^*)-E_0\left\{\sum_{i=1}^{n}\delta(i,\theta_0^*)\right\}\right]\left[\sum_{i=1}^{n}\xi(i,\theta_0^*)-E_0\left\{\sum_{i=1}^{n}\xi(i,\theta_0^*)\right\}\right]^t\right)$$

is asymptotically equivalent to

$$\frac{1}{n}\sum_{i=1}^{n} E_0\left[\sum_{k=1}^{G} g^*_{hk}(Y_{ik},\theta^*_0) - \frac{1}{n}\sum_{s=1}^{n} E_0\left\{\sum_{k=1}^{G} g^*_{hk}(Y_{sk},\theta^*_0)\right\}\right]^{\otimes 2}.$$

When put together, the variance $E_0\left\{\frac{1}{\sqrt{n}}\dot{l}(\theta^*_0)\right\}^{\otimes 2}$ is asymptotically equivalent to

$$\frac{1}{n}\sum_{i=1}^{n} E_0\left\{\xi(i,\theta^*_0) - E_0\{\xi(i,\theta^*_0)\} - \sum_{k=1}^{G} \hat{g}^*_{hk}(Y_{ik},\theta^*_0)\right\}^{\otimes 2}.$$

The exact evaluation of $\sum_{k=1}^{G} g^*_{hk}(Y_{ik},\theta^*_0)$ is computationally impractical. The Monte Carlo sampling algorithm can be used. Let $\sigma^{(1)},\cdots,\sigma^{(N)}$ be a sample of permutations following the distribution

$$P\{\sigma^{(t)}(i_k) = (i_1,\cdots,i_G) \mid \hat{\theta}\},$$

where $\sigma^{(t)} = (\sigma^{(t)}_1,\cdots,\sigma^{(t)}_G)$ and $\sigma^{(t)}_k$ is set to identity permutation. We approximate $g^*_{h_sk}(Y_{ik},\theta^*_0)$ by

$$\hat{g}^*_{h_sk}(Y_{ik},\hat{\theta}) = \frac{1}{N}\sum_{t=1}^{N} \hat{h}_s(Y_{\sigma^{(t)}_1(i)1},\cdots,Y_{\sigma^{(t)}_G(i)G} \mid \hat{\theta}).$$

where

$$\hat{h}_s(Y_{\sigma^{(t)}_1(i)1},\cdots,Y_{\sigma^{(t)}_G(i)G} \mid \hat{\theta}) = h_s(Y_{\sigma^{(t)}_1(i)1},\cdots,Y_{\sigma^{(t)}_G(i)G} \mid \hat{\theta}) - \frac{1}{n}\sum_{j=1}^{n}\xi(i,\hat{\theta}).$$

Let $\hat{g}^*(Y_{ik},\hat{\theta}) = (\hat{g}^*_{h_1}(Y_{ik},\hat{\theta}),\cdots,\hat{g}^*_{h_G}(Y_{ik},\hat{\theta}))^t$. A consistent estimate of the variance is

$$\frac{1}{n}\sum_{i=1}^{n}\left\{\xi(i,\hat{\theta}) - \frac{1}{n}\sum_{j=1}^{n}\xi(i,\hat{\theta}) - \sum_{k=1}^{G}\hat{g}^*_{hk}(Y_{ik},\hat{\theta})\right\}^{\otimes 2}.$$

5.5.3 The maximum semiparametric likelihood approach under misspecified model

For data $(Y_i, X_i), i = 1,\cdots,n$ generated from (5.9), model (5.10) can be fit to the data. The model fitted is asymptotically determined by

$$(\theta^*_0, h^*_0) = \arg\min_{\theta,h} E_0\left\{-\log\frac{p(Y \mid X,\theta,h)}{p_0(Y \mid X)}\right\}$$

The model $p(y \mid x,\theta^*_0,h^*_0)$ is closest to the true model in Kullback-Leibler divergence. Estimation of the parameters (θ^*_0,h^*_0) can be obtained by maximizing the log-likelihood (3.18). The algorithm for maximizing the likelihood discussed in the previous sections can be applied. The asymptotic behavior of the maximum likelihood estimator is different from that with the correct model.

Chen et al (2015)[35] studied the maximum likelihood estimator for the misspecified semiparametric odds ratio model. The basic results obtained show that the maximum likelihood estimator for (θ^*_0,h^*_0) is consistent and θ^*_0 estimator is asymptotically normally distributed. The variance of the maximum likelihood estimator is no longer consistently estimated by the inverse of the observed data profile information matrix. A sandwich estimate can be used to estimate the variance consistently. More details can be found in Chen et al (2015)[35] and Chapter 10.

5.6 Applications

In this section, we apply the results obtained so far to a number of epidemiological problems with biased sampling designs.

5.6.1 Gene-environment independence in case-control genetic association study

Gene-environment interaction is one of the important topics in modern genetic association studies of disease phenotypes. Let D denote the disease status, G denote the genotype, and E denote the environment risk factor. Let Z denote other confounders to be adjusted. The population model for studying the gene-environment interaction is often conveniently set by the following logistic regression model

$$\log \frac{p(D=1 \mid G,E,Z)}{p(D=0 \mid G,E,Z)} = \beta_0 + \beta_1 G + \beta_2 E + \beta_3 GE + \beta_4 Z.$$

A case-control study design is usually employed because of the rare disease phenotype. Conventional statistical analysis of the case-control design performs a prospective analysis to estimate the interaction parameter β_3 as well as other parameters. Such an analysis is straightforward.

In many practical situations, it may be reasonable to assume genotype and the environment risk factor are independent in the general population. This information may be used to gain efficiency in estimating the interaction effect β_3. To see how such gains can occur, consider the conditional decomposition of the population model for the data,

$$p(D,G,E,Z) = p(D \mid G,E,Z)p(G \mid Z)p(E \mid Z)p(Z).$$

The odds ratio function with reference $(D_0, G_0, E_0, Z_0) = (0,0,0,0)$ of the joint density is

$$\eta(D;G;E;Z) = \eta\{D;(G,E,Z)\}\eta\{G;(E,Z) \mid D_0\}\eta(E;Z \mid D_0,G_0),$$

where

$$\begin{aligned}
\log \eta\{D;(G,E,Z)\} &= \beta_1 DG + \beta_2 DE + \beta_3 DGE + \beta_4 DZ,\\
\log \eta\{G;(E,Z) \mid D_0\} &= \log \frac{p(D_0 \mid G,E,Z)p(D_0 \mid G_0,E_0,Z_0)}{p(D_0 \mid G,E_0,Z_0)p(D_0 \mid G_0,E,Z)} + \log \frac{p(G \mid Z)p(G_0 \mid Z_0)}{p(G \mid Z_0)p(G_0 \mid Z)},\\
\log \eta(E;Z \mid D_0,G_0) &= \log \frac{p(D_0 \mid G_0,E,Z)p(D_0 \mid G_0,E_0,Z_0)}{p(D_0 \mid G_0,E,Z_0)p(D_0 \mid G_0,E_0,Z)} + \log \frac{p(E \mid Z)p(E_0 \mid Z_0)}{p(E \mid Z_0)p(E_0 \mid Z)}.
\end{aligned}$$

From the population model for the disease phenotype, we have

$$\begin{aligned}
\log & \frac{p(D_0 \mid G,E,Z)p(D_0 \mid G_0,E_0,Z_0)}{p(D_0 \mid G,E_0,Z_0)p(D_0 \mid G_0,E,Z)}\\
&= \frac{\{1+\exp(\beta_0+\beta_1 G)\}\{1+\exp(\beta_0+\beta_2 E+\beta_4 Z)\}}{\{1+\exp(\beta_0+\beta_1 G+\beta_2 E+\beta_3 GE+\beta_4 Z)\}\{1+\exp(\beta_0)\}},\\
\log \frac{p(D_0 \mid G_0,E,Z)p(D_0 \mid G_0,E_0,Z_0)}{p(D_0 \mid G_0,E,Z_0)p(D_0 \mid G_0,E_0,Z)} &= \frac{\{1+\exp(\beta_0+\beta_2 E)\}\{1+\exp(\beta_0+\beta_4 Z)\}}{\{1+\exp(\beta_0+\beta_2 E+\beta_4 Z)\}\{1+\exp(\beta_0)\}}.
\end{aligned}$$

Under the case-control sampling design, the sampling density becomes

$$p(D,G,E,Z \mid S=1) = \frac{\eta(D;G;E;Z)dP_0(D \mid S=1)dP_0(G)dP_0(E)dP_0(Z)}{\int \eta(D;G;E;Z)dP_0(D \mid S=1)dP_0(G)dP_0(E)dP_0(Z)},$$

where integral with respect to a discrete variable is understood as summation, and $P_0(D \mid S=1)$, $P_0(G)$, $P_0(E)$, and $P_0(Z)$ denote respectively $P(D \mid G_0, E_0, Z_0, S=1)$, $P(G \mid D_0, E_0, Z_0)$, $P_0(E \mid D_0, G_0, Z_0)$, and $P_0(Z \mid D_0, G_0, E_0)$.

From the identifiability results discussed in previous sections, under the case-control design, β_0 may be identifiable when $\beta_3 \neq 0$, or $\beta_3 = 0$ and one of the following conditions holds: $\beta_1 \neq 0$ and $\beta_2 \neq 0$, or $\beta_1 \neq 0$ and $\beta_4 \neq 0$, or $\beta_2 \neq 0$ and $\beta_4 \neq 0$. However, the identifiability of β_0 is weak in the sense that β_0 become non-identifiable when the conditions do not hold.

To facilitate the parameter estimation, we need to further model the odds ratio functions of $p(G \mid Z)$ and $p(E \mid Z)$ respectively. One convenient choice is to model the log-odds ratio function by bilinear functions, i.e.,

$$
\begin{aligned}
\log \frac{p(G \mid Z)p(G_0 \mid Z_0)}{p(p \mid Z_0)p(G_0 \mid Z)} &= \gamma_1 GZ, \\
\log \frac{p(E \mid Z)p(E_0 \mid Z_0)}{p(E \mid Z_0)p(E_0 \mid Z)} &= \gamma_2 EZ.
\end{aligned}
$$

With these models in place, we can in principle maximize the likelihood

$$
\prod_{i=1}^{n} p(D_i, G_i, E_i, Z_i \mid S_i = 1)
$$

to obtain estimators for parameters $(\beta_0, \beta_1, \beta_2, \beta_3, \beta_4, \gamma_1, \gamma_2)$. In practice, however, the maximization can be difficult when conditions on the non-identifiability of β_0 are close to being true. For example, $\beta_1, \beta_2, \beta_3$ are all close to 0, or $\beta_2, \beta_3, \beta_4$ are all close to 0. To circumvent the problem, approximation to the odds ratio functions has been proposed. The approximation relies on the following Taylor expansion,

$$
\log\left(1+e^{\beta_0+\Delta}\right) \approx \log\left(1+e^{\beta_0}\right) + \frac{e^{\beta_0}}{1+e^{\beta_0}}\Delta + \frac{e^{\beta_0}}{\left(1+e^{\beta_0}\right)^2}\frac{\Delta^2}{2} + \frac{e^{\beta_0}\left(1-e^{\beta_0}\right)}{\left(1+e^{\beta_0}\right)^3}\frac{\Delta^3}{6}. \quad (5.11)
$$

The two complex odds ratio functions can be approximated respectively by

$$
\begin{aligned}
\log \frac{p(D_0 \mid G, E, Z)p(D_0 \mid G_0, E_0, Z_0)}{p(D_0 \mid G, E_0, Z_0)p(D_0 \mid G_0, E, Z)} &\approx -\left\{G_0\beta_3 + p_0(1-p_0)\beta_1\beta_2\right\} GE \\
&\quad -p_0(1-p_0)\beta_1\beta_4 GZ, \\
\log \frac{p(D_0 \mid G_0, E, Z)p(D_0 \mid G_0, E_0, Z_0)}{p(D_0 \mid G_0, E, Z_0)p(D_0 \mid G_0, E_0, Z)} &\approx -p_0(1-p_0)\beta_2\beta_4 EZ.
\end{aligned}
$$

With those approximation, the odds ratio function for the case-control sample can be approximated by

$$
\begin{aligned}
\eta(D;G;E;Z) \approx\; & \beta_1 DG + \beta_2 DE + \beta_3 DGE + \beta_4 DZ - \left\{G_0\beta_3 + p_0(1-p_0)\beta_1\beta_2\right\} GE \\
& \left\{\gamma_1 - p_0(1-p_0)\beta_1\beta_4\right\} GZ + \left\{\gamma_2 - p_0(1-p_0)\beta_2\beta_4\right\} EZ.
\end{aligned}
$$

Using this approximation allows for identification of β_0, but efficiency gain in estimating $\beta_1, \beta_2, \beta_3$ is not materialized because three additional parameters $(\beta_0, \gamma_1, \gamma_2)$ need to be estimated. On the other hand, if we known G and Z and/or E and Z are independent in the general population, efficiency gain in estimating parameter $\beta = (\beta_0, \cdots, \beta_4)$ can be achieved by further analysis. For example, suppose we know that $\gamma_1 = \gamma_2 = 0$. The likelihood analysis provides us with estimates for $g(\beta) = (\beta_1, \beta_2, \beta_3, \beta_4, p_0\beta_3 + p_0(1-p_0)\beta_1\beta_2, -p_0(1-p_0)\beta_1\beta_4, -p_0(1-p_0)\beta_2\beta_4)$ and their variance-covariance estimates. Let $U = (U_1, \cdots, U_7)$

denote the maximum likelihood estimators and V denote the covariance matrix. We estimate β by

$$\hat{\beta} = \arg\min \{U - g(\beta)\}^t V^{-1} \{U - g(\beta)\}.$$

The variance of $\hat{\beta}$ can be estimated by

$$\left[\left\{\frac{\partial g(\hat{\beta})}{\partial \beta}\right\}^t V^{-1} \frac{\partial g(\hat{\beta})}{\partial \beta}\right]^{-1}.$$

If neither γ_1 nor γ_2 is known to be 0, we need to use higher orders of expansion in the approximation of complex odds ratio functions first and then apply the weighted least-square combination approach to obtaining β estimator.

5.6.2 Secondary traits analysis in a case-control genetic association study

Secondary outcome analysis is a problem frequently raised in case-control design. We discuss the topic here in the context of a genetic association study. The basic problem has a close relationship with the problem discussed in the previous section although it appears differently from the previous problem. Let D denote the case-control status, Y denote the secondary outcome, G denote the genotype, and Z denote the covariate to be adjusted. The primary objective is to examine the association of the secondary outcome Y and the genotype G. Let the population model for the secondary outcome be $p(y \mid G, Z)$. The population model for the disease outcome is assumed to follow a logistic regression model, i.e.,

$$\log \frac{p(D=1 \mid Y, G, Z)}{p(D=0 \mid Y, G, Z)} = \beta_0 + \beta_1 Y + \beta_2 G + \beta_3 Z.$$

The population model for the data can be represented as

$$p(D, Y, G, Z) = p(D \mid Y, G, Z) p(Y \mid G, Z) p(G \mid Z) p(Z).$$

Similar to the discussion in the previous section, the odds ratio function for the joint density is

$$\eta(D; Y; G; Z) = \eta\{D; (Y, G, Z)\}\eta\{Y; (G, Z) \mid D_0\}\eta(G; Z \mid D_0, Y_0),$$

where the reference point is $(D_0, Y_0, G_0, Z_0) = (0, 0, 0, 0)$ and

$$\begin{aligned}
\log \eta\{D; (Y, G, Z)\} &= \beta_1 DY + \beta_2 DG + \beta_3 DZ, \\
\log \eta\{Y; (G, Z) \mid D_0\} &= \log \frac{p(D_0 \mid Y, G, Z) p(D_0 \mid Y_0, G_0, Z_0)}{p(D_0 \mid Y, G_0, Z_0) p(D_0 \mid Y_0, G, Z)} \\
&\quad + \log \frac{p(Y \mid G, Z) p(Y_0 \mid G_0, Z_0)}{p(Y \mid G_0, Z_0) p(Y_0 \mid G, Z)}, \\
\log \eta(G; Z \mid D_0, Y_0) &= \log \frac{p(D_0 \mid Y_0, G, Z) p(D_0 \mid Y_0, G_0, Z_0)}{p(D_0 \mid Y_0, G, Z_0) p(D_0 \mid Y_0, G_0, Z)} + \log \frac{p(G \mid Z) p(G_0 \mid Z_0)}{p(G \mid Z_0) p(G_0 \mid Z)}.
\end{aligned}$$

The odds ratio functions related to the population disease model have the following forms

$$\begin{aligned}
\log \frac{p(D_0 \mid Y, G, Z) p(D_0 \mid Y_0, G_0, Z_0)}{p(D_0 \mid Y, G_0, Z_0) p(D_0 \mid Y_0, G, Z)} &= \frac{\{1 + \exp(\beta_0 + \beta_1 Y)\}\{1 + \exp(\beta_0 + \beta_2 G + \beta_3 Z)\}}{\{1 + \exp(\beta_0 + \beta_1 Y + \beta_2 G + \beta_3 Z)\}\{1 + \exp(\beta_0)\}}, \\
\log \frac{p(D_0 \mid Y_0, G, Z) p(D_0 \mid Y_0, G_0, Z_0)}{p(D_0 \mid Y_0, G, Z_0) p(D_0 \mid Y_0, G_0, Z)} &= \frac{\{1 + \exp(\beta_0 + \beta_2 G)\}\{1 + \exp(\beta_0 + \beta_3 Z)\}}{\{1 + \exp(\beta_0 + \beta_2 G + \beta_3 Z)\}\{1 + \exp(\beta_0)\}}.
\end{aligned}$$

Under the case-control sampling design, the sampling density becomes

$$p(D,Y,G,Z \mid S=1) = \frac{\eta(D;Y;G;Z)dP_0(D \mid S=1)dP_0(Y)dP_0(G)dP_0(Z)}{\int \eta(D;Y;G;Z)dP_0(D \mid S=1)dP_0(Y)dP_0(G)dP_0(Z)},$$

where integral with respect to a discrete variable is understood as summation, and $P_0(D \mid S=1)$, $P_0(Y)$, $P_0(G)$, and $P_0(Z)$ denote respectively $P(D \mid Y_0, G_0, Z_0, S=1)$, $P(Y \mid D_0, G_0, Z_0)$, $P_0(G \mid D_0, Y_0, Z_0)$, and $P_0(Z \mid D_0, Y_0, G_0)$. If any two of the three parameters $(\beta_1, \beta_2, \beta_3)$ are 0, β_0 is not identifiable. Otherwise, β_0 is potentially identifiable. The identifiability of β_0 is weak and is therefore need special care in estimation when any two of the three parameters $(\beta_1, \beta_2, \beta_3)$ are close to 0.

In the secondary outcome analysis, the key population model is $p(Y \mid G, Z)$. When Y is a binary outcome, a logistic regression model is usually employed. When Y is continuous, a normal linear model is usually employed. In either case, the model can be represented as an odds ratio model with odds ratio function

$$\log \eta\{Y;(G,Z)\} = \gamma_1 YG + \gamma_2 YZ.$$

The population level odds ratio function between G and Z can be modeled by

$$\log \eta(G;Z) = \gamma_3 GZ.$$

With these models for odds ratio functions, the likelihood can in principle be maximized to obtain the parameter estimators. To avoid possible difficulty in the likelihood maximization, we can again use the approximation (5.11) to simplify the odds ratio functions. Using the Taylor expansion, the two complex odds ratio functions can be approximated respectively by

$$\begin{aligned}
\log \frac{p(Y_0 \mid Y,G,Z)p(D_0 \mid Y_0,G_0,Z_0)}{p(D_0 \mid Y,G_0,Z_0)p(D_0 \mid Y_0,G,Z)} &\approx -p_0(1-p_0)\beta_1\beta_2 YG - p_0(1-p_0)\beta_1\beta_3 YZ,\\
\log \frac{p(D_0 \mid Y_0,G,Z)p(D_0 \mid Y_0,G_0,Z_0)}{p(D_0 \mid Y_0,G,Z_0)p(D_0 \mid Y_0,G_0,Z)} &\approx -p_0(1-p_0)\beta_2\beta_3 GZ.
\end{aligned}$$

The odds ratio function for the joint population model can be approximated by

$$\begin{aligned}
\eta(D;Y;G;Z) &\approx \beta_1 DY + \beta_2 DG + \beta_3 DZ + \{\gamma_1 - p_0(1-p_0)\beta_1\beta_2\} YG\\
&\quad \{\gamma_2 - p_0(1-p_0)\beta_1\beta_3\} YZ + \{\gamma_3 - p_0(1-p_0)\beta_2\beta_3\} GZ.
\end{aligned}$$

Using this approximation, we are able to estimate parameters

$$(\beta_1, \beta_2, \beta_3, \gamma_1 - p_0(1-p_0)\beta_1\beta_2, \gamma_2 - p_0(1-p_0)\beta_1\beta_3, \gamma_3 - p_0(1-p_0)\beta_2\beta_3).$$

Note, however, that our primary interest in the secondary analysis is to estimate γ_1 or $(\gamma_1, \gamma_2, \gamma_3)$. We are not able to find a γ_1 estimator without further restrictions because the three additional parameters we estimated contain 4 additional unknown parameters. This problem can be resolved if we can assume $\gamma_3 = 0$. A more general alternative way to avoid this problem is to use higher order Taylor expansion to approximate the odds ratio functions. One such higher order expansion has

$$\begin{aligned}
\log \frac{p(Y_0 \mid Y,G,Z)p(D_0 \mid Y_0,G_0,Z_0)}{p(D_0 \mid Y,G_0,Z_0)p(D_0 \mid Y_0,G,Z)} &\approx -p_0(1-p_0)\beta_1\beta_2 YG - p_0(1-p_0)\beta_1\beta_3 YZ,\\
&\quad -\frac{G_0(1-p_0)(1-2p_0)}{2}\{\beta_1 Y(\beta_2 G + \beta_3 Z)^2 + (\beta_1 Y)^2(\beta_2 G + \beta_3 Z)\}\\
\log \frac{p(D_0 \mid Y_0,G,Z)p(D_0 \mid Y_0,G_0,Z_0)}{p(D_0 \mid Y_0,G,Z_0)p(D_0 \mid Y_0,G_0,Z)} &\approx -p_0(1-p_0)\beta_2\beta_3 GZ\\
&\quad -\frac{G_0(1-p_0)(1-2p_0)}{2}\{(\beta_2 G)^2\beta_3 Z + \beta_2 G(\beta_3 Z)^2\}
\end{aligned}$$

The new approximation for the joint odds ratio function is

$$
\begin{aligned}
\eta(D;Y;G;Z) \approx\ & \beta_1 DY + \beta_2 DG + \beta_3 DZ + \{\gamma_1 - p_0(1-p_0)\beta_1\beta_2\} YG \\
& \{\gamma_2 - p_0(1-p_0)\beta_1\beta_3\} YZ + \{\gamma_3 - p_0(1-p_0)\beta_2\beta_3\} GZ \\
& -\frac{1}{2}p_0(1-p_0)(1-2p_0)\Big(\beta_1\beta_2^2 YG^2 + 2\beta_1\beta_2\beta_3 YGZ \\
& +\beta_1\beta_3^2 YZ^2 + \beta_1^2\beta_2 Y^2 G + \beta_1^2\beta_3 Y^2 Z + \beta_2^2\beta_3 G^2 Z + \beta_2\beta_3^2 GZ^2\Big)
\end{aligned}
$$

The likelihood analysis yields estimates for

$$
\begin{aligned}
g(\beta,\gamma) =\ & (\beta_1, \beta_2, \beta_3, \gamma_1 - p_0(1-p_0)\beta_1\beta_2, \gamma_2 - p_0(1-p_0)\beta_1\beta_3, \gamma_3 - p_0(1-p_0)\beta_2\beta_3, \\
& -\frac{1}{2}p_0(1-p_0)(1-2p_0)\beta_1\beta_2^2, -p_0(1-p_0)(1-2p_0)\beta_1\beta_2\beta_3, \\
& -\frac{1}{2}p_0(1-p_0)(1-2p_0)\beta_1\beta_3^2, -\frac{1}{2}p_0(1-p_0)(1-2p_0)\beta_1^2\beta_2, \\
& -\frac{1}{2}p_0(1-p_0)(1-2p_0)\beta_1^2\beta_3, -\frac{1}{2}p_0(1-p_0)(1-2p_0)\beta_2^2\beta_3, -\frac{1}{2}p_0(1-p_0)(1-2p_0)\beta_2\beta_3^2\Big),
\end{aligned}
$$

and their variance-covariance estimates. Again, let U denote the maximum likelihood estimators and V denote the covariance matrix. Parameter estimators for $\beta = (\beta_0, \beta_1, \beta_2, \beta_3)$ and $\gamma = (\gamma_1, \gamma_2, \gamma_3)$ can be obtained by

$$
(\hat\beta, \hat\gamma) = \arg\min_{(\beta,\gamma)} \{U - g(\beta,\gamma)\}^t V^{-1} \{U - g(\beta,\gamma)\}.
$$

The variance matrix of $(\hat\beta, \hat\gamma)$ can be estimated by

$$
\left[\left\{\frac{\partial g(\hat\beta,\hat\gamma)}{\partial(\beta,\gamma)}\right\}^t V^{-1}\frac{\partial g(\hat\beta,\hat\gamma)}{\partial(\beta,\gamma)}\right]^{-1}.
$$

5.6.3 Case-only design to study interactions

Case-only design is used in epidemiology to study the interaction of two risk factors when they are independent in the population. This design uses similar principle for efficiency gain by exploiting gene-environment independence in the general population. Let Y_1 denote the case status, Y_2 denote the risk factor, Y_3 be the stratification variable, and Y_4 be the confounding variables to be controlled. The population model is

$$
p(y_1, y_2, y_3, y_4) = p(y_1 \mid y_2, y_3, y_4)p(y_2 \mid y_4)p(y_3 \mid y_4)p(y_4).
$$

Since only cases are sampled, i.e.,

$$
P(S = 1 \mid y_1, y_2, y_3, y_4) = p_0 1_{\{y_1=1\}},
$$

the sampling model can be reduced to the population model conditional on case status, i.e.,

$$
p(y_2, y_3, y_4 \mid y_1 = 1) = \frac{\eta(y_2; y_3; y_4 \mid y_1 = 1)\prod_{k=2}^4 dF_k(y_k \mid y_{j0}, j \neq 1, k, y_1 = 1)}{\int \eta(y_2; y_3; y_4 \mid y_1 = 1)\prod_{k=2}^4 dF_k(y_k \mid y_{j0}, j \neq 1, k, y_1 = 1)}.
$$

The joint odds ratio function can be written as

$$
\eta(y_2; y_3; y_4 \mid y_1 = 1) = \eta\{y_2; (y_3, y_4) \mid y_1 = 1)\eta(y_3; y_4 \mid y_2 = y_{20}, y_1 = 1),
$$

where

$$\log \eta\{y_2;(y_3,y_4)\mid y_1=1) = \log \frac{p(y_2,y_3,y_4\mid y_1=1)p(y_{20},y_{30},y_{40}\mid y_1=1)}{p(y_2,y_{30},y_{40}\mid y_1=1)p(y_{20},y_3,y_4\mid y_1=1)}$$
$$= \log \frac{p(y_1=1\mid y_2,y_3,y_4)p(y_1=1\mid y_{20},y_{30},y_{40})}{p(y_1=1\mid y_{20},y_3,y_4)p(y_1=1\mid y_2,y_{30},y_{40})} + \log \frac{p(y_2\mid y_4)p(y_{20}\mid y_{40})}{p(y_{20}\mid y_4)p(y_2\mid y_{40})},$$
$$\eta(y_3;y_4\mid y_2=y_{20},y_1=1) = \log \frac{p(y_3,y_4\mid y_{20},y_1=1)p(y_{30},y_{40}\mid y_{20},y_1=1)}{p(y_3,y_{40}\mid y_{20},y_1=1)p(y_{30},y_4\mid y_{20},y_1=1)}$$
$$= \log \frac{p(y_1=1\mid y_{20},y_3,y_4)p(y_1=1\mid y_{20},y_{30},y_{40})}{p(y_1=1\mid y_{20},y_{30},y_4)p(y_1=1\mid y_{20},y_3,y_{40})} + \log \frac{p(y_3\mid y_4)p(y_{30}\mid y_{40})}{p(y_{30}\mid y_4)p(y_3\mid y_{40})}.$$

Suppose that the population outcome model is

$$p(y_1=1\mid y_2,y_3,y_4) = \frac{\exp(\beta_0+\beta_1y_2+\beta_2y_3+\beta_3y_2y_3+\beta_4y_4)}{1+\exp(\beta_0+\beta_1y_2+\beta_2y_3+\beta_3y_2y_3+\beta_4y_4)}.$$

Let the reference point be $(y_{20},y_{30},y_{40})=(0,0,0)$. It follows that

$$\log \frac{p(y_1=1\mid y_2,y_3,y_4)p(y_1=1\mid y_{20},y_{30},y_{40})}{p(y_1=1\mid y_{20},y_3,y_4)p(y_1=1\mid y_2,y_{30},y_{40})} = \beta_3y_2y_3$$
$$+\log \frac{\{1+\exp(\beta_0+\beta_1y_2)\}\{1+\exp(\beta_0+\beta_2y_3+\beta_4y_4)\}}{\{1+\exp(\beta_0+\beta_1y_2+\beta_2y_3+\beta_3y_2y_3+\beta_4y_4)\}\{1+\exp(\beta_0)\}},$$
$$\log \frac{p(y_1=1\mid y_{20},y_3,y_4)p(y_1=1\mid y_{20},y_{30},y_{40})}{p(y_1=1\mid y_{20},y_{30},y_4)p(y_1=1\mid y_{20},y_3,y_{40})}$$
$$= \log \frac{\{1+\exp(\beta_0+\beta_1y_2)\}\{1+\exp(\beta_0+\beta_4y_4)\}}{\{1+\exp(\beta_0+\beta_2y_3+\beta_4y_4)\}\{1+\exp(\beta_0)\}}.$$

Using Taylor expansion, the following approximation is obtained,

$$\log \frac{p(y_1=1\mid y_2,y_3,y_4)p(y_1=1\mid y_{20},y_{30},y_{40})}{p(y_1=1\mid y_{20},y_3,y_4)p(y_1=1\mid y_2,y_{30},y_{40})} \approx$$
$$\{(1-p_0)\beta_3-p_0(1-p_0)\beta_1\beta_2\}\,y_2y_3-p_0(1-p_0)\beta_1\beta_4y_2y_4,$$
$$\log \frac{p(y_1=1\mid y_{20},y_3,y_4)p(y_1=1\mid y_{20},y_{30},y_{40})}{p(y_1=1\mid y_{20},y_{30},y_4)p(y_1=1\mid y_{20},y_3,y_{40})} \approx -p_0(1-p_0)\beta_2\beta_4y_3y_4.$$

If we further assume that

$$\log \frac{p(y_2\mid y_4)p(y_{20}\mid y_{40})}{p(y_{20}\mid y_4)p(y_2\mid y_{40})} = \gamma_1y_2y_4, \tag{5.12}$$
$$\log \frac{p(y_3\mid y_4)p(y_{30}\mid y_{40})}{p(y_{30}\mid y_4)p(y_3\mid y_{40})} = \gamma_2y_3y_4, \tag{5.13}$$

the joint conditional odds ratio function can be approximated by

$$\log \eta\{y_2;(y_3,y_4)\mid y_1=1) \approx \{(1-p_0)\beta_3-p_0(1-p_0)\beta_1\beta_2\}\,y_2y_3$$
$$+(\gamma_1-p_0(1-p_0)\beta_1\beta_4)y_2y_4+(\gamma_3-p_0(1-p_0)\beta_2\beta_4)y_3y_4.$$

When $p_0\approx 0$,

$$\log \eta\{y_2;(y_3,y_4)\mid y_1=1) \approx \beta_3y_2y_3+\gamma_1y_2y_4+\gamma_3y_3y_4.$$

The odds ratio model analysis yields an estimator of the interaction parameter β_3.

If either the odds ratio model (5.12) or (5.13) can be misspecified, the conditional odds ratio model representation may be used, i.e.,

$$p(y_2, y_3 \mid y_4, y_1 = 1) = \frac{\eta(y_2; y_3 \mid y_4, y_1 = 1) dF_2(y_2 \mid y_{30}, y_4, y_1 = 1) dF_3(y_3 \mid y_{20}, y_4, y_1 = 1)}{\int \eta(y_2; y_3 \mid y_4, y_1 = 1) dF_2(y_2 \mid y_{30}, y_4, y_1 = 1) dF_3(y_3 \mid y_{20}, y_4, y_1 = 1)}$$

where

$$\begin{aligned} \log \eta(y_2; y_3 \mid y_4, y_1 = 1) &= \log \frac{p(y_1 = 1 \mid y_2, y_3, y_4) p(y_1 = 1 \mid y_{20}, y_{30}, y_4)}{p(y_1 = 1 \mid y_{20}, y_3, y_4) p(y_1 = 1 \mid y_2, y_{30}, y_4)} \\ = \beta_3 y_2 y_3 + \log & \frac{\{1 + \exp(\beta_0 + \beta_1 y_2 + \beta_4 y_4)\} \{1 + \exp(\beta_0 + \beta_2 y_3 + \beta_4 y_4)\}}{\{1 + \exp(\beta_0 + \beta_1 y_2 + \beta_2 y_3 + \beta_3 y_2 y_3 + \beta_4 y_4)\} \{1 + \exp(\beta_0 + \beta_4 y_4)\}}. \end{aligned}$$

Applying the expansion (5.11) around $\beta_0 + \beta_4 y_4$ to obtain

$$\log \eta(y_2; y_3 \mid y_4, y_1 = 1) \approx \{1 - p(y_4)\} \beta_3 y_2 y_3,$$

where

$$p(y_4) = \frac{\exp(\beta_0 + \beta_4 y_4)}{1 + \exp(\beta_0 + \beta_4 y_4)}.$$

The inference approaches to conditional odds ratio function discussed in Chapter 4 can be applied.

5.7 Summary and literature

Case-control and matched case-control designs have a long history in epidemiological studies (Breslow, 1996)[14]. The literature on the methodological development for such designs was provided in Chapter 1. Parameter identifiability for regression model other than the logistic model for case-control data was discussed in Weinberg and Wacholder(1993)[197] and Rabinowitz (1997)[146]. More general outcome-dependent sampling designs (Chen, 2003; Weaver and Zhou, 2005; Didelez et al, 2010) [26, 196, 56], such as the extreme-value sampling design (Chen and Li, 2011)[33], the length-biased sampling design (Simon, 1980; Jones, 1991; Wang, 1996)[166, 85, 193], and the two-stage sampling design (White, 1982; Lawless et al, 1999; Breslow et al, 2003)[198, 93, 13], have been studied in the statistical literature. The analysis of secondary outcomes in case-control data (Lee et al, 1997; Jiang et al, 2006)[95, 83] is extensively examined in genetic data analysis (Lin and Zeng, 2009; Chen et al, 2013; Ma and Carroll, 2016, among others) [105, 38, 117]. The case-only design (Piegorsch et al, 1994)[140] is for studying interactions of two independent risk factors such as gene-environment (Schmidt and Schaid, 1999; Albert et al, 2001; Tchetgen Tchetgen and Robins, 2010; VanderWeele et al, 2010)[160, 1, 178, 190]. Gene-environment independence is also exploited for gaining estimation efficiency in genetic studies (Umbach and Weinberg, 1997; Chatterjee and Carroll, 2005; Chen et al, 2008; Chen and Chen, 2011)[186, 23, 42, 32].

5.8 Exercises

1. In the analysis of case-control data with non-standard models, one issue encountered is that parameter identifiability requires certain condition to hold. Consider the probit model for case-control data

$$P(Y = 1 \mid X) = \Phi(\beta_0 + \beta_1 X).$$

Identifiability of β_0 requires $\beta_1 \neq 0$ in addition to the conditions on X. Such identifiability is termed weakly identifiable.

   (a) Give an intuitive explanation on why β_0 is not identifiable when $\beta_1 = 0$.

   (b) When $\beta_1 \neq 0$, but is close to 0, the maximum likelihood estimator can be computationally challenging to obtain. Perform a simulation study to demonstrate this problem.

   (c) Suggest a solution to this problem and test it in your simulation study.

2. In the previous problem, Let $Y = 1$ denote a subject having a disease. If we know the disease prevalence rate in the general population, it can be used to help resolve the weak identifiability and increase estimation efficiency of the model parameters.

   (a) Assume the disease prevalence rate $p = P(Y = 1)$ is known. Find the maximum likelihood estimator given the case-control data incorporating the additional information.

   (b) Perform a simulation study of the new estimator and compare with the results in the previous problem.

   (c) Demonstrate that the asymptotic variance of the estimator is smaller than the estimator in the previous problem.

   (d) If we do not know the exact disease prevalence rate. Instead, we have a separate simple random sample from the general population on Y. How the estimator of the parameters are changed?

3. One variant of the last question in the previous problem is that a missing covariate problem. Suppose that $(Y_i, X_i), i = 1, \cdots, n$ is a random sample from the general population, where X is subject to missing values and the missingness of X depends only Y only. Let the observed data be $(Y_i, X_i), i = 1, \cdots, m < n$ and $Y_i, i = m + 1, \cdots, n$.

   (a) Show that, by discarding $Y_i, i = m + 1, \cdots, n$, $(Y_i, X_i), i = 1, \cdots, m < n$ can be treated as a case-control sample if Y is a binary variable. Find the parameter estimator using the complete-cases with the adjustment of the biased sampling.

   (b) Find the likelihood estimator when all the observed data are used.

   (c) Perform a simution study comparing the estimators from the two approaches.

   (d) If the missing data probability is under the control of the data collector, the sampling approach is termed two-stage sampling design. Discuss when such a sampling design may be used.

4. The weak identifiability also occurs in the analysis of case-control data exploiting gene-environment independence for efficiency gain. Let

$$\text{logit}P(Y = 1 \mid G, E) = \beta_0 + \beta_1 G + \beta_2 E + \beta_{12} G \times E.$$

The gene-environmrnt independence assumption states that

$$p(G, E) = P(G)P(E).$$

Let $(Y_i, G_i, E_i), i = 1, \cdots, n$ be a case-control study sample.

(a) Find the retrospective likelihood based on $(p(G, E \mid Y)$ for the parameters taking into consideration the gene-environment independence assumption.

(b) Show that the parameter β_0 is identifiable only when $(\beta_1, \beta_2, \beta_{12}) \neq 0$.

(c) The traditional maximum prospective conditional likelihood based on $P(Y \mid G, E, S = 1)$ in general yields less efficient estimator than that from the maximum retrospective likelihood, where $S = 1$ means a case-control sample.

(d) Show that the maximum prospective likelihood estimator based on $P(Y, G \mid E, S = 1)$ is equivalent to the maximum retrospective likelihood estimator.

(e) Show that when both G and E are binary variables, the maximum retrospective likelihood estimator does not have any efficiency gain over the traditional maximum prospective likelihood estimator.

(f) If we assume the disease is sparse, i.e., $P(Y = 1) \approx 0$, demonstrate that the maximum approximate likelihood estimator is more efficient than that obtained in the previous question.

5. For the length-biased sampling design, show that if the conditional distribution of Y given X does not have a finite mean, the sampling design is impossible. An example is

$$P(y \mid x) = \frac{2}{\pi} \frac{\exp(\beta_0 + \beta_1 x) 1_{\{y \geq 0\}}}{1 + \{y \exp(\beta_0 + \beta_1 x)\}^2}.$$

Show also that a truncated length-biased sampling design is possible, i. e., discard the sample if the length is greater than a pre-specified bound.

6. In the case-only sampling design to study the interaction of risk factors, we generally assume both sparse disease and independence of the two risk factors. For example, in studying gene-environment interactions using case-only sample, we assume gene-environment independence. Suppose we have a case-only sample plus an additional random sample on the risk factors from the general population.

(a) Show that the gene-environment independence assumption is not needed in identifying and estimating the interactions.

(b) Find the maximum likelihood estimator of the interaction parameter.

(c) Propose an inference method for the interaction.

6

Application to Test of Conditional Independence

6.1 An introduction to test of conditional independence

One of the topics of lasting interest in statistics is testing hypotheses on independence, both conditional and unconditional. The problem of testing unconditional independence is a special case of testing conditional independence. Testing conditional independence can be formulated as follows. Let Y_1, Y_2, and X be vectors of random variables. The problem is equivalent to testing independence of Y_1 and Y_2 given X. Testing conditional independence in general is a hard problem as shown in Shah and Peters (2019)[164] because there is no test having nontrivial power against all alternatives when X is continuous. There are, however, a vast number of tests having been proposed in the literature targeting at different situations.

Traditionally, testing conditional independence is often cast into the regression modeling framework. The odds ratio modeling framework provides a more flexible and unified platform for discussing the independence testing with few restrictions. Assume that the conditions for the odds ratio decomposition of the joint conditional density of (Y_1, Y_2) given X hold. That is,

$$f(y_1, y_2 \mid x) = \frac{\eta(y_1; y_2 \mid x)p(y_1 \mid y_{20}, x)p(y_2 \mid y_{10}, x)}{\int\int \eta(y_1; y_2 \mid x)p(y_1 \mid y_{20}, x)p(y_2 \mid y_{10}, x)dy_1dy_2}. \tag{6.1}$$

The conditional independence of Y_1 and Y_2 given X is equivalent to $\eta(y; x \mid z) \equiv 1$. Therefore, testing the conditional independence is the same as testing the hypothesis

$$H_0 : \eta(y; x \mid z) = 1.$$

Suppose that the observed data $(Y_{i1}, Y_{i2}, X_i), i = 1, \cdots, n$ are independent identically distributed across unit i. In testing the hypothesis of conditional independence of Y_{i1} and Y_{i2} given X_i for all i, we usually need to know $\eta(y_1; y_2 \mid x)$ to design a test to achieve the maximum power. In practice, however, we usually do not know the function form of $\eta(y_1; y_2 \mid x)$ if the null hypothesis is not true. Let $d(y_1, y_2, x)$ be a vector of measurable functions such that $E_0\{||d(Y_1, Y_2, X)||_2^2\} < \infty$. Let

$$h(y_1, y_2, x) = d(y_1, y_2, x) - d(y_{10}, y_2, x) - d(y_1, y_{20}, x) + d(y_{10}, y_{20}, x). \tag{6.2}$$

Without loss of generality, we may assume a model for $\eta(y_1; y_2 \mid x)$ under the alternative hypothesis that

$$\log \eta(y_1; y_2 \mid x, \theta) = \theta^t h(y_1, y_2, x), \tag{6.3}$$

Assume also that

$$\int\int \eta(y_1; y_2 \mid x, \theta)f(y_1 \mid y_{20}, x)f(y_2 \mid y_{10}, x)dy_1dy_2 < \infty.$$

Note that if the foregoing integral is not finite, we can always add a high order term of

DOI: 10.1201/9781351049757-6

θ such that the integral is finite while keeping the likelihood score at $\theta = 0$ the same as $h(y_1, y_2, x)$. Testing the hypothesis of conditional independence may be based on different distributional components of the data. Among them, the test may use the likelihood derived from the conditional distributions of Y_1 given Y_2, X, or the conditional distribution of Y_2 given Y_1, X, or the joint conditional distribution of Y_1, Y_2 given X. More details are given in the next few sections.

6.2 Likelihood ratio tests

6.2.1 Prospective, retrospective, and joint likelihood ratio tests

There are three ways to construct likelihood ratio tests for the conditional independence using likelihoods derived from (6.1). The first, termed the prospective likelihood ratio test, is based on the conditional distribution of Y_1 given Y_2 and X. For observed data $(Y_{i1}, Y_{i2}, X_i), i = 1, \cdots, n$. the likelihood is

$$L_p\{\theta, F(y_1 \mid y_{20}, x)\} = \prod_{i=1}^{n} \frac{\exp\{\theta h(Y_{i1}, Y_{i2}, X_i)\} dF(Y_{i1} \mid y_{20}, X_i)}{\int \exp\{\theta h(y_1, Y_{i2}, X_i)\} dF(y_1 \mid y_{20}, X_i)}.$$

The conditional distribution function $F(y_1 \mid y_{20}, x)$ can usually be further decomposed as

$$dF(y_1 \mid y_{20}, x) = \frac{\eta_1(y_1; x \mid y_{20}) dF(y_1 \mid y_{20}, x_0)}{\int \eta_1(y_1; x \mid y_{20}) dF(y_1 \mid y_{20}, x_0)}.$$

When this decomposition is taken into consideration, the prospective likelihood becomes

$$L_p\{\theta, F(y_1 \mid y_{20}, x)\} = \prod_{i=1}^{n} \frac{\exp\{\theta h(Y_{i1}, Y_{i2}, X_i)\} \eta_1(Y_{i1}; X_i \mid y_{20}) dF(Y_{i1} \mid y_{20}, x_0)}{\int \exp\{\theta h(y_1, Y_{i2}, X_i)\} \eta_1(y_1; X_i \mid y_{20}) dF(y_1 \mid y_{20}, x_0)}.$$

To avoid the curse of dimensionality, we model $\eta_1(Y_{i1}; X_i \mid y_{20})$ by a parametric function denote by $\eta_1(Y_{i1}; X_i \mid \alpha)$, where α is the unknown parameter. The functional parameter $F_1(y_1) = F(y_1 \mid y_{20}, x_0)$ can be left unspecified. The prospective likelihood becomes

$$L_p(\theta, \alpha, F_1) = \prod_{i=1}^{n} \frac{\exp\{\theta h(Y_{i1}, Y_{i2}, X_i)\} \eta_1(Y_{i1}; X_i \mid \alpha) dF_1(Y_{i1})}{\int \exp\{\theta h(y_1, Y_{i2}, X_i)\} \eta_1(y_1; X_i \mid \alpha) dF_1(y_1)}.$$

Let $(\hat{\theta}, \hat{\alpha}, \hat{F}_1)$ be the maximum likelihood estimator from $L_p(\theta, \alpha, F_1)$. Let $(\hat{\alpha}_0, \hat{F}_{10})$ be the maximum likelihood estimator when $\theta = 0$. Testing the conditional independence hypothesis is equivalent to testing $\theta = 0$, which can be performed by the likelihood ratio test as

$$2 \log \frac{L_p(\hat{\theta}, \hat{\alpha}, \hat{F}_1)}{L_p(0, \hat{\alpha}_0, \hat{F}_{10})} \sim \chi^2_{df},$$

where the degree of freedom $df = dim(\theta)$.

The second test, termed the retrospective likelihood ratio test, is based on the conditional distribution of Y_2 given Y_1 and X derived from (6.1). The retrospective likelihood is

$$L_r\{\theta, F(y_2 \mid y_{10}, x)\} = \prod_{i=1}^{n} \frac{\exp\{\theta h(Y_{i1}, Y_{i2}, X_i)\} dF(Y_{i2} \mid y_{10}, X_i)}{\int \exp\{\theta h(Y_{i1}, y_2, X_i)\} dF(y_2 \mid y_{10}, X_i)}.$$

By using the odds ratio decomposition for the conditional distribution function $dF(y_2 \mid y_{10}, x)$, i.e.,

$$dF(y_2 \mid y_{10}, x) = \frac{\eta_2(y_2; x \mid y_{10})dF(y_2 \mid y_{10}, x_0)}{\int \eta_2(y_2; x \mid y_{10})dF(y_2 \mid y_{10}, x_0)},$$

the retrospective likelihood can be rewritten as

$$L_r\{\theta, F(y_2 \mid y_{10}, x)\} = \prod_{i=1}^{n} \frac{\exp\{\theta h(Y_{i1}, Y_{i2}, X_i)\}\eta_2(Y_{i2}; X_i \mid y_{10})dF(Y_{i2} \mid y_{10}, x_0)}{\int \exp\{\theta h(Y_{i1}, y_2, X_i)\}\eta_2(y_2; X_i \mid y_{10})dF(y_2 \mid y_{10}, x_0)}.$$

To avoid the curse of dimensionality, we model $\eta_2(Y_{i2}; X_i \mid y_{10})$ by a parametric function denote by $\eta_2(Y_{i2}; X_i \mid \beta)$ with the unknown parameter β. Let $F_2(y_2) = F(y_2 \mid y_{10}, x_0)$ be left unspecified. The retrospective likelihood appears as

$$L_r(\theta, \beta, F_2) = \prod_{i=1}^{n} \frac{\exp\{\theta h(Y_{i1}, Y_{i2}, X_i)\}\eta_2(Y_{i2}; X_i \mid \beta)dF_2(Y_{i2})}{\int \exp\{\theta h(Y_{i1}, y_2, X_i)\}\eta_2(y_2; X_i \mid \beta)dF_2(y_2)}.$$

Let $(\hat{\theta}, \hat{\beta}, \hat{F}_2)$ be the maximum likelihood estimator from $L_r(\theta, \beta, F_2)$. Let $(\hat{\beta}_0, \hat{F}_{20})$ be the maximum likelihood estimator when $\theta = 0$. Testing the conditional independence can now be performed by the likelihood ratio test as

$$2 \log \frac{L_r(\hat{\theta}, \hat{\beta}, \hat{F}_2)}{L_r(0, \hat{\beta}_0, \hat{F}_{20})} \sim \chi^2_{df},$$

where the degree of freedom $df = dim(\theta)$.

The third test, termed the joint likelihood ratio test, is based on the joint conditional distribution of (Y_1, Y_2) given X in (6.1). The joint conditional likelihood is

$$L_j\{\theta, F(y_1 \mid y_{10}, x), F(y_2 \mid y_{10}, x)\} =$$
$$\prod_{i=1}^{n} \frac{\exp\{\theta h(Y_{i1}, Y_{i2}, X_i)\}dF(Y_{i1} \mid y_{20}, X_i)dF(Y_{i2} \mid y_{10}, X_i)}{\int\int \exp\{\theta h(Y_{i1}, y_2, X_i)\}dF(y_1 \mid y_{20}, X_i)dF(y_2 \mid y_{10}, X_i)}.$$

By using both the odds ratio decompositions for the conditional distribution functions $dF(y_1 \mid y_{20}, x)$ and $dF(y_2 \mid y_{10}, x)$ as specified earlier, the joint likelihood can be expressed as

$$L_j\{\theta, \eta_1, \eta_2, F_1, F_2\} =$$
$$\prod_{i=1}^{n} \frac{\exp\{\theta h(Y_{i1}, Y_{i2}, X_i)\}\eta_1(Y_{i1}; X_i \mid y_{20})\eta_2(Y_{i2}; X_i \mid y_{10})dF_1(Y_{i1})dF_2(Y_{i2})}{\int \exp\{\theta h(y_1, y_2, X_i)\}\eta_1(y_1; X_i \mid y_{10})\eta_2(y_2; X_i \mid y_{10})dF_1(y_1)dF_2(y_2)}.$$

Again, to avoid the curse of dimensionality, $\eta_1(Y_{i1}; X_i \mid y_{20})$ and $\eta_2(Y_{i2}; X_i \mid y_{10})$ are respectively modeled by parametric models $\eta_1(Y_{i1}; X_i \mid \alpha)$ and $\eta_2(Y_{i2}; X_i \mid \beta)$. The joint likelihood becomes

$$L_j(\theta, \alpha, \beta, F_1, F_2) =$$
$$\prod_{i=1}^{n} \frac{\exp\{\theta h(Y_{i1}, Y_{i2}, X_i)\}\eta_1(Y_{i1}; X_i \mid \alpha)\eta_2(Y_{i2}; X_i \mid \beta)dF_1(Y_{i1})dF_2(Y_{i2})}{\int \exp\{\theta h(y_1, y_2, X_i)\}\eta_1(y_1; X_i \mid \alpha)\eta_2(y_2; X_i \mid \beta)dF_1(y_1)dF_2(y_2)}.$$

Let $(\hat{\theta}, \hat{\alpha}, \hat{\beta}, \hat{F}_1, \hat{F}_2)$ be the maximum likelihood estimator from $L_j(\theta, \alpha, \beta, F_1, F_2)$. Let $(\hat{\alpha}_0, \hat{\beta}_0, \hat{F}_{10}, \hat{F}_{20})$ be the maximum likelihood estimator when $\theta = 0$. Testing for independence can also be performed by the likelihood ratio test as

$$2 \log \frac{L_j(\hat{\theta}, \hat{\alpha}, \hat{\beta}, \hat{F}_1, \hat{F}_2)}{L_j(0, \hat{\alpha}_0, \hat{\beta}_0, \hat{F}_{10}, \hat{F}_{20})} \sim \chi^2_{df},$$

where the degree of freedom $df = dim(\theta)$.

6.2.2 Likelihood ratio tests based on permutation likelihoods

Test of independence can also be derived from the permutation likelihood approach. Let $\mathbf{Y}_1 = (Y_{11}, \cdots, Y_{n1})$, $\mathbf{Y}_2 = (Y_{12}, \cdots, Y_{n2})$, and $\mathbf{X} = (X_1, \cdots, X_n)$. Let $\mathbf{Y}_1^{ord} = (Y_{(1)1}, \cdots, Y_{(n)1})$ and $\mathbf{Y}_2^{ord} = (Y_{(1)2}, \cdots, Y_{(n)2})$ be respectively the order statistics for $Y_{11}, \cdots, Y_{n1}$ and $Y_{12}, \cdots, Y_{n2}$.

The conditional density for $\mathbf{Y}_1$ given $\mathbf{Y}_2$ and $\mathbf{X}$ is

$$p(\mathbf{Y}_1 \mid \mathbf{Y}_2, \mathbf{X}) = \prod_{i=1}^{n} \frac{\exp\{\theta h(Y_{i1}, Y_{i2}, X_i)\}\eta_1(Y_{i1}; X_i \mid y_{20}) dF(Y_{i1} \mid y_{20}, x_0)}{\int \exp\{\theta h(y_1, Y_{i2}, X_i)\}\eta_1(y_1; X_i \mid y_{20}) dF(y_1 \mid y_{20}, x_0)}.$$

The prospective permutation likelihood is derived from the conditioning on the order statistic $\mathbf{Y}_1^{ord}$ of $\mathbf{Y}_1$ as

$$p(\mathbf{Y}_1 \mid \mathbf{Y}_1^{ord}, \mathbf{Y}_2, \mathbf{X}) = \frac{\prod_{i=1}^{n} \exp\{\theta h(Y_{i1}, Y_{i2}, X_i)\}\eta_1(Y_{i1}; X_i \mid y_{20})}{\sum_{\sigma \in \mathcal{P}_n} \prod_{i=1}^{n} \exp\{\theta h(Y_{\sigma(i)1}, Y_{i2}, X_i)\}\eta_1(Y_{\sigma(i)1}; X_i \mid y_{20})},$$

where $\mathcal{P}_n$ is the collection of all permutations of $\{1, 2, \cdots, n\}$. When η_1 is parametrically modeled by $\eta_1(y_1; x \mid \alpha)$, the prospective permutation likelihood becomes

$$L_{pp}(\theta, \alpha) = \frac{\prod_{i=1}^{n} \exp\{\theta h(Y_{i1}, Y_{i2}, X_i)\}\eta_1(Y_{i1}; X_i \mid \alpha)}{\sum_{\sigma \in \mathcal{P}_n} \prod_{i=1}^{n} \exp\{\theta h(Y_{\sigma(i)1}, Y_{i2}, X_i)\}\eta_1(Y_{\sigma(i)1}; X_i \mid \alpha)}.$$

Let $(\hat{\theta}, \hat{\alpha})$ be the maximum prospective permutation likelihood estimator and $\hat{\alpha}_0$ be the maximum prospective permutation likelihood estimator when $\theta = 0$. The test of independence can be performed using the permutation likelihood ratio test as

$$2 \log \frac{L_{pp}(\hat{\theta}, \hat{\alpha})}{L_{pp}(0, \hat{\alpha}_0)} \sim \chi^2_{df},$$

where $df = dim(\theta)$.

The retrospective permutation likelihood is based on the conditional density for $\mathbf{Y}_2$ given $\mathbf{Y}_1$ and $\mathbf{X}$, i.e.,

$$p(\mathbf{Y}_2 \mid \mathbf{Y}_1, \mathbf{X}) = \prod_{i=1}^{n} \frac{\exp\{\theta h(Y_{i1}, Y_{i2}, X_i)\}\eta_2(Y_{i2}; X_i \mid y_{10}) dF(Y_{i2} \mid y_{10}, x_0)}{\int \exp\{\theta h(Y_{i1}, y_2, X_i)\}\eta_2(y_2; X_i \mid y_{10}) dF(y_2 \mid y_{10}, x_0)}.$$

The corresponding permutation likelihood is derived from the conditioning on the order statistic $\mathbf{Y}_1^{ord}$ of $\mathbf{Y}_1$ as

$$p(\mathbf{Y}_2 \mid \mathbf{Y}_1, \mathbf{Y}_2^{ord}, \mathbf{X}) = \prod_{i=1}^{n} \frac{\exp\{\theta h(Y_{i1}, Y_{i2}, X_i)\}\eta_2(Y_{i2}; X_i \mid y_{10}) dF(Y_{i2} \mid y_{10}, x_0)}{\int \exp\{\theta h(Y_{i1}, y_2, X_i)\}\eta_2(y_2; X_i \mid y_{10}) dF(y_2 \mid y_{10}, x_0)}.$$

When η_2 is parametrically modeled by $\eta_2(y_2; x \mid \beta)$, the retrospective permutation likelihood becomes

$$L_{rp}(\theta, \beta) = \frac{\prod_{i=1}^{n} \exp\{\theta h(Y_{i1}, Y_{i2}, X_i)\}\eta_2(Y_{i2}; X_i \mid \beta)}{\sum_{\sigma \in \mathcal{P}_n} \prod_{i=1}^{n} \exp\{\theta h(Y_{\sigma(i)1}, Y_{i2}, X_i)\}\eta_2(Y_{\sigma(i)2}; X_i \mid \beta)}.$$

Let $(\hat{\theta}, \hat{\beta})$ be the maximum prospective permutation likelihood estimator and $\hat{\beta}_0$ be the maximum prospective permutation likelihood estimator when $\theta = 0$. The test of independence can be performed using the retrospective permutation likelihood ratio test as

$$2 \log \frac{L_{rp}(\hat{\theta}, \hat{\beta})}{L_{rp}(0, \hat{\beta}_0)} \sim \chi^2_{df},$$

where $df = dim(\theta)$.

The joint permutation likelihood is based on

$$p(\mathbf{Y}_1, \mathbf{Y}_2 \mid \mathbf{X}) = \prod_{i=1}^{n} \frac{\exp\{\theta h(Y_{i1}, Y_{i2}, X_i)\}\eta_1(Y_{i1}; X_i \mid y_{20})\eta_2(Y_{i2}; X_i \mid y_{10})dF_1(Y_{i1})dF_2(Y_{i2})}{\int \exp\{\theta h(y_1, y_2, X_i)\}\eta_1(y_1; X_i \mid y_{10})\eta_2(y_2; X_i \mid y_{10})dF_1(y_1)dF_2(y_2)}.$$

The corresponding permutation likelihood is derived from the conditioning on the order statistic $\mathbf{Y}_1^{ord}$ and $\mathbf{Y}_2^{ord}$ as

$$p(\mathbf{Y}_1, \mathbf{Y}_2 \mid \mathbf{Y}_1^{ord}, \mathbf{Y}_2^{ord}, \mathbf{X}) = \frac{\prod_{i=1}^{n} \exp\{\theta h(Y_{i1}, Y_{i2}, X_i)\}\eta_1(Y_{i1}; X_i \mid y_{20})\eta_2(Y_{i2}; X_i \mid y_{10})}{\sum_{\sigma,\tau\in\mathcal{P}_n} \prod_{i=1}^{n} \exp\{\theta h(Y_{\sigma(i)1}, Y_{\tau(i)2}, X_i)\}\eta_1(Y_{\sigma(i)1}; X_i \mid y_{20})\eta_2(Y_{\tau(i)2}; X_i \mid y_{10})}.$$

When η_1 and η_2 are respectively modeled parametrically by $\eta_1(y_1; x \mid \alpha)$ and $\eta_2(y_2; x \mid \beta)$, the joint permutation likelihood becomes

$$L_{jp}(\theta, \alpha, \beta) = \frac{\prod_{i=1}^{n} \exp\{\theta h(Y_{i1}, Y_{i2}, X_i)\}\eta_1(Y_{i1}; X_i \mid \alpha)\eta_2(Y_{i2}; X_i \mid \beta)}{\sum_{\sigma,\tau\in\mathcal{P}_n} \prod_{i=1}^{n} \exp\{\theta h(Y_{\sigma(i)1}, Y_{\tau(i)2}, X_i)\}\eta_1(Y_{\sigma(i)1}; X_i \mid \alpha)\eta_2(Y_{\tau(i)2}; X_i \mid \beta)}.$$

Let $(\hat{\theta}, \hat{\alpha}, \hat{\beta})$ be the maximum joint permutation likelihood estimator and $(\hat{\alpha}_0, \hat{\beta}_0)$ be the maximum joint permutation likelihood estimator when $\theta = 0$. The test of independence can be performed using the joint permutation likelihood ratio test as

$$2 \log \frac{L_{jp}(\hat{\theta}, \hat{\alpha}, \hat{\beta})}{L_{jp}(0, \hat{\alpha}_0, \hat{\beta}_0)} \sim \chi^2_{df},$$

where $df = dim(\theta)$.

6.3 Likelihood score tests

6.3.1 Prospective, retrospective, and joint likelihood score tests

The likelihood ratio tests have counterparts in the form of the likelihood score tests. Let

$$h(\mathbf{Y}_1, \mathbf{Y}_2, \mathbf{X}) = \sum_{i=1}^{n} h(Y_{i1}, Y_{i2}, X_i).$$

The likelihood score from the prospective likelihood is

$$W_{nA}(h) = h(\mathbf{Y}_1, \mathbf{Y}_2, \mathbf{X}) - E_0\left\{h(\mathbf{Y}_1, \mathbf{Y}_2, \mathbf{X}) \mid \mathbf{Y}_2, \mathbf{X}\right\}.$$

Under the null hypothesis, the conditional density at a fixed condition $dF(y_1 \mid y_{20}, x)$ in the density $dF(y_1 \mid y_2, x)$ reduces to $dF(y_1 \mid x)$, which needs to be estimated in order to compute the conditional expectations in the test statistic. One way to estimate this function is to model it again using the semiparametric odds ratio model as

$$dF(y_1 \mid x) = \frac{\eta_1(y_1; x \mid \alpha)dF_1(y_1)}{\int \eta_1(y_1; x \mid \alpha)dF_1(y_1)}.$$

Unknown parameters in the density can be estimated by

$$(\hat{\alpha}, \hat{F}_1) = \arg\max_{\alpha, F_1} \prod_{i=1}^{n} \frac{\eta_1(Y_{i1}; X_i \mid \alpha) dF_1(Y_{i1})}{\int \eta_1(y_1; X_i \mid \alpha) dF_1(y_1)},$$

where $F_1(y_1) = F(y_1 \mid x_0)$. The estimated likelihood score is

$$\hat{W}_{nA}(h) = h(\mathbf{Y}_1, \mathbf{Y}_2, \mathbf{X}) - E\left\{h(\mathbf{Y}_1, \mathbf{Y}_2, \mathbf{X}) \mid \mathbf{Y}_2, \mathbf{X}, \hat{\alpha}, \hat{F}_1\right\}.$$

Note that when $F(y_1 \mid x)$ is known, the variance of the likelihood score can be easily obtained as the variance of $h(Y_1, Y_2, X)$. When $F(y_1 \mid x)$ needs to be estimated, the variance of the likelihood score is no longer that simple. Using the variance of $h(Y_1, Y_2, X)$ in the construction of the likelihood score test will have incorrect test size under the null hypothesis. That is,

$$\begin{aligned} U_{nA}(h) &= \left[h(\mathbf{Y}_1, \mathbf{Y}_2, \mathbf{X}) - E_{(\hat{\alpha}, \hat{F}_1)}\left\{h(\mathbf{Y}_1, \mathbf{Y}_2, \mathbf{X}) \mid \mathbf{Y}_2, \mathbf{X}\right\}\right]^t \\ & V^{-1}_{(\hat{\alpha}, \hat{F}_1)}\left\{h(\mathbf{Y}_1, \mathbf{Y}_2, \mathbf{X}) \mid \mathbf{Y}_2, \mathbf{X}\right\}\left[h(\mathbf{Y}_1, \mathbf{Y}_2, \mathbf{X}) - E_{(\hat{\alpha}, \hat{F}_1)}\left\{h(\mathbf{Y}_1, \mathbf{Y}_2, \mathbf{X}) \mid \mathbf{Y}_2, \mathbf{X}\right\}\right], \end{aligned}$$

does not follow $\chi^2_{dim(\theta)}$ under the null hypothesis. A correction to the variance estimate is needed.

To derive the correction term, let the likelihood score for estimating (α, F_1) be

$$b^t \dot{l}^p_{\alpha} + \dot{l}^p_{F_1}(e) = \sum_{i=1}^{n} b^t \dot{l}^p_{i\alpha} + \dot{l}^p_{iF_1}(e),$$

where, for bounded vector b and bounded variation function e of y_1,

$$\begin{aligned} b^t \dot{l}^p_{i\alpha} + \dot{l}^p_{iF_1}(e) &= b^t \frac{\partial}{\partial \alpha} \log \eta_1(Y_{i1}; X_i \mid \alpha) - E\left\{b^t \frac{\partial}{\partial \alpha} \log \eta_1(Y_{i1}; X_i \mid \alpha) \mid X_i\right\} \\ & \quad + e(Y_{i1}) - E\left\{e(Y_{i1}) \mid X_i\right\}. \end{aligned}$$

Let the information operator for (α, F_1) be

$$(b, e)\Sigma_1(b, e) = E_0\left[\left\{b^t \dot{l}^p_{\alpha} + \dot{l}^p_{F_1}(e)\right\}^2\right].$$

From the likelihood expansion, it follows that

$$b^t \sqrt{n}(\hat{\alpha} - \alpha_0) + \int e(y_1) d\sqrt{n}(\hat{F}_1 - F_{10}) = \sqrt{n}(\dot{l}^p_{\alpha}, \dot{l}^p_{F_1})\left\{\Sigma_1^{-1}(b, e)\right\} + o_p(1),$$

where (α_0, F_{10}) are the true parameters. Note that

$$\begin{aligned} \frac{1}{\sqrt{n}} \hat{W}_{nA}(h) &= \frac{1}{\sqrt{n}} \sum_{i=1}^{n} \left[h(Y_{i1}, Y_{i2}, X_i) - E_0\left\{h(Y_{i1}, Y_{i2}, X_i) \mid Y_{i2}, X_i\right\}\right] \\ & - \frac{1}{\sqrt{n}} \sum_{i=1}^{n} E_0\left[h(Y_{i1}, Y_{i2}, X_i)\{\dot{l}^p_{i\alpha}(\hat{\alpha} - \alpha_0) + \dot{l}^p_{iF_1}(\hat{F}_1 - F_{10})\} \mid Y_{i2}, X_i\right] + o_p(1). \end{aligned}$$

The asymptotic variance of $\hat{W}_{nA}(h)/\sqrt{n}$ can be expressed as

$$V_{\hat{W}_{nA}} = E_0\left[h(Y_1, Y_2, X) - E_0\left\{h(Y_1.Y_2, X) \mid Y_2, X\right\}\right]^2 - (s_1, t_1)\Sigma_1^{-1}(s_1, t_1),$$

where

$$(s_1, t_1) = \left(E_0\left\{h(\mathbf{Y}_1, \mathbf{Y}_2, \mathbf{X}) l_\alpha^{ip}\right\}, E_0\left\{h(\mathbf{Y}_1, \mathbf{Y}_2, \mathbf{X}) l_{F_1}^{ip}\right\}\right).$$

An estimate of the variance can be obtained as

$$\begin{aligned}\hat{V}_{\hat{W}_{nA}} &= \frac{1}{n}\sum_{i=1}^{n}\left[h(Y_{i1}, Y_{i2}, X_i) - E_{(\hat{\alpha}, \hat{F}_1)}\{h(Y_{i1}, Y_{i2}, X_i) \mid Y_{i2}, X_i\}\right]^2 - (\hat{s}_1, \hat{t}_1)\hat{\Sigma}_1^{-1}(\hat{s}_1, \hat{t}_1) \\ &= \frac{1}{n}\sum_{i=1}^{n}\left[h(Y_{i1}, Y_{i2}, X_i) - E_{(\hat{\alpha}, \hat{F}_1)}\{h(Y_{i1}, Y_{i2}, X_i) \mid Y_{i2}, X_i\} - (\hat{s}_1, \hat{t}_1)\hat{\Sigma}_1^{-1}(l_{i\alpha}^{p}, l_{iF_1}^{p})\right]^2,\end{aligned}$$

where $\hat{\Sigma}_1$ is the observed data information operator and

$$\hat{s}_1 = \frac{1}{n}\sum_{i=1}^{n} h(Y_{i1}, Y_{i2}, X_i)\left[\frac{\partial}{\partial\alpha}\log\eta_1(Y_{i1}; X_i \mid \hat{\alpha}) - E_{(\hat{\alpha}, \hat{F}_1)}\left\{\frac{\partial}{\partial\alpha}\log\eta_1(Y_{i1}; X_i \mid \hat{\alpha}) \mid X_i\right\}\right]$$

$$\hat{t}_1(u) = \frac{1}{n}\sum_{i=1}^{n} h(Y_{i1}, Y_{i2}, X_i)\left\{1_{\{Y_{i1} \le u\}} - P_{(\hat{\alpha}, \hat{F}_1)}(Y_{i1} \le u \mid X_i)\right\}.$$

The modified test

$$\begin{aligned}\hat{U}_{nA}(h) &= \left[h(\mathbf{Y}_1, \mathbf{Y}_2, \mathbf{X}) - E_{(\hat{\alpha}, \hat{F}_1)}\{h(\mathbf{Y}_1, \mathbf{Y}_2, \mathbf{X}) \mid \mathbf{Y}_2, \mathbf{X}\}\right]^t \\ &\quad \left\{n\hat{V}_{\hat{W}_{nA}}\right\}^{-1}\left[h(\mathbf{Y}_1, \mathbf{Y}_2, \mathbf{X}) - E_{(\hat{\alpha}, \hat{F}_1)}\{h(\mathbf{Y}_1, \mathbf{Y}_2, \mathbf{X}) \mid \mathbf{Y}_2, \mathbf{X}\}\right],\end{aligned}$$

follows $\chi^2_{dim(\theta)}$ under the null hypothesis.

The likelihood score from the retrospective likelihood is

$$W_{nB}(h) = h(\mathbf{Y}_1, \mathbf{Y}_2, \mathbf{X}) - E_0\{h(\mathbf{Y}_1, \mathbf{Y}_2, \mathbf{X}) \mid \mathbf{Y}_1, \mathbf{X}\}.$$

Under the null hypothesis, the conditional density at a fixed condition $dF(y_2 \mid y_{10}, x)$ reduces to $dF(y_2 \mid x)$, which needs to be estimated. It can be modeled by the following semiparametric odds ratio model,

$$dF(y_2 \mid x) = \frac{\eta_2(y_2; x \mid \beta)dF_2(y_2)}{\int \eta_2(y_2; x \mid \beta)dF_2(y_2)}.$$

Unknown parameters can be estimated by

$$(\hat{\beta}, \hat{F}_2) = \arg\max_{\beta, F_2}\prod_{i=1}^{n}\frac{\eta_2(Y_{i2}; X_i \mid \beta)dF_2(Y_{i2})}{\int \eta_2(y_2; X_i \mid \beta)dF_2(y_2)},$$

where $F_2(y_2) = F(y_2 \mid x_0)$. The estimated likelihood score is

$$\hat{W}_{nB}(h) = h(\mathbf{Y}_1, \mathbf{Y}_2, \mathbf{X}) - E\left\{h(\mathbf{Y}_1, \mathbf{Y}_2, \mathbf{X}) \mid \mathbf{Y}_1, \mathbf{X}, \hat{\beta}, \hat{F}_2\right\}.$$

As argued before,

$$\begin{aligned}U_{nB}(h) &= \left[h(\mathbf{Y}_1, \mathbf{Y}_2, \mathbf{X}) - E_{(\hat{\beta}, \hat{F}_2)}\{h(\mathbf{Y}_1, \mathbf{Y}_2, \mathbf{X}) \mid \mathbf{Y}_1, \mathbf{X}\}\right]^t \\ &\quad V^{-1}_{(\hat{\beta}, \hat{F}_2)}\{h(\mathbf{Y}_1, \mathbf{Y}_2, \mathbf{X}) \mid \mathbf{Y}_1, \mathbf{X}\}\left[h(\mathbf{Y}_1, \mathbf{Y}_2, \mathbf{X}) - E_{(\hat{\beta}, \hat{F}_2)}\{h(\mathbf{Y}_1, \mathbf{Y}_2, \mathbf{X}) \mid \mathbf{Y}_1, \mathbf{X}\}\right],\end{aligned}$$

does not follow $\chi^2_{dim(\theta)}$ under the null hypothesis. A similar correction to the variance estimate can be carried out as for the prospective likelihood score approach.

Let the likelihood score for estimating (β, F_2) be

$$b^t \dot{l}_\beta^r + \dot{l}_{F_2}^r(e) = \sum_{i=1}^n b^t \dot{l}_{i\beta}^r + \dot{l}_{iF_2}^r(e),$$

where, for bounded vector b and bounded variation function e of y_2,

$$\begin{aligned} b^t \dot{l}_{i\beta}^r + \dot{l}_{iF_2}^r(e) &= b^t \frac{\partial}{\partial \beta} \log \eta_2(Y_{i2}; X_i \mid \beta) - E\left\{ b^t \frac{\partial}{\partial \beta} \log \eta_2(Y_{i2}; X_i \mid \beta) \mid X_i \right\} \\ &\quad + e(Y_{i2}) - E\left\{ e(Y_{i2}) \mid X_i \right\}. \end{aligned}$$

Let the information operator for (β, F_2) be

$$(b, e)\Sigma_2(b, e) = E_0\left[\left\{ b^t \dot{l}_\beta^r + \dot{l}_{F_2}^r(e) \right\}^2 \right].$$

From the likelihood expansion, it follows that

$$b^t \sqrt{n}(\hat{\beta} - \beta_0) + \int e(y_2) d\sqrt{n}(\hat{F}_2 - F_{20}) = \sqrt{n}(\dot{l}_\beta^r, \dot{l}_{F_2}^r)\left\{ \Sigma_2^{-1}(b, e) \right\} + o_p(1),$$

where (β_0, F_{20}) are the true parameters. Note that

$$\begin{aligned} \frac{1}{\sqrt{n}} \hat{W}_{nB}(h) &= \frac{1}{\sqrt{n}} \sum_{i=1}^n \left[h(Y_{i1}, Y_{i2}, X_i) - E_0\left\{ h(Y_{i1}, Y_{i2}, X_i) \mid Y_{i1}, X_i \right\} \right] \\ &\quad - \frac{1}{\sqrt{n}} \sum_{i=1}^n E_0\left[h(Y_{i1}, Y_{i2}, X_i)\{ \dot{l}_{i\beta}^r(\hat{\beta} - \beta_0) + \dot{l}_{iF_2}^r(\hat{F}_2 - F_{20}) \} \mid Y_{i1}, X_i \right] + o_p(1). \end{aligned}$$

The asymptotic variance of $\hat{W}_{nB}(h)/\sqrt{n}$ can be expressed as

$$V_{\hat{W}_{nB}} = E_0\left[h(Y_1, Y_2, X) - E_0\left\{ h(Y_1.Y_2, X) \mid Y_1, X \right\} \right]^2 - (s_2, t_2)\Sigma_2^{-1}(s_2, t_2),$$

where

$$(s_2, t_2) = \left(E_0\left\{ h(\mathbf{Y}_1, \mathbf{Y}_2, \mathbf{X}) \dot{l}_\beta^r \right\}, E_0\left\{ h(\mathbf{Y}_1, \mathbf{Y}_2, \mathbf{X}) \dot{l}_{F_2}^r \right\} \right).$$

An estimate of the variance can be obtained as

$$\begin{aligned} \hat{V}_{\hat{W}_{nB}} &= \frac{1}{n} \sum_{i=1}^n \left[h(Y_{i1}, Y_{i2}, X_i) - E_{(\hat{\beta}, \hat{F}_2)}\left\{ h(Y_{i1}, Y_{i2}, X_i) \mid Y_{i1}, X_i \right\} \right]^2 - (\hat{s}_2, \hat{t}_2)\hat{\Sigma}_2^{-1}(\hat{s}_1, \hat{t}_2) \\ &= \frac{1}{n} \sum_{i=1}^n \left[h(Y_{i1}, Y_{i2}, X_i) - E_{(\hat{\beta}, \hat{F}_2)}\left\{ h(Y_{i1}, Y_{i2}, X_i) \mid Y_{i1}, X_i \right\} - (\hat{s}_2, \hat{t}_2)\hat{\Sigma}_2^{-1}(\dot{l}_{i\beta}^r, \dot{l}_{iF_2}^r) \right]^2, \end{aligned}$$

where $\hat{\Sigma}$ is the observed data information operator and

$$\hat{s}_2 = \frac{1}{n} \sum_{i=1}^n h(Y_{i1}, Y_{i2}, X_i)\left[\frac{\partial}{\partial \alpha} \log \eta_2(Y_{i2}; X_i \mid \hat{\beta}) - E_{(\hat{\beta}, \hat{F}_2)}\left\{ \frac{\partial}{\partial \beta} \log \eta_2(Y_{i2}; X_i \mid \hat{\beta}) \mid X_i \right\} \right]$$

$$\hat{t}_2(u) = \frac{1}{n} \sum_{i=1}^n h(Y_{i1}, Y_{i2}, X_i)\left\{ 1_{\{Y_{i2} \le u\}} - P_{(\hat{\beta}, \hat{F}_2)}(Y_{i2} \le u \mid X_i) \right\}.$$

The modified test

$$\begin{aligned} \hat{U}_{nB}(h) &= \left[h(\mathbf{Y}_1, \mathbf{Y}_2, \mathbf{X}) - E_{(\hat{\beta}, \hat{F}_2)}\left\{ h(\mathbf{Y}_1, \mathbf{Y}_2, \mathbf{X}) \mid \mathbf{Y}_1, \mathbf{X} \right\} \right]^t \\ &\quad \left\{ n\hat{V}_{\hat{W}_{nB}} \right\}^{-1} \left[h(\mathbf{Y}_1, \mathbf{Y}_2, \mathbf{X}) - E_{(\hat{\beta}, \hat{F}_2)}\left\{ h(\mathbf{Y}_1, \mathbf{Y}_2, \mathbf{X}) \mid \mathbf{Y}_1, \mathbf{X} \right\} \right], \end{aligned}$$

follows $\chi^2_{dim(\theta)}$ under the null hypothesis.

The third likelihood score test is based on the joint conditional distribution of (Y_1, Y_2) given X derived from (6.1). The likelihood score is

$$W_{nAB}(h) = h(\mathbf{Y}_1, \mathbf{Y}_2, \mathbf{X}) - E_0 \{h(\mathbf{Y}_1, \mathbf{Y}_2, \mathbf{X}) \mid \mathbf{X}\}.$$

Under the null hypothesis, $E_0 \{h(\mathbf{Y}_1, \mathbf{Y}_2, \mathbf{X}) \mid \mathbf{X}\}$ can be estimated by $E_{(\hat{\alpha},\hat{F}_1,\hat{\beta},\hat{F}_2)}$ $\{h(\mathbf{Y}_1, \mathbf{Y}_2, \mathbf{X}) \mid \mathbf{X}\}$ calculated using the distribution

$$dF(y_1) \mid x)dF(y_2 \mid x) = \frac{\eta_1(y_1; x \mid \beta)dF_1(y_1)}{\int \eta_1(y_1; x \mid \beta)dF_1(y_1)} \frac{\eta_2(y_2; x \mid \beta)dF_2(y_2)}{\int \eta_2(y_2; x \mid \beta)dF_2(y_2)}.$$

By similar asymptotic arguments, it can be seen that

$$\begin{aligned} \frac{1}{\sqrt{n}}\hat{W}_{nAB}(h) &= \frac{1}{\sqrt{n}}\sum_{i=1}^{n} [h(Y_{i1}, Y_{i2}, X_i) - E_0 \{h(Y_{i1}, Y_{i2}, X_i) \mid X_i\}] \\ &\quad -\frac{1}{\sqrt{n}}\sum_{i=1}^{n} E_0 \left[h(Y_{i1}, Y_{i2}, X_i)\{\dot{l}^p_{i\alpha}(\hat{\alpha} - \alpha_0) + \dot{l}^p_{iF_1}(\hat{F}_1 - F_{10})\} \mid X_i\right] \\ &\quad -\frac{1}{\sqrt{n}}\sum_{i=1}^{n} E_0 \left[h(Y_{i1}, Y_{i2}, X_i)\{\dot{l}^r_{i\beta}(\hat{\beta} - \beta_0) + \dot{l}^r_{iF_2}(\hat{F}_2 - F_{20})\} \mid X_i\right] + o_p(1). \end{aligned}$$

The asymptotic variance of $\hat{W}_{nAB}(h)/\sqrt{n}$ can be expressed as

$$V_{\hat{W}_{nAB}} = E_0 \left[h(Y_1, Y_2, X) - E_0 \{h(Y_1.Y_2, X) \mid Y_1, X\}\right]^2 - (s_1, t_1)\Sigma_1^{-1}(s_1, t_1) - (s_2, t_2)\Sigma_2^{-1}(s_2, t_2).$$

An estimate of the variance can be obtained as

$$\begin{aligned} \hat{V}_{\hat{W}_{nAB}} &= \frac{1}{n}\sum_{i=1}^{n} \left[h(Y_{i1}, Y_{i2}, X_i) - E_{(\hat{\beta},\hat{F}_2)} \{h(Y_{i1}, Y_{i2}, X_i) \mid Y_{i1}, X_i\}\right]^2 \\ &\quad -(\hat{s}_1, \hat{t}_1)\hat{\Sigma}_1^{-1}(\hat{s}_1, \hat{t}_1) - (\hat{s}_2, \hat{t}_2)\hat{\Sigma}_2^{-1}(\hat{s}_2, \hat{t}_2) \\ &= \frac{1}{n}\sum_{i=1}^{n} \Big[h(Y_{i1}, Y_{i2}, X_i) - E_{(\hat{\beta},\hat{F}_2)} \{h(Y_{i1}, Y_{i2}, X_i) \mid Y_{i1}, X_i\} \\ &\quad -(\hat{s}_1, \hat{t}_1)\hat{\Sigma}_1^{-1}(\dot{l}^p_{i\alpha}, \dot{l}^p_{iF_1}) - (\hat{s}_2, \hat{t}_2)\hat{\Sigma}_2^{-1}(\dot{l}^r_{i\beta}, \dot{l}^r_{iF_2})\Big]^2. \end{aligned}$$

The modified test

$$\begin{aligned} \hat{U}_{nAB}(h) &= \left[h(\mathbf{Y}_1, \mathbf{Y}_2, \mathbf{X}) - E_{(\hat{\alpha},\hat{F}_1,\hat{\beta},\hat{F}_2)} \{h(\mathbf{Y}_1, \mathbf{Y}_2, \mathbf{X}) \mid \mathbf{X}\}\right]^t \\ &\quad \left\{n\hat{V}_{\hat{W}_{nAB}}\right\}^{-1} \left[h(\mathbf{Y}_1, \mathbf{Y}_2, \mathbf{X}) - E_{(\hat{\alpha},\hat{F}_1,\hat{\beta},\hat{F}_2)} \{h(\mathbf{Y}_1, \mathbf{Y}_2, \mathbf{X}) \mid \mathbf{X}\}\right], \end{aligned}$$

follows $\chi^2_{dim(\theta)}$ under the null hypothesis.

Computation of the variance of the estimated likelihood score can be difficult. A simple approach to finding the variance is through the inverse of the profile likelihood information matrix. The second derivative of the profile likelihood for θ at $\theta = 0$ can be approximated as follows. Fix θ at a point near 0 but different from 0. Maximize the likelihood with respect to the rest of the parameters. Find the score for θ evaluated at the new maximum point. Find the approximation of the second derivative matrix by dividing the components of likelihood score by the corresponding components of θ.

6.3.2 Permutation likelihood score tests

The prospective permutation likelihood score for θ is

$$S_{nA}(h) = h(\mathbf{Y}_1, \mathbf{Y}_2, \mathbf{X}) - E_{nA}\left\{h(\mathbf{Y}_1, \mathbf{Y}_2, \mathbf{X}) \mid \mathbf{Y}_1^{ord}, \mathbf{Y}_2, \mathbf{X}, \alpha\right\}, \tag{6.4}$$

where E_{nA} is computed under the distribution

$$P_{nA}(\sigma \mid \mathbf{Y}_1^{ord}, \mathbf{Y}_2, \mathbf{X}, \alpha) = \frac{\prod_{i=1}^n \eta_1(Y_{\sigma(i)1}; X_i \mid \alpha)}{\sum_{\sigma \in \mathcal{P}_n} \prod_{i=1}^n \eta_1(Y_{\sigma(i)1}; X_i \mid \alpha)}.$$

The unknown parameter α in P_{nA} can be estimated by the permutation likelihood approach,

$$\hat{\alpha} = \arg\max_{\alpha} \frac{\prod_{i=1}^n \eta_1(Y_{i1}; X_i \mid \alpha)}{\sum_{\sigma \in \mathcal{P}_n} \prod_{i=1}^n \eta_1(Y_{\sigma(i)1}; X_i \mid \alpha)}. \tag{6.5}$$

The estimated prospective permutation likelihood score is

$$\hat{S}_{nA}(h) = h(\mathbf{Y}_1, \mathbf{Y}_2, \mathbf{X}) - E_{nA}\left\{h(\mathbf{Y}_1, \mathbf{Y}_2, \mathbf{X}) \mid \mathbf{Y}_1^{ord}, \mathbf{Y}_2, \mathbf{X}, \hat{\alpha}\right\}.$$

Note that

$$\begin{aligned} \hat{S}_{nA}(h) &= h(\mathbf{Y}_1, \mathbf{Y}_2, \mathbf{X}) - E_{nA}\left\{h(\mathbf{Y}_1, \mathbf{Y}_2, \mathbf{X}) \mid \mathbf{Y}_1^{ord}, \mathbf{Y}_2, \mathbf{X}, \alpha_0\right\} \\ &\quad - E_{nA}\left\{h(\mathbf{Y}_1, \mathbf{Y}_2, \mathbf{X}) l_\alpha^{pp}(\hat{\alpha} - \alpha_0) \mid \mathbf{Y}_1^{ord}, \mathbf{Y}_2, \mathbf{X}, \alpha_0\right\} + o_p(\sqrt{n}), \end{aligned}$$

where

$$l_\alpha^{pp} = \sum_{i=1}^n \frac{\partial}{\partial \alpha} \log \eta_1(Y_{i1}; X_i \mid \alpha) - \sum_{i=1}^n E_{nA}\left\{\frac{\partial}{\partial \alpha} \log \eta_1(Y_{i1}; X_i \mid \alpha) \mid X_i, \alpha\right\},$$

is the prospective permutation likelihood score for α. The variance of the estimated prospective likelihood score for θ at $\theta = 0$ is

$$\begin{aligned} V_{\hat{S}_{nA}} &= E_{nAB}\left[h(\mathbf{Y}_1, \mathbf{Y}_2, \mathbf{X}) - E_{nA}\left\{h(\mathbf{Y}_1, \mathbf{Y}_2, \mathbf{X}) \mid \mathbf{Y}_1^{ord}, \mathbf{Y}_2, \mathbf{X}, \alpha_0\right\}\right]^2 \\ &\quad - E_{nA}^t\left\{h(\mathbf{Y}_1, \mathbf{Y}_2, \mathbf{X}) l_\alpha^{pp} \mid \mathbf{Y}_1^{ord}, \mathbf{Y}_2, \mathbf{X}, \alpha_0\right\} E_{nA}^{-1}\left\{l_\alpha^{pp}(l_\alpha^{pp})^t \mid \mathbf{Y}_1^{ord}, \mathbf{Y}_2, \mathbf{X}, \alpha_0\right\} \\ &\quad \times E_{nA}\left\{h(\mathbf{Y}_1, \mathbf{Y}_2, \mathbf{X}) l_\alpha^{pp} \mid \mathbf{Y}_1^{ord}, \mathbf{Y}_2, \mathbf{X}, \alpha_0\right\}. \end{aligned}$$

An estimate of the variance denote by $\hat{V}_{\hat{S}_{nA}}$ replaces α_0 by $\hat{\alpha}$ in the above expression. The prospective permutation likelihood score test is

$$T_{nA}(h) = \hat{S}_{nA} \hat{V}_{\hat{S}_{nA}}^{-1} \hat{S}_{nA} \sim \chi^2_{dim(\theta)}.$$

The retrospective permutation likelihood score for θ is

$$S_{nB}(h) = h(\mathbf{Y}_1, \mathbf{Y}_2, \mathbf{X}) - E_{nB}\left\{h(\mathbf{Y}_1, \mathbf{Y}_2, \mathbf{X}) \mid \mathbf{Y}_1, \mathbf{Y}_2^{ord}, \mathbf{X}, \beta\right\},$$

where E_{nB} is computed under the distribution

$$P_{nB}(\tau \mid \mathbf{Y}_1, \mathbf{Y}_2^{ord}, \mathbf{X}, \beta) = \frac{\prod_{i=1}^n \eta_2(Y_{\tau(i)2}; X_i \mid \beta)}{\sum_{\tau \in \mathcal{P}_n} \prod_{i=1}^n \eta_2(Y_{\tau(i)2}; X_i \mid \beta)}.$$

The unknown parameter α in P_{nB} can be estimated by the permutation likelihood approach,

$$\hat{\beta} = \arg\max_{\beta} \frac{\prod_{i=1}^n \eta_2(Y_{i2}; X_i \mid \beta)}{\sum_{\tau \in \mathcal{P}_n} \prod_{i=1}^n \eta_2(Y_{\tau(i)2}; X_i \mid \beta)}.$$

The estimated prospective permutation likelihood score is

$$\hat{S}_{nB}(h) = h(\mathbf{Y}_1, \mathbf{Y}_2, \mathbf{X}) - E_{nB}\left\{h(\mathbf{Y}_1, \mathbf{Y}_2, \mathbf{X}) \mid \mathbf{Y}_1, \mathbf{Y}_2^{ord}, \mathbf{X}, \hat{\beta}\right\}.$$

Similarly, we have

$$\begin{aligned}\hat{S}_{nB}(h) &= h(\mathbf{Y}_1, \mathbf{Y}_2, \mathbf{X}) - E_{nB}\left\{h(\mathbf{Y}_1, \mathbf{Y}_2, \mathbf{X}) \mid \mathbf{Y}_1, \mathbf{Y}_2^{ord}, \mathbf{X}, \beta_0\right\} \\ &\quad -E_{nB}\left\{h(\mathbf{Y}_1, \mathbf{Y}_2, \mathbf{X}) l_\beta^{irp}(\hat{\beta} - \beta_0) \mid \mathbf{Y}_1, \mathbf{Y}_2^{ord}, \mathbf{X}, \beta_0\right\} + o_p(\sqrt{n}),\end{aligned}$$

where

$$l_\beta^{irp} = \sum_{i=1}^n \frac{\partial}{\partial \beta} \log \eta_2(Y_{i2}; X_i \mid \beta) - \sum_{i=1}^n E_{nB}\left\{\frac{\partial}{\partial \beta} \log \eta_2(Y_{i2}; X_i \mid \beta) \mid X_i, \beta\right\},$$

is the retrospective permutation likelihood score for β. The variance of the estimated retrospective likelihood score for θ at $\theta = 0$ is

$$\begin{aligned}V_{\hat{S}_{nB}} &= E_{nB}\left[h(\mathbf{Y}_1, \mathbf{Y}_2, \mathbf{X}) - E_{nB}\left\{h(\mathbf{Y}_1, \mathbf{Y}_2, \mathbf{X}) \mid \mathbf{Y}_1, \mathbf{Y}_2^{ord}, \mathbf{X}, \beta_0\right\}\right]^2 \\ &\quad -E_{nB}^t\left\{h(\mathbf{Y}_1, \mathbf{Y}_2, \mathbf{X}) l_\beta^{irp} \mid \mathbf{Y}_1, \mathbf{Y}_2^{ord}, \mathbf{X}, \beta_0\right\} E_{nB}^{-1}\left\{l_\beta^{irp}(l_\beta^{irp})^t \mid \mathbf{Y}_1, \mathbf{Y}_2^{ord}, \mathbf{X}, \beta_0\right\} \\ &\quad \times E_{nB}\left\{h(\mathbf{Y}_1, \mathbf{Y}_2, \mathbf{X}) l_\beta^{irp} \mid \mathbf{Y}_1, \mathbf{Y}_2^{ord}, \mathbf{X}, \beta_0\right\}.\end{aligned}$$

An estimate of the variance denote by $\hat{V}_{\hat{S}_{nB}}$ replaces β_0 by $\hat{\beta}$ in the above expression. The retrospective permutation likelihood score test is

$$T_{nB}(h) = \hat{S}_{nB} \hat{V}_{\hat{S}_{nB}}^{-1} \hat{S}_{nB} \sim \chi^2_{dim(\theta)}.$$

The joint permutation likelihood score for θ is

$$S_{nAB}(h) = h(\mathbf{Y}_1, \mathbf{Y}_2, \mathbf{X}) - E_{nAB}\left\{h(\mathbf{Y}_1, \mathbf{Y}_2, \mathbf{X}) \mid \mathbf{Y}_1^{ord}, \mathbf{Y}_2^{ord}, \mathbf{X}, \alpha, \beta\right\},$$

where E_{nAB} is computed under the distribution

$$P_{nAB}(\sigma, \tau \mid \mathbf{Y}_1^{ord}, \mathbf{Y}_2^{ord}, \mathbf{X}, \alpha, \beta) = \frac{\prod_{i=1}^n \eta_1(Y_{\sigma(i)1}; X_i \mid \alpha)\eta_2(Y_{\tau(i)2}; X_i \mid \beta)}{\sum_{\sigma,\tau \in \mathcal{P}_n} \prod_{i=1}^n \eta_1(Y_{\sigma(i)1}; X_i \mid \alpha)\eta_2(Y_{\tau(i)2}; X_i \mid \beta)}.$$

The unknown parameter (α, β) in P_{nAB} can be estimated respectively by the permutation likelihood approaches as specified in the prospective and retrospective permutation likelihood score tests. The estimated joint permutation likelihood score is

$$\hat{S}_{nAB}(h) = h(\mathbf{Y}_1, \mathbf{Y}_2, \mathbf{X}) - E_{nB}\left\{h(\mathbf{Y}_1, \mathbf{Y}_2, \mathbf{X}) \mid \mathbf{Y}_1^{ord}, \mathbf{Y}_2^{ord}, \mathbf{X}, \hat{\alpha}, \hat{\beta}\right\}.$$

The asymptotic expansion of the estimated score is

$$\begin{aligned}\hat{S}_{nAB}(h) &= h(\mathbf{Y}_1, \mathbf{Y}_2, \mathbf{X}) - E_{nAB}\left\{h(\mathbf{Y}_1, \mathbf{Y}_2, \mathbf{X}) \mid \mathbf{Y}_1^{ord}, \mathbf{Y}_2^{ord}, \mathbf{X}, \alpha_0, \beta_0\right\} \\ &\quad -E_{nAB}\left[h(\mathbf{Y}_1, \mathbf{Y}_2, \mathbf{X})\left\{l_\alpha^{ipp}(\hat{\alpha} - \alpha_0) + l_\beta^{irp}(\hat{\beta} - \beta_0)\right\} \mid \mathbf{Y}_1^{ord}, \mathbf{Y}_2^{ord}, \mathbf{X}, \alpha_0, \beta_0\right\} + o_p(\sqrt{n}),\end{aligned}$$

where the likelihood scores for α and β are respectively given earlier. The variance of the estimated joint likelihood score for θ at $\theta = 0$ is

$$\begin{aligned} V_{\hat{S}_{nAB}} &= E_{nAB}\left[h(\mathbf{Y}_1,\mathbf{Y}_2,\mathbf{X}) - E_{nAB}\left\{h(\mathbf{Y}_1,\mathbf{Y}_2,\mathbf{X}) \mid \mathbf{Y}_1^{ord},\mathbf{Y}_2^{ord},\mathbf{X},\alpha_0,\beta_0\right\}\right]^2 \\ &\quad -E_{nAB}^t\left\{h(\mathbf{Y}_1,\mathbf{Y}_2,\mathbf{X})\dot{l}_\alpha^{pp} \mid \mathbf{Y}_1^{ord},\mathbf{Y}_2^{ord},\mathbf{X},\alpha_0,\beta_0\right\} \\ \times E_{nAB}^{-1}&\left\{\dot{l}_\alpha^{pp}(\dot{l}_\alpha^{pp})^t \mid \mathbf{Y}_1^{ord},\mathbf{Y}_2^{ord},\mathbf{X},\alpha_0,\beta_0\right\} E_{nAB}\left\{h(\mathbf{Y}_1,\mathbf{Y}_2,\mathbf{X})\dot{l}_\alpha^{pp} \mid \mathbf{Y}_1^{ord},\mathbf{Y}_2^{ord},\mathbf{X},\alpha_0,\beta_0\right\} \\ &\quad -E_{nAB}^t\left\{h(\mathbf{Y}_1,\mathbf{Y}_2,\mathbf{X})\dot{l}_\beta^{rp} \mid \mathbf{Y}_1^{ord},\mathbf{Y}_2^{ord},\mathbf{X},\alpha_0,\beta_0\right\} \\ \times E_{nAB}^{-1}&\left\{\dot{l}_\beta^{rp}(\dot{l}_\beta^{rp})^t \mid \mathbf{Y}_1^{ord},\mathbf{Y}_2^{ord},\mathbf{X},\alpha_0,\beta_0\right\} E_{nAB}\left\{h(\mathbf{Y}_1,\mathbf{Y}_2,\mathbf{X})\dot{l}_\beta^{pp} \mid \mathbf{Y}_1^{ord},\mathbf{Y}_2^{ord},\mathbf{X},\alpha_0,\beta_0\right\}. \end{aligned}$$

An estimate of the variance denote by $\hat{V}_{\hat{S}_{nAB}}$ replaces (α_0, β_0) by $(\hat{\alpha}, \hat{\beta})$ in the foregoing expression. The joint permutation likelihood score test is

$$T_{nAB}(h) = \hat{S}_{nAB}\hat{V}_{\hat{S}_{nAB}}^{-1}\hat{S}_{nAB} \sim \chi^2_{dim(\theta)}.$$

6.4 Semiparametric efficient score test

6.4.1 Semiparametric likelihood formulation

The tests for conditional independence all rely on model assumptions on either Y_1 given X or Y_2 given X, or both. These models are not directly related to the conditional independence to be tested. Consider the semiparametric efficient score for estimating θ that attempts to remove the influence of those nuisance models. Under the null hypothesis, the semiparametric efficient score projects the likelihood scores to the orthogonal complement of the score space generated by the scores from all the regular parametric sub-models for Y_1 given X and Y_2 given X. For the prospective, retrospective, and the joint likelihood score, the projections are the same. The semiparametric efficient score for estimating θ at $\theta = 0$ is

$$\begin{aligned} R(h)(Y_1, Y_2, X) &= h(Y_1, Y_2, X) - E_0\{h(Y_1, Y_2, X) \mid Y_1, X\} \\ &\quad -E_0\{h(Y_1, Y_2, X) \mid Y_2, X\} + E_0\{h(Y_1, Y_2, X) \mid X\}. \end{aligned}$$

The variance of the efficient score is

$$\begin{aligned} E_0\{R(h)\}^2 &= E_0\left[h(Y_1, Y_2, X) - E_0\{h(Y_1, Y_2, X) \mid Y_1, X\}\right]^2 \\ &\quad -E_0\left[E_0\{h(Y_1, Y_2, X) \mid Y_2, X\} - E_0\{h(Y_1, Y_2, X) \mid X\}\right]^2, \end{aligned}$$

which can be alternatively expressed as

$$\begin{aligned} E_0\{R(h)\}^2 &= E_0\left[h(Y_1, Y_2, X) - E_0\{h(Y_1, Y_2, X) \mid Y_2, X\}\right]^2 \\ &\quad -E_0\left[E_0\{h(Y_1, Y_2, X) \mid Y_1, X\} - E_0\{h(Y_1, Y_2, X) \mid X\}\right]^2. \end{aligned}$$

or as

$$\begin{aligned} E_0\{R(h)\}^2 &= E_0\left[h(Y_1, Y_2, X) - E_0\{h(Y_1, Y_2, X) \mid X\}\right]^2 \\ &\quad -E_0\left[E_0\{h(Y_1, Y_2, X) \mid Y_1, X\} - E_0\{h(Y_1, Y_2, X) \mid X\}\right]^2 \\ &\quad -E_0\left[E_0\{h(Y_1, Y_2, X) \mid Y_2, X\} - E_0\{h(Y_1, Y_2, X) \mid X\}\right]^2. \end{aligned}$$

If the conditional expectations in the score expression are known, a test of conditional independence can be constructed as

$$T_{eff} = \sum_{i=1}^{n} R(h)(Y_{i1}, Y_{i2}, X_i) \left[\sum_{i=1}^{n} \{R(h)(Y_{i1}, Y_{i2}, X_i)\}^2 \right]^{-1} \sum_{i=1}^{n} R(h)(Y_{i1}, Y_{i2}, X_i),$$

which follows $\chi^2_{dim(\theta)}$. In practice, however, we seldom know the conditional expectations. As a result, we still need to estimate the conditional expectations.

The semiparametric odds ratio models for Y_1 given X and Y_2 given X can be used to estimate the distributions for Y_1 given X and Y_2 given X, which are in turn used for computing the conditional expectations. The available semiparametric score is an estimate, i.e.,

$$\begin{aligned} \hat{R}(h)(Y_1, Y_2, X) &= h(Y_1, Y_2, X) - E\left\{h(Y_1, Y_2, X) \mid Y_1, X, \hat{\beta}, \hat{F}_2\right\} \\ &\quad - E\left\{h(Y_1, Y_2, X) \mid Y_2, X, \hat{\alpha}, \hat{F}_1\right\} + E\left\{h(Y_1, Y_2, X) \mid X, \hat{\alpha}, \hat{F}_1, \hat{\beta}, \hat{F}_2\right\}. \end{aligned}$$

As in the likelihood score tests, the variance of the estimated score needs to be reevaluated because the variability of the estimated conditional expectations may change the variance of the estimated score from the variance of the original score. These changes have been demonstrated in the previous sections. For the semiparametric efficient score, we demonstrate below that no change in the variance occurs when certain conditions are satisfied.

By Taylor expansion around the true parameter value,

$$\begin{aligned} \hat{R}(h) &= R(h) - E_0\left[h(Y_1, Y_2, X)\left\{\dot{l}^r_\beta(\hat{\beta} - \beta_0) + \dot{l}^r_{F_2}(\hat{F}_2 - F_{20})\right\} \mid Y_1, X\right] \\ &\quad - E_0\left[h(Y_1, Y_2, X)\left\{\dot{l}^p_\alpha(\hat{\alpha} - \alpha_0) + \dot{l}^p_{F_1}(\hat{F}_1 - F_{10})\right\} \mid Y_2, X\right] \\ &\quad + E_0\left[h(Y_1, Y_2, X)\left\{\dot{l}^r_\beta(\hat{\beta} - \beta_0) + \dot{l}^r_{F_2}(\hat{F}_2 - F_{20})\right\} \mid X\right] \\ &\quad + E_0\left[h(Y_1, Y_2, X)\left\{\dot{l}^p_\alpha(\hat{\alpha} - \alpha_0) + \dot{l}^p_{F_1}(\hat{F}_1 - F_{10})\right\} \mid X\right] \\ &\quad + O_p\left(|\hat{\alpha} - \alpha_0|^2 + |\hat{\beta} - \beta_0|^2 + ||\hat{F}_1 - F_{10}||^2 + ||\hat{F}_2 - F_{20}||^2\right), \end{aligned}$$

where the bounded variation norm is used for distribution functions. Note that

$$\begin{aligned} &\sum_{i=1}^{n} E_0\left[h(Y_{i1}, Y_{i2}, X_i)\left\{\dot{l}^p_\alpha(\hat{\alpha} - \alpha_0) + \dot{l}^p_{F_1}(\hat{F}_1 - F_{10})\right\} \mid Y_{i2}, X_i\right] \\ &\quad - \sum_{i=1}^{n} E_0\left[h(Y_{i1}, Y_{i2}, X_i)\left\{\dot{l}^p_\alpha(\hat{\alpha} - \alpha_0) + \dot{l}^p_{F_1}(\hat{F}_1 - F_{10})\right\} \mid X_i\right] \\ &\quad = O_p\left\{\sqrt{n}|\hat{\alpha} - \alpha_0| + \sqrt{n}||\hat{F}_1 - F_{10}||\right\}, \end{aligned}$$

and that

$$\begin{aligned} &\sum_{i=1}^{n} E_0\left[h(Y_{i1}, Y_{i2}, X_i)\left\{\dot{l}^r_\beta(\hat{\beta} - \beta_0) + \dot{l}^r_{F_2}(\hat{F}_2 - F_{20})\right\} \mid Y_{i2}, X_i\right] \\ &\quad - \sum_{i=1}^{n} E_0\left[h(Y_{i1}, Y_{i2}, X_i)\left\{\dot{l}^r_\beta(\hat{\beta} - \beta_0) + \dot{l}^r_{F_2}(\hat{F}_2 - F_{20})\right\} \mid X_i\right] \\ &\quad = O_p\left\{\sqrt{n}|\hat{\beta} - \beta_0| + \sqrt{n}||\hat{F}_2 - F_{20}||\right\}. \end{aligned}$$

Furthermore, $l^p_\alpha(b_1)+l^p_{F_1}(e_1)$ and $l^r_\beta(b_2)+l^r_{F_2}(e_2)$ are mutually orthogonal and are orthogonal to $R(h)$. This means that

$$\begin{aligned}\frac{1}{\sqrt{n}}\sum_{i=1}^n \hat{R}\{h(Y_{i1},Y_{i2},X_i)\} &= \frac{1}{\sqrt{n}}\sum_{i=1}^n R\{h(Y_{i1},Y_{i2},X_i)\}\\ &\quad +O_p\left(|\hat\alpha-\alpha_0|+|\hat\beta-\beta_0|+||\hat F_1-F_{10}||+||\hat F_2-F_{20}||\right)\\ &\quad +O_p\left\{\sqrt{n}\left(|\hat\alpha-\alpha_0|^2+|\hat\beta-\beta_0|^2+||\hat F_1-F_{10}||^2+||\hat F_2-F_{20}||^2\right)\right\}.\end{aligned}$$

When the estimates of the nuisance parameters α, β, F_1, F_2 are at least $n^{1/4}$-consistent, the asymptotic variance for the estimated score remains the same as the original score. As a result of this observation, the conditional independence test based on the semiparametric efficient score can be constructed as

$$\hat T_{eff} = \sum_{i=1}^n \hat R(h)(Y_{i1},Y_{i2},X_i)\left[\sum_{i=1}^n\left\{\hat R(h)(Y_{i1},Y_{i2},X_i)\right\}^2\right]^{-1}\sum_{i=1}^n \hat R(h)(Y_{i1},Y_{i2},X_i).$$

The statistic $\hat T_{eff}$ follows χ^2_{df} with $df = dim(\theta)$ in general.

6.4.2 Permutation likelihood formulation

The observed data likelihood for the odds ratio model can be written as

$$\frac{\prod_{i=1}^n \eta(Y_{i1};Y_{i2}\mid X_i)\eta_1(Y_{i1};X_i)\eta_2(Y_{i2};X_i)p(Y_{i1}\mid y_{20},x_0)p(Y_{i2}\mid y_{10},x_0)}{\left[\int\int \eta(y_1;y_2\mid X_i)\eta_1(y_1;X_i)\eta_2(Y_{i2};X_i)p(y_1\mid y_{20},x_0)p(y_2\mid y_{10},x_0)dy_1dy_2\right]^n}. \tag{6.6}$$

We may assume without loss of generality the following parametri submodels for η_1 and η_2 as

$$\begin{aligned}\log\eta_1(Y_1;X) &= \alpha h_{1a}(Y_1,X),\\ \log\eta_2(Y_2;X) &= \beta h_{2a}(Y_2,X),\end{aligned}$$

where $h_{1a}(y_{10},x)=h_{1a}(y_1,x_0)=0$ for all y_1,x, and $h_{2a}(y_{20},x)=h_{2a}(y_2,x_0)=0$ for all y_2,x. The nuisance score spaces for $(\eta_1, p(y_1\mid y_{20},x_0))$ can be expressed as $H_1=H_{1a}+H_{1b}$ where

$$\begin{aligned}H_{1a} &= \left\{\sum_{i=1}^n [h_{1a}(Y_{i1},X_i)-E\{h_{1a}(Y_{i1},X_i)\mid X_i\}]\mid h_{1a}\in L^2(P_0)\right\},\\ H_{1b} &= \left\{\sum_{i=1}^n [h_{1b}(Y_{i1})-E\{h_{1b}(Y_{i1})\mid X_i\}]\mid h_{1b}\in L^2(P_0)\right\}.\end{aligned}$$

and for $(\eta_2, p(y_2\mid y_{10},x_0))$ can be expressed as $H_2=H_{2a}+H_{2b}$ where

$$\begin{aligned}H_{2a} &= \left\{\sum_{i=1}^n [h_{2a}(Y_{i2},X_i)-E\{h_{2a}(Y_{i2},X_i)\mid X_i\}]\mid h_{2a}\in L^2(P_0)\right\},\\ H_{2b} &= \left\{\sum_{i=1}^n [h_{2b}(Y_{i2})-E\{h_{2b}(Y_{i2})\mid X_i\}]\mid h_{2b}\in L^2(P_0)\right\}.\end{aligned}$$

The orthogonal complement of all the nuisance score space satisfies

$$(H_{1a}+H_{1b}+H_{2a}+H_{2b})^\perp = (H_{1a}+H_{2a})^\perp\cap(H_{1b}+H_{2b})^\perp.$$

The projection of $h(\mathbf{Y}_1, \mathbf{Y}_2, \mathbf{X}) = \sum_{i=1}^n h(Y_{i1}, Y_{i2}, X_i) \in L^2(P_0)$ to $(H_1 + H_2)^\perp$ can be carried out in two stages as

$$\text{Proj}\left\{h \mid (H_1 + H_2)^\perp\right\} = \text{Proj}\left[\text{Proj}\left\{h \mid (H_{1b} + H_{2b})^\perp\right\} \mid (H_{1a} + H_{2a})^\perp \cap (H_{1a} + H_{2a})^\perp\right].$$

The inner projection can be approximately obtained by the permutation likelihood approach as

$$\text{Proj}\left\{h \mid (H_{1b} + H_{2b})^\perp\right\} = \sum_{i=1}^n \left[h(Y_{i1}, Y_{i2}, X_i) - E\left\{h(Y_{i1}, Y_{i2}, X_i) \mid \mathbf{Y}_1^{ord}, \mathbf{Y}_2^{ord}, \mathbf{X}\right\}\right].$$

Similarly, approximately,

$$\begin{aligned} H_{1a} \cap (H_{1b} + H_{2b})^\perp &= \left\{\sum_{i=1}^n \left[h_{1a}(Y_{i1}, X_i) - E\{h_{1a}(Y_{i1}, X_i) \mid \mathbf{Y}_1^{ord}, \mathbf{Y}_2^{ord}, \mathbf{X}\}\right]\right\} \\ H_{2a} \cap (H_{1b} + H_{2b})^\perp &= \left\{\sum_{i=1}^n \left[h_{2a}(Y_{i2}, X_i) - E\{h_{2a}(Y_{i2}, X_i) \mid \mathbf{Y}_1^{ord}, \mathbf{Y}_2^{ord}, \mathbf{X}\}\right]\right\} \end{aligned}$$

Projection of $\text{Proj}\left\{h \mid (H_{1b} + H_{2b})^\perp\right\}$ to $(H_{1a} + H_{2a})^\perp \cap (H_{1a} + H_{2a})^\perp$ can be obtained using the alternating conditional expectation approach, where the alternating conditional expectations are respectively conditional on $(\mathbf{Y}_1^{ord}, \mathbf{Y}_2, \mathbf{X})$ and on $(\mathbf{Y}_1, \mathbf{Y}_2^{ord}, \mathbf{X})$. When $\mathbf{Y}_1$ and $\mathbf{Y}_2$ are independent conditional on $\mathbf{X}$, the projection has an approximately closed-form as

$$\begin{aligned} \text{Proj}\left\{h(\mathbf{Y}_1, \mathbf{Y}_2, \mathbf{X}) \mid (H_1 + H_2)^\perp\right\} &= h(\mathbf{Y}_1, \mathbf{Y}_2, \mathbf{X}) - E\left\{h(\mathbf{Y}_1, \mathbf{Y}_2, \mathbf{X}) \mid \mathbf{Y}_1^{ord}, \mathbf{Y}_2, \mathbf{X}\right\} \\ &\quad - E\left\{h(\mathbf{Y}_1, \mathbf{Y}_2, \mathbf{X}) \mid \mathbf{Y}_1, \mathbf{Y}_2^{ord}, \mathbf{X}\right\} + E\left\{h(\mathbf{Y}_1, \mathbf{Y}_2, \mathbf{X}) \mid \mathbf{Y}_1^{ord}, \mathbf{Y}_2^{ord}, \mathbf{X}\right\}. \end{aligned}$$

Suppose that the odds ratio functions η_1 and η_2 can be respectively modeled parametrically. Assume further the true parameters are α_0, β_0. The conditional independence test derived from the efficient score can be expressed as

$$\begin{aligned} \mathcal{R}(h)(\mathbf{Y}_1, \mathbf{Y}_2, \mathbf{X}) &= h(\mathbf{Y}_1, \mathbf{Y}_2, \mathbf{X}) - E_{nA}\left\{h(\mathbf{Y}_1, \mathbf{Y}_2, \mathbf{X}) \mid \mathbf{Y}_1^{ord}, \mathbf{Y}_2, \mathbf{X}, \alpha_0\right\} \\ &\quad - E_{nB}\left\{h(\mathbf{Y}_1, \mathbf{Y}_2, \mathbf{X}) \mid \mathbf{Y}_1, \mathbf{Y}_2^{ord}, \mathbf{X}, \beta_0\right\} + E_{nAB}\left\{h(\mathbf{Y}_1, \mathbf{Y}_2, \mathbf{X}) \mid \mathbf{Y}_1^{ord}, \mathbf{Y}_2^{ord}, \mathbf{X}, \alpha_0, \beta_0\right\}. \end{aligned}$$

When α and/or β are unknown, the efficient score needs to be estimated. Estimators of α and β can be respectively obtained by maximizing the permutation likelihoods described in the previous sections. With the estimated α and β, the estimated efficient score is

$$\begin{aligned} \hat{\mathcal{R}}(h)(\mathbf{Y}_1, \mathbf{Y}_2, \mathbf{X}) &= h(\mathbf{Y}_1, \mathbf{Y}_2, \mathbf{X}) - E_{nA}\left\{h(\mathbf{Y}_1, \mathbf{Y}_2, \mathbf{X}) \mid \mathbf{Y}_1^{ord}, \mathbf{Y}_2, \mathbf{X}, \hat{\alpha}\right\} \\ &\quad - E_{nB}\left\{h(\mathbf{Y}_1, \mathbf{Y}_2, \mathbf{X}) \mid \mathbf{Y}_1, \mathbf{Y}_2^{ord}, \mathbf{X}, \hat{\beta}\right\} + E_{nAB}\left\{h(\mathbf{Y}_1, \mathbf{Y}_2, \mathbf{X}) \mid \mathbf{Y}_1^{ord}, \mathbf{Y}_2^{ord}, \mathbf{X}, \hat{\alpha}, \hat{\beta}\right\}. \end{aligned}$$

By Taylor expansion around the true parameter value, the estimated score can be expressed as

$$\begin{aligned} \hat{\mathcal{R}}(h)(\mathbf{Y}_1, \mathbf{Y}_2, \mathbf{X}) = \mathcal{R}(h)(\mathbf{Y}_1, \mathbf{Y}_2, \mathbf{X}) &- E_{nA}\left\{h(\mathbf{Y}_1, \mathbf{Y}_2, \mathbf{X}) \dot{l}_\alpha^{pp}(\hat{\alpha} - \alpha_0) \mid \mathbf{Y}_1^{ord}, \mathbf{Y}_2, \mathbf{X}, \alpha_0\right\} \\ &- E_{nB}\left\{h(\mathbf{Y}_1, \mathbf{Y}_2, \mathbf{X}) \dot{l}_\beta^{rp}(\hat{\beta} - \beta_0) \mid \mathbf{Y}_1, \mathbf{Y}_2^{ord}, \mathbf{X}, \beta_0\right\} \\ &+ E_{nAB}\left[h(\mathbf{Y}_1, \mathbf{Y}_2, \mathbf{X}) \left\{\dot{l}_\alpha^{pp}(\hat{\alpha} - \alpha_0) + \dot{l}_\beta^{rp}(\hat{\beta} - \beta_0)\right\} \mid \mathbf{Y}_1^{ord}, \mathbf{Y}_2^{ord}, \mathbf{X}, \alpha_0, \beta_0\right] \\ &+ O_p\left(n|\hat{\alpha} - \alpha_0|^2 + n|\hat{\beta} - \beta_0|^2\right). \end{aligned}$$

Note that

$$E_{nA}\left\{h(\mathbf{Y}_1,\mathbf{Y}_2,\mathbf{X})\dot{l}_{\alpha}^{pp}(\hat{\alpha}-\alpha_0)\mid \mathbf{Y}_1^{ord},\mathbf{Y}_2,\mathbf{X},\alpha_0\right\}$$
$$-E_{nAB}\left[h(\mathbf{Y}_1,\mathbf{Y}_2,\mathbf{X})\left\{\dot{l}_{\alpha}^{pp}(\hat{\alpha}-\alpha_0)\right\}\mid \mathbf{Y}_1^{ord},\mathbf{Y}_2^{ord},\mathbf{X},\alpha_0,\beta_0\right]=O_p\left(\sqrt{n}|\hat{\alpha}-\alpha_0|\right),$$

and that

$$E_{nB}\left\{h(\mathbf{Y}_1,\mathbf{Y}_2,\mathbf{X})\dot{l}_{\beta}^{rp}(\hat{\beta}-\beta_0)\mid \mathbf{Y}_1,\mathbf{Y}_2^{ord},\mathbf{X},\beta_0\right\}$$
$$-E_{nAB}\left[h(\mathbf{Y}_1,\mathbf{Y}_2,\mathbf{X})\left\{\dot{l}_{\beta}^{rp}(\hat{\beta}-\beta_0)\right\}\mid \mathbf{Y}_1^{ord},\mathbf{Y}_2^{ord},\mathbf{X},\alpha_0,\beta_0\right]=O_p\left(\sqrt{n}|\hat{\beta}-\beta_0|\right).$$

It then follows that

$$\begin{aligned}\frac{1}{\sqrt{n}}\hat{\mathcal{R}}(h)(\mathbf{Y}_1,\mathbf{Y}_2,\mathbf{X}) &= \frac{1}{\sqrt{n}}\mathcal{R}(h)(\mathbf{Y}_1,\mathbf{Y}_2,\mathbf{X})+O_p\left(|\hat{\alpha}-\alpha_0|+|\hat{\beta}-\beta_0|\right)\\ &\quad +O_p\left(\sqrt{n}|\hat{\alpha}-\alpha_0|^2+\sqrt{n}|\hat{\beta}-\beta_0|^2\right).\end{aligned}$$

When the estimates of the parameters α, β are at least $n^{1/4}$-consistent, the asymptotic variance for the estimated efficient score from the permutation remains the same. The conditional independence test based on the efficient score from the permutation can be constructed as

$$\hat{T}_{eff}=\hat{\mathcal{R}}(h)(\mathbf{Y}_1,\mathbf{Y}_2,\mathbf{X})E_{nAB}^{-1}\left[\left\{\hat{\mathcal{R}}(h)(\mathbf{Y}_1,\mathbf{Y}_2,\mathbf{X})\right\}^2\mid \mathbf{Y}_1^{ord},\mathbf{Y}_2^{ord},\mathbf{X},\hat{\alpha},\hat{\beta}\right]\hat{\mathcal{R}}(h)(\mathbf{Y}_1,\mathbf{Y}_2,\mathbf{X}).$$

The statistic $\hat{T}_{eff}$ follows χ^2_{df} with $df = dim(\theta)$ in general.

6.5 Doubly robust test of conditional independence

6.5.1 Estimates based on the semiparametric likelihoods

In all the tests for conditional independence, calculation of the p-values is based on certain model assumptions. For the prospective likelihood ratio test, it requires the correct model assumption for $\eta_1(y_1; x)$. The retrospective likelihood ratio test requires the correct model assumption for $\eta_2(y_2; x)$. The joint likelihood score test requires both models to be correct. If the required model assumptions fail to hold, the type I error can be inflated. To protect from the potential inflation of the type I error, it is important to make the test robust against the model misspecification. However, total robustness against the model assumptions may not be achievable in general due to the curse of dimensionality when X is of moderate to high dimension. A weaker version of robustness against the misspecification of the parametric models for $\eta_1(y_1; x)$ and $\eta_2(y_2; x)$, i.e., the double robustness, can be achieved. The doubly robust tests guarantee the control of type I error when either the model for $\eta(y_1; x)$ or the model for $\eta(y_2, x)$ is correctly specified and reasonably accurately estimated.

Suppose that the semiparametric odds ratio models are specified and estimated as stated previously. The estimated efficient score is

$$\begin{aligned}\hat{R}(h)(Y_1,Y_2,X) &= h(Y_1,Y_2,X)-E_{(\hat{\beta},\hat{F}_2)}\{h(Y_1,Y_2,X)\mid Y_1,X\}\\ &\quad -E_{(\hat{\alpha},\hat{F}_1)}\{h(Y_1,Y_2,X)\mid Y_2,X\}+E_{(\hat{\alpha},\hat{F}_1,\hat{\beta},\hat{F}_2)}\{h(Y_1,Y_2,X)\mid X\}.\end{aligned}$$

Unlike in the previous section, we do not assume the model for the η_1 and η_2 are always correctly specified. Instead, we assume that

$$\begin{aligned}(\hat{\alpha}, \hat{F}_1) &\rightarrow (\alpha_*, F_{1*}),\\ (\hat{\beta}, \hat{F}_2) &\rightarrow (\beta_*, F_{2*}),\end{aligned}$$

in probability, where (α_*, F_{1*}) may or may not be (α_0, F_{10}) and (β_*, F_{2*}) may or may not be (β_0, F_{20}). Similarly, use Taylor expansion around $(\alpha_*, \beta_*, F_{1*}, F_{2*})$, we have

$$\begin{aligned}\hat{R}(h) &= R_*(h) - E\left[h(Y_1,Y_2,X)\left\{\dot{l}^r_\beta(\hat{\beta}-\beta_*) + \dot{l}^r_{F_2}(\hat{F}_2 - F_{2*})\right\} \mid Y_1, X, \beta_*, F_{2*}\right]\\ &\quad -E\left[h(Y_1,Y_2,X)\left\{\dot{l}^p_\alpha(\hat{\alpha}-\alpha_*) + \dot{l}^p_{F_1}(\hat{F}_1 - F_{1*})\right\} \mid Y_2, X, \alpha_*, F_{1*}\right]\\ &\quad +E\left[h(Y_1,Y_2,X)\left\{\dot{l}^r_\beta(\hat{\beta}-\beta_*) + \dot{l}^r_{F_2}(\hat{F}_2 - F_{2*})\right\} \mid X, \alpha_*, F_{1*}, \beta_*, F_{2*}\right]\\ &\quad +E\left[h(Y_1,Y_2,X)\left\{\dot{l}^p_\alpha(\hat{\alpha}-\alpha_*) + \dot{l}^p_{F_1}(\hat{F}_1 - F_{1*})\right\} \mid X, \alpha_*, F_{1*}, \beta_*, F_{2*}\right]\\ &\quad +O_p\left(|\hat{\alpha}-\alpha_*|^2 + |\hat{\beta}-\beta_*|^2 + ||\hat{F}_1 - F_{1*}||^2 + ||\hat{F}_2 - F_{2*}||^2\right),\end{aligned}$$

where

$$\begin{aligned}R_*(h) &= h(Y_1,Y_2,X) - E\{h(Y_1,Y_2,X) \mid Y_1, X, \beta_*, F_{2*}\}\\ &\quad -E\{h(Y_1,Y_2,X) \mid Y_2, X, \alpha_*, F_{1*}\} + E\{h(Y_1,Y_2,X) \mid X, \alpha_*, \beta_*, F_{1*}, F_{2*}\}.\end{aligned}$$

When the model for η_1 is correctly specified and η_2 is misspecified, $(\alpha^*, F_{1*}) = (\alpha_0, F_{10})$. The limit of the estimated semiparametric efficient score can be written as

$$\begin{aligned}R_*(h)(Y_1,Y_2,X) &= h(Y_1,Y_2,X) - E\{h(Y_1,Y_2,X) \mid Y_2, X, \alpha_0, F_{10}\}\\ &\quad -E\{h(Y_1,Y_2,X) \mid Y_1, X, \beta_*, F_{2*}\} + E\{E\{h(Y_1,Y_2,X) \mid Y_1, X, \beta_*, F_{2*}\} \mid X, \alpha_0, F_{10}].\end{aligned}$$

It can be seen that $E_0\{R_*(h)(Y_1,Y_2,X)\} = 0$. When the model for η_2 is correctly specified and η_1 is misspecified, then $(\beta_*, F_{2*}) = (\beta_0, F_{20})$. The limit of the estimated semiparametric efficient score can be rewritten as

$$\begin{aligned}R_*(h)(Y_1,Y_2,X) &= h(Y_1,Y_2,X) - E\{h(Y_1,Y_2,X) \mid Y_1, X, \beta_0, F_{20}\}\\ &\quad -E\{h(Y_1,Y_2,X) \mid Y_2, X, \alpha_*, F_{1*}\} + E\{E\{h(Y_1,Y_2,X) \mid Y_2, X, \alpha_*, F_{1*}\} \mid X, \beta_0, F_{20}].\end{aligned}$$

It also follows that $E_0\{R_*(h)(Y_1,Y_2,X)\} = 0$. Note further that, when η_1 is correctly specified and η_2 is misspecified,

$$\begin{aligned}&\sum_{i=1}^n E\left[h(Y_{i1},Y_{i2},X_i)\left\{\dot{l}^p_\alpha(\hat{\alpha}-\alpha_0) + \dot{l}^p_{F_1}(\hat{F}_1 - F_{10})\right\} \mid Y_{i2}, X_i, \alpha_0, F_{10}\right]\\ &\quad -\sum_{i=1}^n E\left[h(Y_{i1},Y_{i2},X_i)\left\{\dot{l}^p_\alpha(\hat{\alpha}-\alpha_0) + \dot{l}^p_{F_1}(\hat{F}_1 - F_{10})\right\} \mid X_i, \alpha_0, F_{10}, \beta_*, F_{2*}\right]\\ &\quad = O_p\left\{n|\hat{\alpha}-\alpha_0| + n||\hat{F}_1 - F_{10}||\right\}.\end{aligned}$$

and

$$\begin{aligned}&\sum_{i=1}^n E\left[h(Y_{i1},Y_{i2},X_i)\left\{\dot{l}^r_\beta(\hat{\beta}-\beta_*) + \dot{l}^r_{F_2}(\hat{F}_2 - F_{2*})\right\} \mid Y_{i1}, X_i, \beta_*, F_{2*}\right]\\ &\quad -\sum_{i=1}^n E\left[h(Y_{i1},Y_{i2},X_i)\left\{\dot{l}^r_\beta(\hat{\beta}-\beta_*) + \dot{l}^r_{F_2}(\hat{F}_2 - F_{2*})\right\} \mid X_i, \alpha_0, F_{10}, \beta_*, F_{2*}\right]\\ &\quad = O_p\left\{\sqrt{n}|\hat{\beta}-\beta_*| + \sqrt{n}||\hat{F}_2 - F_{2*}||\right\}.\end{aligned}$$

When η_2 is correctly specified and η_1 is misspecified,

$$\sum_{i=1}^{n} E\left[h(Y_{i1}, Y_{i2}, X_i)\left\{\dot{l}_{\alpha}^{p}(\hat{\alpha}-\alpha_*) + \dot{l}_{F_1}^{p}(\hat{F}_1 - F_{1*})\right\} \mid Y_{i2}, X_i, \alpha_*, F_{1*}\right]$$
$$-\sum_{i=1}^{n} E\left[h(Y_{i1}, Y_{i2}, X_i)\left\{\dot{l}_{\alpha}^{p}(\hat{\alpha}-\alpha_*) + \dot{l}_{F_1}^{p}(\hat{F}_1 - F_{1*})\right\} \mid X_i, \alpha_*, F_{1*}, \beta_0, F_{20}\right]$$
$$= O_p\left\{\sqrt{n}|\hat{\alpha}-\alpha_*| + \sqrt{n}||\hat{F}_1 - F_{1*}||\right\}.$$

and

$$\sum_{i=1}^{n} E\left[h(Y_{i1}, Y_{i2}, X_i)\left\{\dot{l}_{\beta}^{r}(\hat{\beta}-\beta_*) + \dot{l}_{F_2}^{r}(\hat{F}_2 - F_{2*})\right\} \mid Y_{i1}, X_i, \beta_*, F_{2*}\right]$$
$$-\sum_{i=1}^{n} E\left[h(Y_{i1}, Y_{i2}, X_i)\left\{\dot{l}_{\beta}^{r}(\hat{\beta}-\beta_*) + \dot{l}_{F_2}^{r}(\hat{F}_2 - F_{2*})\right\} \mid X_i, \alpha_0, F_{10}, \beta_*, F_{2*}\right]$$
$$= O_p\left\{n|\hat{\beta}-\beta_0| + n||\hat{F}_2 - F_{20}||\right\}.$$

From the expansion of the likelihood scores, it follows that

$$\begin{aligned}
b_1^t\sqrt{n}(\hat{\alpha}-\alpha_*) + \int e_1(y_1)d\sqrt{n}(\hat{F}_1 - F_{1*}) &= \sqrt{n}(\dot{l}_{\alpha}^{p}, \dot{l}_{F_1}^{p})\left\{\Sigma_{1*}^{-1}(b_1, e_1)\right\} + o_p(1), \\
b_2^t\sqrt{n}(\hat{\beta}-\beta_*) + \int e_2(y_2)d\sqrt{n}(\hat{F}_2 - F_{2*}) &= \sqrt{n}(\dot{l}_{\beta}^{r}, \dot{l}_{F_2}^{r})\left\{\Sigma_{2*}^{-1}(b_2, e_2)\right\} + o_p(1).
\end{aligned}$$

where Σ_{1*} and Σ_{2*} respectively satisfy

$$\begin{aligned}
(b_1, e_1)\Sigma_{1*}(b_1, e_1) &= E_0\left[\sigma_{1*}^{-1}\{\dot{l}_{\alpha}^{p}(b_1) + \dot{l}_{F_1}^{p}(e_1)\}\right]^2, \\
(b_2, e_2)\Sigma_{2*}(b_2, e_2) &= E_0\left[\sigma_{2*}^{-1}\{\dot{l}_{\beta}^{r}(b_2) + \dot{l}_{F_2}^{r}(e_2)\}\right]^2,
\end{aligned}$$

with

$$\begin{aligned}
(b_1, e_1)\sigma_{1*}(b_1, e_1) &= -E_0\left\{\ddot{l}_{\alpha\alpha}^{p}(b_1, b_1) + \ddot{l}_{\alpha F_1}^{p}(b_1, e_1) + \ddot{l}_{F_1,\alpha}^{p}(e_1, b_1) + \ddot{l}_{F_1F_1}^{p}(e_1, e_1)\right\}, \\
(b_2, e_2)\sigma_{2*}(b_2, e_2) &= -E_0\left\{\ddot{l}_{\beta\beta}^{r}(b_2, b_2) + \ddot{l}_{\beta F_2}^{r}(b_2, e_2) + \ddot{l}_{F_2,\beta}^{r}(e_2, b_2) + \ddot{l}_{F_2F_2}^{r}(e_2, e_2)\right\}.
\end{aligned}$$

The asymptotic variance of the doubly robust test score is

$$V_{\hat{R}} = E_0\left\{R_*(h)(Y_{i1}, Y_{i2}, X_i) - (s_{1*}, t_{1*})\Sigma_{1*}^{-1}(\dot{l}_{i\alpha}^{p}, \dot{l}_{iF_1}^{p}) - (s_{2*}, t_{2*})\Sigma_{2*}^{-1}(\dot{l}_{i\beta}^{r}, \dot{l}_{iF_2}^{r})\right\}^2,$$

where

$$\begin{aligned}
(s_1, t_1) &= \Big(E_0\left[E_*\left\{h(\mathbf{Y}_1, \mathbf{Y}_2, \mathbf{X})\dot{l}_{\alpha} \mid \mathbf{Y}_2, \mathbf{X}\right\} - E_*\left\{h(\mathbf{Y}_1, \mathbf{Y}_2, \mathbf{X})\dot{l}_{\alpha} \mid \mathbf{X}\right\}\right], \\
&\quad E_0\left[E_*\left\{h(\mathbf{Y}_1, \mathbf{Y}_2, \mathbf{X})\dot{l}_{F_1} \mid \mathbf{Y}_2, \mathbf{X}\right\} - E_*\left\{h(\mathbf{Y}_1, \mathbf{Y}_2, \mathbf{X})\dot{l}_{F_1} \mid \mathbf{X}\right\}\right]\Big), \\
(s_2, t_2) &= \Big(E_0\left[E_*\left\{h(\mathbf{Y}_1, \mathbf{Y}_2, \mathbf{X})\dot{l}_{\beta} \mid \mathbf{Y}_1, \mathbf{X}\right\} - E_*\left\{h(\mathbf{Y}_1, \mathbf{Y}_2, \mathbf{X})\dot{l}_{\beta} \mid \mathbf{X}\right\}\right], \\
&\quad E_0\left[E_*\left\{h(\mathbf{Y}_1, \mathbf{Y}_2, \mathbf{X})\dot{l}_{F_2} \mid \mathbf{Y}_2, \mathbf{X}\right\} - E_*\left\{h(\mathbf{Y}_1, \mathbf{Y}_2, \mathbf{X})\dot{l}_{F_2} \mid \mathbf{X}\right\}\right]\Big),
\end{aligned}$$

and E_* is computed based on parameter values $(\alpha_*.F_{1*}, \beta_*, F_{2*})$. A robust estimate of the variance is

$$\hat{V}_{\hat{R}} = \sum_{i=1}^{n} \left[\hat{R}\{h(Y_{i1}, Y_{i2}, X_i)\} - (\hat{s}_1, \hat{t}_1)\hat{\Sigma}_1^{-1}(l_{i\alpha}^{p}, l_{iF_1}^{p}) - (\hat{s}_2, \hat{t}_2)\hat{\Sigma}_2^{-1}(l_{i\beta}^{r}, l_{iF_2}^{r}) \right]^2,$$

where $\hat{\Sigma}_1$ and $\hat{\Sigma}_2$ are sample version of Σ_{1*} and Σ_{2*}, and

$$\begin{aligned}
(\hat{s}_1, \hat{t}_1) &= \Big(\hat{E}\left\{ h(\mathbf{Y}_1, \mathbf{Y}_2, \mathbf{X}) \dot{l}_\alpha \mid \mathbf{Y}_2, \mathbf{X} \right\} - \hat{E}\left\{ h(\mathbf{Y}_1, \mathbf{Y}_2, \mathbf{X}) \dot{l}_\alpha \mid \mathbf{X} \right\}, \\
&\quad \hat{E}\left\{ h(\mathbf{Y}_1, \mathbf{Y}_2, \mathbf{X}) \dot{l}_{F_1} \mid \mathbf{Y}_2, \mathbf{X} \right\} - \hat{E}\left\{ h(\mathbf{Y}_1, \mathbf{Y}_2, \mathbf{X}) \dot{l}_{F_1} \mid \mathbf{X} \right\} \Big), \\
(\hat{s}_2, \hat{t}_2) &= \Big(\hat{E}\left\{ h(\mathbf{Y}_1, \mathbf{Y}_2, \mathbf{X}) \dot{l}_\beta \mid \mathbf{Y}_1, \mathbf{X} \right\} - \hat{E}\left\{ h(\mathbf{Y}_1, \mathbf{Y}_2, \mathbf{X}) \dot{l}_\beta \mid \mathbf{X} \right\}, \\
&\quad \hat{E}\left\{ h(\mathbf{Y}_1, \mathbf{Y}_2, \mathbf{X}) \dot{l}_{F_2} \mid \mathbf{Y}_2, \mathbf{X} \right\} - \hat{E}\left\{ h(\mathbf{Y}_1, \mathbf{Y}_2, \mathbf{X}) \dot{l}_{F_2} \mid \mathbf{X} \right\} \Big)
\end{aligned}$$

and $\hat{E}$ is computed under the estimated parameter values $(\hat{\alpha}, \hat{F}_1, \hat{\beta}, \hat{F}_2)$. Note that, when one of the models for η_1 and η_2 is correctly specified, one component in (s_1, t_1) and in (s_2, t_2) is 0. However, since we do not know which one is 0, both estimates are needed. A doubly robust conditional independence test can be constructed as

$$\hat{T}_{DR} = \sum_{i=1}^{n} \hat{R}(h)(Y_{i1}, Y_{i2}, X_i) \hat{V}_{\hat{R}}^{-1} \sum_{i=1}^{n} \hat{R}(h)(Y_{i1}, Y_{i2}, X_i).$$

The statistic $\hat{T}_{DR}$ follows χ^2_{df} with $df = dim(\theta)$ under the null hypothesis of conditional independence.

6.5.2 Permutation-based doubly robust test of independence

The estimated efficient score based on permutations has the form

$$\begin{aligned}
\hat{\mathcal{R}}(h)(\mathbf{Y}_1, \mathbf{Y}_2, \mathbf{X}) &= h(\mathbf{Y}_1, \mathbf{Y}_2, \mathbf{X}) - E_{nA}\left\{ h(\mathbf{Y}_1, \mathbf{Y}_2, \mathbf{X}) \mid \mathbf{Y}_1^{ord}, \mathbf{Y}_2, \mathbf{X}, \hat{\alpha} \right\} \\
-E_{nB}&\left\{ h(\mathbf{Y}_1, \mathbf{Y}_2, \mathbf{X}) \mid \mathbf{Y}_1, \mathbf{Y}_2^{ord}, \mathbf{X}, \hat{\beta} \right\} + E_{nAB}\left\{ h(\mathbf{Y}_1, \mathbf{Y}_2, \mathbf{X}) \mid \mathbf{Y}_1^{ord}, \mathbf{Y}_2^{ord}, \mathbf{X}, \hat{\alpha}, \hat{\beta} \right\}.
\end{aligned}$$

When the models for η_1 and η_2 can be wrong, assume that

$$\hat{\alpha} \to \alpha_* \text{ and } \hat{\beta} \to \beta_*,$$

in probability. By Taylor expansion around (α_*, β_*), the estimated score can be expressed as

$$\begin{aligned}
\hat{\mathcal{R}}(h) &= \mathcal{R}_*(h) - E_{nA}\left\{ h(\mathbf{Y}_1, \mathbf{Y}_2, \mathbf{X}) \dot{l}_\alpha^{pp} \mid \mathbf{Y}_1^{ord}, \mathbf{Y}_2, \mathbf{X}, \alpha_* \right\} (\hat{\alpha} - \alpha_*) \\
&\quad + E_{nAB}\left\{ h(\mathbf{Y}_1, \mathbf{Y}_2, \mathbf{X}) \dot{l}_\alpha^{pp} \mid \mathbf{Y}_1^{ord}, \mathbf{Y}_2^{ord}, \mathbf{X}, \alpha_*, \beta_* \right\} (\hat{\alpha} - \alpha_*) \\
&\quad - E_{nB}\left\{ h(\mathbf{Y}_1, \mathbf{Y}_2, \mathbf{X}) \dot{l}_\beta^{rp} \mid \mathbf{Y}_1, \mathbf{Y}_2^{ord}, \mathbf{X}, \beta_* \right\} (\hat{\beta} - \beta_*) \\
&\quad + E_{nAB}\left\{ h(\mathbf{Y}_1, \mathbf{Y}_2, \mathbf{X}) \dot{l}_\beta^{rp} \mid \mathbf{Y}_1^{ord}, \mathbf{Y}_2^{ord}, \mathbf{X}, \alpha_*, \beta_* \right\} (\hat{\beta} - \beta_*) \\
&\quad + O_p\left(n|\hat{\alpha} - \alpha_*|^2 + n|\hat{\beta} - \beta_*|^2 \right),
\end{aligned}$$

where

$$\begin{aligned}\mathcal{R}_*(h) &= h(\mathbf{Y}_1,\mathbf{Y}_2,\mathbf{X}) - E_{nA}\left\{h(\mathbf{Y}_1,\mathbf{Y}_2,\mathbf{X}) \mid \mathbf{Y}_1^{ord},\mathbf{Y}_2,\mathbf{X},\alpha_*\right\} \\ &\quad -E_{nB}\left\{h(\mathbf{Y}_1,\mathbf{Y}_2,\mathbf{X}) \mid \mathbf{Y}_1,\mathbf{Y}_2^{ord},\mathbf{X},\beta_*\right\} \\ &\quad +E_{nAB}\left\{h(\mathbf{Y}_1,\mathbf{Y}_2,\mathbf{X}) \mid \mathbf{Y}_1^{ord},\mathbf{Y}_2^{ord},\mathbf{X},\alpha_*,\beta_*\right\}.\end{aligned}$$

When $\alpha_* = \alpha_0$,

$$\begin{aligned}\mathcal{R}_*(h) &= h(\mathbf{Y}_1,\mathbf{Y}_2,\mathbf{X}) - E_{nA}\left\{h(\mathbf{Y}_1,\mathbf{Y}_2,\mathbf{X}) \mid \mathbf{Y}_1^{ord},\mathbf{Y}_2,\mathbf{X},\alpha_0\right\} \\ &\quad -E_{nB}\left\{h(\mathbf{Y}_1,\mathbf{Y}_2,\mathbf{X}) \mid \mathbf{Y}_1,\mathbf{Y}_2^{ord},\mathbf{X},\beta_*\right\} \\ &\quad +E_{nA}\left[E_{nB}\left\{h(\mathbf{Y}_1,\mathbf{Y}_2,\mathbf{X}) \mid \mathbf{Y}_1,\mathbf{Y}_2^{ord},\mathbf{X},\beta_*\right\} \mid Y_1^{ord},Y_2^{ord},\alpha_0\right].\end{aligned}$$

It follows that $E_0\{\mathcal{R}_*(h)\} = 0$. On the other hand, when $\beta_* = \beta_0$,

$$\begin{aligned}\mathcal{R}_*(h) &= h(\mathbf{Y}_1,\mathbf{Y}_2,\mathbf{X}) - E_{nB}\left\{h(\mathbf{Y}_1,\mathbf{Y}_2,\mathbf{X}) \mid \mathbf{Y}_1,\mathbf{Y}_2^{ord},\mathbf{X},\beta_0\right\} \\ &\quad -E_{nA}\left\{h(\mathbf{Y}_1,\mathbf{Y}_2,\mathbf{X}) \mid \mathbf{Y}_1^{ord},\mathbf{Y}_2,\mathbf{X},\alpha_*\right\} \\ &\quad +E_{nB}\left[E_{nA}\left\{h(\mathbf{Y}_1,\mathbf{Y}_2,\mathbf{X}) \mid \mathbf{Y}_1^{ord},\mathbf{Y}_2,\mathbf{X},\alpha_*\right\} \mid Y_1^{ord},Y_2^{ord},\beta_0\right].\end{aligned}$$

It also follows that $E_0\{\mathcal{R}_*(h)\} = 0$. Assume further that

$$\begin{aligned}\sqrt{n}(\hat{\alpha}-\alpha_*) &= \left\{-\ddot{l}_\alpha^{pp}(\alpha_*)\right\}^{-1}\sqrt{n}\dot{l}_\alpha^{pp}(\alpha_*) + o_p(1), \\ \sqrt{n}(\hat{\beta}-\beta_*) &= \left\{-\ddot{l}_\beta^{rp}(\beta_*)\right\}^{-1}\sqrt{n}\dot{l}_\beta^{rp}(\beta_*) + o_p(1).\end{aligned}$$

Let

$$\begin{aligned}s_1 &= E_{nA}\left\{h(\mathbf{Y}_1,\mathbf{Y}_2,\mathbf{X})\dot{l}_\alpha^{pp} \mid \mathbf{Y}_1^{ord},\mathbf{Y}_2,\mathbf{X},\alpha_*\right\} \\ &\quad -E_{nAB}\left\{h(\mathbf{Y}_1,\mathbf{Y}_2,\mathbf{X})\dot{l}_\alpha^{pp} \mid \mathbf{Y}_1^{ord},\mathbf{Y}_2^{ord},\mathbf{X},\alpha_*,\beta_*\right\} \\ s_2 &= E_{nB}\left\{h(\mathbf{Y}_1,\mathbf{Y}_2,\mathbf{X})\dot{l}_\beta^{rp} \mid \mathbf{Y}_1,\mathbf{Y}_2^{ord},\mathbf{X},\beta_*\right\} \\ &\quad -E_{nAB}\left\{h(\mathbf{Y}_1,\mathbf{Y}_2,\mathbf{X})\dot{l}_\beta^{rp} \mid \mathbf{Y}_1^{ord},\mathbf{Y}_2^{ord},\mathbf{X},\alpha_*,\beta_*\right\}.\end{aligned}$$

The asymptotic variance of $\hat{\mathcal{R}}(h)$ is

$$V_{\hat{\mathcal{R}}(h)} = E_0\left[\mathcal{R}_*(h) - s_1\left\{-E_0\left(\ddot{l}_\alpha^{pp}\right)\right\}^{-1}\dot{l}_\alpha^{pp} - s_2\left\{-E_0\left(\ddot{l}_\beta^{rp}\right)\right\}^{-1}\dot{l}_\beta^{rp}\right]^2,$$

To obtain an etimate of the variance, apply the arguments in §5.5.2. First, let

$$\begin{aligned}g_{11h}(Y_{i2},X_i \mid \alpha_*) &= \sum_{j\neq i} h(Y_{j1},Y_{i2},X_i)P\{\sigma(i)=j \mid \alpha_*\}, \\ g_{21h}(Y_{i1} \mid \alpha_*) &= \sum_{j\neq i} h(Y_{i1},Y_{j2},X_j)P\{\sigma(j)=i \mid \alpha_*\}.\end{aligned}$$

Likewise, let

$$\begin{aligned}g_{12h}(Y_{i2} \mid \beta_*) &= \sum_{j\neq i} h(Y_{j1},Y_{i2},X_j)P\{\tau(j)=i \mid \beta_*\}, \\ g_{22h}(Y_{i1},X_i \mid \beta_*) &= \sum_{j\neq i} h(Y_{i1},Y_{j2},X_i)P\{\tau(i)=j \mid \beta_*\},\end{aligned}$$

and

$$\begin{aligned} g_{13h}(Y_{i1} \mid \alpha_*, \beta_*) &= \sum_{j \neq i,k} \sum_{k \neq i,j} h(Y_{i1}, Y_{j2}, X_k) P\{\sigma(k) = i \mid \alpha_*\} P\{\tau(k) = j \mid \beta_*\}, \\ g_{23h}(Y_{i2} \mid \alpha_*, \beta_*) &= \sum_{j \neq i,k} \sum_{k \neq i,j} h(Y_{j1}, Y_{i2}, X_k) P\{\sigma(k) = i \mid \alpha_*\} P\{\tau(k) = j \mid \beta_*\}, \\ g_{33h}(X_i \mid \alpha_*, \beta_*) &= \sum_{j \neq i,k} \sum_{k \neq i,j} h(Y_{j1}, Y_{k2}, X_i) P\{\sigma(i) = j \mid \alpha_*\} P\{\tau(i) = k \mid \beta_*\}. \end{aligned}$$

Define

$$\begin{aligned} \mathcal{R}_{i*}(h)(\alpha_*, \beta_*) &= h(Y_{i1}, Y_{i2}, X_i) - g_{11h}(Y_{i2}, X_i \mid \alpha_*) - g_{21h}(Y_{i1} \mid \alpha_*) \\ &\quad - g_{12h}(Y_{i2} \mid \beta_*) - g_{22h}(Y_{i1}, X_i \mid \alpha_*) \\ &\quad + g_{13h}(Y_{i1} \mid \alpha_*, \beta_*) + g_{23h}(Y_{i2} \mid \alpha_*, \beta_*) + g_{33h}(X_i \mid \alpha_*, \beta_*). \end{aligned}$$

Furthermore, let

$$\begin{aligned} g_{1sp}(Y_{i1} \mid \alpha_*) &= \sum_{j \neq i} \frac{\partial l^{pp}}{\partial \alpha_s}(Y_{i1} \mid X_j, \alpha_*) P\{\sigma(j) = i \mid \alpha_*\}, \\ g_{2sp}(X_i \mid \alpha_*) &= \sum_{j \neq i} \frac{\partial l^{pp}}{\partial \alpha_s}(Y_{j1} \mid X_i, \alpha_*) P\{\sigma(i) = j \mid \alpha_*\}. \end{aligned}$$

Let

$$\begin{aligned} g_{1tr}(Y_{i2} \mid \beta_*) &= \sum_{j \neq i} \frac{\partial l^{rp}}{\partial \beta_t}(Y_{i2} \mid X_j, \beta_*) P\{\tau(j) = i \mid \alpha_*\}, \\ g_{2tr}(X_i \mid \beta_*) &= \sum_{j \neq i} \frac{\partial l^{rp}}{\partial \beta_t}(Y_{j2} \mid X_i, \beta_*) P\{\tau(i) = j \mid \alpha_*\}, \end{aligned}$$

where $s = 1, \cdots, dim(\alpha), t = 1, \cdots, dim(\beta)$. Let

$$\begin{aligned} g_{1p}(Y_{i1} \mid \alpha_*) &= (g_{11p}(Y_{i1} \mid \alpha_*), \cdots, g_{1dim(\alpha)p}(Y_{i1} \mid \alpha_*))^t, \\ g_{2p}(X_i \mid \alpha_*) &= (g_{11p}(X_i \mid \alpha_*), \cdots, g_{1dim(\alpha)p}(X_i \mid \alpha_*))^t, \\ g_{1r}(Y_{i2} \mid \beta_*) &= (g_{21r}(Y_{i2} \mid \beta_*), \cdots, g_{2\,\mathrm{dim}(\beta)r}(Y_{i2} \mid \beta_*))^t, \\ g_{2r}(X_i \mid \beta_*) &= (g_{21r}(X_i \mid \beta_*), \cdots, g_{2\,\mathrm{dim}(\beta)r}(X_i \mid \beta_*))^t. \end{aligned}$$

An estimate of the variance can be obtained as

$$\begin{aligned} \hat{V}_{\hat{\mathcal{R}}(h)} &= \sum_{i=1}^{n} \Big[\mathcal{R}_{i*}(h)(\hat{\alpha}, \hat{\beta}) - \hat{s}_1 \left\{ -\ddot{l}_{\alpha}^{pp}(\hat{\alpha}) \right\}^{-1} \{ g_{1p}(Y_{i1} \mid \hat{\alpha}) + g_{2p}(X_i \mid \hat{\alpha}) \} \\ &\quad - \hat{s}_2 \left\{ -\ddot{l}_{\beta}^{rp}(\hat{\beta}) \right\}^{-1} \left\{ g_{1r}(Y_{i2} \mid \hat{\beta}) + g_{2r}(X_i \mid \hat{\beta}) \right\} \Big]^2, \end{aligned}$$

where

$$\begin{aligned} s_1 &= E_{nA} \left\{ h(\mathbf{Y}_1, \mathbf{Y}_2, \mathbf{X}) \dot{l}_{\alpha}^{pp} \mid \mathbf{Y}_1^{ord}, \mathbf{Y}_2, \mathbf{X}, \hat{\alpha} \right\} \\ &\quad - E_{nAB} \left\{ h(\mathbf{Y}_1, \mathbf{Y}_2, \mathbf{X}) \dot{l}_{\alpha}^{pp} \mid \mathbf{Y}_1^{ord}, \mathbf{Y}_2^{ord}, \mathbf{X}, \hat{\alpha}, \hat{\beta} \right\} \\ s_2 &= E_{nB} \left\{ h(\mathbf{Y}_1, \mathbf{Y}_2, \mathbf{X}) \dot{l}_{\beta}^{rp} \mid \mathbf{Y}_1, \mathbf{Y}_2^{ord}, \mathbf{X}, \hat{\beta} \right\} \\ &\quad - E_{nAB} \left\{ h(\mathbf{Y}_1, \mathbf{Y}_2, \mathbf{X}) \dot{l}_{\beta}^{rp} \mid \mathbf{Y}_1^{ord}, \mathbf{Y}_2^{ord}, \mathbf{X}, \hat{\alpha}, \hat{\beta} \right\}. \end{aligned}$$

A doubly robust test of independence can be obtained as

$$\hat{\mathcal{T}}_{DR} = \hat{\mathcal{R}}(h)(\mathbf{Y}_1, \mathbf{Y}_2, \mathbf{X})\hat{V}^{-1}_{\hat{\mathcal{R}}(h)}\hat{\mathcal{R}}(h)(\mathbf{Y}_1, \mathbf{Y}_2, \mathbf{X}).$$

The statistic $\hat{\mathcal{T}}_{DR}$ follows χ^2_{df} with $df = dim(\theta)$ in general.

6.5.3 Bootstrap implementation of the doubly robust tests

The doubly robust tests can be difficult to implement due to the complications in obtaining the variance estimate. A bootstrapping approach may be used to circumvent this difficulty. For the observed data $(Y_{i1}, Y_{i2}, X_i), i = 1, \cdots, n$, we may apply the following algorithm to obtain the variance estimate when the semiparametric efficient score is employed.

1. Take bootstrap samples $(Y_{i1}^{(k)}, Y_{i2}^{(k)}, X_i^{(k)}), i = 1, \cdots, n$, where $k = 1, \cdots, N$ index the bootstrap samples.
2. Fit the nuisance models respectively to the bootstrap samples using the semiparametric likelihood approach to obtain estimates $(\hat{\alpha}^{(k)}, \hat{F}_1^{(k)})$ and $(\hat{\beta}^{(k)}, \hat{F}_2^{(k)})$ for $k = 1, \cdots, N$.
3. Compute the doubly robust score as

$$\begin{aligned} R^{(k)}(h) = \frac{1}{\sqrt{n}}\sum_{i=1}^{n}\Big(& h(Y_{i1}^{(k)}, Y_{i2}^{(k)}, X_i^{(k)}) - E\left\{h(Y_{i1}^{(k)}, Y_{i2}^{(k)}, X_i^{(k)}) \mid Y_{i1}^{(k)}, X_i^{(k)}, \hat{\alpha}^{(k)}, \hat{F}_1^{(k)}\right\} \\ & - E\left\{h(Y_{i1}^{(k)}, Y_{i2}^{(k)}, X_i^{(k)}) \mid Y_{i2}^{(k)}, X_i^{(k)}, \hat{\beta}^{(k)}, \hat{F}_2^{(k)}\right\} \\ & + E\left[E\left\{h(Y_{i1}^{(k)}, Y_{i2}^{(k)}, X_i^{(k)}) \mid Y_{i1}^{(k)}, X_i^{(k)}, \hat{\alpha}^{(k)}, \hat{F}_1^{(k)}\right\} \mid Y_{i2}^{(k)}, X_i^{(k)}, \hat{\beta}^{(k)}, \hat{F}_2^{(k)}\right]\Big), \end{aligned}$$

   for $k = 1, \cdots, N$.
4. Compute the empirical variance of the bootstrapped doubly robust statistic as

$$\hat{V}_{bt} = \frac{1}{N-1}\sum_{k=1}^{N}\left\{R^{(k)}(h) - \frac{1}{N}\sum_{k=1}^{N} R^{(k)}(h)\right\}^2.$$

The bootstrap-based doubly robust test is

$$\mathcal{T}_{DRbt} = \frac{1}{\sqrt{n}}\sum_{i=1}^{n}\hat{\mathcal{R}}(h)(Y_i, X_i, Z_i)\hat{V}_{bt}^{-1}\frac{1}{\sqrt{n}}\sum_{i=1}^{n}\hat{\mathcal{R}}(h)(Y_i, X_i, Z_i),$$

which asymptotically follows χ^2_{df} with degrees of freedom $df = dim(\theta)$.

When the permutation likelihoods are used in forming the test statistics, steps 2 and 3 in the foregoing algorithm can be changed as follow.

2. Fit the nuisance models respectively to the bootstrap samples using the permutation likelihood approach to obtain estimates $\hat{\alpha}^{(k)}$ and $\hat{\beta}^{(k)}$ for $k = 1, \cdots, N$.
3. Compute the doubly robust score as

$$\begin{aligned} R^{(k)}(h) = {} & h(\mathbf{Y}_1^{(k)}, \mathbf{Y}_2^{(k)}, \mathbf{X}^{(k)}) - E_{nA}\left\{h(\mathbf{Y}_1^{(k)}, \mathbf{Y}_2^{(k)}, \mathbf{X}^{(k)}) \mid \mathbf{Y}_1^{(k)ord}, \mathbf{Y}_2^{(k)}, \mathbf{X}^{(k)}, \hat{\alpha}\right\} \\ & - E_{nB}\left\{h(\mathbf{Y}_1^{(k)}, \mathbf{Y}_2^{(k)}, \mathbf{X}^{(k)}) \mid \mathbf{Y}_1^{(k)}, \mathbf{Y}_2^{(k)ord}, \mathbf{X}^{(k)}, \hat{\beta}\right\} \\ & + E_{nAB}\left\{h(\mathbf{Y}_1^{(k)}, \mathbf{Y}_2^{(k)}, \mathbf{X}^{(k)}) \mid \mathbf{Y}_1^{(k)ord}, \mathbf{Y}_2^{(k)ord}, \mathbf{X}, \hat{\alpha}, \hat{\beta}\right\}. \end{aligned}$$

   for $k = 1, \cdots, N$.

The rest of steps can be carried out similarly.

6.6 Connections to classical tests of conditional independence

A number of tests for conditional independence hypothesis have been proposed in the literature. Among them, the most well-known test to the epidemiologists is perhaps the Mantel-Haeszel test for independence of row and column variables in stratified contingency tables. A little less well-known test that extends the Mantel-Haeszel test to continuous variables is the general Mantel test. These tests assume the log-conditional odds ratio function has the bilinear form, i.e.,

$$\log \eta(y_1; y_2 \mid x) = \theta(x)(y_1 - y_{10})(y_2 - y_{20}),$$

where one choice for $\theta(x)$ is $\theta(x) = \theta_0 + \theta_1 x$. The hypothesis corresponds $H_0 : \theta_0 = 0$ when we know $\theta_1 = 0$, or $H_0 : \theta_0 = \theta_1 = 0$ when we do not know if $\theta_1 = 0$. The log-likelihood scores for θ corresponding to the two cases, i.e, the h functions, are respectively

$$\begin{aligned} h_1(Y_1, Y_2, X) &= (Y_1 - y_{10})(Y_2 - y_{20}), \\ h_2(Y_1, Y_2, X) &= (1, X)(Y_1 - y_{10})(Y_2 - y_{20}). \end{aligned}$$

Note that h_2 is a vector function of 2-dimension.

The efficient scores under the null hypothesis are respectively,

$$\begin{aligned} R(h_1) &= \{Y_1 - E_0(Y_1 \mid X)\}\{Y_2 - E_0(Y_2 \mid X)\}, \\ R(h_2) &= (1, X)\{Y_1 - E_0(Y_1 \mid X)\}\{Y_2 - E_0(Y_2 \mid X)\}. \end{aligned}$$

The variances of the efficient scores are respectively

$$\begin{aligned} E_0 \{R(h_1)\}^2 &= \text{Var}_0(Y_1 \mid X)\text{Var}_0(Y_2 \mid X), \\ E_0 \{R(h_2)\}^2 &= (1, X)^t(1, X)\text{Var}_0(Y_1 \mid X)\text{Var}_0(Y_2 \mid X). \end{aligned}$$

where E_0 and Var_0 denote expectation and variance computed under the null hypothesis. To construct a test based on the efficient score, we need to obtain the conditional means and variances of Y_1 and Y_2 respectively.

For the stratified sampling design, let $(Y_{kj1}, Y_{kj2}), j = 1, \cdots, n_k$ be the observed outcomes for (Y_1, Y_2) in the stratum $X = x_k$, for $k = 1, \cdots, K$. The means and variances can be estimated separately within each stratum. For stratum k,

$$\hat{E}(Y_1 \mid X = x_k) = \frac{1}{n_k} \sum_{j=1}^{n_k} Y_{kj1}, \qquad \hat{V}(Y_1 \mid X = x_k) = \frac{1}{n_k - 1} \sum_{j=1}^{n_k} \left\{Y_{kj1} - \hat{E}(Y_1 \mid X = k)\right\}^2,$$

$$\hat{E}(Y_2 \mid X = x_k) = \frac{1}{n_k} \sum_{j=1}^{n_k} Y_{kj2}, \qquad \hat{V}(Y_2 \mid X = x_k) = \frac{1}{n_k - 1} \sum_{j=1}^{n_k} \left\{Y_{kj2} - \hat{E}(Y_2 \mid X = k)\right\}^2.$$

In the first case, the test for conditional independence can be carried out by $S_1/\sqrt{V_1} \sim N(0, 1)$ under the null hypothesis, where

$$\begin{aligned} S_1 &= \sum_{k=1}^{K} \sum_{j=1}^{n_k} \{Y_{kj1} - \hat{E}(Y_1 \mid X = x_k)\}\{Y_{kj2} - \hat{E}(Y_2 \mid X = x_k)\}, \\ V_1 &= \sum_{k=1}^{K} \hat{V}(Y_1 \mid X = x_k)\hat{V}(Y_2 \mid X = x_k). \end{aligned}$$

In the second case, let

$$\begin{aligned} S_2 &= \sum_{k=1}^{K}(1,x_k)\sum_{j=1}^{n_k}\{Y_{kj1}-\hat{E}(Y_1 \mid X=x_k)\}\{Y_{kj2}-\hat{E}(Y_2 \mid X=x_k)\}, \\ V_2 &= \sum_{k=1}^{K}(1,x_k)^t(1,x_k)\hat{V}(Y_1 \mid X=x_k)\hat{V}(Y_2 \mid X=x_k). \end{aligned}$$

The semiparametric efficient test is $S_2V_2^{-1}S_2 \sim \chi_2^2$. The validity of the tests requires the sample size in each stratum to be $O(n^\gamma)$ with $\gamma > 1/2$.

The Mantel-Haeszel test for conditional independence for the stratified $I \times J \times K$ contingency tables can be obtained as follows. Let Y_1 and Y_2 be respectively indicator functions of row and column categories. More specifically, $Y_1 = (1_{\{R=1\}}, \cdots, 1_{\{R=I-1\}})$ and $Y_2 = (1_{\{C=1\}}, \cdots, 1_{\{C=J-1\}})$, where R is the row variable and C is the column variable and the reference categories are $R = I$ and $C = J$. The log-conditional odds ratio function is

$$\log \eta(Y_1; Y_2 \mid X) = \sum_{i=1}^{I-1}\sum_{j=1}^{J-1}\theta_{ij}(1_{\{R=i\}} - 1_{\{R=I\}})(1_{\{C=j\}} - 1_{\{C=J\}}).$$

The null hypothesis is $H_0 : \theta_{ij} = 0, i = 1, \cdots, I-1; j = 1, \cdots, J-1$. The efficient scores for estimating $\theta_{ij}, i = 1, \cdots, I-1, j = 1, \cdots, J-1$ under the null hypothesis are

$$S = \left(\sum_{k=1}^{K}\left(n_{ijk} - \frac{n_{i+k}n_{+jk}}{n_{++k}}\right), i = 1, \cdots, I-1, j = 1, \cdots, J-1\right),$$

where n_{ijk} is the cell counts for $Y_1 = i$, $Y_2 = j$ in stratum k and a $+$ in the subscript means a sum over that subscript index. The variance matrix is

$$V = \sum_{k=1}^{K}\hat{V}(Y_1 \mid X_k) \otimes \hat{V}(Y_2 \mid X_k),$$

where $\otimes$ is the Kronecker product, and

$$\begin{aligned} \hat{V}(Y_1 \mid X=x_k) &= \operatorname{diag}\left(\frac{n_{i+k}}{n_{++k}}, i = 1, \cdots, I-1\right) - \left(\frac{n_{i+k}n_{j+k}}{n_{++k}^2}, i, j = 1, \cdots, I-1\right), \\ \hat{V}(Y_2 \mid X=x_k) &= \operatorname{diag}\left(\frac{n_{+jk}}{n_{++k}}, i = 1, \cdots, J-1\right) - \left(\frac{n_{+ik}n_{+jk}}{n_{++k}^2}, i, j = 1, \cdots, J-1\right). \end{aligned}$$

The test has $SV^{-1}S \sim \chi^2_{(I-1)(J-1)}$.

Stratified sample corresponds to have categorical X, which allows for consistent estimation of the conditional expectations in the efficient score. For problems without stratification structure, in particular, with continuous covariate vector X, modeling the conditional expectations of Y_1 given X and Y_2 given X is needed for constructing the independence test. The development in the previous sections used semiparametric odds ratio model for estimating the conditional expectations. Regression models are often used instead to directly model those expectations. For example, nonlinear regression models may be specified as follow

$$\begin{aligned} Y_1 &= g_1(X) + \epsilon_1, \\ Y_2 &= g_2(X) + \epsilon_2, \end{aligned}$$

where $E(\epsilon_k) = 0$, $V(\epsilon_k) = \sigma_k^2$, for $k = 1, 2$. Let $g_1(X)$ be modeled parametrically as $g_1(X, \alpha)$ and $g_2(X)$ be modeled parametrically as $g_2(X, \beta)$. Denote the estimator of α and β respectively by $\hat{\alpha}$ and $\hat{\beta}$. If both models are specified corrected and

$$\hat{\alpha} - \alpha_0 = o_p(n^{1/4}), \text{ and } \hat{\beta} - \beta_0 = o_p(n^{1/4}),$$

then the test statistic

$$T_{eff} = \frac{\sqrt{n}\sum_{i=1}^n \{Y_{i1} - g_1(X_i, \hat{\alpha})\}\left\{Y_{i2} - g_2(X_i, \hat{\beta})\right\}}{\left[\sum_{i=1}^n \{Y_{i1} - g_1(X_i, \hat{\alpha})\}^2 \sum_{i=1}^n \left\{Y_{i2} - g_2(X_i, \hat{\beta})\right\}^2\right]^{1/2}} \sim \chi_1^2.$$

In practice, there is no guarantee that these two parametric models are correctly specified. If one model is correctly specified and the other model may not be correctly specified, the variance in the denominator needs to be modified to avoid the inflation of the type I error due to incorrect asymptotic variance. Let

$$\begin{aligned} r_{i1} &= Y_{i1} - g_1(X_i, \alpha_0), \\ r_{i2} &= Y_{i2} - g_2(X_i, \beta_0). \end{aligned}$$

Let $\hat{r}_{i1}$ and $\hat{r}_{i2}$ be the estimated residual when α_0, β_0 in r_{i1}, r_{i2} are replaced by $\hat{\alpha}, \hat{\beta}$ respectively. Suppose that

$$\begin{aligned} \sqrt{n}(\hat{\alpha} - \alpha_0) &= \frac{1}{\sqrt{n}}\sum_{i=1}^n S_A(Y_{i1}, X_i) + o_p(1), \\ \sqrt{n}(\hat{\beta} - \beta_0) &= \frac{1}{\sqrt{n}n}\sum_{i=1}^n S_B(Y_{i2}, X_i) + o_p(1). \end{aligned}$$

we have that

$$\sum_{i=1}^n \hat{r}_{i1}\hat{r}_{i2} = \sum_{i=1}^n \{r_{i1}r_{i2} - t_A S_A(Y_{i1}, X_i) - t_B S_B(Y_{i2}, X_i)\},$$

where

$$\begin{aligned} t_A &= \frac{1}{n}\sum_{i=1}^n r_{i1}\frac{\partial}{\partial\alpha} g_1(X_i, \alpha_0), \\ t_B &= \frac{1}{n}\sum_{i=1}^n r_{i2}\frac{\partial}{\partial\beta} g_2(X_i, \beta_0). \end{aligned}$$

The modified variance for the estimated score is

$$V = \sum_{i=1}^n \{r_{i1}r_{i2} - t_A S_A(Y_{i1}, X_i) - t_B S_B(Y_{i2}, X_i)\}^2.$$

A doubly robust test of conditional independence in this case is

$$T_{DR} = \frac{\sum_{i=1}^n \hat{r}_{i1}\hat{r}_{i2}}{\left[\sum_{i=1}^n \{\hat{r}_{i1}\hat{r}_{i2} - \hat{t}_A S_A(Y_{i1}, X_i) - \hat{t}_B S_B(Y_{i2}, X_i)\}^2\right]^{1/2}},$$

follows $N(0,1)$ where

$$\hat{t}_A = \frac{1}{n}\sum_{i=1}^{n} r_{i1}\frac{\partial}{\partial\alpha}g_1(X_i,\hat{\alpha}),$$
$$\hat{t}_B = \frac{1}{n}\sum_{i=1}^{n} r_{i2}\frac{\partial}{\partial\beta}g_2(X_i,\hat{\beta}).$$

Comparing T_{eff} with T_{DR}, it can be seen that T_{DR} is asymptotically equivalent to T_{eff} when both model for conditional expectation $E(Y_1 \mid X)$ and $E(Y_2 \mid X)$ are correctly specified and reasonably well estimated. When only one model is correct and the other model is wrong, T_{DR} is still valid, i.e., has asymptotically correct size, while T_{eff} fails. When both models are wrong, T_{DR} fails also. This property will be illustrated in simulation studies in the next section.

6.7 Likelihood ratio test using the R package SPORM

The likelihood ratio tests for conditional independence can be carried out for any test function $h(y,x,z)$ satisfying the basic requirements and any odds ratio functions η_1 and η_2 where the maximum likelihood approaches apply. This generality is restricted in practice by the availability of software. Currently, the `R` package `SPORM` can be used for testing the conditional independence under the following restrictions.

Suppose that we want to use the likelihood ratio test for the conditional independence of Y and X given Z. The test function in `SPORM` is limited to

$$h(y,x,z) = (y-y_0)\otimes(x-x_0).$$

For the odds ratio functions η_1 and η_2, the `R` package `SPORM` can be used under the following restrictions.

1. For the prospective likelihood ratio test,
$$\log\eta(y;z \mid x_0) = \alpha^t(y-y_0)\otimes(z-z_0).$$
2. For the retrospective likelihood ratio test,
$$\log\eta(x;z \mid x_0) = \beta^t(x-x_0)\otimes(z-z_0).$$
3. For the joint likelihood ratio test, both models in 1 and 2 are assumed.

In these cases, either the function `splkhfix` or `pmlkhfix` may be used for the tests. The output of these functions contains the maximum log-likelihood (the third one on the output list) which can be used to construct the likelihood ratio test.

Consider a more specific example where $y = (y_1,y_2)$, $x = (x_1,x_2,x_3)$, and $z = (z_1,z_2,z_3,z_4)$. We set

$$h(y,x,z) = \Big((y_1-y_{10})(x_1-x_{10}), (y_1-y_{10})(x_2-x_{20}),$$
$$(y_2-y_{20})(x_2-x_{20}), (y_2-y_{20})(x_3-x_{30})\Big).$$

for a fixed point (y_0, x_0, z_0). Suppose that

$$\log \eta_1(y; z) = \Big((y_1 - y_{10})(z_1 - z_{10}), (y_1 - y_{10})(z_3 - z_{30}),$$
$$(y_2 - y_{20})(z_2 - z_{20}), (y_2 - y_{20})(z_3 - z_{30})\Big)\alpha,$$

and that

$$\log \eta_2(x; z) = \Big((x_1 - x_{10})(z_2 - z_{20}), (x_1 - x_{10})(z_3 - z_{30}), (x_2 - x_{20})(z_1 - z_{10}),$$
$$(x_3 - x_{30})(z_2 - z_{20}), (x_3 - x_{30})(z_4 - z_{40})\Big)\beta,$$

Let the data matrix with the sequence of variables in the order $(y_1, y_2, x_1, x_2, x_3, z_1, z_2, z_3, z_4)$. Let the number of observations be n and the data matrix be `dat[1:n, 1:9]`. The joint model have the fixed structure matrix as

$$M = \begin{pmatrix} \mathbf{0} & \mathbf{0} & 1 & 1 & 0 & 1 & 0 & 1 & 0 \\ \mathbf{0} & \mathbf{0} & 0 & 1 & 1 & 0 & 1 & 1 & 0 \\ 1 & 0 & \mathbf{0} & \mathbf{0} & \mathbf{0} & 0 & 1 & 1 & 0 \\ 1 & 1 & \mathbf{0} & \mathbf{0} & \mathbf{0} & 1 & 0 & 0 & 0 \\ 0 & 1 & \mathbf{0} & \mathbf{0} & \mathbf{0} & 0 & 1 & 0 & 1 \\ 1 & 0 & 0 & 1 & 0 & \mathbf{0} & \mathbf{0} & \mathbf{0} & \mathbf{0} \\ 0 & 1 & 1 & 0 & 1 & \mathbf{0} & \mathbf{0} & \mathbf{0} & \mathbf{0} \\ 1 & 1 & 1 & 0 & 0 & \mathbf{0} & \mathbf{0} & \mathbf{0} & \mathbf{0} \\ 0 & 0 & 0 & 0 & 1 & \mathbf{0} & \mathbf{0} & \mathbf{0} & \mathbf{0} \end{pmatrix},$$

where the bold face blacks show the variable groups. The null model have the fixed structure matrix as

$$M_0 = \begin{pmatrix} \mathbf{0} & \mathbf{0} & 0 & 0 & 0 & 1 & 0 & 1 & 0 \\ \mathbf{0} & \mathbf{0} & 0 & 0 & 0 & 0 & 1 & 1 & 0 \\ 0 & 0 & \mathbf{0} & \mathbf{0} & \mathbf{0} & 0 & 1 & 1 & 0 \\ 0 & 0 & \mathbf{0} & \mathbf{0} & \mathbf{0} & 1 & 0 & 0 & 0 \\ 0 & 0 & \mathbf{0} & \mathbf{0} & \mathbf{0} & 0 & 1 & 0 & 1 \\ 1 & 0 & 0 & 1 & 0 & \mathbf{0} & \mathbf{0} & \mathbf{0} & \mathbf{0} \\ 0 & 1 & 1 & 0 & 1 & \mathbf{0} & \mathbf{0} & \mathbf{0} & \mathbf{0} \\ 1 & 1 & 1 & 0 & 0 & \mathbf{0} & \mathbf{0} & \mathbf{0} & \mathbf{0} \\ 0 & 0 & 0 & 0 & 1 & \mathbf{0} & \mathbf{0} & \mathbf{0} & \mathbf{0} \end{pmatrix}.$$

For the prospective likelihood ratio test using the semiparametric likelihoods, first compute the loglikelihood for y given x, z under the alternative hypothesis by

```
fit1=splkhfix(y=dat[, 1:2],x=dat[,3:9],fixstruct=M[1:2,3:9]).
```

The maximum log-likelihood value in the output list is `fit1[[3]]`. Next, compute the loglikelihood for $y \mid x, z$ under the null hypothesis by

```
fit0=splkhfix(y=dat[, 1:2],x=dat[,3:9],fixstruct=M0[1:2,3:9]).
```

The maximum log-likelihood value is `fit0[[3]]`. The likelihood ratio test statistic can be computed as

```
2(fit1[[3]]-fit0[[3]])
```

which asymptotically follows a χ^2 with degree of freedom 4, the non-zero cells in the matrix $M[1:2, 3:5]$.

For the retrospective likelihood ratio test using the semiparametric likelihoods, first compute the loglikelihood for $x \mid y, z$ under the alternative hypothesis by

```
fit1=splkhfix(y=dat[, 3:5],x=cbind(dat[,1:2], dat[,6:9]),
          fixstruct=cbind(M[3:5,1:2], M[3:5, 6:9])).
```

The maximum log-likelihood value in the output is `fit1[[3]]`. Next, compute the loglikelihood for $y \mid x, z$ under the null hypothesis by

```
fit0=splkhfix(y=dat[, 3:5],x=cbind(dat[,1:2], dat[,6:9]),
          fixstruct=cbind(M[3:5,1:2], M[3:5, 6:9])).
```

The maximum log-likelihood value is `fit0[[3]]`. The likelihood ratio test statistic can be computed as

```
2(fit1[[3]]-fit0[[3]])
```

which asymptotically follows a χ^2 with degree of freedom 4, the non-zero cells in the matrix $M[3:5, 1:2]$.

For the joint likelihood ratio test, since the current SPORM package does not support the joint model estimated by maximizing the joint semiparametric likelihood due to its computation burden, the test cannot be performed using the semiparametric likelihood. However, the joint likelihood ratio test can be performed in SPORM using the permutation likelihood approach. The joint model under the alternative hypothesis can be fit by

```
fit1=pmlkhfix(dat=dat,group=c(2,3,4), fixstruct=M).
```

The joint model under the null hypothesis can be fit by

```
fit1=pmlkhfix(dat=dat,group=c(2,3,4), fixstruct=M0).
```

The likelihood ratio test statistic can be computed as

```
2(fit1[[3]]-fit0[[3]])
```

which asymptotically follows a χ^2 with degree of freedom 4, the non-zero cells in the matrix $M[3:5, 1:2]$. In addition, the prospective likelihood ratio test and the retrospective likelihood ratio test using the permutation likelihood approach can be obtained similarly as in the case of semiparametric likelihood approach.

6.8 A simulation study of the doubly robust tests

In this section, we perform some simulation studies to evaluate the performance of different tests for conditional independence. The evaluation is done in two steps. In the first step, we check the type I errors of the tests under different situations. The second step examines the power of those tests whose type I errors are asymptotically the same as the nominal error rate when conditional independence holds.

When Y_1 and Y_2 are independent conditional on X, we simulated the data as follows. First, generate X follows a standard normal distribution. Then generate (Y_1, Y_2) as follows.

$$
\begin{aligned}
Y_1 &= \alpha_{10} X + \alpha_{20} \exp(X) + \epsilon_1, \\
Y_2 &= \beta_{10} X^2 + \beta_{20} \exp(X) + \epsilon_2,
\end{aligned}
$$

TABLE 6.1
Parameter setup and model correctness.

$(\alpha_{10}, \alpha_{20})$	(β_0, β_1)	Model $Y_1 \mid X$	Model $Y_2 \mid X$
(1,0)	(1,0)	correct	correct
(1,0)	(1,1)	correct	wrong
(1,1)	(1,0)	wrong	correct
(1,1)	(1,1)	wrong	wrong

where ϵ_1 and ϵ_2 are independent standard normals. The exponential functions represent severe model misspecification. A mild model misspecification we will use replaces them by sine functions. In the data generation, we set the parameter values as listed in Table 6.1. The sample size of each simulated data is 200.

TABLE 6.2
Type I error of the tests from the simulations with severe model misspecification.

Model $Y_1 \mid X$	Model $Y_2 \mid X$	test	TIER=0.1	TIER=0.05	TIER=0.01
correct	correct	Prospective	**0.100**	**0.047**	**0.009**
		Retrospective	**0.100**	**0.047**	**0.009**
		Efficient	**0.099**	**0.048**	**0.010**
		Doubly Robust	**0.100**	**0.047**	**0.009**
correct	wrong	Prospective	**0.107**	**0.050**	**0.006**
		Retrospective	1.000	1.000	1.000
		Efficient	0.007	0.002	0.000
		Doubly Robust	**0.109**	**0.050**	**0.008**
wrong	correct	Prospective	0.982	0.944	0.819
		Retrospective	**0.109**	**0.053**	**0.010**
		Efficient	0.031	0.009	0.001
		Doubly Robust	**0.105**	**0.051**	**0.007**
wrong	wrong	Prospective	0.949	0.884	0.705
		Retrospective	1.000	1.000	0.988
		Efficient	0.454	0.347	0.185
		Doubly Robust	0.377	0.258	0.077

TIER: The nominal type I error rate. Bold values are those having correct models that guarantee asymptotically the correct type I error rates under the null hypothesis.

In testing the conditional independence, the nuisance models for finding the conditional expectations $E(Y_1 \mid X)$ and $E(Y_2 \mid X)$ are as follow.

$$\begin{aligned} Y_1 &= \alpha_0 + \alpha_1 X + \varepsilon_1, \\ Y_2 &= \beta_0 + \beta_1 X^2 + \varepsilon_2. \end{aligned}$$

We compared the following tests of conditional independence. The prospective likelihood score test, the retrospective likelihood score test, the semiparametric efficient test, and the doubly robust test. The nominal type I errors were set up at 10%, 5%, and 1% respectively. The number of replicates in the simulation is set to $10,000$. The simulation results for type I error rates (TIER) are listed in Table 6.2.

The results show clearly that the doubly robust test has the best performance in keeping the type I close to the nominal levels. Note that the type I error can be inflated or deflated when the nuisance models are incorrectly specified. The former is usually due to the bias in the test score, and the latter is usually due to the under-estimated variance of the test score. When both nuisance models are incorrect, the doubly robust test fails also.

TABLE 6.3
Type I error of the tests from the simulations with mild model misspecification.

Model $Y_1 \mid X$	Model $Y_2 \mid X$	test	TIER=0.1	TIER=0.05	TIER=0.01
correct	correct	Prospective	**0.107**	**0.055**	**0.010**
		Retrospective	**0.105**	**0.050**	**0.010**
		Efficient	**0.100**	**0.051**	**0.011**
		Doubly Robust	**0.102**	**0.051**	**0.011**
correct	wrong	Prospective	**0.101**	**0.049**	**0.009**
		Retrospective	1.000	1.000	0.997
		Efficient	0.057	0.026	0.002
		Doubly Robust	**0.102**	**0.049**	**0.009**
wrong	correct	Prospective	0.102	0.049	0.009
		Retrospective	**0.102**	**0.047**	**0.009**
		Efficient	0.102	0.049	0.009
		Doubly Robust	**0.106**	**0.052**	**0.009**
wrong	wrong	Prospective	0.129	0.062	0.013
		Retrospective	1.000	1.000	1.000
		Efficient	0.153	0.087	0.020
		Doubly Robust	0.211	0.127	0.038

TIER: The nominal type I error rate. Bold values are those having correct models that guarantee asymptotically the correct type I error rates under the null hypothesis.

The foregoing simulation results are for models deviated from the correct model by the exponential function. If we change the exponential function into a sine function, the results can be less biased. See Table 6.3.

TABLE 6.4
Power of the tests from the simulations with severe model misspecification.

Model $Y_1 \mid X$	Model $Y_2 \mid X$	test	TIER=0.1	TIER=0.05	TIER=0.01
correct	correct	Prospective	**0.504**	**0.374**	**0.164**
		Retrospective	**0.632**	**0.506**	**0.258**
		Efficient	**0.881**	**0.803**	**0.584**
		Doubly Robust	**0.879**	**0.794**	**0.561**
correct	wrong	Prospective	**0.304**	**0.202**	**0.068**
		Retrospective	1.00	1.00	1.00
		Efficient	0.360	0.191	0.034
		Doubly Robust	**0.762**	**0.647**	**0.395**
wrong	correct	Prospective	0.988	0.963	0.871
		Retrospective	**0.279**	**0.179**	**0.061**
		Efficient	0.558	0.404	0.159
		Doubly Robust	**0.759**	**0.664**	**0.410**

TIER: The nominal type I error rate. Bold values are those having correct models that guarantee asymptotically the correct type I error rates under the null hypothesis.

To check the power of the tests, we change the data generation slightly. We keep all the models the same except that we make ϵ_1 and ϵ_2 correlated. We first simulate independent normally distributed ϵ_1^* and ϵ_2^* and then generate ϵ_1 and ϵ_2 such that

$$\begin{aligned} \epsilon_1 &= \frac{\epsilon_1 + \rho\epsilon_2}{\sqrt{1+\rho^2}}, \\ \epsilon_2 &= \frac{\rho\epsilon_1 + \epsilon_2}{\sqrt{1+\rho^2}}, \end{aligned}$$

TABLE 6.5
Power of the tests from simulations with mild misspecification.

Model $Y_1 \mid X$	Model $Y_2 \mid X$	test	TIER=0.1	TIER=0.05	TIER=0.01
correct	correct	Prospective	**0.516**	**0.392**	**0.175**
		Retrospective	**0.642**	**0.514**	**0.263**
		Efficient	**0.882**	**0.813**	**0.607**
		Doubly Robust	**0.884**	**0.805**	**0.577**
correct	wrong	Prospective	**0.512**	**0.381**	**0.172**
		Retrospective	1.00	1.00	1.00
		Efficient	0.788	0.666	0.386
		Doubly Robust	**0.854**	**0.766**	**0.520**
wrong	correct	Prospective	0.405	0.291	0.125
		Retrospective	**0.430**	**0.306**	**0.126**
		Efficient	0.862	0.776	0.552
		Doubly Robust	**0.860**	**0.774**	**0.529**

TIER: The nominal type I error rate. Bold values are those having correct models that guarantee asymptotically the correct type I error rates under the null hypothesis.

where ρ is adjusted to control the level of dependence. We set $\rho = 0.1$ in the simulation.

The simulation results are listed in Table 6.4 for the case of serious model misspecification and in Table 6.5 for the case of mild model misspecification. The power of the tests for those whose type I error rates are asymptotically correct under the null hypothesis are labeled in bold values. Strictly speaking, power comparisons make sense only among those cases. For those tests, it can be seen that the simple conditional tests are less powerful than the semiparametric efficient test when both the nuisance models are correctly specified. The doubly robust test is slightly less powerful than the semiparametric efficient test. The doubly robust test remains more powerful than the respective conditional tests with correct type I error rate under the null hypothesis when one of the nuisance model is correctly specified, but the other nuisance model is incorrectly specified. Interestingly, the semiparametric efficient test is less powerful than the doubly robust test under misspecified nuisance models. The case with misspecification of both nuisance models is excluded because comparison of power in this case makes little sense.

6.9 Summary and literature

Testing conditional independence is one of the major themes in statistics. Some well-known early tests include the Mantel-Heaszel test and Mantel test (Mantel and Haenszel, 1959; Mantel 1963)[121, 120]. Dawid (1979a, 1979b,1980)[50, 51, 52] extensively discussed conditional independence in statistical analysis. Various tests of conditional independence (Rosenbaum, 1984; Stute, 1991; Upton, 2000; Ernst, 2004; Chen, 2007; Byrne and Dawid, 2014)[154, 187, 169, 60, 28, 18] and measures of conditional dependence have been proposed (Chen, 2007; Su and White, 2007; Szekely et al, 2007; 2014; Song, 2009; Szekely and Rizzo, 2014)[28, 170, 174, 171, 168, 173]. Recent attentions focused more on testing conditional independence with high-dimensional covariates (Berret et al, 2020; Shah and Peters, 2020)[5, 164] and measuring conditional dependence with complex data(Gretton, et al, 2005; Fukumizu etal, 2008; Kojadinovic and Holmes, 2009; Pfister et al, 2018; Chatterjee, 2020; Geenens and de Micheaux, 2020)[73, 67, 89, 138, 24, 68].

6.10 Exercises

1. For testing independence, the scaled sample correlation coefficient $\sqrt{n-2}\hat{\rho}$ is often used, where

$$\hat{\rho} = \frac{\sum_{i=1}^{n}(X_i - \bar{X})(Y_i - \bar{Y})}{\sqrt{\sum_{i=1}^{n}(X_i - \bar{X})^2 \sum_{i=1}^{n}(Y_i - \bar{Y})^2}},$$

which converges to the theoretical correlation coefficient

$$\rho = \frac{\text{cov}(Y, X)}{\sqrt{\text{var}(Y)\text{var}(X)}}.$$

It is known that uncorrelated random variables (i.e., $\rho = 0$) may not be independent. For example, if $X \sim N(0,1)$ and $Y = X^2$, then Y and X are uncorrelated. Design a test that can at least partially resolve this issue, i.e., the test has non-trivial power.

2. An extension of the test statistic $\hat{\rho}$ to test conditional independence is the partial correlation coefficient. The partial correlation coefficient of Y and X given Z is defined as

$$\rho_{YX|Z} = \frac{\text{cov}(Y, X \mid Z)}{\sqrt{\text{var}(Y \mid Z)\text{var}(X \mid Z)}}.$$

A sample version of the partial correlation coefficient can be obtained as follows. First, regress Y on Z to obtain the residual R_1 and regress X on Z to obtain residual R_2. Then compute the correlation coefficient of R_1 and R_2. Aside from the problem encountered previously in using the sample correlation coefficient to test independence, the partial correlation coefficient may not converges to 0 when the two marginal regression models are wrong even if $\rho_{YX|Z} = 0$. What would happen if one regression model is correct, the other is incorrect?

3. In a recent paper on testing conditional independence, Berret et al (2020)[5] proposed a method corresponding to sampling from $\mathbf{Y}_2$ given $\mathbf{X}$ for the permutation likelihood of $f(\mathbf{Y}_2 \mid \mathbf{Y}_2^{ord}, \mathbf{X})$. Under the semiparametric odds ratio model for $f(y \mid x)$, when $\eta_2(y_2; x)$ is known, the Monte Carlo approach to calculating p-values for the conditional independence test is exact. However, when $\eta_2(y_2; x)$ is known up to an unknown parameter β, the Monte Carlo approach for computing p-values for the test of conditional independence using the estimated $\eta_2(y_2; x)$ is generally invalid even asymptotically.

   (a) Explain the reason for the invalidity of p-value calculation.

   (b) Propose a way that may be used to correct the problem to make the calculation approximately right.

4. Suppose that we know $f(y_2 \mid x)$ exactly,

   (a) Derive a retrospective likelihood score test for the conditional independence of Y_1 and Y_2 given X.

   (b) Derive the doubly robust test of conditional independence.

   (c) Perform a simulation study to compare the performance of the two tests.

5. Suppose that we want to use the SPORM package to perform the likelihood ratio tests for the specific example in section 6.7 with the following models

$$
\begin{aligned}
h(y,x,z) &= \Big((y_1-y_{10})(x_1-x_{10}), (y_1-y_{10})(x_2-x_{20}), (y_1-y_{10})(x_3-x_{30}),\\
&\quad (y_2-y_{20})(x_1-x_{10}), (y_2-y_{20})(x_2-x_{20}), (y_2-y_{20})(x_3-x_{30})\Big),
\end{aligned}
$$

$$
\begin{aligned}
\log\eta_1(y;z) &= \Big((y_1-y_{10})(z_1-z_{10}), (y_1-y_{10})(z_2-z_{20}), (y_1-y_{10})(z_3-z_{30}),\\
&\quad (y_2-y_{20})(z_2-z_{20}), (y_2-y_{20})(z_4-z_{40})\Big)\alpha,
\end{aligned}
$$

and that

$$
\begin{aligned}
\log\eta_2(x;z) &= \Big((x_1-x_{10})(z_2-z_{20}), (x_1-x_{10})(z_3-z_{30}),\\
&\quad (x_2-x_{20})(z_1-z_{10}), (x_2-x_{20})(z_2-z_{20}), (x_2-x_{20})(z_3-z_{30}),\\
&\quad (x_3-x_{30})(z_2-z_{20}), (x_3-x_{30})(z_4-z_{40})\Big)\beta,
\end{aligned}
$$

(a) Determine the structure matrices M and M_0.
(b) Write R code for computing the prospective likelihood ratio test.
(c) Write R code for computing the retrospective likelihood ratio test.
(d) Write R code for computing the joint likelihood ratio test.

7

Application to Network Detection and Estimation

7.1 Network and Gaussian graphical model

A graphical network consists of a set of nodes $\mathcal{N}$ and a set of edges $\mathcal{E} \subset \mathcal{N} \times \mathcal{N}$ that denote the connection between any pair of nodes. Let Y_j denote the random variables associated with node j. The edges of a network depict the dependence structure of the node variables $Y_j, j \in \mathcal{N}$. The connection can be directed or undirected. A directed graph depicts the partial conditional independence structure of the nodes. An undirected graph depicts the fully conditional dependence of the nodes. The undirected graph is also called association network. Two nodes j and k in an undirected network are not directly connected if and only if Y_j and Y_k are independent conditional on $\{Y_l, l \neq j, k\}$, or in short,

$$Y_j \perp\!\!\!\perp Y_k \mid \{Y_l, l \neq j, k\}. \tag{7.1}$$

For any node j, let $ne(j)$ denote the neighborhood of j, i.e., the set of nodes other than j that directly connect with node j. Then

$$f(y_j \mid y_{-j}) = f\{y_j \mid y_k, k \in ne(j)\}.$$

Note that, for a subset of nodes S that do not contain either node j or k,

$$Y_j \perp\!\!\!\perp Y_k \mid \{Y_j, j \in S\} \tag{7.2}$$

does not necessarily implies (7.1) because partially conditional independence does not necessarily imply fully conditional independence. On the other hand, for any set of nodes S, if (7.2) holds implies that (7.1) is true, the network is said to have the partial faithfulness in Buhlmann et al (2010)'s[17] terminology. If the partial faithfulness holds for any node in a network, the network is called a partially faithful network. The partial faithfulness property of a network can have a big impact on the approaches to identify the network.

When $(Y_j, j \in \mathcal{N})$ follows a joint normal distribution, the dependence structure determined by the joint normal distribution is termed a Gaussian network on $\mathcal{N}$. Suppose that the joint density for the network nodes $(Y_j, j \in \mathcal{N})$ is

$$f(y \mid \mu, \Sigma) = \frac{1}{(2\pi)^{p/2}|\Sigma|^{1/2}} \exp\left\{-\frac{1}{2}(y-\mu)^t\Sigma^{-1}(y-\mu)\right\}, \tag{7.3}$$

where $p = |\mathcal{N}|$, the number of nodes in the network. Since the network structure is determined purely by the covariance matrix Σ, we will assume from now on that $\mu = 0$. Let $\Omega = (w_{jk}) = -\Sigma^{-1}$, termed precision matrix, or concentration matrix. The conditional density for Y_j conditional on Y_{-j} is

$$f(y_j \mid Y_{-j}) = \frac{\sqrt{-w_{jj}}}{\sqrt{2\pi}} \exp\left\{\frac{w_{jj}}{2}\left(y_j + \sum_{k \neq j} \frac{w_{jk}}{w_{jj}} Y_k\right)^2\right\}. \tag{7.4}$$

DOI: 10.1201/9781351049757-7

The neighborhood of node j in a Gaussian network is determined by the jth column or the jth row of the precision matrix as

$$ne(j) = \{k \mid w_{jk} \neq 0, k \neq j\}.$$

The network connections are determined by

$$\mathcal{N} = \{(j,k) \mid j \neq k, w_{jk} \neq 0\},$$

all the non-zero off-diagonal elements of the precision matrix.

7.2 Gaussian network detection approaches

One of the key problems in network analysis is to infer the network structure, i.e., connections of a network, based on a sample, $\{Y_{ij}, j = 1, \cdots, N, i = 1, \cdots, n\}$. In a Gaussian network, this problem is equivalent to inferring the non-zero elements in the precision matrix. Many approaches have been proposed to inferring the Gaussian network structure, in particular, the sparse Gaussian network in high-dimensional data. These approaches include the neighborhood selection approach by the penalized least square, the graphical lasso to jointly infer the network structure, various other penalty approaches for joint inference, and the screening approaches based on the partial faithfulness assumption. Under appropriate conditions, these methods are shown to yield consistent network detection for sparse Gaussian network. The major approaches for Gausssian network detection are reviewed in the following subsections.

7.2.1 The neighborhood detection approach

The neighborhood detection approach at a given node j is based on the conditional distribution (7.4) which is equivalent to a linear regression model

$$Y_{ij} = \sum_{k \neq j} \beta_{jk} Y_{ik} + \epsilon_{ij},$$

where $\epsilon_{ij}, i = 1, \cdots, n$ are independent and follows $N(0, \sigma_j^2)$, and $\beta_{jk} = -w_{jk}/w_{jj}$ and $\sigma_j^2 = -1/w_{jj}$. The penalized least-square approach is proposed to select the nonzero β_{jk} by minimizing

$$\frac{1}{2} \sum_{i=1}^{n} \left(Y_{ij} - \sum_{k \neq j} \beta_{jk} Y_{ik} \right)^2 + \lambda \sum_{k \neq j} |\beta_{jk}|.$$

The minimizer satisfies the following equations

$$-\sum_{i=1}^{n} Y_{ik} \left(Y_{ij} - \sum_{k \neq j} \beta_{jk} Y_{ik} \right) + \lambda \delta_{jk} = 0,$$

for $k \neq j$, where $\delta_{jk} = \text{sign}(\beta_{jk})$ if $\beta_{jk} \neq 0$ and $|\delta_{jk}| \leq 1$ if $\beta_{jk} = 0$.

For a given λ, the equations can be solved by the following coordinate descending algorithm. For a given current estimate of $(\beta_{jk}, k \neq j, l)$, note that

$$\sum_{i=1}^{n} Y_{il}^2 \beta_{jl} = \sum_{i=1}^{n} Y_{il} \left(Y_{ij} - \sum_{k \neq j,l} \beta_{jk} Y_{ik} \right) - \lambda \delta_{jl}.$$

Update β_{jl} as follows.

$$\beta_{jl} = \begin{cases} \frac{\sum_{i=1}^n Y_{il}\left(Y_{ij}-\sum_{k\neq j,l}\beta_{jk}Y_{ik}\right)-\lambda}{\sum_{i=1}^n Y_{il}^2} & \text{if } \sum_{i=1}^n Y_{il}\left(Y_{ij}-\sum_{k\neq j,l}\beta_{jk}Y_{ik}\right) > \lambda, \\ \frac{\sum_{i=1}^n Y_{il}\left(Y_{ij}-\sum_{k\neq j,l}\beta_{jk}Y_{ik}\right)+\lambda}{\sum_{i=1}^n Y_{il}^2} & \text{if } \sum_{i=1}^n Y_{il}\left(Y_{ij}-\sum_{k\neq j,l}\beta_{jk}Y_{ik}\right) < -\lambda, \\ 0 & \text{otherwise.} \end{cases}$$

The algorithm goes through all the indices $l \neq j$ and repeats the iterations until no change in $\beta_{jk}, k \neq j$ is achieved.

After the neighborhood algorithm is applied to each node in the network, the network structure is estimated by the connections identified. If the estimate of $\beta_{jk} \neq 0$ while the estimate of $\beta_{kj} = 0$ or vice versa, node j and k can be regarded as connected. Meinshausen and Bulhmann (2006)[125] showed the algorithm can consistently identify the sparse network for appropriately chosen tuning parameter value when a set of regularity conditions are satisfied.

7.2.2 The joint detection approach

The neighborhood selection approach to sparse Gaussian network detection is computationally fast. But the algorithm is inefficient. Yuan and Lin (2007)[202] proposed a more efficient approach using convex programming for the positive definite matrix. Their approach maximizes the penalized likelihood based on the normal model.

For appropriately centered and scaled data, the approach of Yuan and Lin (2007)[202] minimizes the following objective function

$$-\log|\Omega| + \frac{1}{n}\sum_{i=1}^{n} Y_i^t \Omega Y_i + \lambda \sum_{j\neq k} |w_{jk}|,$$

with respect to the symmetric positive definite matrix Ω, where $Y_i = (Y_{i1}, \cdots, Y_{iN})^t$ is the centered and scaled data for subject i, and λ is the tuning parameter. Let

$$\hat{\Sigma} = \frac{1}{n}\sum_{i=1}^{n} Y_i Y_i^t,$$

The optimization can be restated as

$$\min_{\Omega>0} \left\{ tr(\hat{\Sigma}\Omega) - \log|\Omega| + \lambda \sum_{j\neq k} |w_{jk}| \right\},$$

which can be equivalently stated as

$$\min_{\Omega>0} \left\{ tr(\hat{\Sigma}\Omega) - \log|\Omega| \right\} \text{ subject to } \sum_{j\neq k} |w_{jk}| \leq t,$$

for some $t \geq 0$. This problem can be transformed into a special case of the convex optimization problem termed the detmax problem. The detmax problem has the general form

$$\min\nolimits_{x\in R^m} \{c^t x - \log|G(x)|\}$$
$$\text{subject to: } G(x) > 0 \text{ and } F(x) \geq 0,$$

where

$$G(x) = G_0 + x_1 G_1 + \cdots + x_m G_m$$
$$F(x) = F_0 + x_1 F_1 + \cdots + x_m F_m,$$

and G_i and F_i are known symmetric matrices for $i = 0, 1, \cdots, m$.

To see the joint penalized estimation problem in terms of the detmax problem, let $I^{(ij)}$ be the symmetric matrix with (i, j) and (j, i) elements being 1 and other elements being 0. For the special case $i = j$, $I^{(ii)}$ denotes the matrix with the ith diagonal element being 1 and all other elements being 0. Then G is defined as Ω having the form

$$G(w_{ij}, i \leq j) = \sum_{i=1}^{p} w_{ii} I^{(ii)} + \sum_{i=1}^{p-1} \sum_{j=i+1}^{p} w_{ij} I^{(ij)},$$

where $x = (w_{ii}, i = 1, \cdots, p; w_{ij}, i = 1, \cdots, p-1; j = i+1, \cdots, p)$ and $m = p(p+1)/2$. Since

$$tr\{\hat{\Sigma}\Omega\} = \sum_{i=1}^{p} \hat{\sigma}_{ii} w_{ii} + \sum_{i=1}^{p-1} \sum_{j=i+1}^{p} 2\hat{\sigma}_{ij} w_{ij},$$

where $\hat{\Sigma} = (\hat{\sigma}_{ij})$, it follows that $c = (\hat{\sigma}_{ii}, i = 1, \cdots, p, 2\hat{\sigma}_{ij}, i = 1, \cdots, p-1; j = i+1, \cdots, p)$. The penalty term can be transformed into

$$t - \sum_{i=1}^{p-1} \sum_{j=i+1}^{p} 2 s_{ij} w_{ij} \geq 0, \text{and } s_{ij} w_{ij} \geq 0,$$

where $s_{ij} = \text{sign}(w_{ij})$ and t is the tuning parameter value. Yuan and Lin (2007) proposed the following algorithm to solve the minimization problem. Initially, we may take $\hat{\Sigma}$ and $\hat{\Omega} = \hat{\Sigma}^{-1}$.

1. For current estimate Ω^{old}, compute $S = (s_{ij}) = \text{sign}(\Omega^{old})$.
2. Solve the maxdet problem to obtain Ω^{new}.
3. If $\Omega^{new} = \Omega^{old}$, then stop. Otherwise, set $\Omega^{old} = \Omega^{new}$ and $s_{ij} = -s_{ij}$ for all $w_{ij} = 0$. Repeat steps 2-3.

The joint analysis may also take other forms which make the minimization problem easier to solve. See Cai and Yuan (2010)[20] for more details.

7.2.3 The screening approach based on partial correlation coefficient

When the partial faithfulness condition holds, screening approaches may be used to detect the network, starting from the marginal screening. Since marginal independence implies the full conditional independence when the partial faithfulness condition holds, marginally disconnected nodes remain disconnected in the association network. For Gaussian network, partial faithfulness implies that, for any two node variables, if they are uncorrelated conditional on a subset of other node variables, they are uncorrelated conditional on all other node variables. In other words, they are disconnected in the network.

The screening for a Gaussian network can be done in two different, but related, ways. The first way is, for a given node, to run a regression using the node variable as the outcome and other node variables as covariates. For example, for node j, the regression model is

$$Y_{ij} = \beta_0^{(S_j)} + \sum_{k \in S_j} \beta_{jk}^{(S_j)} Y_{ik} + \epsilon_{ij}.$$

The screening algorithm can be carried out as follows

1. For level 1 screening, starting from node 1, run regression models with one node variable from nodes $2, \cdots, p$. Let

$$ne(1,1) = \{k \mid k \in \{2, \cdots, p\}, \beta_{1k}^{(\{k\})} \neq 0\}.$$

For node j, run regression models with one node variable from nodes $j+1, \cdots, p$. Let

$$ne(j,1) = \{k \mid k \in \{j+1, \cdots, p\}, \beta_{1k}^{(\{k\})} \neq 0\} \cup \{k \mid j \in ne(k,1), k < j\}.$$

Define the set of level 1 nodes $L_1 = \{j \mid |ne(j,1)| > 1\}$. Stop screening if $L_1 = \emptyset$.

2. For level $t+1$ screening, consider the set of nodes L_t only. For $j \in L_t$, $|ne(j,t)| > t$. If any of the nodes in $ne(j,t)$ is unconnected to j in the regression with $t+1$ covariates from $ne(j,t)$, that node variable is excluded from $ne(j,t+1)$, i.e., $ne(j,t+1) \subset ne(j,t)$ such that none of the regression with Y_j as the outcome and $t+1$ covariates from $ne(j,t)$ has zero coefficient for variable Y_k. Define the set of level $t+1$ nodes $L_{t+1} = \{j \mid |ne(j,t+1)| > t+1\}$. Stop screening if $L_{t+1} = \emptyset$. Otherwise, repeat this step.

The screening can also be done by testing the partial correlation coefficients, starting from the unconditional correlation coefficients. For example, Buhlmann et al (2010)[17] proposed the following algorithm to screen based on partial correlation coefficients. Define the active sets of a node j as

$$\begin{aligned} A_1(j) &= \{k \mid cor(Y_j, Y_k) \neq 0, k \neq j\} \\ A_t(j) &= \{k \mid cor(Y_j, Y_k \mid Y_l, l \in S), \forall S \subset A_{t-1}(j) - \{k\}, |S| = t-1\}, \text{ for } t = 2, 3, \cdots. \end{aligned}$$

where $cor(Y_j, Y_k \mid Y_l, l \in S)$ is the partial correlation coefficient conditional on $\{Y_l, l \in S\}$. A population level screening is to find $A_t(j)$ for each j and stop as soon as $|A_t(j)| = t$ for a given node. The network is the union of the final active sets for all the nodes. The sample-based screening makes use of

$$\sqrt{n - |S| - 3}\; Z(Y_j, Y_k \mid Y_l, l \in S) \sim N(0,1),$$

where Z is Fisher transformation of the partial correlation coefficient, i.e.,

$$Z(Y_j, Y_k \mid Y_l, l \in S) = \frac{1}{2} \log \frac{1 + \rho(Y_j, Y_k \mid Y_l, l \in S)}{1 - \rho(Y_j, Y_k \mid Y_l, l \in S)}.$$

The partial correlation coefficient $\rho(Y_j, Y_k \mid Y_l, l \in S)$ can be obtained based on an iterative formula as

$$\rho(Y_j, Y_k \mid Y_l, l \in S) = \frac{\rho(Y_j, Y_k \mid Y_l, l \in S - \{m\}) - \rho(Y_j, Y_m \mid Y_l, l \in S - \{m\})\rho(Y_k, Y_m \mid Y_l, l \in S - \{m\})}{\sqrt{\{1 - \rho^2(Y_j, Y_m \mid Y_l, l \in S - \{m\})\}\{1 - \rho^2(Y_k, Y_m \mid Y_l, l \in S - \{m\})\}}},$$

for any given $m \in S$. For an observed sample, ρ can be replaced by sample estimates of the correlation coefficients. For a given level α, the screening tests if the correlation coefficients are 0 by a pre-adjusted test level for multiple tests based on the sparsity assumption.

7.3 Network modeling by the semiparametric odds ratio model

Gaussian network models the distribution of the random variables corresponding to the $\mathcal{N}$ nodes by a joint normal distribution. The joint normal model may not be appropriate for an arbitrary set of random variables, particularly when they are discrete or a mix of continuous and discrete random variables. When the normal distributional assumption is violated, network detected under the normal model assumption only picks up the linear relationship among random variables.

Consider the general dependence structure of random variables. Let $Y_1, \cdots, Y_G$ denote vectors of random variables corresponding to the clusters of nodes of an association network. For the generality of the framework, we allow Y_k to be a vector of random variables representing a group of nodes with d_k nodes for $k = 1, \cdots, G$, where $p = \sum_{k=1}^{G} d_k$. Let $f(y_1, \cdots, y_G)$ denote the density with respect to a known measure for the distribution of $(Y_1, \cdots, Y_G)$. Assume the positivity condition (Besag, 1974)[7] holds for the node distribution, i.e., if the marginal densities $f_k(y_k) > 0$ for $k = 1, \cdots, G$, then the joint density $f(y_1, \cdots, y_G) > 0$. For any density function satisfying the positivity condition, it has the following odds ratio representation,

$$f(y_1, \cdots, y_G) = \frac{\eta(y_1; \cdots; y_G) \prod_{k=1}^{G} f_k(y_k \mid y_{-k0})}{\int \cdots \int \eta(y_1; \cdots; y_G) \prod_{k=1}^{G} f_k(y_k \mid y_{-k0}) dy_k}, \tag{7.5}$$

where

$$\eta(y_1; \cdots; y_G) = \prod_{k=1}^{G-1} \eta_k\{y_k; (y_G, \cdots, y_{(k+1)}, y_{(k-1)0}, \cdots, y_{10})\} \tag{7.6}$$

and $f_k(y_k \mid y_{-k0})$ is the conditional density of Y_k given $Y_{-k} = y_{-k0}$ for $k = 1, \cdots, G$. Note that $\eta_k\{y_k; (y_G, \cdots, y_{(k+1)}, y_{(k-1)0}, \cdots, y_{10})$ is the odds ratio function of y_k versus $(y_G, \cdots, y_{(k+1)}, y_{(k-1)0}, \cdots, y_{10})$ with reference point $y_0 = (y_{10}, \cdots, y_{G0})$ for the function $f(y_{10}, \cdots, y_{(k-1)0}, y_k, \cdots, y_G)$. The odds ratio representation of a density plays a key role in characterizing an association network.

Theorem 7.1. Under the positivity condition, an association network is uniquely determined by the collection of odds ratio functions in (7.6).

Proof: For any two clusters of nodes y_j and y_k, there is no connection between the two clusters of nodes if and only if Y_j and Y_k are independent conditional on $Y_{-(j,k)}$. This is equivalent to

$$f(y_j, y_k \mid y_{-(j,k)}) = f(y_j|y_{-(j,k)}) f(y_k \mid y_{-(j,k)}).$$

For a network satisfies the positivity condition, it is equivalent to the conditional odds ratio

$$\eta(y_j; y_k \mid y_{-(j,k)}) = \frac{f(y_j, y_k \mid y_{-(j,k)}) f(y_{j0}, y_{k0} \mid y_{-(j,k)})}{f(y_{j0}, y_k \mid y_{-(j,k)}) f(y_j, y_{k0} \mid y_{-(j,k)})} = 1.$$

It is not difficult to see from (7.5) that $\eta(y_j; y_k \mid y_{-(j,k)})$ is uniquely determined by the odds ratio functions in (7.6) through

$$\eta(y_j; y_k \mid y_{-(j,k)}) = \frac{\eta(y_1; \cdots; y_j; \cdots; y_k; \cdots; y_G) \eta(y_1; \cdots; y_{j0}; \cdots; y_{k0}; \cdots; y_G)}{\eta(y_1; \cdots; y_{j0}; \cdots; y_k; \cdots; y_G) \eta(y_1; \cdots; y_j; \cdots; y_{k0}; \cdots; y_G)},$$

As a result, the connections are uniquely determined by the odds ratio functions in (7.5).

Theorem 7.1 suggests that we may concentrate on studying the collection of odds ratio function for the detection of an association network. For a Gaussian network, the odds ratio functions reduce to

$$\log \eta(y_1; \cdots ; y_G) = \sum_{j=1}^{G-1} \sum_{k=j+1}^{G} \theta_{jk} y_j y_k,$$

where $\theta_{jk} = -w_{jk}$ and w_{jk} is the element of the jth column and the kth row of the precision matrix Ω. Similar odds ratio functions also appear in the Ising models for the network where the node variables are all binary. In general, if any node variable conditional on the rest node variables follows a generalized linear model with a canonical link function, the corresponding odds ratio function is bilinear. Such a network has the same odds ratio functions as those in the Gaussian network model.

The semiparametric odds ratio model provides a general framework for studying the association network. For example, when the log-odds ratio function is the bilinear function

$$\log \eta_k\{y_k; (y_G, \cdots , y_{(k+1)}, y_{(k-1)0}, \cdots , y_{10}) = \sum_{j=k+1}^{G} \theta_{kj}(y_k - y_{k0})(y_j - y_{j0}),$$

The semiparametric odds ratio model specifies

$$p(y_1, \cdots , y_G \mid \theta, F_1, \cdots , F_G) = \frac{\eta(y_1; \cdots ; y_G \mid \theta) \prod_{k=1}^{G} dF_k(y_k)}{\int \cdots \int \eta(y_1; \cdots ; y_G \mid \theta) \prod_{k=1}^{G} dF_k(y_k)}, \tag{7.7}$$

where $F_k(y_k)$ is an abbreviation of $F_k(y_k \mid y_{-k0})$ which is the distribution function of $f_k(y_k \mid y_{-k0})$ and the integral in the denominator is constrained to be finite. Such a model extends both the Gaussian network model and the Ising model. The parameter θ is composed of $\theta_{12}, \cdots , \theta_{1G}, \theta_{23}, \cdots , \theta_{2G}, \cdots , \theta_{(G-1)G}$, which are the only set of parameters used to characterize the network. Network detection corresponds to estimation and inference on this set of parameters.

7.4 Network detection by penalized likelihoods based on permutations

7.4.1 Penalized pairwise pseudo-likelihood approach

The pairwise pseudo-likelihood can be applied to individual nodes and is computationally fast when the sample size is not very large. The penalized pairwise pseudo-likelihood can be used to carry out node-wise network selection. For a given node Y_j, the model for that node is

$$f(y_j \mid y_{-j}) = \frac{\eta(y_j; y_{-j} \mid \theta) f(y_j \mid y_{-j0})}{\int \eta(y_j; y_{-j} \mid \theta) f(y_j \mid y_{-j0}) dy_j}.$$

The pairwise pseudo-likelihood for the observed data $(Y_{ik}, k = 1, \cdots , G; i = 1, \cdots , n)$ is

$$L_j(\theta) = \prod_{i=1}^{n-1} \prod_{k=i+1}^{n} \frac{\eta(Y_{ij}; Y_{i(-j)} \mid \theta)\eta(Y_{kj}; Y_{k(-j)} \mid \theta)}{\eta(Y_{ij}; Y_{i(-j)} \mid \theta)\eta(Y_{kj}; Y_{k(-j)} \mid \theta) + \eta(Y_{ij}; Y_{k(-j)} \mid \theta)\eta(Y_{kj}; Y_{i(-j)} \mid \theta)}.$$

The l^1-penalized negative pairwise pseudo-likelihood for node j is

$$l_{j\lambda}(\theta) = l_j(\theta) + \lambda|\theta|,$$

where

$$l_j(\theta) = -\log L_j(\theta) = \sum_{i=1}^{n-1} \sum_{k=i+1}^{n} -\log\left\{1 + \frac{\eta(Y_{ij}; Y_{k(-j)} \mid \theta)\eta(Y_{kj}; Y_{i(-j)} \mid \theta)}{\eta(Y_{ij}; Y_{i(-j)} \mid \theta)\eta(Y_{kj}; Y_{k(-j)} \mid \theta)}\right\}.$$

Minimizing $l_{j\lambda}(\theta)$ is carried out by the coordinate descending method. Let d_l denote a vector of the same dimension as that of θ with the lth component being 1 and the rest of components being 0. Then, for a given δ,

$$\begin{aligned} l_{j\lambda}(\theta + \delta d_l) - l_{j\lambda}(\theta) &= l_j(\theta + \delta d_l) - l_j(\theta) + \lambda(|\theta_l + \delta d_l| - |\theta_l|) \\ &\approx d_l^t \dot{l}_j(\theta)\delta + \frac{1}{2} d_l^t \ddot{l}_j(\theta) d_l \delta^2 + \lambda(|\theta_l + \delta| - |\theta_l|). \end{aligned}$$

Minimizing the right-hand side with respect to δ has a closed-form solution as

$$\theta_l + \delta = \begin{cases} -\frac{d_l^t \dot{l}_j(\theta) - \lambda}{d_l^t \ddot{l}_j(\theta) d_l} & \text{if } d_l^t \dot{l}_j(\theta) > \lambda, \\ -\frac{d_l^t \dot{l}_j(\theta) + \lambda}{d_l^t \ddot{l}_j(\theta) d_l} & \text{if } d_l^t \dot{l}_j(\theta) < -\lambda, \\ 0 & \text{otherwise.} \end{cases}$$

The algorithm cycles through the coordinate of θ until no more variables are selected and the parameters in the model with selected variables have little change between iterations.

Consider more specifically the log-bilinear odds ratio model. The pairwise pseudo-likelihood for node j is

$$L_j(\theta) = \prod_{i=1}^{n-1} \prod_{k=i+1}^{n} \pi_{ik,j}(\theta),$$

where we assume $d_k = 1, k = 1, \cdots, G$ and

$$\pi_{ik,j}(\theta) = \frac{\exp\left\{\sum_{l=1,l\neq j}^{G} \theta_{jl}(Y_{ij} - Y_{kj})(Y_{il} - Y_{kl})\right\}}{1 + \exp\left\{\sum_{l=1,l\neq j}^{G} \theta_{jl}(Y_{ij} - Y_{kj})(Y_{il} - Y_{kl})\right\}}$$

The penalized pairwise pseudo-likelihood for node j is

$$\begin{aligned} l_{j\lambda}(\theta) = & \sum_{i=1}^{n-1} \sum_{k=i+1}^{n} \Bigg(\sum_{l=1,l\neq j}^{G} \theta_{jl}(Y_{ij} - Y_{kj})(Y_{il} - Y_{kl}) \\ & - \log\left[1 + \exp\left\{\sum_{l=1,l\neq j}^{G} \theta_{jl}(Y_{ij} - Y_{kj})(Y_{il} - Y_{kl})\right\}\right]\Bigg) - \lambda \sum_{l=1,l\neq j}^{G} |\theta_{jl}|, \end{aligned}$$

The coordinate ascending method can be applied to maximizing the penalized pairwise pseudo-likelihood. For a given index $t \neq j$,

$$\sum_{i=1}^{n-1} \sum_{k=i+1}^{n} \frac{(Y_{ij} - Y_{kj})(Y_{it} - Y_{kt})}{1 + \exp\left\{\sum_{l=1,l\neq j}^{G} \theta_{jl}(Y_{ij} - Y_{kj})(Y_{il} - Y_{kl})\right\}} - \lambda\delta_{jt} = 0,$$

where $\delta_{jt} = \text{sign}(\theta_{jt})$ if $\delta_{jt} \neq 0$, otherwise, $|\delta_{jt}| \leq 1$. Let the current estimate of θ be $\theta^{(c)}$.

The equation can be expanded into

$$
\begin{aligned}
&\sum_{i=1}^{n-1}\sum_{k=i+1}^{n}\sum_{l=1,l\neq j}^{G}(Y_{ij}-Y_{kj})(Y_{il}-Y_{kl})\{1-\pi_{ik,j}(\theta^{(c)})\}-\lambda\delta_{jt}^{new}\\
=&\sum_{i=1}^{n-1}\sum_{k=i+1}^{n}\sum_{l=1,l\neq j}^{G}\{(Y_{ij}-Y_{kj})(Y_{il}-Y_{kl})\}^2\{1-\pi_{ik,j}(\theta^{(c)})\}\pi_{ik,j}(\theta^{(c)})(\theta_{jt}-\theta_{jt}^{(c)}),
\end{aligned}
$$

The solution to the equation is

$$
\theta_{jt}=\begin{cases}(A-\lambda)/B & \text{if } A>\lambda,\\ (A+\lambda)/B & \text{if } A<-\lambda,\\ 0 & \text{otherwise,}\end{cases}
$$

where

$$
\begin{aligned}
A =& \sum_{i=1}^{n-1}\sum_{k=i+1}^{n}\sum_{l=1,l\neq j}^{G}(Y_{ij}-Y_{kj})(Y_{il}-Y_{kl})\{1-\pi_{ik,j}(\theta^{(c)})\}\\
&+\sum_{i=1}^{n-1}\sum_{k=i+1}^{n}\sum_{l=1,l\neq j}^{G}\{(Y_{ij}-Y_{kj})(Y_{il}-Y_{kl})\}^2\{1-\pi_{ik,j}(\theta^{(c)})\}\pi_{ik,j}(\theta^{(c)})\theta_{jt}^{(c)},\\
B =& \sum_{i=1}^{n-1}\sum_{k=i+1}^{n}\sum_{l=1,l\neq j}^{G}\{(Y_{ij}-Y_{kj})(Y_{il}-Y_{kl})\}^2\{1-\pi_{ik,j}(\theta^{(c)})\}\pi_{ik,j}(\theta^{(c)}).
\end{aligned}
$$

The current estimate for θ_{jt} may then be replaced by θ_{jt}^{new}. The above steps are then iterated until convergence. The neighborhood selection approach is completed by going through all the nodes. Once the network structure and the strength of the connections are obtained, they can be used as initial values for the penalized permutation likelihood approach.

For each node, the algorithm detects nodes that connect with it. The collection of the directly connected nodes is termed the neighborhood of the node. For the estimated neighborhood, it may occur that node A is in the neighborhood of node B, but node B is not in the neighborhood of A. In this case, we may either exclude node A from the neighborhood of B, or include B in the neighborhood of node A. Asymptotically, there is not difference in the two treatments.

7.4.2 Penalized permutation likelihood approach

The penalized pairwise pseudo-likelihood approach to network detection is inefficient at two different levels. At the individual node level, the method is inefficient because of the use of the pseudo-likelihood approach. At the network level, it is inefficient because of the node-wise detection approach used. To overcome these problems, we consider the network detection approach directly using permutation likelihood in this section.

Let $Y_{ik}, k=1,\cdots,G; i=1,\cdots,n$ be the observed data. Let $Y_{(1)k},\cdots,Y_{(n)k}$, $k=1,\cdots,G$ denote the corresponding order statistics. For a fixed θ, it can be seen that the order statistics are sufficient statistics for the baseline distribution functions $F_k(y_k)$, $k=1,\cdots,G$. The conditional likelihood given the order statistics has the following simple form:

$$
L_n(\theta)=\frac{\prod_{i=1}^{n}\eta(Y_{i1};\cdots;Y_{iG}\mid\theta)}{\sum_{\sigma\in\mathcal{P}_{G-1}}\prod_{i=1}^{n}\eta(Y_{\sigma_1(i)1};\cdots;Y_{\sigma_{G-1}(i)(G-1)},Y_{iG}\mid\theta)}, \tag{7.8}
$$

where $\mathcal{P}_{G-1}$ is the collection of permutations $\sigma = (\sigma_1, \cdots, \sigma_{G-1})$, and σ_k is a permutation on the indices $\{1, \cdots, n\}$ for $k = 1, \cdots, G-1$. Let

$$M_n(\theta) = \sum_{\sigma \in \mathcal{P}_{G-1}} \exp\left\{\sum_{i=1}^{n} \log \frac{\eta(Y_{\sigma_1(i)1}; \cdots; Y_{\sigma_{G-1}(i)(G-1)}, Y_{iG} \mid \theta)}{\eta(Y_{i1}; \cdots; Y_{iG} \mid \theta)}\right\}. \tag{7.9}$$

Maximizing the penalized log-permutation likelihood

$$l_{n\lambda}(\theta) = \log L_n(\theta) - \sqrt{n}\lambda||\theta||_1,$$

is equivalent to minimizing

$$m_{n\lambda}(\theta) = \log M_n(\theta) + \lambda||\sqrt{n}\theta||_1.$$

Directly computation of the permutation likelihood is very challenging. Monte Carlo approximation based on a simple random sample of uniformly distributed permutations does not yield sufficient accuracy in approximating $M_n(\theta)$. To overcome the difficulties, the Metropolis sampling algorithm for sampling permutations from the distribution,

$$P(\Gamma = \sigma \mid Y, \theta_b) = \frac{\prod_{i=1}^{n} \eta(Y_{\sigma_1(i)1}; \cdots; Y_{\sigma_{G-1}(i)(G-1)}, Y_{iG} \mid \theta_b)}{\sum_{\tau \in \mathcal{P}_{G-1}} \prod_{i=1}^{n} \eta(Y_{\tau_1(i)1}; \cdots; Y_{\tau_{G-1}(i)(G-1)}, Y_{iG} \mid \theta_b)}, \tag{7.10}$$

to approximate $M_n(\theta)$ is used instead. This approach has been shown to yield a good approximation to the permutation likelihood When there is no penalty.

The computational approach to the permutation likelihood by Monte Carlo simulation can be summarized as follows. Let Γ be a random element taking values in $\mathcal{P}_{G-1}$ and $S_n(\Gamma|\theta) = \sum_{i=1}^{n} a\{i, \Gamma(i)|\theta\}$, where

$$a\{i, \Gamma(i)|\theta\} = \log \frac{\eta(Y_{\Gamma_1(i)1}; \cdots; Y_{\Gamma_{G-1}(i)(G-1)}, Y_{iG} \mid \theta)}{\eta(Y_{i1}; \cdots; Y_{iG} \mid \theta)}.$$

It can be seen that

$$\frac{M(\theta)}{M(\theta_b)} = E_{\mathcal{P}(\theta_b)}\left[\exp\left\{S_n(\Gamma \mid \theta) - S_n(\Gamma \mid \theta_b)\right\}\right], \tag{7.11}$$

where $E_{\mathcal{P}(\theta_b)}$ is the conditional expectation computed under the distribution in (7.10). By Taylor expansion of the exponent in (7.11) around θ_b and the normal approximation in computing the summation over all permutations, it can be derived that

$$\frac{M(\theta)}{M(\theta_b)} \approx \exp\left[\mu_n^t(\theta_b)\sqrt{n}(\theta - \theta_b) + \frac{1}{2}\sqrt{n}(\theta - \theta_b)^t\{\Omega_n(\theta_b) + \Sigma_n(\theta_b)\}\sqrt{n}(\theta - \theta_b)\right], \tag{7.12}$$

where

$$\begin{aligned}
\mu_n(\theta_b) &= E_{\mathcal{P}_{\theta_b}}\left\{\frac{1}{\sqrt{n}}\frac{\partial}{\partial\theta}S_n(\Gamma \mid \theta_b)\right\}, \\
\Sigma_n(\theta_b) &= E_{\mathcal{P}_{\theta_b}}\left\{\frac{1}{\sqrt{n}}\frac{\partial}{\partial\theta}S_n(\Gamma \mid \theta_b) - \mu_n(\theta_b)\right\}^2, \\
\Omega_n(\theta_b) &= E_{\mathcal{P}_{\theta_b}}\left\{\frac{1}{n}\frac{\partial^2}{\partial\theta^2}S_n(\Gamma \mid \theta_b)\right\}.
\end{aligned}$$

When the dimension of θ is fixed while $n \to \infty$, it can be shown that, under some regularity conditions, the approximation is accurate in the neighborhood of θ_b, i.e.,

$$\left|\frac{M(\theta)}{M(\theta_b)} - \exp\left[\mu_n^t(\theta_b)\sqrt{n}(\theta - \theta_b) + \frac{1}{2}\sqrt{n}(\theta - \theta_b)^t\{\Omega_n(\theta_b) + \Sigma_n(\theta_b)\}\sqrt{n}(\theta - \theta_b)\right]\right| = o_p(1),$$

when $||\sqrt{n}(\theta - \theta_b)||_2 = O(1)$. The penalized log-permutation likelihood approximately reduces to minimizing

$$\mu_n^t(\theta_b)\sqrt{n}(\theta - \theta_b) + \frac{1}{2}\sqrt{n}(\theta - \theta_b)^t\{\Omega_n(\theta_b) + \Sigma_n(\theta_b)\}\sqrt{n}(\theta - \theta_b) + \lambda\sqrt{n}||\theta||_1. \quad (7.13)$$

This is a quadratic programming having solutions satisfying

$$\mu_n(\theta_b) + \{\Omega_n(\theta_b) + \Sigma_n(\theta_b)\}\sqrt{n}(\theta - \theta_b) + \lambda\delta = 0,$$

where $\delta = (\delta_1, \cdots, \delta_p)$ with $||\delta||_\infty \le 1$ and $\delta_j = \text{sign}(\theta_j)$ if $\theta_j \ne 0$.

The quadratic programming problem can be solved by the coordinate descending algorithm. Let j be the current index of the coordinate and $\theta_{(-j)}^{(c)}$ and $\theta_{(-j)b}$ be respectively the current estimate for θ components other than θ_j, and the vector of θ_b without the jth component. The solution for θ_j satisfies

$$\begin{aligned}&\mu_{nj}(\theta_b) + \{\Omega_n(\theta_b) + \Sigma_n(\theta_b)\}_{j(-j)}\sqrt{n}(\theta_{(-j)}^{(c)} - \theta_{(-j)b})\\ &+\{\Omega_n(\theta_b) + \Sigma_n(\theta_b)\}_{jj}\sqrt{n}(\theta_j - \theta_{jb}) + \lambda\delta_j = 0,\end{aligned}$$

where $\mu_{nj}(\theta_b)$ is the jth component of $\mu_n(\theta_b)$, and $\{\Omega_n(\theta_b) + \Sigma_n(\theta_b)\}_{j(-j)}$ and $\{\Omega_n(\theta_b)\Sigma_n(\theta_b)\}_{jj}$ are respectively the jth row vector without the jth element and the (j, j) element of $\{\Omega_n(\theta_b) + \Sigma_n(\theta_b)\}$. Let

$$\Delta_j = \mu_{nj}(\theta_b) - \{\Omega_n(\theta_b) + \Sigma_n(\theta_b)\}_{jj}\sqrt{n}\theta_{bj} + \{\Omega_n(\theta_b) + \Sigma_n(\theta_b)\}_{j(-j)}\sqrt{n}\{\theta_{(-j)}^{(c)} - \theta_{b(-j)}\}.$$

Equation (7.4.2) can be further rewritten as

$$\{\Omega_n(\theta_b) + \Sigma_n(\theta_b)\}_{jj}\sqrt{n}\theta_j = -\Delta_j - \lambda\delta_j.$$

The solution to the equation can be expressed as

$$\theta_j = \begin{cases} -\frac{\Delta_j + \lambda}{\{\Omega_n(\theta_b) + \Sigma_n(\theta_b)\}_{jj}\sqrt{n}} & \text{if } \Delta_j < -\lambda, \\ -\frac{\Delta_j - \lambda}{\{\Omega_n(\theta_b) + \Sigma_n(\theta_b)\}_{jj}\sqrt{n}} & \text{if } \Delta_j > \lambda, \\ 0 & \text{otherwise.} \end{cases}$$

This solution is used to update the current estimate of θ_j. The algorithm goes through all the indices of θ and repeats until no change in θ estimate is achieved. The final θ obtained is then used to update θ_b.

To approximate $\Omega_n(\theta_b)$, $\Sigma_n(\theta_b)$, and $\mu_n(\theta_b)$, a Metropolis sampling algorithm is implemented to sample Γ from the distribution (7.10). Let $\sigma^{(k)}, k = 1, \cdots, N$ be such a sample. We estimate the quantities as follow.

$$\begin{aligned}\mu_n(\theta_b) &\approx \hat{\mu}_{n,N}(\theta_b) = \frac{1}{N}\sum_{k=1}^{N}\frac{1}{\sqrt{n}}\frac{\partial}{\partial\theta}S_n(\sigma^{(k)} \mid \theta_b),\\ \Sigma_n(\theta_b) &\approx \hat{\Sigma}_{n,N}(\theta_b) = \frac{1}{N-1}\sum_{k=1}^{N}\left\{\frac{1}{\sqrt{n}}\frac{\partial}{\partial\theta}S_n(\sigma^{(k)} \mid \theta_b) - \hat{\mu}_{n,N}(\theta_b)\right\}^2,\\ \Omega_n(\theta_b) &\approx \hat{\Omega}_{n,N}(\theta_b) = \frac{1}{N}\sum_{k=1}^{N}\frac{1}{n}\frac{\partial^2}{\partial\theta^2}S_n(\sigma^{(k)} \mid \theta_b).\end{aligned}$$

For a given θ_b, the following Metropolis algorithm for generating a sample of permutations following (7.10) can be applied to the approximation.

1. Denote the current sampled permutation by σ_A. Randomly sample two indices from $\{1, \cdots, n\}$ to form a permutation that switches the two indices. Applying the switching to σ^A to obtain permutation σ^B.
2. Update σ^A to σ^B with probability $\min\{1, \pi(\sigma^A, \sigma^B)\}$ and keep σ^A unchanged with probability $1 - \min\{1, \pi(\sigma^A, \sigma^B)\}$, where

$$\pi(\sigma^A, \sigma^B) = \frac{\prod_{i=1}^n \eta(Y_{\sigma_1^B(i)1}; \cdots ; Y_{\sigma_{G-1}^B(i)(G-1)}, Y_{iG} \mid \theta_b)}{\prod_{i=1}^n \eta(Y_{\sigma_1^A(i)1}; \cdots ; Y_{\sigma_{G-1}^A(i)(G-1)}, Y_{iG} \mid \theta_b)}.$$

3. Generate a Markov chain by repeating the foregoing steps.

We set the observed data as the initial step, i.e., σ_A starts from identity permutation. To generate a sample of permutations for compute the mean and variances, we sample permutation after a burn-in period and keep the sampling interval sufficiently apart to suppress the correlation among sampled permutations.

Starting from $\theta = 0$, the algorithm for finding the penalized permutation likelihood estimator can be summarized as follows.

1. Draw a random sample of permutations following (7.10) for the current θ_b by the Metropolis algorithm.
2. Estimate $\mu_n(\theta_b)$, $\Sigma_n(\theta_b)$, and $\Omega_n(\theta_b)$ based on the sample drewn. Update θ_b by solving the quadratic program (7.13).
3. Repeat the two steps until convergence which is defined by a criterion on the change of θ.

Due to the sampling variation, we choose the selected network as the average of the last few steps after the algorithm stabilized to oscillation. As a result, the connection detected in a network can be a fraction. To avoid such fractions, we may choose to define a detected connection between two nodes to be those with the fraction bigger than 0.5.

7.4.3 Hybrid algorithm for increasing the computation speed and estimation efficiency

The penalized permutation likelihood approach is efficient and computationally manageable for low and moderate high-dimension problem. For a high-dimensional problem, the repeated Metropolis sampling approach can substantially slow down the computation. Note that the pairwise pseudo-likelihood approach is fast but inefficient. To tackle the computational problem with high-dimensional data, the pairwise pseudo-likelihood approach may be combined with the permutation likelihood approach. The idea is to use the pairwise pseudo-likelihood approach to perform a neighborhood selection approach to obtain a selected network first. This network selection approach is consistent for the sparse network under some regularity conditions. Once the network is selected and the parameter estimated. The estimated network can be used as the starting point for refining the network detection by the penalized permutation likelihood approach.

Let the penalized pseudo-likelihood estimator of θ be $\tilde{\theta}$. Let $\sigma^{(k)}, k = 1, \cdots, N$ be a sample of permutations from the estimated model using the pairwise pseudo-likelihood selection and estimation approach. Denote the θ estimate by θ_a. The log-permutation likelihood can be approximated by

$$l_a(\theta) = \sum_{i=1}^n \log \eta(Y_{i1}; \cdots ; Y_{iG} \mid \theta) - \log \frac{1}{N} \sum_{k=1}^N \prod_{i=1}^n \frac{\eta(Y_{\sigma_1^{(k)}1}; \cdots , Y_{\sigma_{G-1}^{(k)}(G-1)}; Y_{iG} \mid \theta)}{\eta(Y_{\sigma_1^{(k)}1}; \cdots , Y_{\sigma_{G-1}^{(k)}(G-1)}; Y_{iG} \mid \theta_a)}.$$

The log-penalized approximate permutation likelihood under l^1-penalty is

$$l_{a\lambda} = l_a(\theta) - \lambda||\theta||_1.$$

Maximizing the penalized likelihood can be carried out using the coordinate ascending algorithm. The score equation for finding the maximizer is

$$\sum_{i=1}^{n}\left[\frac{\partial}{\partial\theta_j}\log\eta(Y_{i1};\cdots;Y_{iG}\mid\theta) - E_{\sigma,N}\left\{\frac{\partial}{\partial\theta_j}\log\eta(Y_{\sigma_1(i)1};\cdots;Y_{\sigma_{G-1}(i)(G-1)};Y_{iG}\mid\theta)\right\}\right] + \lambda\delta_j = 0,$$

where $\delta_j = sign(\theta_j)$ if $\theta_j \neq 0$, and $|\delta_j \leq 1$ if $\theta_j = 0$, and $E_{\sigma,N}$ is the conditional expectation computed under the distribution

$$P(\sigma^{(k)}) = \frac{\prod_{i=1}^{n}\frac{\eta(Y_{\sigma_1^{(k)}(i)1};\cdots;Y_{\sigma_{G-1}^{(k)}(i)(G-1)};Y_{iG}|\theta)}{\eta(Y_{\sigma_1^{(k)}(i)1};\cdots;Y_{\sigma_{G-1}^{(k)}(i)(G-1)};Y_{iG}|\theta_a)}}{\sum_{k=1}^{N}\prod_{i=1}^{n}\frac{\eta(Y_{\sigma_1^{(k)}(i)1};\cdots;Y_{\sigma_{G-1}^{(k)}(i)(G-1)};Y_{iG}|\theta)}{\eta(Y_{\sigma_1^{(k)}(i)1};\cdots;Y_{\sigma_{G-1}^{(k)}(i)(G-1)};Y_{iG}|\theta_a)}}.$$

In implementing the coordinate ascending algorithm, the approximate θ_a may need to be updated if the weights $P(\theta^{(k)})$ becomes overly concentrated on a few sampled permutations.

7.5 Network detection by penalized semiparametric likelihood approach

An alternative approach to the network detection is through the penalized semiparametric likelihood approach. In principle, this approach can be used either as a node-wise selection approach or a joint selection approach. Since the denominator in the joint likelihood is difficult to compute, we concentrate on the node-wise selection approach.

7.5.1 Penalized likelihood for neighborhood selection

For node j, consider the nodes that potentially connect to node j directly. To simplify notation, let $Z = Y_j$ and $X = Y_{-j}$. Denote the parametric odds ratio function by $\eta(z; x \mid \theta), \theta \in \Theta$. For a fixed point x_0, let F denote the distribution function of $f(z \mid x_0)$. Given an independent sample $(Z_i, X_i), i = 1, \cdots, n$, the likelihood for (θ, F) under the semiparametric odds ratio model is

$$L(\theta, F) = \prod_{i=1}^{n}\frac{\eta(Z_i; X_i \mid \theta)dF(Z_i \mid x_0)}{\int\eta(z; X_i \mid \theta)dF(z \mid x_0)}.$$

Let $||\theta||_1 = \sum_{j=1}^{r}|\theta_j|$, where r is the dimension of the parameter θ. The following objective function

$$l_\lambda(\theta, F) = l(\theta, F) + \lambda_n||\theta||_1 \tag{7.14}$$

is minimized with respect to (θ, F) in the penalized semiparametric likelihood approach for some $\lambda_n \geq 0$, where

$$l(\theta, F) = -\frac{1}{n}\sum_{i=1}^{n}\log\eta(Z_i, X_i|\theta) - \frac{1}{n}\sum_{i=1}^{n}\log dF(Z_i) + \frac{1}{n}\sum_{i=1}^{n}\log\int\eta(z; X_i|\theta)dF(z),$$

is the negative log-likelihood given the observed data $(Z_i, X_i), i = 1, \cdots, n$, and x_0 is suppressed. Similar to the procedure in obtaining the maximum semiparametric likelihood estimator for (θ, F), we also restricted F to be discrete distributions in the maximum penalized semiparametric likelihood approach. It can be shown that the maximum penalized likelihood estimator for F is obtained in the collection of F_n for Z, and $\mathcal{F}_n$ with probability masses on the observed Z points only.

To better facilitate the computation, let P_n and Q_n be the empirical measures respectively generated by Z and X. Define

$$\mathcal{H}_{P_n} = \left\{ H_n(z) = \log \frac{dF_n(z)}{dP_n(z)} \mid F_n \in \mathcal{F}_n, P_n \ll F_n \ll P_n \right\}.$$

We use $\mathcal{H}_n$ for $\mathcal{H}_{P_n}$ for simplicity. Under the new parametrization, (10.10) can be converted into minimizing

$$l_\lambda(\theta, H) = l(\theta, H) + \lambda_n ||\theta||_1,$$

with $(\theta, H) \in \Theta \times \mathcal{H}_n$ where

$$l(\theta, H) = -\frac{1}{n} \sum_{i=1}^{n} \log \eta(Z_i; X_i|\theta) - \frac{1}{n} \sum_{i=1}^{n} H(Z_i) + \frac{1}{n} \sum_{i=1}^{n} \log \int \eta(z; X_i|\theta) e^{H(z)} dP_n(z).$$

Note that, for brevity, we still use l_λ and l for the new functions. Given θ, P, and Q, let $H(z \mid \theta, P, Q)$ denote the minimizer of

$$-\int H(z)dP(z) + \int \left\{ \log \int \eta(z; x \mid \theta) e^{H(z)} dP(z) \right\} dQ(x),$$

in a given set of H when the minimizer exists. The minimizer, when exists, satisfies

$$H(z \mid \theta, P, Q) = -\log \int \frac{\eta(z; x \mid \theta)}{\int \eta(z; x \mid \theta) \exp\{H(z \mid \theta, P, Q)\} dP(z)} dQ(x), \tag{7.15}$$

where $\int \exp\{H(z \mid \theta, P, Q)\}dP(z) = 1$. Let $H_n(z \mid \theta) = H(z \mid \theta, P_n, Q_n)$, the minimizer for $H \in \mathcal{H}_n$. Denote the profile likelihood for θ by

$$l_{pn}(\theta) = l\{\theta, H_n(\cdot \mid \theta)\}.$$

If Θ is a convex set and for any fixed $(z, x) \in \mathcal{Z} \times \mathcal{X}$,

$$\begin{aligned}
&-\log \frac{\eta\{z; x \mid u\theta_1 + (1-u)\theta_2\} e^{uH_1(z) + (1-u)H_2(z)}}{\int \eta\{z; x \mid u\theta_1 + (1-u)\theta_2\} e^{uH_1(z)+(1-u)H_2(z)} dP(z)} \\
&\leq \quad -u \log \frac{\eta(z; x \mid \theta_1) e^{H_1(z)}}{\int \eta(z; x \mid \theta_1) e^{H_1(z)} dP(z)} - (1-u) \log \frac{\eta(z; x \mid \theta_2) e^{H_2(z)}}{\int \eta(z; x \mid \theta_2) e^{H_2(z)} dP(z)},
\end{aligned}$$

for any $0 < u < 1$, $\theta_1, \theta_2 \in \Theta$ and $H_1, H_2 \in \mathcal{H}_n$, then the profile likelihood for θ, $l_{pn}(\theta)$, is a convex function of θ. This is because $l(\theta, H - c) = l(\theta, H)$ for any c, $(\theta, H) \in \Theta \times \mathcal{H}_n$, and

$$\begin{aligned}
l_{pn}(u\theta_1 + (1-u)\theta_2) &= l[u\theta_1 + (1-u)\theta_2, H_n\{\cdot \mid u\theta_1 + (1-u)\theta_2\}] \\
&\leq l\{u\theta_1 + (1-u)\theta_2, uH_n(\cdot \mid \theta_1) + (1-u)H_n(\cdot \mid \theta_2) - c(u, n)\} \\
&\leq ul\{\theta_1, H_n(\cdot \mid \theta_1)\} + (1-u)l\{\theta_2, H_n(\cdot \mid \theta_2)\} \\
&\leq ul_{pn}(\theta_1) + (1-u)l_{pn}(\theta_2),
\end{aligned}$$

where $c(u,n) = a\{uH_n(\cdot \mid \theta_1) + (1-u)H(\cdot \mid \theta_2)\}$ and $a(H) = \log \int e^{H(z)} dP_n(z)$. This means the joint minimization problem can be transformed into a convex programming for θ as

$$\min_{\theta \in \Theta} \{l_{pn}(\theta) + \lambda_n ||\theta||_1\}, \tag{7.16}$$

which is equivalent to

$$\min_{\theta \in \Theta} l_{pn}(\theta) \text{ subject to } ||\theta||_1 \leq C(\lambda_n), \tag{7.17}$$

for some constant $C(\lambda_n) \geq 0$. The minimizer exists if $l_{pn}(\theta)$ is a continuous function of θ on the compact set $\Theta \cap \{\theta \mid ||\theta||_1 < C(\lambda_n)\}$.

7.5.2 Algorithms for finding the maximum penalized likelihood estimator

The θ estimator can in principle be computed using one of the available methods for convex programming (Boyd and Vanderberghe, 2004)[12]. However, the profile likelihood for θ does not have a closed-form expression. This makes direct application of such methods difficult. When the original objective function is a convex function of (θ, H_n), minimizing the objective function jointly with respect to (θ, H_n) can be simpler. Coordinate descent methods (Tseng, 2001)[184] have been shown to be numerically stable in solving large scale problems (Friedman et al, 2007; Wu and Lange, 2008; Tseng and Yun, 2009)[64, 199, 185]. A block coordinate descent algorithm (Fu, 1998; Tseng, 2001)[65, 184] for solving the minimization problem may be implemented as follows.

1. For a fixed θ, update H by the iterative algorithm:

$$H^{new}(z) = -\log t_H(z) - \log \int t_H^{-1}(z) dP_n(z),$$

   where $t_H(z) = \int [\eta(z; x \mid \theta) / \{\int \eta(z; x \mid \theta) \exp\{H(z)\} dP_n(z)] dQ_n(x)$.

2. For a fixed H, update θ by solving the following quadratic program

$$\arg\min_{\theta_j} \left\{ \frac{\partial l(\theta_c, H_c)}{\partial \theta_j} (\theta_j - \theta_{jc}) + \frac{\delta}{2} \frac{\partial^2 l(\theta_c, H_c)}{\partial^2 \theta_j} (\theta_j - \theta_{jc})^2 + \lambda |\theta_j| \right\},$$

   where $\theta_c, \theta_{jc}, H_c$ denote the current estimates of θ, θ_j, H, and $\delta \geq 1$ is the step size selected to ensure the objective function decreases.

3. Repeat steps 1 and 2 until the algorithm converges.

Chen (2015)[31] showed that Step 1 for updating H guarantees to decrease the objective function. As a result, an arbitrarily fixed number of iterations may be used in this step. The minimization problem in Step 2 has a closed-form solution as follows.

(a) If $\delta \frac{\partial^2 l(\theta_c, H_c)}{\partial^2 \theta_{jc}} \theta_{jc} - \frac{\partial l(\theta_c, H_c)}{\partial \theta_j} > \lambda$, then

$$\theta_j = \theta_{jc} - \delta^{-1} \left\{ \frac{\partial l(\theta_c, H_c)}{\partial \theta_j} + \lambda \right\} \Big/ \frac{\partial^2 l(\theta_c, H_c)}{\partial^2 \theta_j}.$$

(b) If $\delta \frac{\partial^2 l(\theta_c, H_c)}{\partial^2 \theta_{jc}} \theta_{jc} - \frac{\partial l(\theta_c, H_c)}{\partial \theta_j} < -\lambda$, then

$$\theta_j = \theta_{jc} - \delta^{-1} \left\{ \frac{\partial l(\theta_c, H_c)}{\partial \theta_j} - \lambda \right\} \Big/ \frac{\partial^2 l(\theta_c, H_c)}{\partial^2 \theta_j}.$$

(c) If $-\lambda \le \delta \frac{\partial^2 l(\theta_c, H_c)}{\partial^2 \theta_{jc}} \theta_{jc} - \frac{\partial l(\theta_c, H_c)}{\partial \theta_j} \le \lambda$, then $\theta_j = 0$.

The initial values for θ can be conveniently set to 0.

To determine the tuning parameter value, Chen and Chen (2008; 2012)[39, 40] proposed an extended BIC (EBIC) approach for the generalized linear model, which yields selection consistency. For the semiparametric odds ratio model,

$$EBIC(\lambda) = 2l\{\hat{\theta}_\lambda, H_n(\cdot \mid \hat{\theta}_\lambda)\} + ||\hat{\theta}_\lambda||_0 \frac{\log(n)}{n} + 2\gamma ||\hat{\theta}_\lambda||_0 \frac{\log(r)}{n},$$

where $\hat{\theta}_\lambda$ is the penalized likelihood estimator under a fixed λ value, $||\hat{\theta}_\lambda||_0$ is the number of nonzero components in $\hat{\theta}_\lambda$, and $\gamma > 1 - 1/(2k)$ and $r = O(n^k)$, is the dimension of θ. The EBIC tends to select models with many false positive parameters, likelihood cross-validation is an alternative choice.

Combination of neighborhoods detected at individual nodes forms the network. For two nodes a and b, if b is detected in the neighborhood of a, but a is not detected in the neighborhood of b, the two nodes are assumed to be connected. In fact, whether including such connections in the network or not does not make any difference asymptotically. Under a set of regularity condition, The penalized likelihood selection approach is consistent for sparse network selection. More details can be found in Chapter 10.

7.6 Network selection using package SPORM

Many software packages exist for network detection under the Gaussian model, such as the GGM. The R package `SPORM` has several functions which can be used in combination for network detection under the semiparametric odds ratio model. The maximum penalized pairwise pseudo-likelihood approach for node-wise detection of network connections is implemented in the `R` function `pwpenlkh`. The function has the following form

```
pwpenlkh(dat, group, lambda, niter, eps),
```

where `dat` denotes the data matrix, `group` denotes the variable groups in the model, `lambda` denotes the penalty parameter values. One network is determined for each value of `lambda`. `niter` sets the maximum number of iterations and `eps` is the convergence criterion. The output of the function is primarily a series of matrices representing structures of the network detected, each corresponding to one penalty value given in `lambda`. Possible status of the network connections are coded as follow: 0 means not connected; 1 means weakly connected, i.e., in two different node-wise models, one detected as connected and the other detected as unconnected; 2 denotes strongly connected, i.e., both node-wise models detected as connected; A `NA` status means the connection is not included in the consideration of the detection algorithm. Under the semiparametric odds ratio model, only connections between groups of variables are modeled and detected. Connections within a group are not modeled or detected. If all the connections between pairs of nodes are considered, the group is a vector of length equal to the number of variables and the number of variables in each group is 1. Only the diagonal elements are `NA`s. The output of a set of selected network is the second object in the output list. The total number of networks detected is the same as the length of the vector of the penalty parameter `lambda`.

The maximum penalized semiparametric-likelihood approach for node-wise detection of network connections is implemented in the `R` function `sppenlkh`. The function has the following form

```
sppenlkh(dat, group, lambda, niter, eps),
```

The argument of this function are the same as those in `pwpenlkh`. The output is also the same.

The maximum penalized permutation likelihood approach for joint selection of network connections is implemented in the `R` function `pmpenlkh`. The function has the following form

```
pmpenlkh(dat, group, lambda, niter, eps, nburnin, nsamp, nintv, maxcyc).
```

In addition to the arguments similar to the previous two functions, four other arguments are used for controlling the Monte Carlo sample. `nburnin` specifies the length of the sampling sequence to be run before the samples are taken. `nsamp` is the sample size to be taken. `nintv` is the length of interval between two samples, and `maxcyc` is the cycle length of the candidate permutation move taken in the Metropolis sampling algorithm. Default values are given so that a user does not have to choose them.

The foregoing three functions are used to select connections for a given penalty value. For a series of networks corresponding to different penalty values, which network is chosen can be carried out by the following R function `networkselect`. It can be used to fit a pre-determined set of networks using a particular model-fitting method and to select a particular network from the set of networks based on a given criterion. The function has the following form

```
networkselect(dat, group, network, lambda, method, criterion),
```

where `dat` denotes the data used for the network selection, `group` specifies the number of clusters and the cluster sizes, `network` contains the network structures corresponding to `lambda` values determined by one of the previous three penalized network selection approaches specified in `method`. Those methods are the pairwise likleihood (`pw`), the semi-parametric likelihood (`sp`), or the permutation likelihood (`pm`). The `criterion` specifies the model selection approach such as the BIC. The output of the function is the selected network structure.

7.7 Software usage cases

To illustrate the use of the `SPORM` package for network detection, data from a nine-dimensional joint normal model are simulated. The variables are divided into three groups of size $(2, 3, 4)$. The true strength of the connections between nodes of the groups are as follows.

$$\Theta = \begin{pmatrix} 0 & 0 & -0.564 & 2.45 & 1.94 & 0.750 & -2.11 & -0.291 & -1.19 \\ 0 & 0 & -3.84 & 5.08 & -1.67 & 0.377 & -5.07 & 0.656 & -3.12 \\ -0.564 & -3.84 & 0 & 0 & 0 & 3.57 & -8.17 & 1.54 & -7.37 \\ 2.45 & 5.08 & 0 & 0 & 0 & -4.63 & 14.3 & -1.19 & 11.4 \\ 1.94 & -1.67 & 0 & 0 & 0 & -0.957 & -0.778 & -0.352 & -0.293 \\ 0.750 & 0.377 & 3.57 & -4.63 & -0.957 & 0 & 0 & 0 & 0 \\ -2.11 & -5.07 & -8.17 & 14.3 & -0.778 & 0 & 0 & 0 & 0 \\ -0.291 & 0.656 & 1.54 & -1.19 & -0.352 & 0 & 0 & 0 & 0 \\ -1.19 & -3.12 & -7.37 & 11.4 & -0.293 & 0 & 0 & 0 & 0 \end{pmatrix},$$

To illustrate, the variable clusters are `group=c(2,3,4)`, and a set of penalty parameter values are chosen as `lambda=c(10, 50, 200, 500, 800)`. Using the penalized pairwise pseudo-likelihood approach to select the network, apply the following statement

```
fit1=pwpenlkh(dat=dat,group=group,lambda=lambda).
```

The selected networks can be found in

```
network=fit1[[4]].
```

The five selected networks corresponding respectively to the five penalty values are

$$\texttt{network[,,1]} = \begin{pmatrix} \mathbf{0} & \mathbf{0} & 2 & 2 & 2 & 2 & 2 & 2 & 2 \\ \mathbf{0} & \mathbf{0} & 2 & 2 & 2 & 2 & 2 & 2 & 2 \\ 2 & 2 & \mathbf{0} & \mathbf{0} & \mathbf{0} & 2 & 2 & 2 & 2 \\ 2 & 2 & \mathbf{0} & \mathbf{0} & \mathbf{0} & 2 & 2 & 2 & 2 \\ 2 & 2 & \mathbf{0} & \mathbf{0} & \mathbf{0} & 2 & 2 & 2 & 2 \\ 2 & 2 & 2 & 2 & 2 & \mathbf{0} & \mathbf{0} & \mathbf{0} & \mathbf{0} \\ 2 & 2 & 2 & 2 & 2 & \mathbf{0} & \mathbf{0} & \mathbf{0} & \mathbf{0} \\ 2 & 2 & 2 & 2 & 2 & \mathbf{0} & \mathbf{0} & \mathbf{0} & \mathbf{0} \\ 2 & 2 & 2 & 2 & 2 & \mathbf{0} & \mathbf{0} & \mathbf{0} & \mathbf{0} \end{pmatrix},$$

$$\texttt{network[,,2]} = \begin{pmatrix} \mathbf{0} & \mathbf{0} & 2 & 2 & 2 & 2 & 2 & 2 & 2 \\ \mathbf{0} & \mathbf{0} & 2 & 2 & 0 & 0 & 2 & 2 & 2 \\ 2 & 2 & \mathbf{0} & \mathbf{0} & \mathbf{0} & 2 & 0 & 2 & 2 \\ 2 & 2 & \mathbf{0} & \mathbf{0} & \mathbf{0} & 2 & 2 & 2 & 2 \\ 2 & 0 & \mathbf{0} & \mathbf{0} & \mathbf{0} & 2 & 2 & 2 & 2 \\ 2 & 0 & 2 & 2 & 2 & \mathbf{0} & \mathbf{0} & \mathbf{0} & \mathbf{0} \\ 2 & 2 & 0 & 2 & 2 & \mathbf{0} & \mathbf{0} & \mathbf{0} & \mathbf{0} \\ 2 & 2 & 2 & 2 & 2 & \mathbf{0} & \mathbf{0} & \mathbf{0} & \mathbf{0} \\ 2 & 2 & 2 & 2 & 2 & \mathbf{0} & \mathbf{0} & \mathbf{0} & \mathbf{0} \end{pmatrix},$$

$$\texttt{network[,,3]} = \begin{pmatrix} \mathbf{0} & \mathbf{0} & 2 & 2 & 1 & 1 & 2 & 2 & 2 \\ \mathbf{0} & \mathbf{0} & 2 & 0 & 0 & 0 & 2 & 2 & 2 \\ 2 & 2 & \mathbf{0} & \mathbf{0} & \mathbf{0} & 2 & 0 & 2 & 2 \\ 2 & 0 & \mathbf{0} & \mathbf{0} & \mathbf{0} & 2 & 2 & 2 & 2 \\ 1 & 0 & \mathbf{0} & \mathbf{0} & \mathbf{0} & 0 & 2 & 2 & 2 \\ 1 & 0 & 2 & 2 & 0 & \mathbf{0} & \mathbf{0} & \mathbf{0} & \mathbf{0} \\ 2 & 2 & 0 & 2 & 2 & \mathbf{0} & \mathbf{0} & \mathbf{0} & \mathbf{0} \\ 2 & 2 & 2 & 2 & 2 & \mathbf{0} & \mathbf{0} & \mathbf{0} & \mathbf{0} \\ 2 & 2 & 2 & 2 & 2 & \mathbf{0} & \mathbf{0} & \mathbf{0} & \mathbf{0} \end{pmatrix},$$

$$\texttt{network[,,4]} = \begin{pmatrix} \mathbf{0} & \mathbf{0} & 1 & 1 & 1 & 1 & 1 & 0 & 1 \\ \mathbf{0} & \mathbf{0} & 0 & 0 & 0 & 0 & 2 & 0 & 0 \\ 1 & 0 & \mathbf{0} & \mathbf{0} & \mathbf{0} & 2 & 0 & 0 & 0 \\ 1 & 0 & \mathbf{0} & \mathbf{0} & \mathbf{0} & 2 & 2 & 0 & 2 \\ 1 & 0 & \mathbf{0} & \mathbf{0} & \mathbf{0} & 0 & 0 & 0 & 2 \\ 1 & 0 & 2 & 2 & 0 & \mathbf{0} & \mathbf{0} & \mathbf{0} & \mathbf{0} \\ 1 & 0 & 0 & 2 & 0 & \mathbf{0} & \mathbf{0} & \mathbf{0} & \mathbf{0} \\ 0 & 0 & 0 & 0 & 0 & \mathbf{0} & \mathbf{0} & \mathbf{0} & \mathbf{0} \\ 1 & 0 & 0 & 2 & 2 & \mathbf{0} & \mathbf{0} & \mathbf{0} & \mathbf{0} \end{pmatrix},$$

and

$$\texttt{network[,,5]} = \begin{pmatrix} \mathbf{0} & \mathbf{0} & 0 & 0 & 1 & 0 & 1 & 0 & 0 \\ \mathbf{0} & \mathbf{0} & 0 & 0 & 0 & 0 & 0 & 0 & 0 \\ 0 & 0 & \mathbf{0} & \mathbf{0} & \mathbf{0} & 0 & 0 & 0 & 0 \\ 0 & 0 & \mathbf{0} & \mathbf{0} & \mathbf{0} & 0 & 0 & 0 & 2 \\ 1 & 0 & \mathbf{0} & \mathbf{0} & \mathbf{0} & 0 & 0 & 0 & 0 \\ 0 & 0 & 0 & 0 & 0 & \mathbf{0} & \mathbf{0} & \mathbf{0} & \mathbf{0} \\ 1 & 0 & 0 & 0 & 0 & \mathbf{0} & \mathbf{0} & \mathbf{0} & \mathbf{0} \\ 0 & 0 & 0 & 0 & 0 & \mathbf{0} & \mathbf{0} & \mathbf{0} & \mathbf{0} \\ 0 & 0 & 0 & 2 & 0 & \mathbf{0} & \mathbf{0} & \mathbf{0} & \mathbf{0} \end{pmatrix}.$$

Use the network detected as input to call the function `networkselect` as follows

```
fit2=networkselect(dat=dat,group=group,lambda=lambda,network=network,
                   method='pw',criterion='BIC').
```

The output provides the selected network as `fit2[[2]]` is the same as `network[,,1]`. The BIC vector ouput is $(15275.89, 20130.66, 26808.80, 59871.69, 171961.83)$.

To select the network using the penalized semiparametric likelihood, similar statements can be applied.

```
fit1=sppenlkh(dat=dat,group=group,lambda=lambda).
                    network=fit1[[4]].
fit2=networkselect(dat=dat,group=group,lambda=lambda,network=network,
                   method='sp',criterion='BIC').
```

For selection using the permutation likelihood approach, the following statements are applied

```
fit1=pmpenlkh(dat=dat,group=group,lambda=lambda,nsamp=2e4,niter=50),
                    network=fit1[[4]],
fit2=networkselect(dat=dat,group=group,lambda=lambda,network=network,
                   method='pm',criterion='BIC').
```

The difference of the penalized permutation likelihood approach from the other two approaches is that the permutation likelihood approach uses the joint likelihood and is not a node-wise (or block-wise) selection. The selected network coded as either 0 or 1 only.

7.8 Summary and literature

Graphical models for undirected network have been extensively studied (Besag, 1974; Lauritzen, 1996; Edwards, 2000)[7, 92, 59]. Recent surge of interest in graphical models mostly concerns with the detection of network structure in large networks using the Gaussian modeling approach (Dempster, 1972; Li and Gui, 2006; Meinshausen and Buhlmann, 2006; Yuan and Lin, 2007; Lam and Fan, 2009; Buhlmann et al, 2010; Cai and Yuan, 2012; Li et al, 2012; Cai et al, 2013; Danaher et al, 2014)[54, 97, 125, 91, 202, 20, 96, 19, 48]. Non-Gaussian models for network detection have been proposed (Ravikumar et al, 2010; Liu et al, 2012; Xue and Zou, 2012; Loh and Wainwright, 2013; Fellinghauer et al, 2013, Voorman et al, 2014; Liu et al, 2017; Ning et al, 2017)[148, 110, 200, 115, 61, 192, 113, 130]. Network using screening approaches were studied in Friedman et al (2007)[64], Mazumder and Hastie (2012)[123], Buhlmann et al (2010) [17], and Liang et al (2015)[98].

7.9 Exercises

1. The directed acyclic graph is often used to describe causal relationship between variables, for example, the left graph in the following diagram.

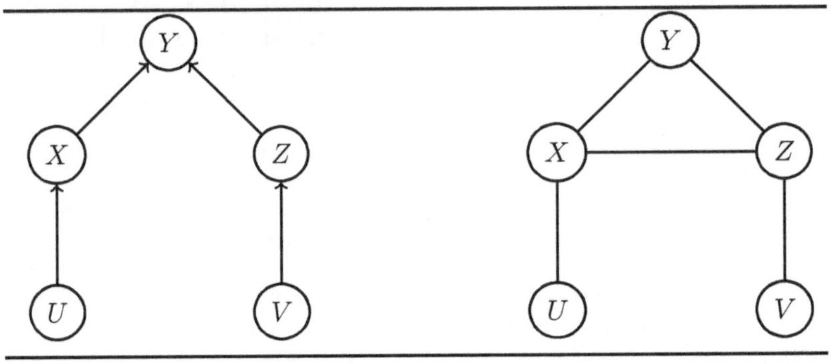

Under the Markov condition, i.e., conditional on the parental nodes (nodes that connects and points to the current node), the current node is independent of the rest nodes in the graph, the distribution of the directed acyclic graph can be expressed as

$$p(y, x, z) = p(y \mid x, z)p(x|u)p(z|v)p(u)p(v).$$

An undirected graph (right graph of the above diagram) describes the conditional distribution of any given node conditional on the rest nodes.

   (a) Show that the directed acyclic graph on the left leads to the undirected graph (termed moral graph) on the right.
   (b) If the directed acyclic graph has also a connection between U and Z and directed to Z, write down the joint distribution corresponding to the directed acyclic graph. What is the moral graph of the directed acyclic graph?
   (c) If the additional connection between U and Z pointing to U instead, what is the moral graph of the directed acyclic graph?

2. Suppose that a causal network containing nodes U, V, X, Y, Z can be expressed as a directed acyclic graph with the Markov property. Suppose also that the network has been detected to have the moralized graph as displayed on the right-hand side of the above diagram. Find all possible directed acyclic graphs that have the undirected graph as their moralized graph.

3. For an undirected network $(\mathcal{N}, \mathcal{E})$, where $\mathcal{N} = \{Y_1, \cdots, Y_p\}$ is the collection of nodes and $\mathcal{E}$ is the collection of edges.

   (a) Show that a biased sampling design with sampling probability

   $$P(S = 1 \mid Y_1, \cdots, Y_p) \propto \prod_{i=1}^{p} \pi_i(Y_i),$$

   changes a Gaussian network into a non-Gaussian network, where $\pi_i(Y_i) > 0$ for $i = 1, \cdots, p$.
   (b) Show that the network structure is not altered by the biased sampling design.

(c) If the undirected network is a moral graph of a directed acyclic graph, does the biased sampling alter the structure of the directed acyclic graph?
Hint: Consider the following directed acyclic graph,

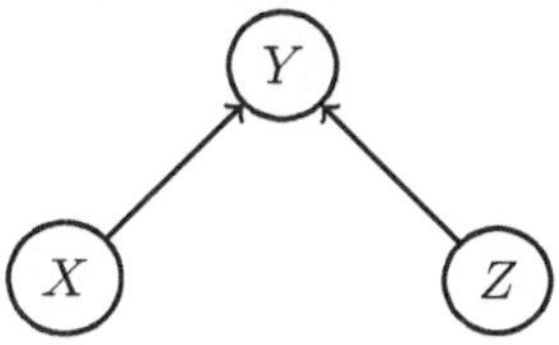

and the biased sampling design $P(S = 1 \mid Y, X, Z) = \pi(Y) > 0$.

4. The partial faithfulness of a network refers to the situation where for any two nodes Y and X, if Y and X are independent conditional on a given set of nodes Z, then Y and X are independent conditional on Z and any other nodes in the network.
   (a) Find a specific Gaussian network where the partial faithfulness does not hold.
   (b) Based on the example, provide some general conditions under which the partial faithfulness does not hold.
   (c) Explain why the partial faithfulness can be a reasonable assumption in practice.
   (d) Discuss the impact of the partial faithfulness on the network detection.
5. One of the advantages of the semiparametric odds ratio model over the Gaussian network model is that the former can be used directly to model the response of network to intervention. For example, consider the connection of nodes Y and X given another node Z. The conditional odds ratio model for the network connection may appear as

$$\log \eta(y; x \mid z) = \gamma(z)(y - y_0)(x - x_0).$$

Unlike the Gaussian model, the connection between nodes Y and X can be moderated by the values other nodes take. If we have

$$\gamma(z) = \gamma_0 + \gamma_1 z,$$

the nodes Y and X are disconnected when $z = -\gamma_0/\gamma_1$, and connected otherwise.
   (a) Assume Z is an intervention variable that acts to modify the strength of the network connections, build a semiparametric odds ratio model that allows Z to impact some of the connections but not others.
   (b) If Z is already a node in the network, intervention modifies the value of Z, show that the semiparametric model allows for evaluation of the effect without change the structure of the network model.
   (c) How can a Gaussian network adjust to such an intervention?

8

Application to Missing Data Problems

8.1 A brief introduction to the missing data problem

Missing data are often an unavoidable nuisance in statistical analysis. There are a vast number of literatures on methods for handling missing values in statistical analysis. These methods can be roughly classified into three types. The weighted estimation equation approach and its extensions, the maximum likelihood approach, and the imputation approach. The difficulties of applying these approaches in practice depend on the missing data mechanisms as well as the complexity of the missing data patterns.

The missing data mechanisms are classified into three categories according to Rubin (1976)[156] and Little and Rubin (2002)[109]. The first category is termed missing completely at random (MCAR) where missingness of a value is independent of the data. Under the MCAR, the complete cases that exclude those data points having missing values can still be viewed as a random sample from the general population. As a result, the complete-case analysis is valid under this type of missing data mechanisms. The complete-case analysis does not use the full data and may therefore be inefficient. The second category is termed missing at random (MAR) where missingness of a value depends on the observed values, but not on the unobserved values. Under the MAR, the complete cases can no longer be viewed as a random sample from the general population. As a result, the complete-case analysis can be biased as well as inefficient. The third category termed missing not at random (MNAR) where missingness depends on the unobserved data. The last category is often loosely called non-ignorable missing due to the need to consider the missing data mechanism model in the maximum likelihood approach while the missing data mechanism need not be modeled in general in applying the maximum likelihood approach when the missing data are MAR.

The weighted estimating equation approach requires us to model the missing data mechanism first and then use the inverse of the estimated missing data probability to weight the complete cases. The missing data mechanism model can be difficult to specify correctly in observational studies under MAR when the missing data patterns are not monotone. On the other hand, neither the maximum likelihood approach nor imputation approach needs to model the missing data mechanism under MAR. However, in the case of missing covariates in regression analysis, the maximum likelihood approach requires us to model the covariate distribution which is not needed when covariates are fully observed. The covariate model can be hard to specify correctly and even harder to estimate jointly with the regression model. The semiparametric odds ratio model provides a flexible modeling approach that is fairly robust against the model misspecification. Both models for the missing data mechanism and for the covariate distribution can be misspecified. For data missing by design, the missing data mechanism is often known without the danger of misspecification. For simple missing data patterns, the covariate model in the maximum likelihood approach can be totally nonparametric and free of misspecification. To take advantages of these situations, the doubly robust estimating approach may be considered. The doubly robust approach

DOI: 10.1201/9781351049757-8

can benefit from a more robust model for the covariate distribution when the missing data patterns are complex.

The imputation approach is usually set in the Bayesian framework where we need to specify a complete full data model plus the priors for the model parameters. Under MAR, the missing data mechanism model can be avoided. The Monte Carlo Markov chain sampling approach is usually employed. The key to the success of the imputation approach is to design fast and efficient sampling algorithm within the modeling framework. In addition to sampling from the posterior distribution, fast and efficient sampling from the conditional distribution of missing data given the observed data is also important.

8.2 The missing covariate problem in regression analysis

8.2.1 Missing covariates in parametric regression

Consider specifically the missing covariate problem where the outcome is Y and the covariate is X. Assume that the conditional density of Y given X is modeled by $f(y \mid x, \theta)$ with unknown parameter θ. We are interested in estimating the parameter θ. Suppose that X is a vector of p random variables denote by $(X_1, \cdots, X_p)$. Assume further that Y is fully observed and X is subject to missing values. Let $R = (R_1, \cdots, R_p)$ be the missing data indicators corresponding respectively to $X = (X_1, \cdots, X_p)$, where $R_{ij} = 1$ if X_{ij} is observed, 0 otherwise.

Let the observed data be $(R_{i1}, \cdots, R_{ip}, Y_i, X_{i1}, \cdots, X_{ip}), i = 1, \cdots, n$. Let $\bar{R}_{ij} = 1 - R_{ij}$. The observed covariate data are $R_{i1}(X_{i1}), \cdots, R_{ip}(X_{ip}), i = 1, \cdots, n$ and the missing data are $\bar{R}_{i1}(X_{i1}), \cdots, \bar{R}_{ip}(X_{ip}), i = 1, \cdots, n$, where $R(X) = X$ if $R = 1$ and missing if $R = 0$. The likelihood for the observed data is

$$\prod_{(r_1, \cdots, r_p)} \left[\int \cdots \int \prod_{i=1}^{n} P(R_1 = r_1, \cdots, R_p = r_p \mid Y_i, X_{i1}, \cdots, X_{ip}) \right.$$
$$\left. f(Y_i \mid X_{i1}, \cdots, Y_{ip}, \theta) g(X_{i1}, \cdots, X_{ip}) d\bar{r}_1(X_{i1}) \cdots d\bar{r}_p(X_{ip}) \right]^{1_{\{R_{i1}=r_1, \cdots, R_{ip}=r_p\}}}, \quad (8.1)$$

where $\bar{r} = 1 - r$ and $d\bar{r}_k(x_{ik})$ represents an integral argument only when $r_k = 0$. When data are missing at random, i.e.,

$$\begin{aligned} & P(R_1 = r_1, \cdots, R_p = r_p \mid Y_i, X_{i1}, \cdots, X_{ip}) \\ & \quad = \quad P\{R_1 = r_1, \cdots, R_p = r_p \mid Y_i, r_1(X_{i1}), \cdots, r_p(X_{ip})\}, \end{aligned}$$

the missing data probability can be moved out of the integral and be separated from the marginal likelihood

$$\prod_{(r_1, \cdots, r_p)} \left[\int \cdots \int \prod_{i=1}^{n} f(Y_i \mid X_{i1}, \cdots, X_{ip}, \theta) \right.$$
$$\left. g(X_{i1}, \cdots, X_{ip}) d\bar{r}_1(X_{i1}) \cdots d\bar{r}_p(X_{ip}) \right]^{1_{\{R_{i1}=r_1, \cdots, R_{ip}=r_p\}}}, \quad (8.2)$$

if the parameters in the missing data mechanism model are variation-independent of the parameters in the marginal likelihood.

To obtain a well-behaved maximum likelihood estimator of the regression parameters, it is usually needed to model the covariate distribution. On the other hand, misspecification of the covariate distribution may bias the regression model estimation. In specifying the covariate model, it is more desirable to put as few constraints on the model as possible. While a nonparametric model for the covariate distribution is feasible for some special cases of missing covariates (Chen, 2002; 2004)[25, 27], it is not feasible in general due to the curse of dimensionality. As a compromise, the semiparametric odds ratio model can be a good choice for the covariate modeling. There are two ways to model the covariate distribution using the semiparametric odds ratio model. The first decomposes the joint density into consecutive conditional densities as

$$g(x_1, \cdots, x_p) = \prod_{j=1}^{p} g_j(x_j \mid x_{j-1}, \cdots, x_1),$$

and then employs the odds ratio decomposition

$$g_j(x_j \mid x_{j-1}, \cdots, x_1) = \frac{\eta_j\{x_j; (x_{j-1}, \cdots, x_1)\} g_j(x_j \mid x_{(j-1)0}, \cdots, x_{10})}{\int \eta_j\{x_j; (x_{j-1}, \cdots, x_1)\} g_j(x_j \mid x_{(j-1)0}, \cdots, x_{10}) dx_j}, \tag{8.3}$$

to model η_j parametrically as $\eta_j(x_j \mid x_{j-1}, \cdots, x_1, \gamma_j)$ for $j = 2, \cdots, p$. All the baseline functions $g_j(x_j \mid x_{(j-1)0}, \cdots, x_{10})$, $j = 1, \cdots, p$ are unspecified. The second applies the odds ratio decomposition of the joint density

$$g(x_1, \cdots, x_p) = \frac{\eta(x_1; \cdots; x_p) \prod_{j=1}^{p} g(x_j \mid x_{-(j)0})}{\int \cdots \int \eta(x_1; \cdots; x_p) \prod_{j=1}^{p} g(x_j \mid x_{-(j)0}) dx_j},$$

models the collection of odds ratio functions parametrically, i.e., $\eta(x_1; \cdots; x_p) = \eta(x_1; \cdots; x_p \mid \gamma)$, and leaves the fully conditioned baseline distributions $g(x_j \mid x_{-(j)0}), j = 1, \cdots, p$ unspecified.

In maximizing the likelihood (8.2), the baseline distribution functions are restricted to the discrete distributions. It can be shown that this way of handling the baseline functions yield a consistent estimator of the regression parameter if the odds ratio functions are specified correctly. Let the full data likelihood be

$$L(\theta, \gamma, G) = \prod_{i=1}^{n} f(Y_i \mid X_{i1}, \cdots, X_{ip}, \theta) g(X_{i1}, \cdots, X_{ip} \mid \gamma, G) \tag{8.4}$$

where $G = (G_1, \cdots, G_p)$ is the collection of baseline distribution functions and g is the odds ratio representation of the covariate density either in the consecutive conditional form or in the joint density form. Under the consecutive conditional odds ratio model, the likelihood score equations for the fully observed data are

$$\begin{aligned}
\dot{l}_\theta(\theta, \gamma, G) &= \sum_{i=1}^{n} \frac{\partial}{\partial \theta} \log f(Y_i \mid X_{i1}, \cdots, X_{ip}, \theta), \\
\dot{l}_\gamma(\theta, \gamma, G) &= \sum_{i=1}^{n} \Bigg(\frac{\partial}{\partial \gamma} \log \eta_j\{X_{ij}; (X_{i(j-1)}, \cdots, X_{i1}) \mid \gamma\} \\
&\quad - E\left[\frac{\partial}{\partial \gamma} \log \eta_j\{X_{ij}; (X_{i(j-1)}, \cdots, X_{i1}) \mid \gamma\} \mid X_{i(j-1)}, \cdots, X_{i1} \right] \Bigg), \\
\dot{l}_{G_j}(\theta, \gamma, G)(h_j) &= \sum_{i=1}^{n} \left[h_j\{X_{ij}) - E\{h_j(X_{ij}) \mid X_{i(j-1)}, \cdots, X_{i1}\} \right],
\end{aligned}$$

where $h_j, j = 1, \cdots, p$ are bounded variation functions on their respective domain. Under the joint semiparametric odds ratio model for the covariates, the second and third equations become

$$\begin{aligned} \dot{l}_\gamma(\theta, \gamma, G) &= \sum_{i=1}^{n} \left[\frac{\partial}{\partial \gamma} \log \eta(X_{i1}; \cdots; X_{ip} \mid \gamma) - E\left\{ \frac{\partial}{\partial \gamma} \log \eta(X_{i1}; \cdots; X_{ip} \mid \gamma) \right\} \right], \\ \dot{l}_{G_j}(\theta, \gamma, G)(h_j) &= \sum_{i=1}^{n} [h_j\{X_{ij}) - E\{h_j(X_{ij})\}] = 0. \end{aligned}$$

For the missing covariate problem with MAR missing data, the likelihood scores for the incompletely observed data are

$$\begin{aligned} \dot{l}_\theta^{obs}(\theta, \gamma, G) &= E\left\{ \dot{l}_\theta(\theta, \gamma, G)(Y, X) \mid Y, R(X) \right\}, \\ \dot{l}_\gamma^{obs}(\theta, \gamma, G) &= E\left\{ \dot{l}_\gamma(\theta, \gamma, G)(Y, X) \mid Y, R(X) \right\}, \\ \dot{l}_{G_j}^{obs}(\theta, \gamma, G)(h_j) &= E\left\{ \dot{l}_{G_j}(\theta, \gamma, G)(Y, X) \mid Y, R(X) \right\}. \end{aligned}$$

One challenge in carrying out the likelihood maximization is to compute the conditional expectations of missing data given the observed data. One advantage of the semiparametric odds ratio models is that the integrals in the conditional expectation all reduce to summations over the observed data values. However, the number of items in the summations is combinatorial and increases in terms of the multiplication of the numbers of categories of the groups. To make the computation feasible, approximation to the summations needs to be considered. The Monte Carlo approach is viable when the number of items are too big to compute.

8.2.2 Missing covariates in the Cox regression model

The proportional hazards regression model is frequently used for the analysis of survival time data. When covariates are fully observed, the partial likelihood can be used to estimate and make inference on the relative risk parameters. Missing values can be much more difficult to handle within the partial likelihood framework when covariates are incompletely observed. Usually the semiparametric likelihood for the proportional hazards model is employed to handle the missing values instead (Chen and Little, 1999)[34]. Under the assumption that the censoring is independent of the survival time given the observed covariates, the semiparametric likelihood for estimating the parameters in the proportional hazards model can be separated from the likelihood for the missing data mechanism model. Let T and C be respectively the survival time and the censoring time. Let $Y = \min(T, C)$ and $\delta = 1_{\{T \leq C\}}$. Let X be the covariates. Assume that T given X follows the proportional hazards regression model with hazards density

$$\lambda(t \mid x) = \lambda_0(t) \exp(\beta X).$$

Let the observed data subject to censoring with fully observed covariate be $(Y_i, \delta_i, X_i), i = 1, \cdots, n$. The likelihood for estimating β and the cumulative baseline hazard Λ_0 is

$$\prod_{i=1}^{n} \{d\Lambda_0(Y_i) \exp(\beta X_i)\}^{\delta_i} \exp\{-\Lambda_0(Y_i) \exp(\beta X_i)\} g(X_i).$$

Let the observed data be $(R_i, Y_i, \delta_i, R_i(X_i)), i = 1, \cdots, n$. Under the MAR missing data mechansim and the independence censoring assumption, the semiparametric likelihood becomes

$$\prod_r \left[\int \cdots \int \prod_{i=1}^n \{d\Lambda_0(Y_i)\exp(\beta X_i)\}^{\delta_i} \exp\{-\Lambda_0(Y_i)\exp(\beta X_i)\} g(X_i) d\bar{r}(X_i) \right]^{1_{\{R_i=r\}}}.$$

By following the convention of maximizing the semiparametric likelihood with Λ_0 restricted to discrete distributions, the maximum likelihood estimator for Λ_0 has probability mass only on the observed failure times. Let

$$\Lambda(t) = \sum_{j=1}^n \lambda_j 1_{\{Y_j \le t\}}.$$

The full data likelihood scores for β and λ are

$$\begin{aligned} \dot{l}_\beta(\beta, \Lambda, \gamma, G) &= \sum_{i=1}^n \delta_i X_i - \sum_{i=1}^n \sum_{j=1}^n \lambda_j 1_{\{Y_j \le t\}} X_i e^{\beta X_i}, \\ \dot{l}_{\lambda_j}(\beta, \Lambda, \gamma, G) &= \frac{\delta_j}{\lambda_j} - \sum_{i=1}^n 1_{\{Y_j \le Y_i\}} \exp(\beta X_i). \end{aligned}$$

The likelihood score for parameters (γ, G) under either the consecutive conditional odds ratio model or the joint odds ratio model has the same forms as those in the parametric regression model.

When the covariates are subject to missing values and the missingness depends on the observed data only, the observed data likelihood scores for β and Λ_0 are

$$\begin{aligned} \dot{l}_\beta^{obs}(\beta, \Lambda, \gamma, G) &= \sum_{i=1}^n \delta_i E\{X_i \mid Y_i, \delta_i, R_i(X_i)\} \\ &\quad - \sum_{i=1}^n \sum_{j=1}^n \lambda_j 1_{\{Y_j \le t\}} E\left\{X_i e^{\beta X_i} \mid Y_i, \delta_i, R_i(X_i)\right\}, \\ \dot{l}_{\lambda_j}^{obs}(\beta, \Lambda, \gamma, G) &= \frac{\delta_j}{\lambda_j} - \sum_{i=1}^n 1_{\{Y_j \le Y_i\}} E\{\exp(\beta X_i) \mid Y_i, \delta_i, R_i(X_i)\}. \end{aligned}$$

The observed data likelihood scores for (γ, G) are similar to those in the parametric regression. The slight difference is that the observed data are $(Y_i, \delta_i, R_i(X_i))$ instead. A similar computational problem to that in the parametric regression models is encountered when missing data patterns are not simple.

8.3 Monte Carlo algorithm for the likelihood maximization

8.3.1 The general formulation

Consider the likelihood maximization for a regression model with incompletely observed covariates.

$$\begin{aligned} \frac{L^{obs}(\theta, \gamma, G)}{L^{obs}(\theta_b, \gamma_b, G_b)} &= E\left\{ \frac{L(\theta, \gamma, G)}{L(\theta_b, \gamma_b, G_b)} \mid Y, R, R(X), \theta_b, \gamma_b, G_b \right\} \\ &= E\left[\exp\{l(\theta, \gamma, G) - l(\theta_b, \gamma_b, G_b)\} \mid Y, R, R(X), \theta_b, \gamma_b, G_b\right\} \\ &\ge \exp\left[E\left\{l(\theta, \gamma, G) - l(\theta_b, \gamma_b, G_b) \mid Y, R, R(X), \theta_b, \gamma_b, G_b\right\}\right]. \end{aligned}$$

Similar to the justification of the EM algorithm in increasing the likelihood, if we find a new value of the parameters, denote by $(\theta_{new}, \gamma_{new}, G_{new})$ such that

$$E\{l(\theta_{new}, \gamma_{new}, G_{new}) \mid Y, R, R(X), \theta_b, \gamma_b, G_b\} \geq E\{l(\theta_b, \gamma_b, G_b) \mid Y, R, R(X), \theta_b, \gamma_b, G_b\},$$

we then have

$$\frac{L^{obs}(\theta, \gamma, G)}{L^{obs}(\theta_b, \gamma_b, G_b)} \geq 1,$$

i.e., the observed data likelihood increases. It is better to maximize the objective function

$$Q(\theta, \gamma, G \mid \theta_b, \gamma_b, G_b) = E\{l(\theta, \gamma, G) \mid Y, R, R(X), \theta_b, \gamma_b, G_b\}.$$

when the objective function can be accurately computed. However, when the objective function cannot be accurately computed, maximizing the approximate objective function may not be a good strategy. If the accuracy can only be guaranteed locally, it is better just to find a local update of the parameters in each updating step.

To update the parameters, note that

$$\begin{aligned} l(\theta, \gamma, G) - l(\theta_b, \gamma_b, G_b) &\approx \dot{l}(\theta_b, \gamma_b, G_b)(\theta - \theta_b, \gamma - \gamma_b, G_b - G) \\ &+ \frac{1}{2}(\theta - \theta_b, \gamma - \gamma_b, G - G_b)\ddot{l}(\theta_b, \gamma_b, G_b)(\theta - \theta_b, \gamma - \gamma_b, G - G_b), \end{aligned}$$

where $\dot{l}$ and $\ddot{l}$ are first and second derivatives of the full data log-likelihood. The one-step update of the parameters has

$$\begin{aligned} (\theta, \gamma, G) &= (\theta_b, \gamma_b, G_b) - \left[E\left\{\ddot{l}(\theta_b, \gamma_b, G_b) \mid Y, R, R(X), \theta_b, \gamma_b, G_b\right\}\right]^{-1} \\ &\quad \times E\left\{\dot{l}(\theta_b, \gamma_b, G_b) \mid Y, R, R(X), \theta_b, \gamma_b, G_b\right\}. \end{aligned}$$

To carry out the computation, we need to compute conditional expectations of the form $E\{h(Y, X) \mid Y, R, R(X), \theta_b, \gamma_b, G_b\}$. If the number of categories in the distribution $p\{\bar{r}(x) \mid Y, R = r, r(X) = r(x), \theta_b, \gamma_b, G_b\}$ is not very large, computation of such conditional expectations can be straightforwardly carried out because of the discrete nature of the conditional distribution

$$p\{\bar{r}(x) \mid Y, R = r, r(X) = r(x), \theta_b, \gamma_b, G_b\} = \frac{f(Y \mid X, \theta_b)p(X \mid \gamma_b, G_b)1_{\{r(X)=r(x)\}}}{\sum_{r(X)=r(x)} f(Y \mid X, \theta_b)p(X \mid \gamma_b, G_b)}. \quad (8.5)$$

If the dimension of $\bar{r}(X)$ is high and the number of the combined categories for $\bar{r}(X)$ is very large, direct computation can be infeasible due to the extremely large number of items to be added in the expectations. For example, if the missing data divide the covariates into say five groups and the number of categories for each group is 200 which is not unlikely when the variables are continuous or the number of categorical variables in each group is relatively large, the total number of terms to be added can be up to $200^5 = 3.2 \times 10^{11}$. To overcome the exponential increase in the number of terms in the integrals, Monte Carlo approximation may be applied.

Due to the discrete nature of the conditional distribution (8.5), the Monte Carlo approximation can be relatively easily implemented using the Gibbs sampler. The Gibbs sampler for drawing from the distribution (8.5) can be implemented as follows. Initially, starting from the observed data point with the missing values filled except one missing value, draw the missing value from the conditional distribution and then move to the next component having missing value. In general, assume the current sample is $(x_{1(l)}, \cdots, x_{p(l)})$ where

$x_{j(l)} = X_j$ if that component of X is observed. If $R_k = 0$, draw $x_{k(l+1)}$ from the distribution

$$\begin{aligned} p(x_k \mid x_{1(l+1)}, \cdots, x_{(k-1)(l+1)}, x_{(k+1)(l)}, \cdots, x_{p(l)}, y, \theta_b, \gamma_b, G_b) \\ \propto f(y \mid x_{1(l+1)}, \cdots, x_{(k-1)(l+1)}, x_k, x_{(k+1)(l)}, \cdots, x_{p(l)}, \theta_b) \\ \times \eta(x_{1(l+1)}; \cdots; x_{(k-1)(l+1)}; x_k; x_{k(l)}; \cdots; x_{p(l)} \mid \theta_b) dG_{bk}(x_k), \end{aligned}$$

when the covariate distribution is modeled by the joint odds ratio model. Move to the next component with missing values and repeat the sampling draw. At convergence, draw a sample for compute the conditional expectations. In modeling the covariate distribution by the joint semiparametric odds ratio model, we need also to compute unconditional expectations in the form $E\{h(X) \mid \gamma_b, G_b)\}$. For computation of such quantities, we need to sample from

$$\begin{aligned} p(x_k \mid x_{1(l+1)}, \cdots, x_{(k-1)(l+1)}, x_{(k+1)(l)}, \cdots, x_{p(l)}, \gamma_b, G_b) \\ \propto \eta(x_{1(l+1)}; \cdots; x_{(k-1)(l+1)}; x_k; x_{k(l)}; \cdots; x_{p(l)} \mid \theta_b) dG_{bk}(x_k). \end{aligned}$$

This can be done similarly using the Gibbs sampler. Note that, when the covariate distribution is modeled by the consecutive conditional odds ratio models, the sampling distribution is more complex and the sampling algorithm can be more complicated. However, computing $E\{h(X) \mid \gamma_b, G_b\}$ is not needed.

Given unit i, let $(x_{i1}^{(l)}, \cdots, x_{ip}^{(l)}), l = 1, \cdots, N$, be a sample drawn from the Markov chain at convergence. We can estimate

$$E\{h(Y_i, X_i) \mid Y_i, R_i, R_i(X_i), \theta_b, \gamma_b, G_b\} \approx \frac{1}{N} \sum_{l=1}^{N} h(Y_i, x_{i1}^{(l)}, \cdots, x_{ip}^{(l)}).$$

Let $(x_1^{(l)}, \cdots, x_p^{(l)}), l = 1, \cdots, N$, be a sample drawn from the Markov chain at convergence for the unconditional expectations. We can estimate

$$E\{h(X) \mid \theta_b, \gamma_b, G_b\} \approx \frac{1}{N} \sum_{l=1}^{N} h(x_1^{(l)}, \cdots, x_p^{(l)}).$$

In the following, we will use E_b, V_b, and Cov_b to denote mean, variance, and covariance computed based on the distribution with the current parameter values.

8.3.2 Application to parametric regression

Applying the sample drawn to compute the likelihood scores, we have

$$\begin{aligned} \hat{E}_b\left\{\dot{l}_\theta(\theta_b, \gamma_b, G_b) \mid Y_i, R_i, R_i(X_i)\right\} &= \sum_{i=1}^{n} \frac{1}{N} \sum_{l=1}^{N} \frac{\partial}{\partial \theta} \log f(Y_i \mid x_{i1}^{(l)}, \cdots, x_{ip}^{(l)}, \theta_b), \\ \hat{E}_b\left\{\dot{l}_\gamma(\theta_b, \gamma_b, G_b) \mid Y_i, R_i, R_i(X_i)\right\} &= \sum_{i=1}^{n} \left[\frac{1}{N} \sum_{l=1}^{N} \frac{\partial}{\partial \gamma} \log \eta(x_{i1}^{(l)}; \cdots; x_{ip}^{(l)} \mid \gamma_b)\right. \\ &\quad \left. - \frac{1}{N} \sum_{l=1}^{N} \frac{\partial}{\partial \gamma} \log \eta(x_1^{(l)}; \cdots; x_p^{(l)} \mid \gamma_b)\right]. \end{aligned}$$

Since $x_k^{(l)}, l = 1, \cdots, N$ only take values of the observed X_k. Let $z_{jk}, j = 1, \cdots, J_k$ be distinctive observed X_k values with corresponding observed frequencies $n_{jk}, j = 1, \cdots, J_k$.

Let $N_{jk}^{(i)}$ be the frequency of z_{jk} in the simulated sample $x_{ik}^{(l)}, l = 1, \cdots, N$. Let N_{jk} be the frequency of z_{jk} in the simulated sample $x_{k}^{(l)}, l = 1, \cdots, N$.

$$\begin{aligned}\hat{E}_b\left\{\dot{l}_{G_k}(\theta_b, \gamma_b, G_b)(h) \mid Y_i, R_i, R_i(X_i)\right\} &= \sum_{i=1}^{n}\frac{1}{N}\sum_{l=1}^{N} h_k(x_{ik}^{(l)}) - \frac{1}{N}\sum_{l=1}^{N} h_k(x_k^{(l)}) \\ &= \sum_{j=1}^{J_k}\left\{\left(n_{jk} + \sum_{i:R_{ik}=0}\frac{N_{jk}^{(i)}}{N}\right) - \frac{N_{jk}}{N}\right\} h_k(z_{jk}),\end{aligned}$$

for $k = 1, \cdots, G$. The parameter update also requires the second derivative matrix be computed. They are respectively,

$$\begin{aligned}\hat{E}_b\left\{\ddot{l}_{\theta\theta} \mid Y_i, R_i, R_i(X_i)\right\} &= \sum_{i=1}^{n}\frac{1}{N}\sum_{l=1}^{N}\frac{\partial^2}{\partial\theta^2}\log f(Y_i \mid x_{i1}^{(l)}, \cdots, x_{ip}^{(l)}, \theta_b), \\ \hat{E}_b\left\{\ddot{l}_{\gamma\gamma} \mid Y_i, R_i, R_i(X_i)\right\} &= \sum_{i=1}^{n}\left[\frac{1}{N}\sum_{l=1}^{N}\frac{\partial^2}{\partial\gamma^2}\log\eta(x_{i1}^{(l)}; \cdots; x_{ip}^{(l)} \mid \gamma_b)\right. \\ &\qquad \left. - \frac{1}{N}\sum_{l=1}^{N}\frac{\partial^2}{\partial\gamma^2}\log\eta(x_1^{(l)}; \cdots; x_p^{(l)} \mid \gamma_b)\right], \\ \hat{E}_b\left\{\ddot{l}_{G_kG_j} \mid Y_i, R_i, R_i(X_i)\right\} &= \frac{1}{N}\sum_{l=1}^{N} h_k(x_k^{(l)})h_j(x_j^{(l)}) - \frac{1}{N}\sum_{l=1}^{N} h_k(x_k^{(l)})\frac{1}{N}\sum_{l=1}^{N} h_j(x_j^{(l)}) \\ \hat{E}_b\left\{\ddot{l}_{\gamma G_k} \mid Y_i, R_i, R_i(X_i)\right\} &= \frac{1}{N}\sum_{l=1}^{N} h_k(x_k^{(l)})\frac{\partial}{\partial\gamma}\log\eta(x_1^{(l)}; \cdots; x_p^{(l)} \mid \gamma_b), \\ &\qquad -\frac{1}{N}\sum_{l=1}^{N} h_k(x_k^{(l)})\frac{1}{N}\sum_{l=1}^{N}\frac{\partial}{\partial\gamma}\log\eta(x_1^{(l)}; \cdots; x_p^{(l)} \mid \gamma_b).\end{aligned}$$

With the estimated first and second derivatives of the objective function in the EM algorithm, the update of the model parameters can be implemented by the one-step Newton-Raphson algorithm.

Since the observed data information matrix can be relatively easily approximated using the Monte Carlo sample, it is more efficient to update of the model parameters based on the observed likelihood function. Based on the Louis' formula, the additional items to be computed are conditional covariance matrix between the full data score functions given the observed data.

$$\begin{aligned}&\hat{Cov}_b\left\{\dot{l}_\theta, \dot{l}_\gamma \mid Y_i, R_i, R_i(X_i)\right\} \\ &\quad = \sum_{i=1}^{n}\frac{1}{N}\sum_{l=1}^{N}\frac{\partial}{\partial\theta}\log f(Y_i \mid x_{i1}^{(l)}, \cdots, x_{ip}^{(l)}, \theta_b)\frac{\partial}{\partial\gamma}\log\eta(x_{i1}^{(l)}; \cdots; x_{ip}^{(l)} \mid \gamma_b) \\ &\qquad -\hat{E}_b\left\{\dot{l}_\theta(\theta_b, \gamma_b, G_b)\right\}\left\{\frac{1}{N}\sum_{l=1}^{N}\frac{\partial}{\partial\gamma}\log\eta(x_{i1}^{(l)}; \cdots; x_{ip}^{(l)} \mid \gamma_b)\right\},\end{aligned}$$

$$\begin{aligned}\hat{Cov}_b\left\{\dot{l}_\theta, \dot{l}_{G_k} \mid Y_i, R_i, R_i(X_i)\right\} &= \sum_{i=1}^{n}\left[\frac{1}{N}\sum_{l=1}^{N}\frac{\partial}{\partial\theta}\log f(Y_i \mid x_{i1}^{(l)}, \cdots, x_{ip}^{(l)}, \theta_b)h_k(x_{ik}^{(l)})\right. \\ &\qquad \left. -\hat{E}_b\left\{\dot{l}_\theta(\theta_b, \gamma_b, G_b)\right\}\left\{\frac{1}{N}\sum_{l=1}^{N} h_k(x_{ik}^{(l)})\right\},\right.\end{aligned}$$

$$\begin{aligned}\hat{Cov}_b\left(\dot{l}_\gamma, \dot{l}_{G_k} \mid Y_i, R_i, R_i(X_i)\right) &= \sum_{i=1}^{n}\left[\frac{1}{N}\sum_{l=1}^{N}\frac{\partial}{\partial\gamma}\log\eta(x_{i1}^{(l)};\cdots;x_{ip}^{(l)} \mid \gamma_b)h_k(x_{ik}^{(l)}) \right.\\ &\left. -\left\{\frac{1}{N}\sum_{l=1}^{N}\frac{\partial}{\partial\gamma}\log\eta(x_{i1}^{(l)};\cdots;x_{ip}^{(l)} \mid \gamma_b)\right\}\left\{\frac{1}{N}\sum_{l=1}^{N}h_k(x_{ik}^{(l)})\right\}\right]\end{aligned}$$

In addition to the covariance across different parameters, the cross-derivatives can also be obtained similarly. With the estimated second derivative of the observed data likelihood, the update scheme is

$$\begin{aligned}(\theta,\gamma,G) &= (\theta_b,\gamma_b,G_b) - \left[E_b\left\{\ddot{l}(\theta_b,\gamma_b,G_b) \mid Y,R,R(X)\right\}\right.\\ &\left. +V_b\left\{\dot{l}(\theta_b,\gamma_b,G_b) \mid Y,R,R(X)\right\}\right]^{-1} E_b\left\{\dot{l}(\theta_b,\gamma_b,G_b) \mid Y,R,R(X)\right\}.\end{aligned}$$

8.3.3 Application to Cox regression model

For the Cox regression model, computation of the expected values of the first and second derivatives for (γ, G) are the same as those in the parametric regression. For the first and second derivatives with respect to (β, Λ), we need to find the relevant conditional expectations as follow.

$$\begin{aligned}\hat{E}_b\left\{X_{ik} \mid Y_i,\delta_i,R_i(X_i)\right\} &= \frac{1}{N}\sum_{l=1}^{N}x_{ik}^{(l)},\\ \hat{E}_b\left\{\exp(\beta X_i) \mid Y_i,\delta_i,R_i(X_i)\right\} &= \frac{1}{N}\sum_{l=1}^{N}\exp\left(\sum_{j=1}^{p}\beta_j x_{ij}^{(l)}\right),\\ \hat{E}_b\left\{X_{ik}\exp(\beta X_i) \mid Y_i,\delta_i,R_i(X_i)\right\} &= \frac{1}{N}\sum_{l=1}^{N}x_{ik}^{(l)}\exp\left(\sum_{j=1}^{p}\beta_j x_{ij}^{(l)}\right),\\ \hat{E}_b\left\{X_{ik}X_{it}\exp(\beta X_i) \mid Y_i,\delta_i,R_i(X_i)\right\} &= \frac{1}{N}\sum_{l=1}^{N}x_{ik}^{(l)}x_{it}^{(l)}\exp\left(\sum_{j=1}^{p}\beta_j x_{ij}^{(l)}\right),\end{aligned}$$

Since Λ has a closed-form solution, update of parameters can be carried out first on the profile likelihood of (β, γ, G) and then on Λ.

8.4 Imputation approach to missing covariates

8.4.1 A Bayesian framework

The likelihood approach to the missing covariate problem can yield an efficient estimator of the regression parameters. However, it is computationally inefficient when the Monte Carlo EM algorithm is used. In addition, it can be challenging for a general user to implement the approach. A more popular approach imputes the missing values first and then performs analyses on the imputed data. Although this approach may not be as efficient as the maximum likelihood approach, it separates the difficult part of imputing the missing values from the relatively easy part of the analysis of the full data. A general user may focus more on the second part while the imputation part may be handled by an expert or by a streamlined software. Computationally, the imputation approach is computationally more efficient because only one or a few Monte Carlo Markov chains need to be run.

A Bayesian inference framework for the fully observed data is often required for generation of the proper imputation of the missing values. For the regression problem with full covariate data, let the likelihood be

$$L(\theta, \gamma, G) = \prod_{i=1}^{n} f(Y_i \mid X_i, \theta) g(X_i \mid \gamma, G).$$

Assume the prior for parameter θ and (γ, G) be respectively $q_0(\theta)$ and $q(\gamma, G)$. The posterior for (θ, γ, G) is then

$$p(\theta, \gamma, G \mid Y_i, X_i, i = 1, \cdots, n) \propto \prod_{i=1}^{n} f(Y_i \mid X_i, \theta) g(X_i \mid \gamma, G) q_0(\theta) q(\gamma, G).$$

The posterior for θ and (γ, G) given the full data is separated into two parts as

$$p(\theta \mid Y_i, X_i, i = 1, \cdots, n) \propto \prod_{i=1}^{n} f(Y_i \mid X_i, \theta) q_0(\theta),$$

and

$$p(\gamma, G \mid Y_i, X_i, i = 1, \cdots, n) \propto \prod_{i=1}^{n} g(X_i \mid \gamma, G) q(\gamma, G).$$

When covariates are subject to missing values, let the missing data mechanism be

$$P(R_i = r \mid Y_i, X_i) = \pi(r \mid Y_i, X_i, \alpha).$$

Let the prior for α be $s(\alpha)$. The posterior for α is

$$p(\alpha \mid R_i, Y_i, X_i, i = 1, \cdots, n) \propto \prod_{i=1}^{n} \pi(R_i \mid Y_i, X_i, \alpha) s(\alpha).$$

Let $\alpha^*, \theta^*, \gamma^*, G^*$ be drawn respectively from the posterior distributions. Imputation for missing values are drawn from the following distribution

$$\bar{R}_i(X_i) \propto \pi(R_i \mid Y_i, X_i, \alpha^*) f(Y_i \mid X_i, \theta^*) g(X_i \mid \gamma^*).$$

When the semiparametric odds ratio model is used to model the covariate distribution, imputing missing covariate values is equivalent to drawing from the observed values with appropriate weights. The weights are determined by the models. As a result, imputing the missing values can in principle be straightforwardly performed once the draws from the posterior distribution are obtained.

Let covariate X be divided into groups based on the missing data patterns and, for notational simplicity, assume $X_{i1}, \cdots, X_{ip}$ correspond to the missing pattern groups. For a given group X_{ij}, let $X_{i(-j)} = (X_{i1}, \cdots, X_{i(j-1)}, X_{i(j+1)}, \cdots, X_{ip})$. When X_{ij} is missing, an imputed value can be drawn using the Gibbs sampler based on the discrete sampling distribution

$$P(X_{ij} = z_k \mid X_{i(-j)}, Y_i, \alpha^*, \theta^*, \gamma^*, G^*) = \frac{\pi(R_i \mid Y_i, X_{i(-j)}, X_{ij} = z_{kj}, \alpha^*) f(Y_i \mid X_{i(-j)}, X_{ij} = z_{kj}, \theta^*) g(X_{i(-j)}, X_{ij} = z_{kj}, \mid \gamma^*)}{\sum_{k=1}^{K_j} \pi(R_i \mid Y_i, X_{i(-j)}, X_{ij} = z_{kj}, \alpha^*) f(Y_i \mid X_{i(-j)}, X_{ij} = z_{kj}, \theta^*) g(X_{i(-j)}, X_{ij} = z_{kj}, \mid \gamma^*)},$$

where $z_{1j}, \cdots, z_{K_j j}$ are the distinctive observed values for X_j, $j = 1, \cdots, p$. Because of the discrete nature of the distribution, independent random draws from the distribution can be obtained straightforwardly.

8.4.2 Draw (γ, G) from the posterior under the consecutive conditional odds ratio models

When the distribution of the covariates is expressed in terms of consecutive conditional distributions and each conditional distribution is modeled by a semiparametric odds ratio model, the distribution of X_i is

$$g(X_i \mid \gamma, G) = \prod_{j=1}^{p} \frac{\eta_j \left\{X_{ij}; (X_{i(j-1)}, \cdots, X_{i1}) \mid \gamma_j\right\} dG_j(x_{ij})}{\int \eta_j \left\{X_{ij}; (X_{i(j-1)}, \cdots, X_{i1}) \mid \gamma_j\right\} dG_j(x_{ij})}.$$

As before, let $z_{1j}, \cdots, z_{K_j j}$ be the distinctive observed values for X_j with corresponding observed frequencies $n_{1j}, \cdots, n_{K_j j}$, $j = 1, \cdots, p$. Let $p_{kj}, k = 1, \cdots, K_j$ be the probability masses assigned respectively to $z_{kj}, j = 1, \cdots, K_j$, i.e., the discretized

$$G_j(u) = \sum_{k=1}^{K_j} p_{kj} 1_{\{z_{jk} \le u\}}.$$

The likelihood function for γ and discretized G given the full data becomes

$$L(\gamma, G) = \left[\prod_{i=1}^{n}\prod_{j=2}^{p} \frac{\eta_j\{X_{ij}; (X_{i(j-1)}, \cdots, X_{i1}) \mid \gamma_j\}}{\sum_{k=1}^{K_j} \eta_j\{z_{kj}; (X_{i(j-1)}, \cdots, X_{i1}) \mid \gamma_j\} p_{kj}}\right] \left(\prod_{j=1}^{p}\prod_{k=1}^{K_j} p_{kj}^{n_{kj}}\right).$$

Let the prior for γ_j be normal with mean 0 and variance Ω_j, and the prior for G_j be the Dirichlet process $\mathcal{D}(c_j F_j)$. Assume the prior distributions for different $(\gamma_j, G_j), j = 1, \cdots, p$ are multiplicative (independent). The prior distribution for all the parameters is

$$p(\gamma, G) = \prod_{j=2}^{p} \frac{1}{(2\pi)^{d_j/2} |\Omega_j|^{1/2}} \exp\left\{-\frac{1}{2}\gamma_j^t \Omega_j^{-1} \gamma_j\right\} \prod_{j=1}^{p} \mathcal{D}(c_j F_j),$$

where d_j is the dimension of parameter γ_j. For a given j, let

$$G_{nj}(u) = \sum_{j=1}^{K_j} \frac{n_{kj}}{n} 1_{\{z_{kj} \le u\}}.$$

It follows that

$$\prod_{k=1}^{K_j} p_{kj}^{n_{kj}} \mathcal{D}(c_j F_j) = \mathcal{D}\left\{(n + c_j)\left(\frac{n}{n + c_j} G_{nj} + \frac{c_j}{n + c_j} F_j\right)\right\}.$$

As argued in Chen et al (2011)[37], when c_j relative to n is small, the above Dirichlet process can be approximated by the Dirichlet process $\mathcal{D}(nG_{nj})$. The posterior for (γ, G) can be approximated by

$$\begin{aligned} p(\gamma, G \mid data) \approx & \left[\prod_{j=2}^{p} \frac{\prod_{i=1}^{n} \eta_j\{X_{ij}; (X_{i(j-1)}, \cdots, X_{i1}) \mid \gamma_j\}}{\prod_{i=1}^{n} \sum_{k=1}^{K_j} \eta_j\{z_{kj}; (X_{i(j-1)}, \cdots, X_{i1}) \mid \gamma_j\} p_{kj}}\right] \\ & \left\{\prod_{j=1}^{p}\prod_{k=1}^{K_j} p_{kj}^{nG_{nj}(z_{kj})}\right\} \prod_{j=2}^{p} \exp\left[-\frac{1}{2}\gamma_j^t \Omega_j^{-1} \gamma_j - \frac{1}{2}\log\left\{(2\pi)^{d_j} |\Omega_j|\right\}\right]. \end{aligned}$$

To draw (γ, G) from the posterior distribution, Chen et al (2011)[37] proposed a hybrid Monte Carlo approach (HMC, also known as Hamiltonian Monte Carlo approach) using the molecular dynamic algorithm for generating candidate draws (Duane et al, 1987)[57]. The method can be described as follows. Let

$$\begin{aligned} U(\gamma,\lambda) &= -\log p(\gamma, G \mid data) - \frac{\sum_{j=2}^{p} d_j}{2}\log(2\pi) \\ &\approx \sum_{j=2}^{p}\left[\sum_{i=1}^{n}\log\eta_j\{X_{ij};(X_{i(j-1)},\cdots,X_{i1})\mid\gamma_j\} - \frac{1}{2}\gamma_j^t\Omega_j^{-1}\gamma_j - \frac{1}{2}\log(|\Omega_j|)\right] \\ &\quad -\sum_{j=1}^{p}\sum_{i=1}^{n}\log\left[\sum_{k=1}^{K_j}\eta_j\{z_{kj};(X_{i(j-1)},\cdots,X_{i1})\mid\gamma_j\}\exp(\lambda_{kj})\right] + \sum_{j=1}^{p}\sum_{k=1}^{K_j} n_{kj}\lambda_{kj}, \end{aligned}$$

where $\lambda_{kj} = \log(p_{kj}/p_{K_j j})$ and $\eta_1 \equiv 1$. Define an Hamiltonian function for molecular dynamics as

$$H(\gamma,\lambda,\beta,\mu) = U(\gamma,\lambda) + \frac{1}{2}\sum_{j=2}^{p}\beta_j^t M_j^{-1}\beta_j + \frac{1}{2}\sum_{j=1}^{p}\sum_{k=1}^{K_j-1}\frac{\mu_{kj}^2}{\sigma_{kj}^2},$$

where β_j has the same dimension as γ_j, M_j is a positive definite matrix usually chosen to be a diagonal matrix. By treating $\gamma, \lambda, \beta, \mu$ as functions of time t, the molecular dynamics can be used to generate the candidate draws. The molecular dynamics interprets $U(\gamma,\lambda)$ as the potential energy and the function of β and μ as the kinetic energy of the molecules. The dynamic of the energy in the form of the position (γ,λ) (representing the potential energy) and momentum (β,μ) (representing the kinetic energy) of the molecules satisfies the following differential equations,

$$\begin{aligned} \frac{d\gamma_{lj}(t)}{dt} &= \frac{\partial H}{\partial\beta_{lj}}(\gamma,\lambda,\beta,\mu), \\ \frac{d\lambda_{kj}(t)}{dt} &= \frac{\partial H}{\partial\mu_{kj}}(\gamma,\lambda,\beta,\mu), \\ \frac{d\beta_{lj}(t)}{dt} &= -\frac{\partial H}{\partial\gamma_{lj}}(\gamma,\lambda,\beta,\mu), \\ \frac{d\mu_{kj}(t)}{dt} &= -\frac{\partial H}{\partial\lambda_{kj}}(\gamma,\lambda,\beta,\mu). \end{aligned}$$

These dynamics conserve the total energy of the system. If the differentiation equations could be solved directly for a given time $\tau > 0$, $\gamma_{lj}(\tau), \lambda_{kj}(\tau), \beta_{lj}(\tau)$, and $\mu_{kj}(\tau)$ could serve as candidate draws. The Markov chain Monte Carlo could be carried out as follows. Starting from the current position $(\gamma_{lj}, \lambda_{kj}, \beta_{lj}, \mu_{kj})$, randomly draw $(\beta_{lj}^*, \mu_{kj}^*)$ independently from the standard normal distributions. Solving the differentiation equations to obtain $\{\gamma_{lj}(\tau), \lambda_{kj}(\tau), \beta_{lj}(\tau), \mu_{kj}(\tau)\}$ with the initial condition

$$\{\gamma_{lj}(0), \lambda_{kj}(0), \beta_{lj}(0), \mu_{kj}(0)\} = (\gamma_{lj}, \lambda_{kj}, \beta_{lj}^*, \mu_{kj}^*).$$

With probability

$$\pi = \min\left(1, \exp\left[H(\gamma,\lambda,\beta,\mu) - H\left\{\gamma(\tau),\lambda(\tau),\beta(\tau),\mu(\tau)\right\}\right]\right),$$

to accept $\gamma(\tau), \lambda(\tau), \beta(\tau), \mu(\tau)$ as the new sample. With probability $1-\pi$ to use the old sample. The cycle would be repeated to generate a Markov chain sample.

Note that the differentiation equations usually do not have a closed-form solution. Numeric solution is found instead. Euler's method that directly discretizes the differentiation equations, i.e.,

$$\begin{aligned}
\gamma_{lj}(t+\epsilon) &= \gamma_{lj}(t)+\epsilon\frac{\partial H}{\partial \beta_{lj}}\{\gamma(t),\lambda(t),\beta(t),\mu(t)\},\\
\lambda_{kj}(t+\epsilon) &= \lambda_{kj}(t)+\epsilon\frac{\partial H}{\partial \mu_{kj}}\{\gamma(t),\lambda(t),\beta(t),\mu(t)\},\\
\beta_{lj}(t+\epsilon) &= \beta_{lj}(t)-\epsilon\frac{\partial H}{\partial \gamma_{lj}}\{\gamma(t),\lambda(t),\beta(t),\mu(t)\},\\
\mu_{kj}(t+\epsilon) &= \mu_{kj}(t)-\epsilon\frac{\partial H}{\partial \lambda_{kj}}\{\gamma(t),\lambda(t),\beta(t),\mu(t)\}.
\end{aligned}$$

does not yield a convergent solution (Neal, 2011)[129]. A modification of Euler's approach yields a good approximate solution to the differentiation equations. This method is termed the leap-frog approach. For a chosen step-size and a chosen number of steps L, the leap-frog method computes

$$\begin{aligned}
\beta_{lj}(t+\epsilon/2) &= \beta_{lj}(t)-\frac{\epsilon}{2}\frac{\partial H}{\partial \gamma_{lj}}\{\gamma(t),\lambda(t),\beta(t),\mu(t)\},\\
\mu_{kj}(t+\epsilon/2) &= \mu_{kj}(t)-\frac{\epsilon}{2}\frac{\partial H}{\partial \lambda_{kj}}\{\gamma(t),\lambda(t),\beta(t),\mu(t)\},\\
\gamma_{lj}(t+\epsilon) &= \gamma_{lj}(t)+\epsilon\frac{\partial H}{\partial \beta_{lj}}\{\gamma(t),\lambda(t),\beta(t+\epsilon/2),\mu(t+\epsilon/2)\},\\
\lambda_{kj}(t+\epsilon) &= \mu_{kj}(t)+\epsilon\frac{\partial H}{\partial \mu_{kj}}\{\gamma(t),\lambda(t),\beta(t+\epsilon/2),\mu(t+\epsilon/2)\},\\
\beta_{lj}(t+\epsilon) &= \beta_{lj}(t+\epsilon/2)-\frac{\epsilon}{2}\frac{\partial H}{\partial \gamma_{lj}}\{\gamma(t+\epsilon),\lambda(t+\epsilon),\beta(t+\epsilon/2),\mu(t+\epsilon/2)\},\\
\mu_{kj}(t+\epsilon) &= \mu_{kj}(t+\epsilon/2)-\frac{\epsilon}{2}\frac{\partial H}{\partial \lambda_{kj}}\{\gamma(t+\epsilon),\lambda(t+\epsilon),\beta(t+\epsilon/2),\mu(t+\epsilon/2)\},
\end{aligned}$$

where t goes from 0 to $(L-1)\epsilon$. When the functions of (γ,λ) and (β,μ) are additive in H, the consecutive two steps after the first step and before the last step of the leap-frog method can be combined because

$$\begin{aligned}
\beta_{lj}(t+3\epsilon/2) &= \beta_{lj}(t+\epsilon)-\frac{\epsilon}{2}\frac{\partial H}{\partial \gamma_{lj}}\{\gamma(t+\epsilon),\lambda(t+\epsilon),\beta(t+\epsilon),\mu(t+\epsilon)\},\\
&= \beta_{lj}(t+\epsilon/2)-\frac{\epsilon}{2}\frac{\partial H}{\partial \gamma_{lj}}\{\gamma(t+\epsilon),\lambda(t+\epsilon),\beta(t+\epsilon/2),\mu(t+\epsilon/2)\},\\
&\quad -\frac{\epsilon}{2}\frac{\partial H}{\partial \gamma_{lj}}\{\gamma(t+\epsilon),\lambda(t+\epsilon),\beta(t+\epsilon),\mu(t+\epsilon)\},\\
&= \beta_{lj}(t+\epsilon/2)-\epsilon\frac{\partial H}{\partial \gamma_{lj}}\{\gamma(t+\epsilon),\lambda(t+\epsilon),\beta(t+\epsilon/2),\mu(t+\epsilon/2)\}.
\end{aligned}$$

Similarly,

$$\mu_{kj}(t+3\epsilon/2) = \mu_{kj}(t+\epsilon/2)-\epsilon\frac{\partial H}{\partial \lambda_{kj}}\{\gamma(t+\epsilon),\lambda(t+\epsilon),\beta(t+\epsilon/2),\mu(t+\epsilon/2)\}.$$

These along with

$$\begin{aligned}\gamma_{lj}(t+\epsilon) &= \gamma_{lj}(t) + \epsilon\frac{\partial H}{\partial \beta_{lj}}\{\gamma(t), \lambda(t), \beta(t+\epsilon/2), \mu(t+\epsilon/2)\},\\ \lambda_{kj}(t+\epsilon) &= \mu_{kj}(t) + \epsilon\frac{\partial H}{\partial \mu_{kj}}\{\gamma(t), \lambda(t), \beta(t+\epsilon/2), \mu(t+\epsilon/2)\},\end{aligned}$$

suggest the leap-frog method uses the half step size in updating the first and the last of β, μ and uses the full step size in the rest of the alternating updates.

To implement the HMC algorithm, note that

$$\begin{aligned}\frac{\partial H}{\partial \beta_{lj}}(\gamma, \lambda, \beta, \mu) &= \left(M_j^{-1}\beta_j\right)_l,\\ \frac{\partial H}{\partial \mu_{kj}}(\gamma, \lambda, \beta, \mu) &= \mu_{kj}/\sigma_{kj}^2,\\ \frac{\partial H}{\partial \gamma_{lj}}(\gamma, \lambda, \beta, \mu) &= \sum_{i=1}^{n}\frac{\partial}{\partial \gamma_{lj}}\log \eta_j\{X_{ij}; (X_{i(j-1)}, \cdots, X_{i1}) \mid \gamma_j\} - \left(\Omega_j^{-1}\gamma_j\right)_l\\ &\quad - \sum_{i=1}^{n}\frac{\sum_{k=1}^{K_j}\frac{\partial}{\partial \gamma_{kj}}\eta_j\{z_{kj}; (X_{i(j-1)}, \cdots, X_{i1}) \mid \gamma_j\}\exp(\lambda_{kj})}{\sum_{k=1}^{K_j}\eta_j\{z_{kj}; (X_{i(j-1)}, \cdots, X_{i1}) \mid \gamma_j\}\exp(\lambda_{kj})},\\ \frac{\partial H}{\partial \lambda_{kj}}(\gamma, \lambda, \beta, \mu) &= n_{kj} - \sum_{i=1}^{n}\frac{\eta_j\{z_{kj}; (X_{i(j-1)}, \cdots, X_{i1}) \mid \gamma_j\}\exp(\lambda_{kj})}{\sum_{k=1}^{K_j}\eta_j\{z_{kj}; (X_{i(j-1)}, \cdots, X_{i1}) \mid \gamma_j\}\exp(\lambda_{kj})},\end{aligned}$$

where $(v_j)_l$ denotes the lth component of the vector v_j. The leap-frog algorithm first computes

$$\begin{aligned}\beta_{lj}(\epsilon/2) &= \beta_{lj}(0) - \frac{\epsilon}{2}\frac{\partial H}{\partial \gamma_{lj}}\{\gamma(0), \lambda(0), \beta(0), \mu(0)\},\\ \mu_{kj}(\epsilon/2) &= \mu_{kj}(0) - \frac{\epsilon}{2}\frac{\partial H}{\partial \lambda_{kj}}\{\gamma(0), \lambda(0), \beta(0), \mu(0)\}.\end{aligned}$$

And then alternatingly computes

$$\begin{aligned}\gamma_{lj}(s\epsilon) &= \gamma_{lj}\{(s-1)\epsilon\} + \epsilon\frac{\partial H}{\partial \beta_{lj}}[\gamma\{(s-1)\epsilon\}, \lambda\{(s-1)\epsilon\}, \beta\{(s-1/2)\epsilon\}, \mu\{(s-1/2)\epsilon\}],\\ \lambda_{kj}(s\epsilon) &= \mu_{kj}\{(s-1)\epsilon\} + \epsilon\frac{\partial H}{\partial \mu_{kj}}[\gamma\{(s-1)\epsilon\}, \lambda\{(s-1)\epsilon\}, \beta\{(s-1/2)\epsilon\}, \mu\{(s-1/2)\epsilon\}],\end{aligned}$$

for $s = 1, \cdots, L$ and

$$\begin{aligned}\beta_{lj}\{(s+1/2)\epsilon\} &= \beta_{lj}\{(s-1/2)\epsilon\} - \epsilon\frac{\partial H}{\partial \gamma_{lj}}[\gamma(s\epsilon), \lambda(s\epsilon), \beta\{(s-1/2)\epsilon\}, \mu\{(s-1/2)\epsilon\}],\\ \mu_{kj}\{(s+1/2)\epsilon\} &= \mu_{kj}\{(s-1/2)\epsilon\} - \epsilon\frac{\partial H}{\partial \lambda_{kj}}[\gamma(s\epsilon), \lambda(s\epsilon), \beta\{(s-1/2)\epsilon\}, \mu\{(s-1/2)\epsilon\}],\end{aligned}$$

for $s = 1, \cdots, L-1$. Lastly, computes

$$\begin{aligned}\beta_{lj}(L\epsilon) &= \beta_{lj}\{(L-1/2)\epsilon\} - \frac{\epsilon}{2}\frac{\partial H}{\partial \gamma_{lj}}[\gamma(L\epsilon), \lambda(L\epsilon), \beta\{(L-1/2)\epsilon\}, \mu\{(L-1/2)\epsilon\}],\\ \mu_{kj}(L\epsilon) &= \mu_{kj}\{(L-1/2)\epsilon\} - \frac{\epsilon}{2}\frac{\partial H}{\partial \lambda_{kj}}[\gamma(L\epsilon), \lambda(L\epsilon), \beta\{(L-1/2)\epsilon\}, \mu\{(L-1/2)\epsilon\}].\end{aligned}$$

For a specific implementation, M_j may be taken as identity matrix and $\sigma_{kj} = 1$. Both the step size ϵ and the number of steps L may be adjusted to yield reasonable high acceptance rate in the Metropolis sampling. Methods for the adjustment were proposed in Girolami and Calderhead (2011)[72], and Hofffman and Gelman (2014)[80].

8.4.3 Draw (γ, G) from the posterior under the joint odds ratio model

When the covariates are modeled by the joint semiparametric odds ratio model, the distribution of X_i has the form

$$g(X_i \mid \gamma, G) = \frac{\eta(X_{i1}; \cdots ; X_{ip} \mid \gamma) \prod_{j=1}^p dG_j(x_{ij})}{\int \cdots \int \eta(X_{i1}; \cdots ; X_{ip} \mid \gamma) \prod_{j=1}^p dG_j(x_{ij})}.$$

Prior specification can be done similarly to the consecutive conditional models. The prior for γ follows a normal distribution as $p(\gamma) = N(0, \Omega)$. Prior for G_j is Dirichlet process $\mathcal{D}(c_j F_j)$ for $j = 1, \cdots, p$. The posterior for (γ, G) is

$$\begin{aligned} p(\gamma, G \mid data) \quad \propto \quad & \frac{\prod_{i=1}^n \left\{ \eta(X_{i1}; \cdots ; X_{ip} \mid \gamma) \prod_{j=1}^p dG_j(x_{ij}) \right\}}{\left\{ \int \cdots \int \eta(X_{i1}; \cdots ; X_{ip} \mid \gamma) \prod_{j=1}^p dG_j(x_{ij}) \right\}^n} \\ & \frac{1}{(2\pi)^{d/2} |\Omega|^{1/2}} \exp\left\{ -\frac{1}{2} \gamma^t \Omega^{-1} \gamma \right\} \prod_{j=1}^p \mathcal{D}(c_j F_j), \end{aligned}$$

where d is the dimension of parameter γ. Again, let $z_{1j}, \cdots, z_{K_j j}$ be the distinctive observed values for X_j with corresponding observed frequencies $n_{1j}, \cdots, n_{K_j j}$, $j = 1, \cdots, p$. Let $g_{kj}, k = 1, \cdots, K_j$ be the probability masses assigned respectively to $z_{kj}, j = 1, \cdots, K_j$, i.e., the discretized

$$G_j(u) = \sum_{k=1}^{K_j} g_{kj} 1_{\{z_{jk} \leq u\}}.$$

For a given j, let

$$G_{nj}(u) = \sum_{j=1}^{K_j} \frac{n_{kj}}{n} 1_{\{z_{kj} \leq u\}}.$$

It follows that

$$\prod_{k=1}^{K_j} g_{kj}^{n_{kj}} \mathcal{D}(c_j F_j) = \mathcal{D} \left\{ (n + c_j) \left(\frac{n}{n + c_j} G_{nj} + \frac{c_j}{n + c_j} F_j \right) \right\}.$$

As argued before, when c_j relative to n is small, the Dirichlet process can be approximated by the Dirichlet process $\mathcal{D}(nG_{nj})$. The posterior for (γ, G) can be approximated by

$$\begin{aligned} p(\gamma, G \mid data) \quad \approx \quad & \frac{\prod_{i=1}^n \eta(X_{i1}; \cdots ; X_{ip} \mid \gamma)}{\left\{ \sum_{k_1=1}^{K_1} \cdots \sum_{k_p=1}^{K_p} \eta\{z_{k_1 1}; \cdots ; z_{k_p p} \mid \gamma) g_{k_1 1} \cdots g_{k_p p} \right\}^n} \\ & \left\{ \prod_{j=1}^p \prod_{k=1}^{K_j} g_{kj}^{nG_{nj}(z_{kj})} \right\} \exp\left[-\frac{1}{2} \gamma^t \Omega^{-1} \gamma - \frac{1}{2} \log\left\{ (2\pi)^d |\Omega| \right\} \right]. \end{aligned}$$

Let $\lambda_{kj} = \log(g_{kj}/g_{K_j j})$. Then

$$p(\gamma,\lambda \mid data) \approx \frac{\prod_{i=1}^n \eta(X_{i1};\cdots;X_{ip} \mid \gamma)}{\left\{\sum_{k_1=1}^{K_1}\cdots\sum_{k_p=1}^{K_p}\eta(z_{k_1 1};\cdots;z_{k_p p}\mid\gamma)\exp(\lambda_{k_1 1}+\cdots+\lambda_{k_p p})\right\}^n}$$
$$\left\{\prod_{j=1}^{p}\prod_{k=1}^{K_j-1}\exp(\lambda_{kj}n_{kj})\right\}\exp\left[-\frac{1}{2}\gamma^t\Omega^{-1}\gamma - \frac{1}{2}\log\left\{(2\pi)^d|\Omega|\right\}\right].$$

In principle, sampling from the posterior can be conducted by the Hamiltonian Monte Carlo as in the case of the consecutive conditional semiparametric odds ratio model discussed in the previous section. Let

$$U(\gamma,\lambda) = \sum_{i=1}^n \log\eta(X_{i1};\cdots;X_{ip}\mid\gamma) - \frac{1}{2}\gamma^t\Omega^{-1}\gamma - \frac{1}{2}\log\left\{(2\pi)^d|\Omega|\right\}$$
$$-n\log\left\{\sum_{k_1=1}^{K_1}\cdots\sum_{k_p=1}^{K_p}\eta(z_{k_1 1};\cdots;z_{k_p p}\mid\gamma)\exp(\lambda_{k_1 1}+\cdots+\lambda_{k_p p})\right\} + \sum_{j=1}^{p}\sum_{k=1}^{K_j-1}\lambda_{kj}n_{kj}.$$

Likewise the Hamiltonian function can be set to

$$H(\gamma,\lambda,\beta,\mu) = U(\gamma,\lambda) + \frac{1}{2}\beta^t\Sigma^{-1}\beta + \frac{1}{2}\sum_{j=1}^{p}\sum_{k=1}^{K_j}\frac{\mu_{kj}^2}{\sigma_{kj}^2}.$$

One difficulty in implementing the Hamiltonian Monte Carlo is the need to compute the normalizing function

$$\varphi(\gamma,\lambda) = \sum_{k_1=1}^{K_1}\cdots\sum_{k_p=1}^{K_p}\eta(z_{k_1 1};\cdots;z_{k_p p}\mid\gamma)\exp(\lambda_{k_1 1}+\cdots+\lambda_{k_p p})$$

and its derivatives. They are nontrivial to compute when p is moderate to high. In most cases, it is reasonable to assume that, when $\gamma = 0$, $\eta(z_{k_1 1};\cdots;z_{k_p p}\mid\gamma) = 1$. Under this assumption,

$$\varphi(0,\lambda) = \prod_{j=1}^{p}\left\{\sum_{k=1}^{K_j}\exp(\lambda_{kj})\right\},$$

which is not difficult to compute. For non-zero γ, it can be seen that

$$\log\frac{\varphi(\gamma,\lambda)}{\varphi(0,\lambda)} = \int_0^1 E_u\left\{\frac{d}{du}\log\eta(X_1;\cdots;X_p\mid u\gamma)\right\}du,$$

where E_u is computed under the distribution

$$p(X_1 = z_{k_1 1};\cdots;X_p = z_{k_p p}\mid u\gamma)$$
$$= \frac{\eta(z_{k_1 1};\cdots;z_{k_p p}\mid u\gamma)\exp(\lambda_{k_1 1}+\cdots+\lambda_{k_p p})}{\sum_{k_1=1}^{K_1}\cdots\sum_{k_p=1}^{K_p}\eta(z_{k_1 1};\cdots;z_{k_p p}\mid u\gamma)\exp(\lambda_{k_1 1}+\cdots+\lambda_{k_p p})}.$$

For a given u, sampling from this distribution can be carried out using Gibbs sampler. Given the values of the rest of variables, sampling from a particular variable, say j, can be directly obtained due to the discrete nature of the distribution. More generally, u can be regarded as uniformly distributed on $[0,1]$, a single Gibbs sampler can be run to sample from the distribution

$$p(X_1 = z_{k_1 1}, \cdots, X_p = z_{k_p p}, U = u) = p(z_{k_1 1}; \cdots; z_{k_p p} \mid u\gamma) 1_{[0,1]}(u).$$

In applying the Hamiltonian Monte Carlo, the derivatives of $U(\gamma, \lambda)$ are also needed. The derivatives have the form

$$\begin{aligned} \frac{\partial}{\partial \gamma_j} U(\gamma, \lambda) &= \sum_{i=1}^{n} \frac{\partial}{\partial \gamma_j} \log \eta(X_{i1}; \cdots; X_{ip} \mid \gamma) - \left(\Omega^{-1}\gamma\right)_j \\ &\quad - n E_1 \left\{ \frac{\partial}{\partial \gamma_j} \log \eta(X_{i1}; \cdots; X_{ip} \mid \gamma) \right\}, \\ \frac{\partial}{\partial \lambda_{kj}} U(\gamma, \lambda) &= n_{kj} - n P_1 \left(X_k = z_{kj}\right), \end{aligned}$$

where E_1 and P_1 are computed under the distribution $P(X_1, \cdots, X_p \mid \gamma)$.

To implement the HMC algorithm, note that

$$\begin{aligned} \frac{\partial H}{\partial \beta_j}(\gamma, \lambda, \beta, \mu) &= \left(M^{-1}\beta\right)_j, \\ \frac{\partial H}{\partial \mu_{kj}}(\gamma, \lambda, \beta, \mu) &= \mu_{kj}/\sigma_{kj}^2, \\ \frac{\partial H}{\partial \gamma_j}(\gamma, \lambda, \beta, \mu) &= \frac{\partial}{\partial \gamma_j} U(\gamma, \lambda), \\ \frac{\partial H}{\partial \lambda_{kj}}(\gamma, \lambda, \beta, \mu) &= \frac{\partial}{\partial \lambda_{kj}} U(\gamma, \lambda). \end{aligned}$$

Start from the current position $(\gamma, \lambda, \beta, \mu)$, randomly draw (β^*, μ^*) independently from the standard normal distributions. Set the initial condition as

$$\{\gamma(0), \lambda(0), \beta(0), \mu(0)\} = (\gamma, \lambda, \beta^*, \mu^*).$$

For a given step size ϵ and the number of steps L, implement the leap-frog algorithm to compute first

$$\begin{aligned} \beta_j(\epsilon/2) &= \beta_j(0) - \frac{\epsilon}{2} \frac{\partial H}{\partial \gamma_j} \{\gamma(0), \lambda(0), \beta(0), \mu(0)\}, \\ \mu_{kj}(\epsilon/2) &= \mu_{kj}(0) - \frac{\epsilon}{2} \frac{\partial H}{\partial \lambda_{kj}} \{\gamma(0), \lambda(0), \beta(0), \mu(0)\}. \end{aligned}$$

Then alternatingly compute

$$\begin{aligned} \gamma_j(s\epsilon) &= \gamma_j\{(s-1)\epsilon\} + \epsilon \frac{\partial H}{\partial \beta_j} [\gamma\{(s-1)\epsilon\}, \lambda\{(s-1)\epsilon\}, \beta\{(s-1/2)\epsilon\}, \mu\{(s-1/2)\epsilon\}], \\ \lambda_{kj}(s\epsilon) &= \mu_{kj}\{(s-1)\epsilon\} + \epsilon \frac{\partial H}{\partial \mu_{kj}} [\gamma\{(s-1)\epsilon\}, \lambda\{(s-1)\epsilon\}, \beta\{(s-1/2)\epsilon\}, \mu\{(s-1/2)\epsilon\}], \end{aligned}$$

for $s = 1, \cdots, L$ and

$$\begin{aligned} \beta_j\{(s+1/2)\epsilon\} &= \beta_j\{(s-1/2)\epsilon\} - \epsilon \frac{\partial H}{\partial \gamma_j} [\gamma(s\epsilon), \lambda(s\epsilon), \beta\{(s-1/2)\epsilon\}, \mu\{(s-1/2)\epsilon\}], \\ \mu_{kj}\{(s+1/2)\epsilon\} &= \mu_{kj}\{(s-1/2)\epsilon\} - \epsilon \frac{\partial H}{\partial \lambda_{kj}} [\gamma(s\epsilon), \lambda(s\epsilon), \beta\{(s-1/2)\epsilon\}, \mu\{(s-1/2)\epsilon\}], \end{aligned}$$

for $s = 1, \cdots, L-1$. Finally, compute

$$\begin{aligned}\beta_j(L\epsilon) &= \beta_j\{(L-1/2)\epsilon\} - \frac{\epsilon}{2}\frac{\partial H}{\partial \gamma_j}[\gamma(L\epsilon), \lambda(L\epsilon), \beta\{(L-1/2)\epsilon\}, \mu\{(L-1/2)\epsilon\}],\\ \mu_{kj}(L\epsilon) &= \mu_{kj}\{(L-1/2)\epsilon\} - \frac{\epsilon}{2}\frac{\partial H}{\partial \lambda_{kj}}[\gamma(L\epsilon), \lambda(L\epsilon), \beta\{(L-1/2)\epsilon\}, \mu\{(L-1/2)\epsilon\}].\end{aligned}$$

With probability

$$\pi = \min\left(1, \exp\left[H(\gamma, \lambda, \beta, \mu) - H\left\{\gamma(L\epsilon), \lambda(L\epsilon), \beta(L\epsilon), \mu(L\epsilon)\right\}\right]\right),$$

to accept $\gamma(L\epsilon), \lambda(L\epsilon), \beta(L\epsilon), \mu(L\epsilon)$ as the new sample. With probability $1-\pi$ to use the old sample $(\gamma, \lambda, \beta, \mu)$. Both the step size ϵ and the number of steps L may be adjusted to yield reasonable high acceptance rate in the Metropolis sampling. The cycle would be repeated to generate a Markov chain sample.

In the computation, in addition to $\varphi(\gamma(t), \lambda)$ for $t \in [0, L\epsilon]$,

$$\frac{\partial U}{\partial \gamma_j}\left\{\gamma(s\epsilon), \lambda(s\epsilon)\right\}, \frac{\partial U}{\partial \lambda_{kj}}\left\{\gamma(s\epsilon), \lambda(s\epsilon)\right\},$$

for $s = 0, 1, \cdots, L$ need to be evaluated numerically. This can substantially increase the computation in comparison to the consecutive conditional odds ratio models.

8.4.4 Draw α and θ from their posteriors

Drawing α and θ is relatively simpler in comparison to drawing (γ, G). Let the prior for α be $q(\alpha)$. The posterior for α is

$$p(\alpha \mid R_i, Y_i, X_i, i = 1, \cdots, n) \propto \prod_{i=1}^{n} p(R_i \mid Y_i, X_i, \alpha) q(\alpha).$$

Let the prior for θ be $\xi(\theta)$, the posterior for θ is

$$p(\theta \mid Y_i, X_i, i = 1, \cdots, n) \propto \prod_{i=1}^{n} f(Y_i \mid X_i, \theta)\xi(\theta).$$

In either of the above cases, we may choose normal distribution as the prior, and the posterior can be simulated using Monte Carlo Markov chain approaches. The specific approaches used will depend on the specific models for the missing data mechanism and for the full data respectively.

8.5 Application to nonignorable missing data

8.5.1 Nonparametric identifiability of the full data model with missing at random

Nonparametric identifiability in missing data problem refers to the identifiability of the full data model without imposing any restrictions on the full data distribution. When data are subject to missing values, the distribution of the full data Y may not be nonparametrically identifiable from the observed data without imposing any restrictions on the missing data probability model. This can be seen as follows. Let $Y = (Y_1, \cdots, Y_p)$ and $R = (R_1, \cdots, R_p)$. The full data distribution of Y can be expressed through pattern mixture formula as

$$f(y_1, \cdots, y_p) = \sum_r f\{\bar{r}(y) \mid r(y), R = r\} f\{r(y) \mid R = r\} P(R = r).$$

None of $f\{\bar{r}(y) \mid r(y), R = r\}$ is identifiable when no restriction is imposed on the distribution of R given Y.

The missing at random mechanism imposes the following constraints

$$P(R = r \mid Y) = P\{R = r \mid r(Y)\},$$

for all missing data patterns. The missing data probabilities are identifiable. When the probability of observing all possible configurations of the variables in the complete cases is uniformly bounded away from 0, i.e.,

$$P(R = \mathbf{1} \mid Y) > \sigma > 0, \tag{8.6}$$

for almost all Y, the full data model is nonparametrically identifiable from the observed data. This is because

$$f(y) = \frac{f(y \mid R = \mathbf{1}) P(R = \mathbf{1})}{P(R = \mathbf{1} \mid y)}.$$

In general,

$$f\{\bar{r}(y) \mid r(y)\} = f\{\bar{r}(y) \mid r(y), R = r\},$$

for all missing data patterns under the missing at random model assumption. These may be used to obtain nonparametric identifiability when (8.6) does not hold.

Nonparametric identifiability may fail when (8.6) does not hold even when missing data are missing at random. For example, let $Y = (Y_1, Y_2)$ and $R = (R_1, R_2)$. Assume that $R_1 + R_2 = 1$ and $P(R_1 = 1 \mid Y) = P(R_2 = 1 \mid Y) = 0.5$. The observed data have either only Y_1 or only Y_2 observed. The missing data are missing at random, but condition (8.6) does not hold. It can be easily seen that only the marginal distribution $f_1(y_1)$ and $f_2(y_2)$ are nonparametrically identifiable. One joint distribution that can be determined is

$$f(y_1, y_2) = f_1(y_1) f_2(y_2).$$

The observed data can be perfectly fit to this model. However, there are many other joint distributions can also be perfectly fit to the observed data, such as the distribution functions

$$F(y_1, y_2 \mid \rho) = \Phi_\rho\{F_1(y_1), F_2(y_2)\}$$

where $\rho \in (-1, 1)$, Φ_ρ is a bivariate normal distribution with mean $(0, 0)$ variance $(1, 1)$ and correlation ρ, and F_1 and F_2 are respectively the marginal distribution functions for f_1 and f_2. The joint model $f(y_1, y_2)$ is not nonparametrically identifiable in this case.

8.5.2 Nonparametric identifiability of the full data model with missing not at random

When data are not missing at random, certain restrictions may also be imposed on the missing data mechanism model so that the full data model is nonparametrically identifiable. One of such restrictions is to assume that any missing data indicator, given the rest of missing data indicators and the full data, does not depend on the particular missing variable, i.e.,

$$R_k \perp\!\!\!\perp Y_k \mid R_{-k}, Y_{-k},$$

for all $k = 1, \cdots, p$. In terms of missing data probability, it corresponds to assuming

$$P(R_k = r_k \mid R_{-k}, Y) = P(R_k = r_k \mid R_{-k}, Y_{-k}),$$

for all $k = 1, \cdots, p$. The independence condition was termed the item-wise conditional independent nonresponse in Sadinle and Reiter (2017)[157] or the no self-censoring model in Malinsky et al (2021)[118]. When the conditional independence holds, Sadinle and Reiter (2017)[157] showed that the full data model can be nonparametrically identifiable. The idea is to recursively construct a nonparametric identifiable full data distribution that generates the observed data probability for each missing patterns, satisfies the item-wise conditional independence assumption, and depends only on the observed data probabilities.

Let $f\{R, R(Y)\}, R \in \mathcal{R}$ be the observed data probabilities, where $\mathcal{R}$ is the collection of all missing data patterns that have $f\{R, R(Y)\} > 0$. Assume that, for any missing data pattern r, $f\{R = r, r(Y)\} > 0$ for almost all $r(Y)$ or 0 for almost all $r(Y)$. Define a joint distribution function

$$g(R, Y) = \prod_{R^* \succeq R, R^* \in \mathcal{R}} q\{R^*, R^*(Y)\},$$

where $R^* \succeq R$ means every variable observed in the missing data patetern R is also observed in the missing data pattern R^*, and

$$q\{R^*, R^*(Y)\} = \frac{f\{R^*, R^*(Y)\}}{\int \prod_{R^+ \succ R^*, R^+ \in \mathcal{R}} q\{R^+, R^+(Y)\} d\bar{R}^*(Y)}.$$

Here, $R^+ \succ R^*$ means every variable observed in the missing data pattern R^* is also observed in the missing data pattern R^+ and at least one more variable is observed in R^+ and is not observed in R^*. If there is no $R^* \in \mathcal{R}$ satisfying $R^* \succ R$, the denominator in the expression is set to 1. From the definition of $g(R, Y)$, it can be seen that

$$g(R, Y) = \frac{f\{R, R(Y)\}}{\int \prod_{R^+ \succ R, R^+ \in \mathcal{R}} q\{R^+, R^+(Y)\} d\bar{R}(Y)} \prod_{R^* \succ R, R^* \in \mathcal{R}} q\{R^*, R^*(Y)\}.$$

It follows that

$$\int g(R, Y) d\bar{R}(Y) = f\{R, R(Y)\}.$$

This suggests that $g(R, Y)$ generates the same observed data probabilities as required. From the construction of $g(R, Y)$, it is easily seen that $g(R, Y)$ depends on $f\{R, R(Y)\}, R \in \mathcal{R}$ only. It remains to show that the construction also satisfies, for any j,

$$P(R_j = 0 \mid R_{-j}, Y) = P(R_j = 0 \mid R_{-j}, Y_{-j}). \tag{8.7}$$

Note, from the assumption that, $f(R = r, r(Y)\} > 0$ for almost all $r(Y)$ or 0 for almost all $r(Y)$ for all missing data patterns r, that

$$P(R_j = 0, R_{-j}, Y) = 0 \iff P(R_j = 0, R_{-j}, Y_{-j}) = 0.$$

For $P(R_j = 0, R_{-j}, Y) > 0$,

$$P(R_j = 0 \mid R_{-j}, Y) = \frac{P(R_j = 0, R_{-j}, Y)}{P(R_j = 0, R_{-j}, Y) + P(R_j = 1, R_{-j}, Y)}.$$

Furthermore,

$$\frac{P(R_j = 0, R_{-j}, Y)}{P(R_j = 1, R_{-j}, Y)} = \frac{\prod_{R^* \succeq (R_j=0, R_{-j}), R^* \in \mathcal{R}} q\{R^*, R^*(Y)\}}{\prod_{R^* \succeq (R_j=1, R_{-j}), R^* \in \mathcal{R}} q\{R^*, R^*(Y)\}}, \tag{8.8}$$

For any R^* in the numerator of the right-hand side of (8.8) that has $R_j^* = 1$, such an R^* also appears in the denominator, and is therefore canceled. For R^* in the numerator of the right-hand side of (8.7) that has $R_j^* = 0$, the term dose not depends on Y_j by construction. This means that $P(R_j = 0, R_{-j}, Y)/P(R_j = 1, R_{-j}, Y)$ does not depend on Y_j, which implies (8.7) holds because the full data density $g(y)$ by construction is the same as the true underlying density $f(y)$ generated the full data, i.e.,

$$g(Y) = \frac{g(R = \mathbf{1}, Y)}{P_g(R = \mathbf{1} \mid Y)} = \frac{f(R = \mathbf{1}, Y)}{P_f(R = \mathbf{1} \mid Y)} = f(Y).$$

The foregoing construction of $g(R, Y), R \in \mathcal{R}$ only guarantees there exists a distribution function that perfectly fit the observed data. That is,

$$f^*(y) = \sum_{r \in \mathcal{R}} g\{R = r, r(y)\}.$$

It does not means such a distribution is unique, i.e., $f^*(y)$ may or may not be the same as the true underlying distribution $f(y)$ that generates the full data. Consider the example given in the previous section, the missing data mechanism also satisfies the item-wise independence nonresponse condition. By the foregoing method for construction,

$$g\{(1,0), Y\} = f\{(1,0), Y_1\} \text{ and } g\{(0,1), Y\} = f\{(0,1), Y_2\}.$$

The full data model is therefore

$$\begin{aligned} g(Y) &= g\{(1,0), Y\} + g\{(0,1), Y\} \\ &= \{f_1(Y_1) + f_2(Y_2)\}/2. \end{aligned}$$

As mentioned in the previous subsection, there are many different distributions that can generated the same observed data. To make the full model uniquely identifiable, one sufficient condition is (8.6).

To see more specifically how the construction was carried out in practice, consider the more general case of missing data with three variables (Y_1, Y_2, Y_3). Assume that the possible missing patterns are

$$\mathcal{R} = \{(1,1,1), (1,1,0), (1,0,1), (0,0,1), (1,0,0)\},$$

i.e., for $r \in \mathcal{R}$, $P\{R = r, r(Y)\} > 0$ for almost all $r(Y)$. By the construction,

$$\begin{aligned} q\{(1,1,1), Y_1, Y_2, Y_3\} &= f\{(1,1,1, Y_1, Y_2, Y_3\}, \\ q\{(1,1,0), Y_1, Y_2\} &= \frac{f\{(1,1,0, Y_1, Y_2\}}{\int q\{(1,1,1, Y_1, Y_2, Y_3\} dY_3}, \\ q\{(1,0,1), Y_1, Y_3\} &= \frac{f\{(1,0,1, Y_1, Y_3\}}{\int q\{(1,1,1, Y_1, Y_2, Y_3\} dY_2}, \\ q\{(0,0,1), Y_3\} &= \frac{f\{(0,0,1, Y_3\}}{\int\int q\{(1,1,1, Y_1, Y_2, Y_3\} q\{(1,0,1), Y_1, Y_3\} dY_1 dY_2}, \\ q\{(1,0,0), Y_1\} &= \frac{f\{(1,0,0, Y_1\}}{\int\int q\{(1,1,1, Y_1, Y_2, Y_3\} q\{(1,1,0), Y_1, Y_2\} q\{(1,0,1), Y_1, Y_3\} dY_2 dY_3}. \end{aligned}$$

The joint distribution of (R, Y) thus constructed has

$$
\begin{aligned}
g\{(1,1,1), Y_1, Y_2, Y_3\} &= q\{(1,1,1, Y_1, Y_2, Y_3\},\\
g\{(1,1,0), Y_1, Y_2, Y_3\} &= q\{(1,1,0), Y_1, Y_2\}q\{(1,1,1, Y_1, Y_2, Y_3\}\\
g\{(1,0,1), Y_1, Y_2, Y_3\} &= q\{(1,0,1), Y_1, Y_3\}q\{(1,1,1, Y_1, Y_2, Y_3\},\\
g\{(0,0,1), Y_1, Y_2, Y_3\} &= q\{(0,0,1), Y_3\}q\{(1,0,1), Y_1, Y_3\}q\{(1,1,1, Y_1, Y_2, Y_3\}\\
g\{(1,0,0), Y_1, Y_2, Y_3\} &= q\{(1,0,0), Y_1\}q\{(1,0,1), Y_1, Y_3\}q\{(1,1,0), Y_1, Y_2\}\\
&\quad q\{(1,1,1, Y_1, Y_2, Y_3\}.
\end{aligned}
$$

8.5.3 Odd ratio model for the item-wise independent nonresponse data

The construction of distribution for the item-wise independent nonresponse data is concise and illustrative. However, the recursive nature makes it difficult to directly model such data. Assume that (8.6) holds. Let $Y = 0$ and $R = \mathbf{1}_p$ be the fixed reference values for (R, Y). The odds ratio representation of $g(R, Y)$ can be obtained as

$$
g(R, Y) = \frac{\eta(R; Y)p(Y \mid R = \mathbf{1}_p)p(R \mid Y = 0)}{\sum_{r \in \mathcal{R}} \int \eta(R; Y)p(Y \mid R = \mathbf{1}_p)p(R \mid Y = 0)dY},
$$

where $\eta(R; Y)$ is the odds ratio function of the density function $g(R, Y)$ with reference point $(R = \mathbf{1}_p, Y = 0)$. The item-wise independent nonresponse condition

$$
R_j \perp\!\!\!\perp Y_j \mid R_{-j}, Y_{-j}, \iff \eta(R_j; Y_j \mid R_{-j}, Y_{-j}) \equiv 1,
$$

for $j = 1, \cdots, p$. From (2.50), it follows that

$$
\begin{aligned}
\eta(R; Y) &= \eta(R_1; Y \mid R_{-1})\eta(R_{-1}; Y \mid R_1 = 1)\\
&= \eta(R_1; Y_1 \mid R_{-1}, Y_{-1})\eta(R_1; Y_{-1} \mid R_{-1}, Y_1 = 0)\eta(R_{-1}; Y \mid R_1 = 1)\\
&= \eta(R_1; Y_{-1} \mid R_{-1}, Y_1 = 0)\eta(R_{-1}; Y \mid R_1 = 1)
\end{aligned}
$$

because $R_1 \perp\!\!\!\perp Y_1 \mid R_{-1}, Y_{-1}$. Note further that

$$
\begin{aligned}
\eta(R_{-1}; Y \mid R_1 = 1) &= \eta(R_2; Y \mid R_1 = 1, R_{-(1,2)})\eta(R_{-(1,2)}; Y \mid R_{(1,2)} = \mathbf{1}_2)\\
&= \eta(R_2; Y_2 \mid R_1 = 1, R_{-(1,2)}, Y_{-2})\eta(R_2; Y_{-2} \mid R_1 = 1, R_{-(1,2)}, Y_2 = 0)\\
&\quad \eta(R_{-(1,2)}; Y \mid R_{(1,2)} = \mathbf{1}_2)\\
&= \eta(R_2; Y_{-2} \mid R_1 = 1, R_{-(1,2)}, Y_2 = 0)\eta(R_{-(1,2)}; Y \mid R_{(1,2)} = \mathbf{1}_2)
\end{aligned}
$$

because $R_2 \perp\!\!\!\perp Y_2 \mid R_{-2}, Y_{-2}$. Continue this type of decomposition for $\eta(R_{-(1,2)}; Y \mid R_{(1,2)} = \mathbf{1}_2)$. In general for j, let $\overline{j} = \{1, \cdots, j-1\}$.

$$
\begin{aligned}
\eta(R_{-\overline{j}}; Y \mid R_{\overline{j}} = 1) &= \eta(R_j; Y \mid R_{-\overline{j}} = \mathbf{1}_{j-1}, R_{-\overline{(j+1)}})\eta(R_{-\overline{(j+1)}}; Y \mid R_{\overline{(j+1)}} = \mathbf{1}_j)\\
&= \eta(R_j; Y_j \mid R_{\overline{j}} = \mathbf{1}_{j-1},, R_{-\overline{(j+1)}}, Y_{-j})\\
&\quad \eta(R_j; Y_{-j} \mid R_{\overline{j}} = 1, R_{-\overline{(j+1)}}, Y_j = 0)\eta(R_{-\overline{(j+1)}}; Y \mid R_{\overline{(j+1)}} = \mathbf{1}_j)\\
&= \eta(R_j; Y_{-j} \mid R_{\overline{j}} = \mathbf{1}_{j-1},, R_{-\overline{(j+1)}}, Y_j = 0)\eta(R_{-\overline{(j+1)}}; Y \mid R_{\overline{(j+1)}} = \mathbf{1}_j)
\end{aligned}
$$

because $R_j \perp\!\!\!\perp Y_j \mid R_{-j}, Y_{-j}$. Put together, the odds ratio function can be expressed as

$$
\eta(R; Y) = \prod_{j=1}^{p} \eta(R_j; Y_{-j} \mid R_{\overline{j}} = \mathbf{1}_{j-1},, R_{-\overline{(j+1)}}, Y_j = 0)
$$

All the conditional independence conditions are satisfied in the last expression for $\eta(R;Y)$. The joint density can now be reduced to

$$g(R,Y)=\frac{\prod_{j=1}^{p}\eta(R_j;Y_{-j}\mid R_{\overline{j}}=\mathbf{1}_{j-1},,R_{-\overline{(j+1)}},Y_j=0)p(Y\mid R=\mathbf{1}_p)p(R\mid Y=0)}{\sum_{r\in\mathcal{R}}\int\prod_{j=1}^{p}\eta(R_j;Y_{-j}\mid R_{\overline{j}}=\mathbf{1}_{j-1},,R_{-\overline{(j+1)}},Y_j=0)p(Y\mid R=\mathbf{1}_p)p(R\mid Y=0)dY},$$

Models for $\eta(R_j;Y_{-j}\mid R_{\overline{j}}=\mathbf{1}_{j-1},,R_{-\overline{(j+1)}},Y_j=0)$ can be parametrically specified without additional restrictions when the missing patterns that do not appear in the observed data are taken care.

To see more clearly the restrictions in the odds ratio models, consider the example given in the previous section where $p=3$. It follows that

$$\begin{aligned}\eta(R;Y) &= \eta\{R_1;(Y_2,Y_3)\mid R_2,R_3,Y_1=0\}\eta\{R_2;(Y_1,Y_3)\mid R_1=1,R_3,Y_2=0\}\\ &\quad \eta\{R_3;(Y_1,Y_2)\mid R_1=1,R_2=1,Y_3=0\}\end{aligned}$$

In the example, three missing patterns $(0,1,1),(0,1,0)$ and $(0,0,0)$ were not observed, which corresponds to

$$g(0,1,1,Y)=0;g(0,1,0,Y)=0;g(0,0,0,Y)=0.$$

From the odds ratio decomposition,

$$\eta\{(0,1,1);Y\}=0;\eta\{(0,1,0);Y\}=0;\eta\{(0,0,0);Y\}=0.$$

It follows that

$$\begin{aligned}0&=\eta\{R_1=0;(Y_2,Y_3)\mid R_2=1,R_3=1,Y_1=0\}\eta\{R_2=1;(Y_1,Y_3)\mid R_1=1,R_3=1,Y_2=0\}\\&\quad\eta\{R_3=1;(Y_1,Y_2)\mid R_1=1,R_2=1,Y_3=0\},\\0&=\eta\{R_1=0;(Y_2,Y_3)\mid R_2=1,R_3=0,Y_1=0\}\eta\{R_2=1;(Y_1,Y_3)\mid R_1=1,R_3=0,Y_2=0\}\\&\quad\eta\{R_3=0;(Y_1,Y_2)\mid R_1=1,R_2=1,Y_3=0\},\\0&=\eta\{R_1=0;(Y_2,Y_3)\mid R_2=0,R_3=0,Y_1=0\}\eta\{R_2=0;(Y_1,Y_3)\mid R_1=1,R_3=0,Y_2=0\}\\&\quad\eta\{R_3=0;(Y_1,Y_2)\mid R_1=1,R_2=1,Y_3=0\}.\end{aligned}$$

Furthermore, for the observed data patterns in $R\in\mathcal{R}$, $g(R,Y)>0$, it follows that

$$\begin{aligned}0&<\eta\{R_1=1;(Y_2,Y_3)\mid R_2=1,R_3=1,Y_1=0\}\eta\{R_2=1;(Y_1,Y_3)\mid R_1=1,R_3=1,Y_2=0\}\\&\quad\eta\{R_3=1;(Y_1,Y_2)\mid R_1=1,R_2=1,Y_3=0\},\\0&<\eta\{R_1=1;(Y_2,Y_3)\mid R_2=1,R_3=0,Y_1=0\}\eta\{R_2=1;(Y_1,Y_3)\mid R_1=1,R_3=0,Y_2=0\}\\&\quad\eta\{R_3=0;(Y_1,Y_2)\mid R_1=1,R_2=1,Y_3=0\},\\0&<\eta\{R_1=1;(Y_2,Y_3)\mid R_2=0,R_3=1,Y_1=0\}\eta\{R_2=0;(Y_1,Y_3)\mid R_1=1,R_3=1,Y_2=0\}\\&\quad\eta\{R_3=1;(Y_1,Y_2)\mid R_1=1,R_2=1,Y_3=0\},\\0&<\eta\{R_1=0;(Y_2,Y_3)\mid R_2=0,R_3=1,Y_1=0\}\eta\{R_2=0;(Y_1,Y_3)\mid R_1=1,R_3=1,Y_2=0\}\\&\quad\eta\{R_3=1;(Y_1,Y_2)\mid R_1=1,R_2=1,Y_3=0\},\\0&<\eta\{R_1=1;(Y_2,Y_3)\mid R_2=0,R_3=0,Y_1=0\}\eta\{R_2=0;(Y_1,Y_3)\mid R_1=1,R_3=0,Y_2=0\}\\&\quad\eta\{R_3=0;(Y_1,Y_2)\mid R_1=1,R_2=1,Y_3=0\}.\end{aligned}$$

Compare the first equation with the first inequality, it follows that

$$\eta\{R_1=0;(Y_2,Y_3)\mid R_2=1,R_3=1,Y_1=0\}\equiv 0.$$

Compare the second equation with the second inequality, it follows that

$$\eta\{R_1 = 0; (Y_2, Y_3) \mid R_2 = 1, R_3 = 0, Y_1 = 0\} \equiv 0.$$

Compare the third equation with the fifth inequality, it follows that

$$\eta\{R_1 = 0; (Y_2, Y_3) \mid R_2 = 0, R_3 = 0, Y_1 = 0\} \equiv 0.$$

The rest of the odds ratio functions appeared in $g(R, Y)$ can be modeled without restrictions as long as the odds ratio function requirement is satisfied.

8.6 Summary and literature

The problem of missing covariates has been extensively studied (Pepe and Fleming, 1991; Little, 1992; Robins et al, 1994; Robins et al, 1995; Lipsitz and Ibrahim, 1996; Reilly and Pepe, 1996; Paik, 1997; Lipsitz et al, 1999; Chen, 2004)[135, 108, 153, 152, 106, 149, 133, 107, 27]. For the proportional hazards regression with missing covariates, methods were proposed in Lin and Ying (1993)[104], Zhou and Pepe (1995)[206], Chen and Little (1999)[34], Martinussen (1999)[122], Herring et al (2001)[75], and Chen (2002)[25]. For nonignorable missing data, parameter identifiability and estimation were studied in Tang et al (2003) [177], Wang et al (2014)[194], Zhao and Shao (2015)[204], Miao and Tchetgen Tchetgen (2016)[126], Shao and Wang (2016)[165], Miao et al (2016)[165], and Sun et al (2018)[172] through an instrumental variable mechanism. Full parameter identifiability and estimation through restrictions on the missing data probability were studied in Sadinle and Reiter (2017)[157] and Malinsky et al (2021)[118].

8.7 Exercises

1. Let $(X_i, \delta_i, Z_i), i = 1, \cdots, n$ be the observed iid censored survival time data, where $X_i = \min(T_i, C_i)$ and $\delta_i = 1_{\{T_i \le C_i\}}$, and T_i and C_i are respectively the survival time and censoring time. Given Z_i, T_i and C_i are independent and T_i given Z_i follows the Cox regression model. In applying the imputation approach to the Cox regression with missing covariates, a Bayesian survival model requires us to specify a prior for the regression parameter and the baseline hazard. Such a prior usually assumes independence of the regression parameter and the baseline hazard. A normal prior is often assumed for the regression parameter and a Gamma process prior is assumed for the baseline(Kalbfleisch, 1978; Ibrahim et al, 2001)[87, 82]. A Gamma process denote by $\mathcal{GP}(\Lambda^*, c)$ is an independent increment process on $[0, \infty)$ such that $\mathcal{GP}(0) = 0$, and for any non-overlapped intervals $[t_1, t_2), \cdots, [t_{k-1}, t_k)$, the joint distribution of $\mathcal{GP}(t_2) - \mathcal{GP}(t_1), \cdots, \mathcal{GP}(t_k) - \mathcal{GP}(t_{k-1})$ are independent following Gamma distributions $\Gamma(\Lambda^*(t_{j+1}) - \Lambda^*(t_j), c), j = 1, \cdots, k-1$ respectively, where Λ^* and c are assumed known. A Gamma distribution $\Gamma(a, b)$ has density

$$f(t \mid a, b) = \frac{a(at)^{b-1}}{\Gamma(b)} \exp(-at),$$

for $t \ge 0, a > 0, b > 0$, where $\Gamma(b)$ is the *Gamma* function at b.

(a) Find the mean and the variance of the $\Gamma(a, b)$ distributed random variable.
(b) If the prior is assumed to be $\mathcal{GP}(c\Lambda^*(t), c)$, find the mean and variance for $\mathcal{GP}(t_j) - \mathcal{GP}(t_{j-1})$.
(c) Suppose that all subjects who are still alive at time τ are censored at time τ, and there is no other censoring occurs, find the posterior for the baseline hazard.

2. Let $s_0 = 0 < s_1 < \cdots < s_K < +\infty$. For a given Λ^*, let

$$\lambda_j^* = \Lambda^*(s_j) - \Lambda^*(s_{j-1}).$$

In a Bayesian model, assume that the baseline cumulative hazard Λ satisfies $\Lambda(s_j) - \Lambda(s_{j-1}), j = 1, \cdots, K$ are independent and

$$\Lambda(s_j) - \Lambda(s_{j-1}) \sim \Gamma\left(c\lambda_j^*, c\right).$$

Suppose that the observed data are only known to fall into one of the intervals. Assume that β are Λ are independent.

(a) Find the likelihood for the observed data.
(b) Find the posterior distribution for $(\beta, \Lambda(s_j) - \Lambda(s_{j-1}), j = 1, \cdots, K)$.
(c) Find the Gibbs sampling distribution for a given component of β.
(d) Propose a sampling approach to the given component of β in the last question.
(e) Find the Gibbs sampling distribution for a given $\Lambda(s_j) - \Lambda(s_{j-1}$.
(f) Show that the distribution in the last question can be approximated by a Gamma distribution.

3. Let Y be the outcome and X be the covariates. Let the observed missing data patterns be as given in Table 8.1.

TABLE 8.1
Missing data patterns.

Variable	Missing pattern			
Y	O	O	M	M
X	O	M	O	M

O=observed, M=missing.

Suppose that the joint distribution of (Y, X) has the odds ratio representation

$$f(y, x) = \frac{\eta(y; x)p(y \mid x_0)p(x \mid y_0)}{\int\int \eta(y; x)p(y \mid x_0)p(x \mid y_0)dydx},$$

where the reference point $(y_0, x_0) = (0, 0)$. Let $R = (R_y, R_x)$ be the missing data indicators for (Y, X). Assume further that

$$P(R = (1, 1) \mid Y, X) \geq \sigma > 0,$$

for all values in the domain of (Y, X).

(a) If the missing data are missing at random, show that $\eta(y; x)$, $p(y \mid x_0)$, and $p(x \mid y_0)$ are all identifiable.

(b) If further

$$\log \eta(y; x) = \theta y x,$$

show that θ is identifiable.

(c) Find the maximum likelihood estimator for θ.

(d) Propose an approach to make inference on the true θ value.

4. For the same missing data problem given in the previous problem, assume that the missing data are not missing at random. More specifically,

$$P\{R = (r_y, r_x) \mid X, Y\} \propto \pi_y(Y)\pi_x(X).$$

(a) Show that $\eta(y; x)$ is still identifiable.

(b) Are $p(y \mid x_0)$ and $p(x \mid y_0)$ still identifiable?

(c) For the log-linear odds ratio model, find an estimator for θ.

(d) If we know the marginal distributions for Y, can you incorporate this piece of information to identify $p(y \mid x_0)$ and $p(x \mid y_0)$.

(e) How to efficiently estimate the identifiable parameters?

5. For the same missing data problem given in the previous problem, assume that the missing data probability satisfies the no self-censoring assumption, i.e.,

$$\begin{aligned} R_y \perp\!\!\!\perp Y &\mid R_x, X, \\ R_x \perp\!\!\!\perp X &\mid R_y, Y. \end{aligned}$$

(a) Let $(R_y = 1, R_x = 1, y_0 = 0, x_0 = 0)$ be the reference point, show that

$$\begin{aligned} P(R_y = r_y, R_x = r_x, Y = y, X = x) &\propto \eta\{r_y; (r_x, x)\}\eta\{y; (r_x, x)\} \\ p(R_y = r_y \mid R_x = 1, x_0)p(y \mid R_x = 1, x_0)&p(R_x = r_x \mid R_y = 1, y_0)p(x \mid R_y = 1, y_0) \end{aligned}$$

(b) How does this missing data mechanism differ from the missing at random missing data mechanism in this example?

(c) For the log-linear odds ratio model, find an estimator for θ.

(d) How to efficiently estimate the identifiable parameters?

9

Other Applications

9.1 Compatibility of conditionally specified models

9.1.1 The compatibility problem

Modeling multivariate data can be challenging when the data are not normally distributed. In practice, fully conditional models are often used instead for their convenience. For example, the chained equations (a.k.a. the fully conditional models) are used for imputing missing values. The fully conditionally models, though easy to specify, may not be compatible with each other, i.e., there may not exist a joint model from which all the conditionally specified models can be derived. Incompatible models may have unfavorable consequence on the imputation (Hobert and Casella, 1998; Liu et al, 2014)[76, 112].

Consider more specifically the models specified under the chained equation approach for imputation. Suppose the data have p variables $(Y_1, \cdots, Y_p)$. As is almost always the case, the variables can be a mix of discrete, counts, and continuous variables. The generalized linear model is often the choice in the conditional model specification. That is, the discrete variable conditional on other variables is modeled by the logit regression model, the count variable conditional on other variables is modeled by Poisson regression, and the continuous variable conditional on other variables is modeled by the linear regression model. The question is whether these conditional models can be derived from a single joint model. Such a joint model is assumed to generate the observed data.

To see the issue more clearly, consider a simple example where we have two variables (Y_1, Y_2) in the data. Suppose that one (Y_1) is a continuous variable and the other (Y_2) is a binary variable. One may model Y_1 conditional on Y_2 by a normal linear regression model, i.e.,

$$Y_1 = \alpha_0 + \alpha_1 Y_2 + \epsilon, \tag{9.1}$$

where $\epsilon \sim N(0, \sigma^2)$, and model Y_2 given Y_1 as a logistic regression model, i.e.,

$$\log \frac{P(Y_2 = 1 \mid Y_1)}{P(Y_2 = 0 \mid Y_1)} = \beta_0 + \beta_1 Y_1. \tag{9.2}$$

The question is whether there exists a joint distribution for (Y_1, Y_2) from which both the normal linear regression model and the logistic regression model can be derived. This simple example can be expanded by adding a new Poisson variable Y_3. The model may then become

$$Y_1 = \alpha_0 + \alpha_1 Y_2 + \alpha_2 Y_3 + \epsilon \tag{9.3}$$

for Y_1 conditional on (Y_2, Y_3), where $\epsilon \sim N(0, \sigma^2)$. For Y_2 given (Y_1, Y_3), the model is

$$\log \frac{P(Y_2 = 1 \mid Y_1, Y_3)}{P(Y_2 = 0 \mid Y_1, Y_3)} = \beta_0 + \beta_1 Y_1 + \beta_2 Y_3. \tag{9.4}$$

For Y_3 given (Y_1, Y_2), we model

$$P(Y_3 = k \mid Y_1, Y_2) = \frac{\lambda^k(Y_1, Y_2)}{k!} \exp\{-\lambda(Y_1, Y_2)\}, \tag{9.5}$$

DOI: 10.1201/9781351049757-9

where

$$\log \lambda(Y_1, Y_2) = \gamma_0 + \gamma_1 Y_1 + \gamma_2 Y_2. \tag{9.6}$$

The question remains if there exists a joint distribution for (Y_1, Y_2, Y_3) from which all the conditional models can be derived.

To add a layer of complication, we may include an interaction term in these models. That is,

$$Y_1 = \alpha_0 + \alpha_1 Y_2 + \alpha_2 Y_3 + \alpha_{12} Y_2 Y_3 + \epsilon, \tag{9.7}$$

and

$$\log \frac{P(Y_2 = 1 \mid Y_1, Y_3)}{P(Y_2 = 0 \mid Y_1, Y_3)} = \beta_0 + \beta_1 Y_1 + \beta_2 Y_3 + \beta_{12} Y_1 Y_3. \tag{9.8}$$

We may also have

$$\log \lambda(Y_1, Y_2) = \gamma_0 + \gamma_1 Y_1 + \gamma_2 Y_2 + \gamma_{12} Y_1 Y_2. \tag{9.9}$$

The conditional models may also be specified as a combination of these models: some with interaction and others without. Checking whether the conditional models are compatible remains a question of interest. Furthermore, we may want to find out the joint model if they are compatible. If the conditional models are not compatible, the question may become how to modify the conditionally specified models such that compatibility may be achieved.

9.1.2 Compatibility of conditional densities

Most of the conditional models can be expressed as a family of conditional densities. Compatibility of conditionally specified models reduces to compatibility of conditional densities. Conditional densities can appear in many forms. For simplicity, we consider the fully conditional densities in this section. Specifically, given a set of conditional densities $g_j(y_j|y_{-j})$, $j = 1, \cdots, t$, we want to know if a joint density exists such that all the given conditional densities can be derived from the joint density.

Let the reference point be $(y_{10}, \cdots, y_{p0})$ and

$$\eta_j(y_j; y_{-j}) = \frac{g_j(y_j|y_{-j}) g_j(y_{j0}|y_{-j0})}{g_j(y_{j0}|y_{-j}) g_j(y_j|y_{-j0})}.$$

The odds ratio representation of $g_j(y_j|y_{-j})$ is

$$g_j(y_j|y_{-j}) = \frac{\eta_j(y_j; y_{-j}) g_j(y_j|y_{-j0})}{\int \eta_j(y_j; y_{-j}) g_j(y_j|y_{-j0}) dy_j}.$$

Let σ be a permutation of the indices $\{1, \cdots, p\}$. We call the following equality,

$$\begin{aligned}
&\prod_{l=2}^{p} \eta_l(y_l; \{y_{p0}, \cdots, y_{(l+1)0}, y_{l-1}, \cdots, y_1\}) \\
&\quad = \prod_{l=2}^{p} \eta_{\sigma(l)}(y_{\sigma(l)}; \{y_{\sigma(p)0}, \cdots, y_{\sigma(l+1)0}, y_{\sigma(l-1)}, \cdots, y_{\sigma(1)}\}),
\end{aligned} \tag{9.10}$$

exchangeable condition with respect to σ. The following condition

$$\int \cdots \int \prod_{j=2}^{p} \eta_j(y_j; \{y_{p0}, \cdots, y_{(j+1)0}, y_{j-1}, \cdots, y_1\}) \prod_{j=1}^{p} \{g_j(y_j|y_{-j0}) dy_j\} < +\infty. \tag{9.11}$$

is called integrable condition. These two conditions are key to checking compatibility of the fully conditional densities.

Theorem 9.1. For a given set of conditional densities $g_j(y_j|y_{-j})$, there exists a joint density $g(y_1, \cdots, y_p)$ with conditional densities being g_j, $j = 1, \cdots, p$, if and only if both the exchangeable conditions for all permutations and the integrable condition hold. Furthermore, the exchangeable conditions for all permutations can be reduced to the exchangeable conditions for those permutation σ_j such that $\sigma_j(\{p, \cdots, 1\}) = \{j, p, \cdots, j+1, j-1, \cdots, 1\}$, $j = 1, \cdots, p-1$.

Proof: First, if $g_j(y_j|y_{-j})$, $j = 1, \cdots, p$ are compatible, there exists a joint density generating all the conditional densities. Let $g(y_1, \cdots, y_p)$ be the joint density. For any given permutation σ of the indices $\{1, \cdots, p\}$, the odds ratio representation of $g(y_1, \cdots, y_p)$ in the order of variables $(y_{\sigma(1)}, \cdots, y_{\sigma(p)})$ is

$$g(y_1, \cdots, y_p) = \frac{\prod_{j=2}^{p} \eta_{\sigma(j)}(y_{\sigma(j)}, \{y_{\sigma(p)0}, \cdots, y_{\sigma(j+1)0}, y_{\sigma(j-1)}, \cdots, y_{\sigma(1)}\}) \prod_{j=1}^{p} g_j(y_j|y_{-j0})}{\int \cdots \int \prod_{l=2}^{p} \eta_{\sigma(l)}(y_{\sigma(l)}, \{y_{\sigma(p)0}, \cdots, y_{\sigma(l+1)0}, y_{\sigma(l-1)}, \cdots, y_{\sigma(1)}\}) \prod_{l=1}^{p}\{g_l(y_l|y_{-l0})dy_l\}}, \tag{9.12}$$

The integrable condition is true because the density is integrable. Let $C(\sigma)$ denotes the denominator of the right-hand side of (9.12). When $y_j = y_{j0}$ for all j in the numerator, it follows that

$$g(y_{10}, \cdots, y_{p0}) = \frac{\prod_{j=1}^{p} g_j(y_{j0}|y_{-j0})}{C(\sigma)},$$

because $\eta_j(y_{j0}; \{y_{p0}, \cdots, y_{(j+1)0}, y_{(j-1)0}, \cdots, y_{10}\}) = 1$. From the equality of any two representations for $g(y_1, \cdots, y_p)$ that $C(\sigma)$ is a constant independent of σ. For any σ such that $\sigma(1, \cdots, p) \neq (1, \cdots, p)$, by comparing its odds ratio representation with the odds ratio representation with $\sigma(1, \cdots, p) = (1, \cdots, p)$, the denominator are the same and thus can be canceled. Furthermore, $\prod_{j=1}^{p} g_j(y_j|y_{-j0})$ in the numerator of the representations can be canceled. It follows that the exchangeable condition holds for all permutations.

To show the reverse is true, assume that (9.10) is true for σ_j, $j = 1, \cdots, p-1$. It follows from the integrable condition that the right-hand side of (9.12) is a density function. Denote the density function by g_{σ_j} for $j = 1, \cdots, p$. It follows from (9.10) that these density functions are all equal. Denote the common density by $g(y_1, \cdots, y_p)$. Since the conditional density for $Y_{\sigma_j(p)} = Y_j$ given the other variables computed from the joint density g_{σ_j} is

$$\begin{aligned}
\frac{g(y_1, \cdots, y_p)}{\int g(y_1, \cdots, y_p)dy_j} &= \frac{g_{\sigma_j}(y_1, \cdots, y_p)}{\int g_{\sigma_j}(y_1, \cdots, y_p)dy_{\sigma_j(p)}} \\
&= \frac{\eta_j(y_j; y_{-j})g_j(y_j|y_{-j0})}{\int \eta_j(y_j; y_{-j})g_j(y_j|y_{-j0})dy_j} \\
&= g_j(y_j|y_{-j}).
\end{aligned}$$

The second equality is true because $\sigma_j(p) = j$ and all the other odds ratio functions and conditional densities at a fixed condition are canceled out from the numerator and denominator. Hence $g_j(y_j|y_{-j})$ corresponds to the conditional densities from the single density $g(y_1, \cdots, y_p)$. This means that the $\{g_j(y_j|y_{-j}), j = 1, \cdots, p\}$ are compatible.

Theorem 9.1 is adapted from Chen (2010)[29]. Compared to the conditions in Theorems 1 and 2 of Hobert and Casella (1998)[76], Theorem 9.1 has two improvements. First, $p-1$ rather than $p! - 1$ equations need to be checked to see if the conditional densities are compatible. The reduction is substantial even for $p = 3$. Second, the conditions in terms of odd ratio functions are simpler because redundancy in terms of the conditional densities at a fixed condition in the conditional density expressions is canceled. Compared with theorem

4 of Wang and Ip (2008)[195], conditions for Theorem 9.1 to hold is also simpler because we need to check $p-1$ equations based on odds ratio functions rather than $p(p-1)/2$ equations based on the conditional densities.

When $p = 2$, condition (9.10) in Theorem 9.1 reduces to verifying if $\eta_1(y_1; y_2) = \eta_2(y_2; y_1)$ holds. When $p = 3$, the exchangeable conditions become

$$\begin{aligned} \eta_3\{y_3; (y_2, y_1)\}\eta_2(y_2; y_1|y_{30}) &= \eta_2\{y_2; (y_1, y_3)\}\eta_3(y_3; y_1|y_{20}) \\ &= \eta_1\{y_1; (y_2, y_3)\}\eta_3(y_3; y_2|y_{10}), \end{aligned}$$

or equivalently

$$\begin{aligned} \eta_3\{y_3; (y_2, y_1)\}\eta_2\{y_2; (y_{30}, y_1)\} &= \eta_2\{y_2; (y_1, y_3)\}\eta_3\{y_3; (y_1, y_{20})\} \\ &= \eta_1\{y_1; (y_3, y_2)\}\eta_3\{y_3; (y_2, y_{10})\}. \end{aligned}$$

Note that, to find incompatibility in the conditional models, we only need to show that one of the equalities does not hold or the integral condition does not converge.

To apply the result of Theorem 9.1 to the simple example defined by (9.1) and (9.2), note that the odds ratio representation for the density Y_1 given Y_2 for model (9.1) is

$$p_1(y_1 \mid y_2) = \frac{\eta_1(y_1; y_2)p_1(y_1 \mid y_{20})}{\int \eta_1(y_1; y_2)p_1(y_1 \mid y_{20})dy_1}$$

where

$$\begin{aligned} \eta_1(y_1; y_2) &= \exp\left\{\frac{\alpha_1}{\sigma^2}(y_1 - y_{10})(y_2 - y_{20})\right\}, \\ p_1(y_1 \mid y_{20}) &= \frac{1}{\sqrt{2\pi}\sigma}\exp\left\{-\frac{1}{2\sigma^2}(y_1 - \alpha_0 - \alpha_1 y_{20})^2\right\}. \end{aligned}$$

The odds ratio representation for the density Y_2 given Y_1 for model (9.2) is

$$p_2(y_2 \mid y_1) = \frac{\eta_2(y_1; y_2)p_2(y_2 \mid y_{10})}{\int \eta_2(y_1; y_2)p_2(y_2 \mid y_{10})dy_2}$$

where

$$\begin{aligned} \eta_2(y_2; y_1) &= \exp\{\beta_1(y_1 - y_{10})(y_2 - y_{20})\}, \\ p_2(y_2 \mid y_{10}) &= \frac{\exp(\beta_0 + \beta_1 y_{10})}{1 + \exp(\beta_0 + \beta_1 y_{10})}. \end{aligned}$$

To check the exchangeable condition, we note that

$$\eta_1(y_1; y_2) = \eta_2(y_2; y_1),$$

when

$$\beta_1 = \frac{\alpha_1}{\sigma^2}.$$

The integrable condition

$$\sum_{y_2=1}^{2} \int \exp\{\beta_1(y_1 - y_{10})(y_2 - y_{20})\}p_1(y_1 \mid y_{20})dy_1 p(y_2 \mid y_{10}) < \infty,$$

is always satisfied. These mean that the two densities are compatible when $\beta_1 = \alpha/\sigma^2$, and the joint density is

$$g(y_1, y_2) = \frac{\exp\{\beta_1(y_1 - y_{10})(y_2 - y_{20})\}p_1(y_1 \mid y_{20})p(y_2 \mid y_{10})}{\sum_{y_2=1}^{2} \int \exp\{\beta_1(y_1 - y_{10})(y_2 - y_{20})\}p_1(y_1 \mid y_{20})dy_1 p(y_2 \mid y_{10})}.$$

To check the compatible of densities defined by (9.3), (9.4), and 9.5 and 9.6, we again use the odds ratio representations. The odds ratio representation for the density Y_1 given Y_2, Y_3 for model (9.3) is

$$p_1(y_1 \mid y_2, y_3) = \frac{\eta_1\{y_1; (y_2, y_3)\} p_1(y_1 \mid y_{20}, y_{30})}{\int \eta_1\{y_1; (y_2, y_3)\} p_1(y_1 \mid y_{20}, y_{30}) dy_1}$$

where

$$\begin{aligned}
\eta_1\{y_1; (y_2, y_3)\} &= \exp\left\{\frac{\alpha_1}{\sigma^2}(y_1 - y_{10})(y_2 - y_{20}) + \frac{\alpha_2}{\sigma^2}(y_1 - y_{10})(y_3 - y_{30})\right\}, \\
p_1(y_1 \mid y_{20}, y_{30}) &= \frac{1}{\sqrt{2\pi}\sigma} \exp\left\{-\frac{1}{2\sigma^2}(y_1 - \alpha_0 - \alpha_1 y_{20} - \alpha_2 y_{30})^2\right\}.
\end{aligned}$$

The odds ratio representation for the density Y_2 given Y_1, Y_3 for model (9.4) is

$$p_2(y_2 \mid y_1, y_3) = \frac{\eta_2\{y_2; (y_1, y_3)\} p_2(y_2 \mid y_{10}, y_{30})}{\int \eta_2\{y_2; (y_1, y_3)\} p_2(y_2 \mid y_{10}, y_{30}) dy_2}$$

where

$$\begin{aligned}
\eta_2\{y_2; (y_1, y_3)\} &= \exp\{\beta_1(y_1 - y_{10})(y_2 - y_{20}) + \beta_2(y_3 - y_{30})(y_2 - y_{20})\}, \\
p_2(y_2 \mid y_{10}, y_{30}) &= \frac{\exp(\beta_0 + \beta_1 y_{10} + \beta_2 y_{30})}{1 + \exp(\beta_0 + \beta_1 y_{10} + \beta_2 y_{30})}.
\end{aligned}$$

The odds ratio representation for the density Y_3 given Y_1, Y_2 for model 9.5 and 9.6 is

$$p_3(y_2 \mid y_1, y_2) = \frac{\eta_3\{y_3; (y_1, y_2)\} p_3(y_3 \mid y_{10}, y_{20})}{\int \eta_3\{y_3; (y_1, y_2)\} p_3(y_3 \mid y_{10}, y_{20}) dy_3}$$

where

$$\begin{aligned}
\eta_3\{y_3; (y_1, y_2)\} &= \exp\{\gamma_1(y_3 - y_{30})(y_1 - y_{10}) + \gamma_2(y_3 - y_{30})(y_2 - y_{20})\}, \\
p_3(y_3 \mid y_{10}, y_{20}) &= \frac{\lambda^{y_3}(y_{10}, y_{20})}{y_3!} \exp\{-\lambda(y_{10}, y_{20})\}.
\end{aligned}$$

The exchangeable conditions are satisfied when

$$\frac{\alpha_1}{\sigma^2} = \beta_1, \quad \frac{\alpha_2}{\sigma^2} = \gamma_1, \quad \beta_2 = \gamma_2.$$

The integrable condition is always satisfied and the joint density can be represented as

$$g(y_1, y_2, y_3) = \frac{\eta(y_1; y_2; y_3) p_1(y_1 \mid y_{20}, y_{30}) p_2(y_2 \mid y_{10}, y_{30}) p_3(y_3 \mid y_{10}, y_{20})}{\sum_{y_2=0}^{1} \sum_{y_3=0}^{\infty} \int \eta(y_1; y_2; y_3) p_1(y_1 \mid y_{20}, y_{30}) p_2(y_2 \mid y_{10}, y_{30}) p_3(y_3 \mid y_{10}, y_{20}) dy_1},$$

where

$$\eta(y_1; y_2; y_3) = \exp\{\beta_1(y_1 - y_{10})(y_2 - y_{20}) + \gamma_1(y_3 - y_{30})(y_1 - y_{10}) + \gamma_2(y_3 - y_{30})(y_2 - y_{20})\}.$$

When interactions are introduced into the conditional models, the odds ratio functions for the three models respectively become

$$\begin{aligned}
\log \eta_1\{y_1; (y_2, y_3)\} &= \frac{\alpha_1}{\sigma^2}(y_1 - y_{10})(y_2 - y_{20}) + \frac{\alpha_2}{\sigma^2}(y_1 - y_{10})(y_3 - y_{30}) \\
&\quad + \frac{\alpha_{12}}{\sigma^2}(y_1 - y_{10})(y_2 y_3 - y_{20} y_{30}), \\
\log \eta_2\{y_2; (y_1, y_3)\} &= \beta_1(y_1 - y_{10})(y_2 - y_{20}) + \beta_2(y_3 - y_{30})(y_2 - y_{20}) \\
&\quad + \beta_{12}(y_2 - y_{20})(y_1 y_3 - y_{10} y_{30}), \\
\log \eta_3\{y_3; (y_1, y_2)\} &= \gamma_1(y_3 - y_{30})(y_1 - y_{10}) + \gamma_2(y_3 - y_{30})(y_2 - y_{20}) \\
&\quad + \gamma_{12}(y_3 - y_{30})(y_1 y_2 - y_{10} y_{20}).
\end{aligned}$$

For the exchangeable conditions to hold, we need to have

$$\frac{\alpha_{12}}{\sigma^2} = \beta_{12} = \gamma_{12},$$
$$\frac{\alpha_1}{\sigma^2} = \beta_1, \quad \frac{\alpha_2}{\sigma^2} = \gamma_1, \quad \beta_2 = \gamma_2.$$

The integrable condition is always true. Under these conditions, the joint density has odds ratio function

$$\begin{aligned}\log \eta(y_1; y_2; y_3) &= (\beta_1 + \beta_{12}y_{30})(y_1 - y_{10})(y_2 - y_{20}) + (\gamma_2 + \gamma_{12}y_{10})(y_2 - y_{20})(y_3 - y_{30}) \\ &\quad + (\gamma_1 + \gamma_{12}y_{20})(y_1 - y_{10})(y_3 - y_{30}) + \gamma_{12}(y_1 - y_{10})(y_2 - y_{20})(y_3 - y_{30})\end{aligned}$$

If one of the model does not have the interaction and the corresponding interaction is not 0 in other models, then the models are not compatible.

9.1.3 Modifications to incompatible conditionally specified models based on the joint model construction

The conditionally specified models considered in the previous section are compatible when parameters are appropriately restricted. The constructed joint model can be very useful in reparametrizing the conditional models. In the case that the conditionally specified models are incompatible, the construction approach using different variable orders may yield different joint models, each of which is partially compatible with the conditionally specified models. Those constructed joint models can also be useful in practice for resolving model conflicts. The following conditionally specified normal models are used to illustrate the idea for resolving model conflicts in conditionally specified models.

Let $Y_1, \cdots, Y_p$ be a set of variables to be modeled by the conditionally modeling approach. For the simplicity of presentation, assume that $Y_1, \cdots, Y_p$ are all continuous variables. Other cases with all discrete variables, or a mix of discrete and continuous variables can be handled in the same way. The fully conditional models are specified as

$$Y_j = \beta_{j0} + \sum_{k \neq j} \beta_{jk} Y_k + \epsilon_j,$$

where $\epsilon_j \sim N(0, \sigma_j)$, $j = 1, \cdots, p$. The conditional densities under these models are respectively

$$g_j(y_j | y_{-j}) = \frac{1}{\sqrt{2\pi}\sigma_j} \exp\left\{ -\frac{1}{2\sigma_j^2} \left(y_j - \beta_{j0} - \sum_{k \neq j} \beta_{jk} y_k \right)^2 \right\},$$

for $j = 1, \cdots, p$. The odds ratio function and the conditional density at a fixed condition for the jth model are respectively

$$\eta_j(y_j; y_{-j}) = \exp\left\{ \sum_{k \neq j} \frac{\beta_{jk}}{\sigma_j^2} (y_j - y_{j0})(y_k - y_{k0}) \right\},$$

and $g_j(y_j | y_{-j0})$. A joint model obtained based on Theorem 9.1 should have a density proportional to

$$\exp\left\{ \sum_{j=2}^{p} \sum_{k=1}^{j-1} \frac{\beta_{jk}}{\sigma_j^2} (y_j - y_{j0})(y_k - y_{k0}) \right\} \prod_{j=1}^{p} \frac{1}{\sqrt{2\pi}\sigma_j} \exp\left\{ -\frac{1}{2\sigma_j^2} \left(y_j - \beta_{j0} - \sum_{k \neq j} \beta_{jk} y_{k0} \right)^2 \right\}.$$

To ensure the function defines a joint density, we need also the integral of the function over the sample space is finite. This entails the precision matrix

$$\Omega = \begin{pmatrix} \frac{1}{\sigma_1^2} & -\frac{\beta_{21}}{\sigma_2^2} & \cdots & -\frac{\beta_{p1}}{\sigma_p^2} \\ -\frac{\beta_{21}}{\sigma_2^2} & \frac{1}{\sigma_2^2} & \cdots & -\frac{\beta_{p2}}{\sigma_p^2} \\ \vdots & \vdots & \cdots & \vdots \\ -\frac{\beta_{p1}}{\sigma_p^2} & -\frac{\beta_{p2}}{\sigma_p^2} & \cdots & \frac{1}{\sigma_p^2} \end{pmatrix}$$

be positive definite. See §2.7 for the derivation. It can be derived from the joint density that the conditional density for Y_j given Y_{-j} is

$$g_j^*(y_j|y_{-j}) =$$

$$\frac{1}{\sqrt{2\pi}\sigma_j}\exp\left\{-\frac{1}{2\sigma_j^2}\left(y_j-\beta_{j0}-\sum_{k=1}^{j-1}\beta_{jk}y_k-\sum_{k=j+1}^{t}\beta_{jk}y_{k0}-\sum_{k=j+1}^{p}\beta_{kj}\frac{\sigma_j^2}{\sigma_k^2}(y_k-y_{k0})\right)^2\right\}.$$

If the conditionally specified models are compatible, g_j^* and g_j must define the same model with possibly different parametrization. This implies that the conditionally specified models are compatible if

$$\beta_{kj}/\sigma_k^2 = \beta_{jk}/\sigma_j^2$$

for all $k \neq j$. In this case, the joint models determined by different variable orders become identical. In practice, if compatible models are fit to the data and no additional restriction is imposed on the parameters, the consistent estimator of the parameters based on the conditionally specified models automatically satisfies the constraints. However, taking the constraints into consideration in estimation will increase the estimation efficiency.

Conditionally specified normal models may also include interaction and/or second order terms. This seemly harmless and natural addition to the conditional models, however, can instantly destroy compatibility of the conditionally specified models. The question now becomes how to amend the conditionally specified models to make them compatible. To illustrate the idea, suppose that the conditionally specified normal model includes one high order term in the model for Y_1 given $Y_2, \cdots, Y_t$. The other conditionally specified models remain unchanged. That is,

$$Y_1 = \beta_{10} + \sum_{k=2}^{t}\beta_{1k}Y_k + \alpha_{12}Y_2^2 + \epsilon_1.$$

It is easy to see that the odds ratio function obtained from this model is

$$\log\eta_1\{y_1;(y_2,\cdots,y_t)\} = \sum_{k=2}^{t}\frac{\beta_{1k}}{\sigma_1^2}(y_1-y_{10})(y_k-y_{k0}) + \frac{\alpha_{12}}{\sigma_1^2}(y_1-y_{10})(y_2^2-y_{20}^2),$$

and the odds ratio functions from the other conditional models are

$$\log\eta_j\{y_j;(y_l, l\neq j)\} = \sum_{k\neq j}\frac{\beta_{jk}}{\sigma_j^2}(y_j-y_{j0})(y_k-y_{k0}),$$

for $j = 2, \cdots, t$. All the conditional densities at a fixed condition that are obtained from the odds ratio decomposition of the conditional densities are normal densities.

The foregoing conditionally specified models are incompatible because the conditional odds ratios functions of Y_1 and Y_2 given the rest of variables from g_1 and g_2 respectively are not equal when $\alpha_{12} \neq 0$. It is hard to modify the conditionally specified models directly in an attempt to correct the conflicts in those models. By applying Theorem 1 to this problem, we can construct many joint models which are partially compatible with the conditionally specified models. Note that if the second order terms in the model Y_1 given Y_{-1} is very important to keep, we can select Y_1 before Y_2 in the order of the odds ratio function combination to keep this high order association term there. To resolve the incompatibility, we can redefine $\log \eta_2$ as

$$\log \eta_2^*(y_2; y_{-2}) = \sum_{k \neq 2} \frac{\beta_{2k}}{\sigma_2^2}(y_2 - y_{20})(y_k - y_{k0}) + \frac{\alpha_{12}}{\sigma_1^2}(y_1 - y_{10})(y_2^2 - y_{20}^2),$$

In addition, we need to modify $g_2(y_2|y_{-20})$ to make it different from the normal density so that the integral in the denominator of the odds ratio representation of the joint density is finite. This can be done easily by modify $g_2(y_2|y_{-20})$ to

$$g^*(y_2|y_{-20}) \propto \exp(-y_2^4/\sigma^2).$$

As a result, $g(y_2|y_{-2})$ is modified to

$$g^*(y_2|y_{-2}) = \frac{\eta_2^*\{Y_2; Y_{-2})g^*(y_2|y_{-20})}{\int \eta_2^*\{Y_2; Y_{-2})g^*(y_2|y_{-20})dy_2}.$$

The modified conditional models can now be seen compatible because each of the modified conditional models can be obtained as the conditional models of the joint model $p^*(y_1, \cdots, y_t)$, which is

$$\frac{\prod_{l=1, l\neq 2}^{t-1} \eta_l(y_l; \{y_t, \cdots, y_{(l+1)}, y_{(l-1)0}, \cdots, y_{10}\}) \times \eta_2^*(y_2; \{y_t, \cdots, y_3, y_{10}\}) \prod_{l \neq 2} g_l(y_l|y_{-l}^0) g_2^*(y_2|y_{-20})}{\int \cdots \int \prod_{l=1, l\neq 2}^{t-1} \eta_l(y_l, \{y_t, \cdots, y_{(l+1)}, y_{(l-1)0}, \cdots, y_{10}\}) \eta_2^*(y_2; \{y_t, \cdots, y_3, y_{10}\}) \times \prod_{l \neq 2} g_l(y_l|y_{-l}^0) g_2^*(y_2|y_{-20}) \prod_{l=1}^{t-1} dy_l},$$

when $\beta_{kj}/\sigma_k^2 = \beta_{jk}/\sigma_j^2$ for all $k \neq j$. Note also that different orders of the odds ratio combination in the foregoing expression do not alter the joint model. Rather, they determine different parameterizations of the joint model. The joint model can also be expressed as

$$p^*(y_1, \cdots, y_t) = \frac{\prod_{l=1}^{t-1} \eta_l(y_l, \{y_t, \cdots, y_{(l+1)}, y_{(l-1)0}, \cdots, y_{10}\}) \prod_{l \neq 2} g_l(y_l|y_{-l}^0) g_2^*(y_2|y_{-20})}{\int \cdots \int \prod_{l=1}^{t-1} \eta_l(y_l, \{y_t, \cdots, y_{(l+1)}, y_{(l-1)0}, \cdots, y_{10}\}) \prod_{l \neq 2} g_l(y_l|y_{-l}^0) g_2^*(y_2|y_{-20}) dy_l\}}.$$

However, the order of the odds ratio combination in this expression cannot be changed arbitrarily without changing the joint model.

It can be seen that p^* is a joint model that is compatible with all of the conditionally specified models except $g_2(y_2|y_{-2})$. Note that we have enforced the change of $g_2(y_2|y_{-20})$ here. It is also possible to allow observed data to determine what distribution to use. It can also be seen from the above arguments that when the conditionally specified models are close to compatible, the modification of the originally conditionally specified models can be kept small in making the conditional models compatible. As a result, most features of the originally conditionally specified models are retained. On the other hand, if the conditionally specified models are in a high degree of conflicts, many modifications to the conditional models may be needed to make the models compatible. Even if some features in the conditionally specified models are eliminated in the modifications, however, Theorem 1 gives us some flexibility in determining which features to keep. Theorem 1 also suggests other modifications of the example. For example, if the order of Y_1 and Y_2 in the odds ratio combination of the last displayed expression is switched, the constructed joint model would be the joint normal model with the second order term removed. This construction may be unfavorable given the importance of the second order term in the conditional model.

The foregoing example provides us with a general strategy to modify conditionally specified models to achieve compatibility and to obtain the joint models. In summary, when the conditionally specified models are incompatible, the following guidelines may be applied to the modification of the incompatible conditional densities to achieve compatibility.

1. We first modify the odds ratio functions of the conditionally specified models. In doing so, we may choose the important features in the models and try to maximally keep them in the joint model.
2. To make the integral in the denominator of the potential joint density finite, we may need to modify the conditional density at a fixed condition also.

From a practical point of view, we may modify the conditional densities at a fixed condition to be unspecified. Such a modification can generate many interesting models to fit the data. The interpretation of those models can also be attractive in practice.

9.1.4 Summary and literature

Model compatibility was discussed in the context of Markov random field (Besag, 1974; Besag and Kooperberg, 1995; Tjelmeland and Besag, 1998; Arnold et al, 2001; Besag et al, 2001)[7, 10, 183, 2, 9]. This problem becomes more acute in the Gibbs sampler (Arnold and Press, 1989; Joe, 1997; Hobert and Casella, 1998; Arnold et al, 2001; van Dyk and Park, 2008)[3, 84, 76, 2, 189] and the missing data imputation (Van Buuren et al, 2006; Liu et al, 2014)[188, 112]. An effective solution to this problem can be found under the positivity condition(Hobert and Casella, 1998; Wang and Ip, 2008; Chen, 2010)[76, 195, 29]. For distributions that do not satisfy the positivity condition, compatible conditions can be much more complicated(Moussouris, 1974; Berti et al, 2014; Kuo and Wang, 2018)[127, 6, 90].

9.2 Semiparametric estimation of multivariate density

9.2.1 The semiparametric odds ratio model for density estimation

For high-dimensional data, nonparametric estimation of the density is virtually impossible due to the curse of dimensionality. Parametric estimation can be too restrictive and may not have the flexibility to capture the complex shapes of a high-dimensional density function. One compromised approach is to consider semiparametric modeling approach. One problem with the semiparametric approach is the lack of a modeling framework that provides adequate flexible shapes and is also relatively easy to handle in practice. The semiparametric odds ratio modeling framework is mostly fit for the needs.

Let $(Y_1, \cdots, Y_G)$ be the multivariate random variables whose joint density is to be estimated. Assume the positivity condition, the joint density can be expressed as

$$f(y_1, \cdots, y_G) = \frac{\eta(y_1; \cdots; y_G) \prod_{j=1}^{G} f(y_j \mid y_{-j0})}{\int \cdots \int \eta(y_1; \cdots; y_G) \prod_{j=1}^{G} \{f(y_j \mid y_{-j0}) dy_j\}},$$

where the integral with respect to y_j is interpreted as summation over all possible categories if Y_j is a categorical variable. In the semiparametric density estimation using the semiparametric odds ratio model, we assume $\eta(y_1; \cdots, y_G)$ is parametrically model by $\eta(y_1; \cdots, y_G \mid \theta)$. An alternative semiparametric odds ratio model for density estimation is the consecutive conditional model where

$$f(y_1, \cdots, y_G) = \prod_{j=1}^{G} \frac{\eta\{y_j; (y_{j-1}, \cdots, y_1)\} f(y_j \mid y_{(j-1)0}, \cdots, y_{10})}{\int \eta\{y_j; (y_{j-1}, \cdots, y_1)\} f(y_j \mid y_{(j-1)0}, \cdots, y_{10}) dy_j},$$

where $\eta\{y_j; (y_{j-1}, \cdots, y_1)$ is parametrically modeled by $\eta\{y_j; (y_{j-1}, \cdots, y_1, \theta_j)$ for $j = 2, \cdots, G$. Comparing the two methods of modeling the multivariate density, the first method does not need to select an order of variables in the model. However, the normalizing constant may be computationally intensive to obtain. The second method requires to pre-select an order of the variables for the model. Different order may lead to different estimates. The advantage for the second method is that the normlizing constant in each conditional model is easy to compute.

The proposed density estimation is divided into two stages. In the first stage, the semiparametric odds ratio model is fit to the observed data. In this stage, the baseline conditional distributions are all discretized distributions. One of the maximum likelihood approaches, namely, the pairwise pseudo-likelihood approach, the maximum semiparametric likelihood approach, or the permutation likelihood approach, is used to obtain estimates for θ. For the joint likelihood model, fully conditional models need to be used in the pairwise pseudo-likelihood approach. This leads to redundant estimates of some of the parameter values. The final estimator will be the average of the repeated estimations. The baseline functions and the normalized constant can be estimated using a weighting approach similar to the approach used for estimating the baseline functions and normalized constant in the permutation likelihood approach. In the maximum semiparametric likelihood approach, the fully conditional models also have to be used when the dimension p is moderate to high. In these cases, the parameter estimates may be repeated for some of the parameters. Average over the repeated estimate will be used. The baseline functions can be obtained in each fully conditional model directly. The normalized constant may still need to be estimated by the weighting approach as in the permutation likelihood described later. In the permutation likelihood, θ can be obtained directly. The baseline functions and the normalizing constant are obtained by the weighting approach.

Once a consistent estimator of θ is obtained using either the pairwise pseudo-likelihood approach or the permutation likelihood approach, the baseline functions can be explicitly estimated by the weighting approach. Denote the θ estimator by $\hat{\theta}$. The baseline estimators are respectively

$$\hat{F}_j(t) = \frac{\sum_{i=1}^{n} \eta^{-1}(Y_{i1}; \cdots, Y_{iG} \mid \hat{\theta}) 1_{\{Y_{ij} \le t\}}}{\sum_{i=1}^{n} \eta^{-1}(Y_{i1}; \cdots, Y_{iG} \mid \hat{\theta})},$$

for $j = 1, \cdots, G$. The normalized constant of the joint density, i.e., the denominator in the first odds ratio model, can be estimated by

$$c = \left\{ \sum_{i=1}^{n} \eta^{-1}(Y_{i1}; \cdots, Y_{iG} \mid \hat{\theta}) \right\}^{-1}.$$

The density can thus be approximated by a probability mass function,

$$\hat{f}(y_1, \cdots, y_G) = \frac{\eta(y_1; \cdots; y_G \mid \hat{\theta}) \prod_{k=1}^G d\hat{F}_k(y_k)}{c}.$$

When the consecutive conditional model is used, the three likelihood approach can be directly obtained. In this case, the maximum semiparametric likelihood estimator of the odds ratio parameter θ and the baseline functions are preferred because of the higher estimation efficiency. In this case, the estimated probability mass function is

$$\hat{f}(y_1, \cdots, y_G) = \prod_{j=1}^{G} \frac{\eta\{Y_j; (Y_{j-1}, \cdots, Y_1) \mid \hat{\theta}_j\} d\hat{F}_j(Y_j \mid y_{(j-1)0}, \cdots, y_{10})}{\sum_{Y_j} \eta\{Y_j; (Y_{j-1}, \cdots, Y_1) \mid \hat{\theta}_j\} d\hat{F}_j(Y_j \mid y_{(j-1)0}, \cdots, y_{10})},$$

9.2.2 Smoothed estimators for the semiparametric density

The second stage of the density estimation is to smooth the baseline functions for the continuous random variables. For this purpose, one of the smoothing approaches, such as the kernel smoothing, is applied to the estimated baseline functions. Let $\hat{f}_{ik}$, $i = 1, \cdots, n_k$ be the estimated probability mass for the distribution function F_k, $k = 1, \cdots, G$. The kernel smoothed estimator of the density function of F_k is

$$\hat{f}_k(y) = \frac{1}{h_{nk}} \sum_{i=1}^{n_k} Q\left(\frac{y - Y_{ik}}{h_{nk}}\right) \hat{f}_{ik},$$

where h_{nk} is the bandwidth and Q is the kernel function satisfying $\int y^2 |Q(y)| dy < +\infty$, and

$$\int Q(y) dy = 1, \int y Q(y) dy = 0, \int y^2 Q(y) dy = v.$$

It follows from the restrictions on the kernal function that

$$\begin{aligned} \int \hat{f}_k(y) dy &= \sum_{i=1}^{n_k} \hat{f}_{ik} = 1, \\ \int y \hat{f}_k(y) dy &= \sum_{i=1}^{n_k} y_{ik} \hat{f}_{ik}, \\ \int y^2 \hat{f}_k(y) dy &= \sum_{i=1}^{n_k} y_{ik}^2 \hat{f}_{ik} + h_{nk}^2 v. \end{aligned}$$

Assume that S is the set of indices of variables that are continuous and S^c is the set of discrete variable indices. When the joint semiparametric odds ratio model is taken, the kernel smoothed estimator of the joint density is

$$\hat{f}_s(y_1, \cdots, y_G) = \frac{\eta(y_1; \cdots; y_G \mid \hat{\theta}) \prod_{k \in S} \hat{f}_k(y_k) \prod_{k \in S^c} d\hat{F}_k(y_k)}{\sum \cdots \sum \int \cdots \int \eta(y_1; \cdots; y_G \mid \hat{\theta}) \prod_{k \in S} \left\{\hat{f}_k(y_k) dy_k\right\} \prod_{k \in S^c} d\hat{F}_k(y_k)},$$

where the integrals in the denominator are respectively for indices in S and the sums are respectively for indices in S^c. Note further that

$$\begin{aligned} &\sum \cdots \sum \int \cdots \int \eta(y_1; \cdots; y_G \mid \hat{\theta}) \prod_{k \in S} \left\{\hat{f}_k(y_k) dy_k\right\} \prod_{k \in S^c} d\hat{F}_k(y_k) \\ &= \sum_{i_1=1}^{n_1} \cdots \sum_{i_G=1}^{n_G} \int \cdots \int \eta(y_{i_1 1} + u_1 h_{n1}; \cdots; y_{i_G G} + h_{nG} u_G \mid \hat{\theta}) \prod_{k \in S} \{Q(u_k) du_k\} \prod_{k=1}^{G} \hat{f}_{ik}, \end{aligned}$$

where $u_k = 0$ if $k \in S^c$. Assume that η is differentiable with respect to the continuous variables up to the third order with bounded derivatives. By Taylor expansion,

$$\begin{aligned}
&\eta(y_{i_1 1} + u_1 h_{n1}; \cdots; y_{i_G G} + h_{nG} u_G \mid \hat{\theta}) \\
&= \eta(y_{i_1 1}; \cdots; y_{i_G G} \mid \hat{\theta}) + \sum_{k=1}^{G} \frac{\partial}{\partial y_k} \eta(y_{i_1 1}; \cdots; y_{i_G G} \mid \hat{\theta}) u_k h_{nk} \\
&\quad + \frac{1}{2} \sum_{k=1}^{G} \sum_{l=1}^{G} \frac{\partial^2}{\partial y_k \partial y_l} \eta(y_{i_1 1}; \cdots; y_{i_G G} \mid \hat{\theta}) u_k u_l h_{nk}^2 + o_p\left(\sum_{k=1}^{n} u_k^2 h_{nk}^2\right).
\end{aligned}$$

It follows that

$$\begin{aligned}
&\sum_{i_{j_1}} \cdots \sum_{i_{j_{|S^c|}}} \int \cdots \int \eta(y_1; \cdots; y_G \mid \hat{\theta}) \prod_{k \in S} \left\{ \hat{f}_k(y_k) dy_k \right\} \prod_{k \in S^c} d\hat{F}_k(y_{i_k k}) \\
&= \sum_{i_1=1}^{n_1} \cdots \sum_{i_G=1}^{n_G} \eta(y_{i_1 1}; \cdots; y_{i_G G} \mid \hat{\theta}) \prod_{k=1}^{G} \hat{f}_{i_k k} \\
&\quad + \frac{1}{2} \sum_{i_1=1}^{n_1} \cdots \sum_{i_G=1}^{n_G} \sum_{k \in S} v_k h_{nk}^2 \frac{\partial^2}{\partial y_k^2} \eta(y_{i_1 1}; \cdots; y_{i_G G} \mid \hat{\theta}) \prod_{k=1}^{G} \hat{f}_{i_k k} + o_p\left(\sum_{k \in S} h_{nk}^2\right) \\
&= c + \frac{1}{2} \sum_{k \in S} v_k h_{nk}^2 c_{2k} + o_p\left(\sum_{k \in S} h_{nk}^2\right),
\end{aligned}$$

where $j_1, \cdots, j_{|S^c|} \in S^c$, and

$$c_{2k} = \sum_{i_1=1}^{n_1} \cdots \sum_{i_G=1}^{n_G} \frac{\partial^2}{\partial y_k^2} \eta(y_{i_1 1}; \cdots; y_{i_G G} \mid \hat{\theta}) \prod_{k=1}^{G} \hat{f}_{i_k k}.$$

When $\max_{k \in S} h_{nk} \to 0$, the smoothed density estimator is approximately

$$\hat{f}_s(y_1, \cdots, y_G) \approx \frac{\eta(y_1; \cdots; y_G \mid \hat{\theta}) \prod_{k \in S} \hat{f}_k(y_k) \prod_{k \in S^c} d\hat{F}_k(y_k)}{c + \sum_{k \in S} v_k h_{nk}^2 c_{2k}/2},$$

When the consecutive conditional models are used for the density estimation, the smoothed density estimator appears as

$$\hat{f}_s(y_1, \cdots, y_G) = \prod_{k=1}^{G} \frac{\eta\{y_k; (y_{k-1}, \cdots, y_1) \mid \hat{\theta}_k\} \hat{f}_k(y_k)}{\int \eta\{y_k; (y_{k-1}, \cdots, y_1) \mid \hat{\theta}_k\} \hat{f}_k(y_k) dy_k},$$

where $\hat{f}_k(y_k) = d\hat{F}_k(y_k)$ in the numerator and $\hat{f}_k(y_k)dy_k = d\hat{F}_k(y_k)$ in the denominator when Y_k is a discrete variable, i.e., $k \in S^c$. For $k \in S$,

$$\begin{aligned}
&\int \eta\{y_k; (y_{k-1}, \cdots, y_1) \mid \hat{\theta}_k\} \hat{f}_k(y_k) dy_k \\
&= \int \eta\{y_k; (y_{k-1}, \cdots, y_1) \mid \hat{\theta}_k\} \frac{1}{h_{nk}} \sum_{i=1}^{n_k} Q\left(\frac{y_k - Y_{ik}}{h_{nk}}\right) \hat{f}_{ik} dy_k \\
&= \sum_{i=1}^{n_k} \eta\{Y_{ik}; (y_{k-1}, \cdots, y_1) \mid \hat{\theta}_k\} \hat{f}_{ik} \\
&\quad + \frac{v_k h_{nk}^2}{2} \sum_{i=1}^{n_k} \frac{\partial^2}{\partial y_k^2} \eta\{Y_{ik}; (y_{k-1}, \cdots, y_1) \mid \hat{\theta}_k\} \hat{f}_{ik} + o_p(h_{nk}^2).
\end{aligned}$$

9.2.3 Summary and literature

Estimation of multiple densities using density ratio models was studied in Fokianos (2004), Cheng and Chu (2004), Qin and Zhang (2005), Aubin and Leoni-Aubin (2008), Hazelton and Turlach (2010) [62, 43, 145, 4, 74]. The semiparametric approach is extended to the spectral density estimation (Fokianos and Savvides, 2008; de Carvalho and Davison, 2014)[63, 53] and the penalized likelihood estimation approach (Yang, 2009)[201]. Chen et al (2020)[41] incorporated log-concave restriction in the semiparametric density estimation. Methods for multivariate density estimation (Terrell and Scott, 1992; Loader, 1996; Part et al, 2002; Slaoui, 2018)[182, 114, 134, 167] may be used in combination with the flexible semiparametric odds ratio model for the density estimation.

9.3 Exercises

1. Let Y denote a binary outcome and (X_1, X_2) be two covariates. Let the observed data be $(Y_i, X_{i1}, X_{i2}), i = 1, \cdots, n$. In specifying conditional models, assume that

$$P(Y = 1 \mid X_1, X_2) = \frac{\exp(\beta_0 + \beta_1 X_1 + \beta_2 X_2 + \beta_{12} X_1 X_2)}{1 + \exp(\beta_0 + \beta_1 X_1 + \beta_2 X_2 + \beta_{12} X_1 X_2)}.$$

   An alternative way to model the data is to treat (X_1, X_2) given Y as a two-sample problem. Assume that the conditional distribution is a bivariate normal with mean $\mu = (\mu_1, \mu_2)^t$ and variance-covariance matrix Σ.

   (a) Under what conditions, the two conditional models are compatible?

   (b) If the logistic regression model does not have the interaction term, i.e., $\beta_{12} = 0$, are the two models compatible?

   (c) If we specify X_1 given Y and X_2, and X_2 given Y and X_1 both as normal linear regression models, under what conditions, the three models are compatible?

2. In the previous problem, if we model the binary outcome given the covariates as a probit regression model instead, i.e.,

$$P(Y = 1 \mid X_1, X_2) = \Phi\left(\beta_0 + \beta_1 X_1 + \beta_2 X_2 + \beta_{12} X_1 X_2\right),$$

   where Φ is the standard normal distribution function. Answer the three questions respectively.

3. Suppose that the parameter θ in the semiparametric odds ratio model is estimated by maximizing the permutation likelihood. Denote the estimator by $\hat{\theta}$. Instead of the weighting approach, the maximum likelihood estimators for the baseline functions are obtained. Discuss the efficiency of the likelihood approach in comparison to the weighting approach to estimating the baseline functions.

4. Let $Y = (Y_1, \cdots, Y_p)$ be the random vector with density $f(y_1, \cdots, y_p), (y_1, \cdots, y_p) \in \Omega_1$ with respect to a known measure. Let $X = (X_1, \cdots, X_q)$ be the random vector with density $g(x_1, \cdots, y_q), (x_1, \cdots, x_q) \in \Omega_2$ with respect to a known measure. Assume the following positivity condition holds, i.e., the joint density for (Y, X),

$$p(y_1, \cdots, y_p, x_1, \cdots, x_q) > 0$$

if

$$f(y_1, \cdots, y_p), (y_1, \cdots, y_p) > 0 \text{ and } g(x_1, \cdots, y_q) > 0.$$

The odds ratio representation of the joint density with respect to a reference point (y_0, x_0) has the form

$$p(y, x) = \frac{\eta\{(y_1, \cdots, y_p); (x_1, \cdots, x_q)\} p(y \mid x_0) p(x \mid y_0)}{\int\int \eta\{(y_1, \cdots, y_p); (x_1, \cdots, x_q)\} p(y \mid x_0) p(x \mid y_0) dy dx},$$

where $y = (y_1, \cdots, y_p)$ and $x = (x_1, \cdots, x_q)$. In a semiparametric estimation of the joint density, both $p(y \mid x_0)$ and $p(x \mid y_0)$ are left unspecified and $\eta(y; x)$ is parametric modeled. For the case $p = 2, q = 2$,

(a) find a kernel estimator of the semiparametric density $p(y, x)$,

(b) perform an analysis of the asymptotic bias of the estimator.

(c) What are advantages and disadvantages of this approach in comparison to the density estimation approach based on the fully semiparametric odds ratio model?

10

Theoretical Results on Estimation and Inference

10.1 Variation independence in the joint odds ratio representation of a density

In the odds ratio representation of a joint conditional density $f(y_1, y_2 \mid x)$, the models for $f(y_1|y_{20}, x)$ and $f(y_2|y_{10}, x)$ can be specified independent of each other without creating any conflicts. To see this, note that it is straightforward that either the pair $f(y_1|y_{20}, x)$ and the marginal conditional density $g(y_2|x)$ or the pair $f(y_2|y_{10}, x)$ and the marginal conditional density $h(y_1|x)$ can be specified freely without constraints, where $g(y_2|x)$ and $h(y_1|x)$ are densities of Y_2 given X and of Y_1 given X respectively. They satisfy

$$\frac{\eta(y_1, y_{10}, y_2, y_{20}|x)f(y_2|y_{10}, x)}{\int \eta(y_1, y_{10}, y_2, y_{20}|x)f(y_2|y_{10}, x)dy_2}h(y_1|x) = \frac{\eta(y_1, y_{10}, y_2, y_{20}|x)f(y_1|y_{20}, x)}{\int \eta(y_1, y_{10}, y_2, y_{20}|x)f(y_1|y_{20}, x)dy_1}g(y_2|x). \tag{10.1}$$

Suppose that $f(y_2|y_{10}, x)$ is specified, then

$$g(y_2|x) = f(y_2|y_{10}, x)\int \frac{\eta(y_1, y_{10}, y_2, y_{20}|x)}{\int \eta(y_1, y_{10}, y_2, y_{20}|x)f(y_2|y_{10}, x)dy_2}h(y_1|x)dy_1 \tag{10.2}$$

and

$$\begin{aligned}
&\frac{f(y_1|y_{20}, x)}{\int \eta(y_1, y_{10}, y_2, y_{20}|x)f(y_1|y_{20}, x)dy_1}\\
&= \frac{f(y_2|y_{10}, x)}{\int \eta(y_1, y_{10}, y_2, y_{20}|x)f(y_2|y_{10}, x)dy_2}h(y_1|x)\left\{g(y_2|x)\right\}^{-1}\\
&= \frac{h(y_1|x)}{\int \eta(y_1, y_{10}, y_2, y_{20}|x)f(y_2|y_{10}, x)dy_2}\Big/\int \frac{\eta(y_1, y_{10}, y_2, y_{20}|x)}{\int \eta(y_1, y_{10}, y_2, y_{20}|x)f(y_2|y_{10}, x)dy_2}h(y_1|x)dy_1.
\end{aligned}$$

We obtain from the last equation that

$$\begin{aligned}
h(y_1|x) &= f(y_1|y_{20}, x)\frac{\int \eta(y_1, y_{10}, y_2, y_{20}|x)f(y_2|y_{10}, x)dy_2}{\int \eta(y_1, y_{10}, y_2, y_{20}|x)f(y_1|y_{20}, x)dy_1}\\
&\quad\times\int \frac{\eta(y_1, y_{10}, y_2, y_{20}|x)}{\int \eta(y_1, y_{10}, y_2, y_{20}|x)f(y_2|y_{10}, x)dy_2}h(y_1|x)dy_1.
\end{aligned}$$

Since $\int h(y_1|x)dy_1 = 1$, it follows from the above expression that

$$\begin{aligned}
&\int\int f(y_1|y_{20}, x)\eta(y_1, y_{10}, y_2, y_{20}|x)dy_1\ f(y_2|y_{10}, x)dy_2\\
&= \frac{\int \eta(y_1, y_{10}, y_2, y_{20}|x)f(y_1|y_{20}, x)dy_1}{\int \frac{\eta(y_1, y_{10}, y_2, y_{20}|x)}{\int \eta(y_1, y_{10}, y_2, y_{20}|x)f(y_2|y_{10}, x)dy_2}h(y_1|x)dy_1}.
\end{aligned}$$

DOI: 10.1201/9781351049757-10

It then follows that

$$h(y_1|x) = \frac{f(y_1|y_{20}, x) \int \eta(y_1, y_{10}, y_2, y_{20}|x) f(y_2|y_{10}, x) dy_2}{\int \int f(y_1|y_{20}, x) \eta(y_1, y_{10}, y_2, y_{20}|x) dy \; f(y_2|y_{10}, x) dy_2}. \tag{10.3}$$

Further, (10.3) implies

$$f(y_1|y_{20}, x) = \frac{h(y_1|x) \int \int f(y_1|y_{20}, x) \eta(y_1, y_{10}, y_2, y_{20}|x) dy_1 \; f(y_2|y_{10}, x) dy_2}{\int \eta(y_1, y_{10}, y_2, y_{20}|x) f(y_2|y_{10}, x) dy_2}.$$

Since $\int f(y_1|y_{20}, x) dy_1 = 1$, it follows that

$$\begin{aligned} f(y_1|y_{20}, x) &= \frac{h(y_1|x)}{\int \eta(y_1, y_{10}, y_2, y_{20}|x) f(y_2|y_{10}, x) dy_2} \\ &\quad \left\{ \int \frac{h(y_1|x)}{\int \eta(y_1, y_{10}, y_2, y_{20}|x) f(y_2|y_{10}, x) dy_2} dy_1 \right\}^{-1}. \end{aligned} \tag{10.4}$$

Equations (10.3) and (10.4) suggest that there is one-to-one correspondence between $h(y_1|x)$ and $f(y_1|y_{20}, x)$. Since models $f(y_2|y_{10}, x)$ and $h(y_1|x)$ can be specified without any constraint. It can be concluded that $f(y_2|y_{10}, x)$ and $f(y_1|y_{20}, x)$ can be specified without constraints.

By replacing $h(y_1|x)$ in (10.2) by (10.3), we obtain similar relationship between $f(y_1|y_{20}, x)$ and $g(y_2|x)$ as

$$g(y_2|x) = \frac{f(y_2|y_{10}, x) \int \eta(y_1, y_{10}, y_2, y_{20}|x) f(y_1|y_{20}, x) dy_1}{\int \int f(y_1|y_{20}, x) \eta(y_1, y_{10}, y_2, y_{20}|x) dy_1 \; f(y_2|y_{10}, x) dy_2}. \tag{10.5}$$

and

$$\begin{aligned} f(y_2|y_{10}, x) &= \frac{g(y_2|x)}{\int \eta(y_1, y_{10}, y_2, y_{20}|x) f(y_1|y_{20}, x) dy_1} \\ &\quad \left\{ \int \frac{g(y_2|x)}{\int \eta(y_1, y_{10}, y_2, y_{20}|x) f(y_1|y_{20}, x) dy_1} dy_2 \right\}^{-1}. \end{aligned} \tag{10.6}$$

Combining the results, we obtain the following representation

$$f(y_1, y_2|x) = \frac{\eta(y_1, y_{10}, y_2, y_{20}|x) f(y_1|y_{20}, x) f(y_2|y_{10}, x)}{\int \int f(y_1|y_{20}, x) \eta(y_1, y_{10}, y_2, y_{20}|x) dy_1 \; f(y_2|y_{10}, x) dy_2},$$

where $f(y_2|y_{10}, x)$ and $f(y_1|y_{20}, x)$ can be specified without constraints.

10.2 The semiparametric efficient score for semiparametric odds ratio model

For the semiparametric model where no parametric form is assumed for either $f(y_2|y_{10}, x)$ or $f(y_1|y_{20}, x)$, the semiparametric efficient score for estimating γ has the following form (Bickel et al, 1993) [11],

$$\frac{\partial}{\partial \gamma} \log \eta(Y_1, Y_2|X, \gamma) - \text{Proj}\left(\frac{\partial}{\partial \gamma} \log \eta(Y_1, Y_2|X, \gamma) \middle| \overline{H_1 + H_2} \right),$$

where $\overline{H_1 + H_2}$ is the closed linear space generated by $H_1 + H_2 = \{h_1 + h_2 | h_1 \in H_1, h_2 \in H_2\}$, and H_1 and H_2 are respectively the mean-zero square-integrable functions in the measurable spaces generated by (Y_1, X) and (Y_2, X). By Neumann projection theorem (Bickel et al, 1993, page 436–453) [11], the efficient score can be expressed as

$$\lim_{k\to\infty} [\{I - \text{Proj}(\cdot|H_1)\}\{I - \text{Proj}(\cdot|H_2)\}]^k \frac{\partial}{\partial\gamma} \log \eta(Y_1, Y_2|X, \gamma),$$

or equivalently as

$$\lim_{k\to\infty} [\{I - \text{Proj}(\cdot|H_2)\}\{I - \text{Proj}(\cdot|H_1)\}]^k \frac{\partial}{\partial\gamma} \log \eta(Y_1, Y_2|X, \gamma),$$

where I is the identity map and the limit is in $L^2(P_0)$ norm with P_0 being the true distribution of (Y_1, Y_2, X). Note that $\text{Proj}(\cdot|H_1) = E(\cdot|Y_1, X)$, also denoted by $E^{(Y_1 X)}$, and that $\text{Proj}(\cdot|H_2) = E(\cdot|Y_2, X)$, also denoted by $E^{(Y_2,X)}$. When expanded, the semiparametric efficient score can be rewritten as

$$\begin{gathered}\textstyle\sum_{k=0}^{\infty}(E^{(Y_2,X)}E^{(Y_1,X)})^k \left[\frac{\partial}{\partial\gamma}\log\eta(Y_1,Y_2|X,\gamma) - E\left\{\frac{\partial}{\partial\gamma}\log\eta(Y_1,Y_2|X,\gamma)|Y_2,X\right\}\right] \\ -E\left(\textstyle\sum_{k=0}^{\infty}(E^{(Y_2,X)}E^{(Y_1,X)})^k \left[\frac{\partial}{\partial\gamma}\log\eta(Y_1,Y_2|X,\gamma) - E\left\{\frac{\partial}{\partial\gamma}\log\eta(Y_1,Y_2|X,\gamma)|Y_2,X\right\}\right]|Y_1,X\right).\end{gathered}$$

An alternative expansion of the semiparametric efficient score is

$$\begin{gathered}\textstyle\sum_{k=0}^{\infty}(E^{(Y_1,X)}E^{(Y_2,X)})^k \left[\frac{\partial}{\partial\gamma}\log\eta(Y_1,Y_2|X,\gamma) - E\left\{\frac{\partial}{\partial\gamma}\log\eta(Y_1,Y_2|X,\gamma)|Y_1,X\right\}\right] \\ -E\left(\textstyle\sum_{k=0}^{\infty}(E^{(Y_1,X)}E^{(Y_2,X)})^k \left[\frac{\partial}{\partial\gamma}\log\eta(Y_1,Y_2|X,\gamma) - E\left\{\frac{\partial}{\partial\gamma}\log\eta(Y_1,Y_2|X,\gamma)|Y_1,X\right\}\right]\middle|Y_2,X\right).\end{gathered}$$

Under the hypothesis of no association,

$$\begin{aligned} E^{(Y_1,X)}E^{(Y_2,X)}\left\{\frac{\partial}{\partial\gamma}\log\eta(Y_1,Y_2|X,\gamma)\right\} &= E^{(Y_2,X)}E^{(Y_1,X)}\left\{\frac{\partial}{\partial\gamma}\log\eta(Y_1,Y_2|X,\gamma)\right\} \\ &= E\left\{\frac{\partial}{\partial\gamma}\log\eta(Y_1,Y_2|X,\gamma)\middle|X\right\}. \end{aligned}$$

The semiparametric efficient score reduces to

$$\begin{aligned} S &= \frac{\partial}{\partial\gamma}\log\eta(Y_1,Y_2|X,\gamma) - E\left\{\frac{\partial}{\partial\gamma}\log\eta(Y_1,Y_2|X,\gamma)\middle|Y_2,X\right\} \\ &\quad -E\left\{\frac{\partial}{\partial\gamma}\log\eta(Y_1,Y_2|X,\gamma)\middle|Y_1,X\right\} + E\left\{\frac{\partial}{\partial\gamma}\log\eta(Y_1,Y_2|X,\gamma)\middle|X\right\}. \quad (10.7)\end{aligned}$$

It follows from the above expression that the semiparametric information bound is $E(SS') = E(V_X)$, where

$$\begin{aligned} V_X &= \text{Var}\left\{\frac{\partial}{\partial\gamma}\log\eta(Y_1,Y_2|X,\gamma)\middle|X\right\} - \text{Var}\left[E\left\{\frac{\partial}{\partial\gamma}\log\eta(Y_1,Y_2|X,\gamma)\middle|Y_2,X\right\}\middle|X\right] \\ &\quad -\text{Var}\left[E\left\{\frac{\partial}{\partial\gamma}\log\eta(Y_1,Y_2|X,\gamma)\middle|Y_1,X\right\}\middle|X\right]. \quad (10.8)\end{aligned}$$

The efficient influence function is $[E\{V_X\}]^{-1}S$.

10.3 Theoretical properties of the semiparametric maximum likelihood estimator

We study the asymptotic behavior of the maximum semiparametric likelihood estimator under the semiparametric odds ratio model for the case $p = 3$. For $p > 3$, similar proof methods can be applied. Assume that the parametric odds ratio function may or may not include the true density. The parametric odds ratio model has $\gamma \in B$ and B is a subset of R^p. Denote $\eta_1\{y_1; (y_2, y_3) \mid \gamma_1\}\eta_2(y_2, y_3|y_{10}, \gamma_2)$ by $\eta(y_1; y_2; y_3 \mid \gamma)$. Let

$$f(y_1, y_2 \mid y_3, \gamma, Q_1, Q_2) = \frac{\eta(y_1; y_2; y_3 \mid \gamma) q_1(y_1) q_2(y_2)}{\int\int \eta(y_1; y_2; y_3 \mid \gamma) dQ_1(y_1) dQ_2(y_2)},$$

where $q_k = dQ_k/d\mu_k$ with μ_k being Lesbegue or count measures in $\mathcal{Y}_k$ for $k = 1, 2$, and $y_k \in \mathcal{Y}_k, k = 1, 2, 3$ and $\mathcal{Y}_k$, $k = 1, 2, 3$, are bounded sets in Euclidean spaces. Let $P_{\mathcal{Y}_k}$ be collection of all distribution functions absolutely continuous with respect to μ_k on $\mathcal{Y}_k$ for $k = 1, 2$. Let $f_0(y_1, y_2 \mid y_3)$ be the true conditional density generating the data having support $\mathcal{Y}_1 \times \mathcal{Y}_2 \times \mathcal{Y}_3$. Let E_0 and Var_0 denote the expectation and variance computed under f_0. Assume the following regularity conditions hold.

1. Assume that the Kullback-Leibler information exists and is finite. That is

   $$E_0\left\{\left| \log \frac{f(y_1, y_2 \mid y_3, \gamma, Q_1, Q_2)}{f_0(y_1, y_2 \mid y_3)} \right|\right\} < +\infty,$$

   for all γ, Q_1, Q_2 satisfiying

   $$\int\int \eta(y_1; y_2; y_3 \mid \gamma) dQ_1(y_1) dQ_2(y_2) < +\infty.$$

   Furthermore,

   $$\max_{\gamma, Q_1, Q_2} E_0\left\{\log \frac{f(y_1, y_2 \mid y_3, \gamma, Q_1, Q_2)}{f_0(y_1, y_2 \mid y_3)}\right\}$$

   exists and the maximizer is unique. Denote the maximizer by γ_0, Q_{10}, Q_{20}.

2. Assume B is a convex compact set in R^p and γ_0 is an inner point of B. Parameter $\gamma \in B$ is identifiable from $\eta(y_1; y_2; y_3|\gamma)$. There exists $c_1 > 0$ such that, for any $y_k \in \mathcal{Y}_k$, k=1,2,3, $\gamma \in B$,

   $$0 < c_1 \leq \eta(y_1; y_2; y_3|\gamma) \leq c_1^{-1} < +\infty.$$

3. Assume $\eta(y_1; y_2; y_3|\gamma)$ has bounded continuous derivatives up to the third order with respect to γ. That is,

   $$\sup\left\{|\dot{\eta}_s(y_1; y_2; y_3|\gamma)|, |\ddot{\eta}_{st}(y_1; y_2; y_3|\gamma)|, |\eta^{(3)}_{rst}(y_1; y_2; y_3|\gamma)|, \forall r, s, t \in \gamma\right\} < \infty,$$

   where the supremum is taken under $\{y_k \in \mathcal{Y}_k, k = 1, 2, 3, \gamma \in B\}$.

4. Assume $\eta(y_1; y_2; y_3|\gamma)$ and its first derivatives with respect to $\gamma \in B$ are bounded variation functions with respect to y_k, $k = 1, 2, 3$, for any $\gamma \in B$, and the total variations are uniformly bounded. That is,

   $$\sup\{||\eta(y_1; y_2; y_3|\gamma)||_{BV(\mathcal{Y}_1)}, ||\dot{\eta}_s(y_1; y_2; y_3|\gamma)||_{BV(\mathcal{Y}_1)}, \forall s \in \gamma\} < \infty,$$

with the supremum being taken under $\{y_k \in \mathcal{Y}_k, k = 2, 3, \gamma \in B\}$ and

$$\sup\left\{||\eta(y_1; y_2; y_3|\gamma)||_{BV(\mathcal{Y}_2)}, ||\dot{\eta}_s(y_1; y_2; y_3|\gamma)||_{BV(\mathcal{Y}_2)}, \forall s \in \gamma\right\} < \infty,$$

with the supremum being taken under $\{y_k \in \mathcal{Y}_k, k = 1, 3, \gamma \in B\}$, where $||h(z)||_{BV(\mathcal{Z})}$ denotes the bounded variation norm for $h(z)$ on $\mathcal{Z}$, and $s \in \gamma$ means s is a component of γ.

5. Let H_0 be a bounded set in R^p. Let H_k be the collection of functions of y_k with bounded variations on the support of Q_{k0} and $\int h_k(y_k)dQ_{k0}(y_k) = 0$ for all $h_k \in H_k$, $k = 1, 2$. For any nonzero $(h_0, h_1, h_2) \in H_0 \times H_1 \times H_2$,

$$-h_0^t E_0\left[\frac{\partial^2}{\partial\gamma^2}\log\eta(Y_1; Y_2; Y_3 \mid \gamma_0) - E_{(\gamma_0, Q_{10}, Q_{20})}\left\{\frac{\partial^2}{\partial\gamma^2}\log\eta(Y_1; Y_2; Y_3 \mid \gamma_0) \mid Y_3\right\}\right] h_0$$
$$+E_0\left[V_{(\gamma_0, Q_{10}, Q_{20})}\left\{h_0^T\frac{\partial}{\partial\gamma}\log\eta(Y_1; Y_2; Y_3 \mid \gamma_0) + h_1(Y_1) + h_2(Y_2) \mid Y_3\right\}\right] > 0.$$

From condition 1, (r_0, Q_{10}, Q_{20}) maximize $E_0[\log\{\eta(Y_1; Y_2; Y_3 \mid \gamma)dQ_1(Y_1)dQ_2(Y_2)/\int\int\eta(y_1; y_2; Y_3 \mid \gamma)dQ_1(y_1)dQ_2(y_2)\}]$, it follows that for any $h_1(y_1) \in H_1$,

$$E_0\left\{h_1(Y_1) - \frac{\int\int h_1(y_1)\eta(y_1; y_2; Y_3 \mid \gamma_0)dQ_{10}(y_1)dQ_{20}(y_2)}{\int\int\eta(y_1; y_2; Y_3 \mid \gamma)dQ_{10}(y_1)dQ_{20}(y_2)}\right\} = 0.$$

It follows that

$$\int h_1(y_1)\left[f_{10}(y_1) - q_{10}(y_1)\int\frac{\int\eta(y_1; y_2; y_3 \mid \gamma_0)dQ_{20}(y_2)}{\int\int\eta(y_1; y_2; y_3 \mid \gamma)dQ_{10}(y_1)dQ_{20}(y_2)}f_{30}(y_3)d\mu_3(y_3)\right]d\mu_1(y_1) = 0,$$

which means

$$f_{10}(y_1) = q_{10}(y_1)\int\frac{\int\eta(y_1; y_2; y_3 \mid \gamma_0)dQ_{20}(y_2)}{\int\int\eta(y_1; y_2; y_3 \mid \gamma)dQ_{10}(y_1)dQ_{20}(y_2)}f_{30}(y_3)d\mu_3(y_3),$$

where $f_{k0}(y_k)$ denotes the true marginal density of Y_k with respect to μ_k, $k = 1, 2, 3$. Similarly,

$$f_{20}(y_2) = q_{20}(y_2)\int\frac{\int\eta(y_1; y_2; y_3 \mid \gamma_0)dQ_{10}(y_1)}{\int\int\eta(y_1; y_2; y_3 \mid \gamma)dQ_{10}(y_1)dQ_{20}(y_2)}f_{30}(y_3)d\mu_3(y_3).$$

Finally,

$$E_0\left\{\frac{\partial}{\partial\gamma}\log\eta(Y_1; Y_2; Y_3 \mid \gamma_0)\right\} = E_0\left\{\frac{\int\int\frac{\partial}{\partial\gamma}\log\eta(y_1; y_2; Y_3 \mid \gamma_0)\eta(y_1; y_2; Y_3 \mid \gamma_0)dQ_1(y_1)dQ_2(y_2)}{\int\int\eta(y_1; y_2; Y_3 \mid \gamma_0)dQ_1(y_1)dQ_2(y_2)}\right\}.$$

The second derivative at $(\gamma_0, Q_{10}, Q_{20})$ is non-positive definite, i.e.,

$$h_0^t E_0\left[\frac{\partial^2}{\partial\gamma^2}\log\eta(Y_1; Y_2; Y_3 \mid \gamma_0) - E_{(\gamma_0, Q_{10}, Q_{20})}\left\{\frac{\partial^2}{\partial\gamma^2}\log\eta(Y_1; Y_2; Y_3 \mid \gamma_0) \mid Y_3\right\}\right] h_0$$
$$-E_0\left[V_{(\gamma_0, Q_{10}, Q_{20})}\left\{h_0^T\frac{\partial}{\partial\gamma}\log\eta(Y_1; Y_2; Y_3 \mid \gamma_0) + h_1(Y_1) + h_2(Y_2) \mid Y_3\right\}\right] \leq 0,$$

where $h_0 \in H_0$. Condition 5 is a strengthened version of this inequality.

If the true density is in the odds ratio model, i.e., the odds ratio model is correctly specified, conditions 1 and 5 can be replaced by the following simple conditions. Denote the true parameter value by $(\gamma_0, Q_{10}, Q_{20})$.

1a .There exists $(\gamma_0, Q_{10}, Q_{20})$ such that $f_0(y_1, y_2 \mid x) = f(y_1, y_2 \mid x, \gamma_0, Q_{10}, Q_{20})$. The Kullback-Leibler information exists and is finite. That is

$$E_0 \left\{ \left| \log \frac{f(y_1, y_2 \mid y_3, \gamma, Q_1, Q_2)}{f_0(y_1, y_2 \mid y_3)} \right| \right\} < +\infty.$$

for all γ, Q_1, Q_2 satisfiying

$$\int \int \eta(y_1; y_2; y_3 \mid \gamma) dQ_1(y_1) dQ_2(y_2) < +\infty.$$

5a . For almost all $Y_3 \in \mathcal{Y}_3$,

$$\mathrm{Var}_0 \left\{ \frac{\partial \log \eta}{\partial \gamma}(Y_1; Y_2; Y_3 | \gamma_0) \middle| Y_3 \right\} > 0.$$

Conditions 1a and 5a imply conditions 1 and 5 hold. Note that, under those conditions,

$$E_0 \left\{ \log \frac{f(y_1, y_2 \mid y_3, \gamma, Q_1, Q_2)}{f_0(y_1, y_2 \mid y_3)} \right\} \le 0$$

by Jensen's inequality and the equality holds only when $f(y_1, y_2 \mid y_3, \gamma, Q_1, Q_2) = f_0(y_1, y_2 \mid y_3)$, which implies $(\gamma, Q_1, Q_2) = (\gamma_0, Q_{10}, Q_{20})$ following identifiability. This means condition 1 holds. Further,

$$E_0 \left\{ \frac{\partial^2}{\partial \gamma^2} \log \eta(Y_1; Y_2; Y_3 \mid \gamma_0) \right\} = E_{(\gamma_0, Q_{10}, Q_{20})} \left\{ \frac{\partial^2}{\partial \gamma^2} \log \eta(Y_1; Y_2; Y_3 \mid \gamma_0) \mid Y_3 \right\}.$$

If, for some (h_0, h_1, h_2),

$$E_0 \left[V_{(\gamma_0, Q_{10}, Q_{20})} \left\{ h_0^T \frac{\partial}{\partial \gamma} \log \eta(Y_1; Y_2; Y_3 \mid \gamma_0) + h_1(Y_1) + h_2(Y_2) \mid Y_3 \right\} \right] = 0,$$

then

$$\begin{aligned} 0 &= h_0^T \left[\frac{\partial}{\partial \gamma} \log \eta(Y_1; Y_2; Y_3 | \gamma_0) - E_{(\gamma_0, Q_{10}, Q_{20})} \left\{ \frac{\partial}{\partial \gamma} \log \eta(Y_1; Y_2; Y_3 | \gamma_0) \middle| Y_3 \right\} \right] \\ &\quad + h_1(Y_1) - E_{(\gamma_0, Q_{10}, Q_{20})} \{h_1(Y_1) | Y_3\} + h_2(Y_2) - E_{(\gamma_0, Q_{10}, Q_{20})} \{h_2(Y_2) | Y_3\}. \end{aligned}$$

Since $\eta(Y_1; Y_{20}; Y_{30} | \gamma_0) = 1$, it follows that $\frac{\partial}{\partial \gamma} \log \eta(Y_1; Y_{20}; Y_{30} | \gamma_0) = 0$. This implies that $h_1(Y_1)$ is a constant. It follows from $\int h_1(Y_1) dQ_{10}(Y_1) = 0$ that $h_1(Y_1) = 0$. Similarly, we can show that $h_2(Y_2) = 0$. By plugging in the results back, we see that

$$h_0^T \left[\frac{\partial}{\partial \gamma} \log \eta(Y_1; Y_2; Y_3 | \gamma_0) - E_{(\gamma_0, Q_{10}, Q_{20})} \left\{ \frac{\partial}{\partial \gamma} \log \eta(Y_1; Y_2; Y_3 | \gamma_0) \middle| Y_3 \right\} \right] = 0.$$

It follows from condition 5a,

$$\mathrm{Var}_{(\gamma_0, Q_{10}, Q_{20})} \left\{ \frac{\partial}{\partial \gamma} \log \eta(y_1; Y_2; y_3 | \gamma_0) \middle| Y_3 \right\} > 0,$$

that $h_0 = 0$. This means condition 5 holds.

For any $\delta > 0$, let $\mathcal{Q}_{k\delta} = \{Q_k \in P_{\mathcal{Y}_k} | \sup_{y_k \in \mathcal{Y}_k} |Q_k(y_k) - Q_{k0}(y_k)| < \delta\}$, $k = 1, 2$. Before showing the consistency of the maximum likelihood estimator, we state a few lemmas that are used in the proof of consistency. Proofs of these lemmas are omitted.

Lemma 10.2.1. Suppose that $F_n(y) \to F(y)$ weakly, and that $\varphi(y,h)$ is uniformly bounded and equicontinuous with respect to y over $h \in H$. That is, for any $\epsilon > 0$, there exists a finite set of $h_k \in H$, $k = 1, \cdots, K_\epsilon$ such that

$$\max_{h\in H}\left\{\min_k |\varphi(y,h) - \varphi(y,h_k)|\right\} < \epsilon.$$

Then,

$$\lim_{n\to\infty}\sup_{h\in H}\left|\int \varphi(y,h)dF_n(y) - \varphi(y,h)dF(y)\right| = 0.$$

Lemma 10.2.2. Suppose that $\sup_{y\in\mathcal{D}}|F_n(y) - F(y)| \to 0$, $\sup_{y\in\mathcal{D},h\in H}|\varphi_n(y,h) - \varphi(y,h)| \to 0$, and $\sup_{h\in H}||\varphi(y,h)||_{BV(y)} < +\infty$ where $\mathcal{D}$ is a measurable set and $||\cdot||_{BV(y)}$ denotes the variation norm in $y \in \mathcal{Y}$. Then

$$\lim_{n\to\infty}\sup_{h\in H}\left|\int_{\mathcal{D}} \varphi_n(y,h)dF_n(y) - \int_{\mathcal{D}} \varphi(y,h)dF(y)\right| = 0.$$

Lemma 10.2.3. Under conditions 1–4, for a single observation, $\{\dot{l}(\gamma, Q_1, Q_2)(h), \gamma \in B, Q_k \in \mathcal{Q}_{k\delta}, k = 1,2, h \in H_0 \times H_1 \times H_2\}$ is a P_0-Donsker class and $\{\ddot{l}(\gamma, Q_1, Q_2)(h)(h^*), \gamma \in B, Q_k \in \mathcal{Q}_{k\delta}, k = 1,2, h^*, h \in H_0 \times H_1 \times H_2\}$ is a P_0-Glivenko-Cantelli class.

Lemma 10.2.4. Let $b^{-1}(y_3) = \int\int \eta(y_1;y_2;y_3|\gamma_0)dQ_{10}(y_1)dQ_{20}(y_2)$,

$$\begin{aligned} Q_{n10}(y_1) &= \frac{1}{n}\sum_{i=1}^{n}\frac{1_{\{Y_{i1}\le y_1\}}}{\frac{1}{n}\sum_{j=1}^{n} b(Y_{j3})\int \eta(Y_{i1};y_2;Y_{j3}|\gamma_0)dQ_{20}(y_2)}, \\ Q_{n20}(y_2) &= \frac{1}{n}\sum_{i=1}^{n}\frac{1_{\{Y_{i2}\le y_2\}}}{\frac{1}{n}\sum_{j=1}^{n} b(Y_{j3})\int \eta(y_1;Y_{i2};Y_{j3}|\gamma_0)dQ_{10}(y_1)}. \end{aligned}$$

Then $Q_{nk0}(y_k) \to Q_{k0}(y_k)$, uniformally over $y_k \in \mathcal{Y}_k$ P_0-almost sure for $k = 1,2$. Further,

$$\sup_{y_k\in\mathcal{Y}_k, k=1,2,3}|P(y_1,y_2 \mid y_3,\gamma_0,Q_{n10},Q_{n20}) - P(y_1,y_2 \mid y_3,\gamma_0,Q_{10},Q_{20})| \to 0,$$

P_0-almost sure, where $P(y_1,y_2 \mid y_3,\gamma,Q_1,Q_2)$ is the distribution function of $f(y_1,y_2 \mid y_3,\gamma,Q_1,Q_2)$.

Theorem 10.2.1. Under the regularity conditions 1-5, the maximum likelihood estimator of γ from maximizing

$$\prod_{i=1}^{n}\frac{\eta_1\{Y_{i1};(Y_{i2},Y_{i3}) \mid \gamma_1\}\eta_2\{Y_{i2};Y_{i3} \mid Y_{10},\gamma_2\}dQ_1(Y_{i1})dQ_2(Y_{i2})}{\int\int \eta_1\{y_1;(y_2,Y_{i3}) \mid \gamma_1\}\eta_2\{y_2;Y_{i3} \mid y_{10},\gamma_2\}dQ_1(y_1)dQ_2(y_2)}, \tag{10.9}$$

exists and is almost surely converges to γ_0, which is the value of γ minimizing the Kullback-Leibler divergence. Furthermore, the maximum likelihood estimator of γ is asymptotically normally distributed. That is,

$$\sqrt{n}(\hat{\gamma} - \gamma_0) \to N(0,V),$$

where the asymptotic variance, satisfying $h_0^T V h_0 = P_0\{\dot{l}_{(\gamma_0,Q_{10},Q_{20})}(Y_1,Y_2,Y_3)\sigma^{-1}(h_0,0,0)\}^2$ with σ defined in the proof later, can be consistently estimated by the corresponding sampling version of the sandwich estimator.

Proof of the theorem are divided into three steps. They are proof of existence and consistency, proof of normality, and proof of consistency of the variance estimate.

Proof of existence and consistency of the maximum likelihood estimator: The likelihood has the form

$$\prod_{i=1}^{n} \frac{\eta(Y_{i1}; Y_{i2}; Y_{i3}|\gamma) dQ_1(Y_{i1}) dQ_2(Y_{i2})}{\int\int \eta(y_1; y_2; Y_{i3}|\gamma) dQ_1(y_1) dQ_2(y_2)}.$$

Direct maximization of the likelihood over all distributions for Q_1 and/or Q_2 does not exist. We follow the convention of maximizing the likelihood over all discrete distributions for Q_1 and Q_2. It can be seen that, for any discrete Q_1 and Q_2, there are Q_1 and Q_2 having probability masses only at the observed Y_1 and Y_2 values that make the likelihood no smaller. This means the maximizer, if exists, is achieved with Q_1 and Q_2 having probability masses only at the observed Y_1 and Y_2 values. The likelihood is then a multivariate function of $(\gamma; q_{k1}, k = 1, \cdots, N_1; q_{j2}, j = 1, \cdots, N_2)$ with $\sum_{k=1}^{N_1} q_{k1} = 1$ and $\sum_{j=1}^{N_2} q_{j2} = 1$. From Assumptions 2 and 3, it follows that the likelihood, as a function of $(\gamma, q_{11}, \cdots, q_{(N_1-1)1}, q_{12}, \cdots, q_{(N_2-1)2})$ with $\gamma \in B$ and the others parameters are respectively in

$$\mathcal{Q}_1 = \left\{ (q_{11}, \cdots, q_{(N_1-1)1}) \middle| \sum_{k=1}^{N_1} q_{k1} = 1 \right\} \text{ and } \mathcal{Q}_2 = \left\{ (q_{12}, \cdots, q_{(N_2-1)2}) \middle| \sum_{k=1}^{N_2} q_{k2} = 1 \right\},$$

is continuous. It follows from that $B \times \mathcal{Q}_1 \times \mathcal{Q}_2$ is a compact set that the maximum of the likelihood exists.

From assumption 2, we see that when $dQ_1(Y_{i1}) = 0$ or $dQ_2(Y_{i2}) = 0$ for any $i = 1, \cdots, n$, the likelihood becomes zero. This implies that, for any fixed γ, the maximum of the restricted likelihood over $Q_k, k = 1, 2$, exists and is attained respectively at inner points of $\mathcal{Q}_1 \times \mathcal{Q}_2$. The retricted maximum likelihood estimators of $Q_k, k = 1, 2$, satisfy the score equations derived from

$$P_n l(\gamma, Q_1, Q_2) + \lambda \left(\sum_{k=1}^{N_1} q_{k1} - 1 \right) + \mu \left(\sum_{k=1}^{N_2} q_{k2} - 1 \right),$$

where $P_n l(\gamma, Q_1, Q_2)$ is the log-likelihood. Denote the maximum restricted likelihood estimator by $(\hat{\gamma}, \hat{Q}_1, \hat{Q}_2)$. The score equations for $\hat{Q}_1$ are

$$\frac{\sum_{i=1}^{n} 1_{\{Y_{i1}=y_{k1}\}}}{n q_{k1}} - \frac{1}{n} \sum_{i=1}^{n} \frac{\sum_{j=1}^{N_2} \eta(y_{k1}; y_{j2}; Y_{i3}|\hat{\gamma}) d\hat{Q}_2(y_{j2})}{\sum_{l=1}^{N_1} \sum_{j=1}^{N_2} \eta(y_{l1}; y_{j2}, Y_{i3}|\hat{\gamma}) d\hat{Q}_1(y_{l1}) d\hat{Q}_2(y_{j2})} + \lambda = 0,$$

for $k = 1, \cdots, N_1$. It follows from $\sum_{k=1}^{N_1} q_{k1} = 1$ that $\lambda = 0$. From this, we see that $\hat{Q}_1(y_1) = \sum_{k=1}^{N_1} q_{k1} 1_{\{y_{k1} \le y_1\}}$, where

$$q_{k1} = \frac{1}{n} \sum_{i=1}^{n} \frac{1_{\{Y_{i1}=y_{k1}\}}}{D_1(y_{k1}, \hat{\gamma}, \hat{Q}_1, \hat{Q}_2)},$$

and

$$D_1(y_1, \hat{\gamma}, \hat{Q}_1, \hat{Q}_2) = \frac{1}{n} \sum_{i=1}^{n} \frac{\sum_{j=1}^{N_2} \eta(y_1; y_{j2}; Y_{i3}|\hat{\gamma}) d\hat{Q}_2(y_{j2})}{\sum_{l=1}^{N_1} \sum_{j=1}^{N_2} \eta(y_{l1}; y_{j2}, Y_{i3}|\hat{\gamma}) d\hat{Q}_1(y_{l1}) d\hat{Q}_2(y_{j2})}.$$

Note that the existence of $\hat{Q}_1$ in the inner part of the set guarantees the maximizer $q_{k1} > 0$

for all $k = 1, \cdots, N_1$ and satisfies $\hat{Q}_1(y_1) = \sum_k q_{k1} 1_{\{y_{k1} \le y_1\}}$. From the expression, we see that

$$\frac{d\hat{Q}_1}{dQ_{n10}}(y_{k1}) = \frac{1}{D_1(y_{k1}, \hat{\gamma}, \hat{Q}_1, \hat{Q}_2)} \left\{ \frac{1}{\frac{1}{n}\sum_{j=1}^n b(Y_{j3}) \int \eta(y_{k1}; y_2; Y_{j3}|\gamma_0) dQ_{20}(y_2)} \right\}^{-1}.$$

Similar derivation can be carried out for $\hat{Q}_2$ to obtain

$$\frac{d\hat{Q}_2}{dQ_{n20}}(y_{k2}) = \frac{1}{D_2(y_{k2}, \hat{\gamma}, \hat{Q}_1, \hat{Q}_2)} \left\{ \frac{1}{\frac{1}{n}\sum_{j=1}^n b(Y_{j3}) \int \eta(y_1; y_{k2}; Y_{j3}|\gamma_0) dQ_{10}(y_1)} \right\}^{-1},$$

where D_2 is defined similarly as D_1. Since γ belongs to a compact set, $\hat{\gamma}$ has a convergent subsequence. Denote the limit by γ. From Helly selection Theorem (Fuchino and Plewik, 1999), there also exists a subsequence of $\hat{Q}_k$ that converges to a subdistribution Q_k on the continuous points of Q_k, for $k = 1, 2$. Since $\mathcal{Y}_k$ is compact, which ensure Q_k is a distribution function. Hence, $\hat{Q}_k$ converges weakly to Q_k, for $k = 1, 2$.

By repeatedly applying Lemma 1, it follows that $D_1(y_1, \hat{\gamma}, \hat{Q}_1, \hat{Q}_2)$ converges uniformly to

$$D_1(y_1, \gamma, Q_1, Q_2) = \int \frac{\int \eta(y_1; y_2; y_3|\gamma) dQ_2(y_2)}{\int\int \eta(y_1; y_2, y_3|\gamma) dQ_1(y_1) dQ_2(y_2)} f_{30}(y_3) d\mu_3(y_3).$$

Similarly, $D_2(y_2, \hat{\gamma}, \hat{Q}_1, \hat{Q}_2)$ converges uniformly to $D_2(y_2, \gamma, Q_1, Q_2)$, defined similarly as D_1. These imply that $Q_1(y_1)$ satisfies

$$Q_1(y_1) = \int_{-\infty}^{y_1} \left\{ \int \frac{\int \eta(y_1; y_2; y_3|\gamma) dQ_2(y_2)}{\int\int \eta(y_1; y_2, y_3|\gamma) dQ_1(y_1) dQ_2(y_2)} f_{30}(y_3) d\mu_3(y_3) \right\}^{-1} f_{10}(y_1) d\mu_1(y_1)$$

and $\hat{Q}_1(y_1)$ converges uniformly over $\mathcal{Y}_1$ to $Q_1(y_1)$. Similar definition and convergence result hold for $Q_2(y_2)$.

Further, it follows that

$$-\frac{1}{n}\sum_{i=1}^n \log D_1(Y_{i1}, \hat{\gamma}, \hat{Q}_1, \hat{Q}_2) \to -\int \{\log D_1(y_1, \gamma, Q_1, Q_2)\} f_{10}(y_1) d\mu_1(y_1).$$

and

$$-\frac{1}{n}\sum_{i=1}^n \log D_2(Y_{i2}, \hat{\gamma}, \hat{Q}_1, \hat{Q}_2) \to -\int \{\log D_2(y_2, \gamma, Q_1, Q_2)\} f_{20}(y_2) d\mu_2(y_2).$$

It now follows that

$$\begin{aligned} P_n\left\{l(\hat{\gamma}, \hat{Q}_1, \hat{Q}_2) - l(\gamma_0, Q_{1n0}, Q_{2n0})\right\} &= P_n\left\{\log \frac{d\hat{Q}_1}{dQ_{n10}}(Y_1)\right\} + P_n\left\{\log \frac{d\hat{Q}_2}{dQ_{n20}}(Y_2)\right\} \\ +P_n\left\{\log \frac{\eta(Y_1; Y_2; Y_3 \mid \hat{\gamma})}{\eta(Y_1; Y_2; Y_3 \mid \gamma_0)}\right\} &- P_n\left\{\log \frac{\int\int \eta(y_1; y_2, Y_3 \mid \hat{\gamma}) d\hat{Q}_1(y_1) d\hat{Q}_2(y_2)}{\int\int \eta(y_1; y_2, Y_3 \mid \gamma_0) dQ_{n10}(y_1) dQ_{n20}(y_2)}\right\} \\ &\to P_0\{l(\gamma, Q_1, Q_2) - l(\gamma_0, Q_{10}, Q_{20})\} \ge 0. \end{aligned}$$

It follows from Assumption 1, i.e., $(\gamma_0, Q_{10}, Q_{20})$ is the unique maximizer of the Kullback-Leibler information, that $(\gamma, Q_1, Q_2) = (\gamma_0, Q_{10}, Q_{20})$. Since every sequence has a convergent subsequence and every convergent subsequence has the same limit, the sequence $(\hat{\gamma}, \hat{Q}_1, \hat{Q}_2)$ converges to the common limit $(\gamma_0, Q_{10}, Q_{20})$.

Proof of asymptotic normality: Note that the maximum likelihood estimator satisfies the score equation,

$$P_n \dot{l}_{(\hat{\gamma},\hat{Q}_1,\hat{Q}_2)}(Y_1, Y_2, Y_3)(h_0, h_{1n}, h_{2n}) = 0,$$

where

$$\begin{aligned} &\dot{l}_{(\gamma,Q_1,Q_2)}(Y_1, Y_2, Y_3)(h_0, h_1, h_2) \\ &= h_0^T \left[\frac{\partial}{\partial \gamma} \log \eta(Y_1; Y_2; Y_3 \mid \gamma) - E\left\{ \frac{\partial}{\partial \gamma} \log \eta(Y_1; Y_2; Y_3 \mid \gamma) \middle| Y_3 \right\} \right] \\ &\quad + h_1(Y_1) - E\{h_1(Y_1)|Y_3\} + h_2(Y_2) - E\{h_2(Y_2)|Y_3\}, \end{aligned}$$

with E being the expectation taken under the odds ratio model with parameter (γ, Q_1, Q_2), h_0 is bounded, $h_{1n} \in H_{1n}$ is a collection of bounded variation functions with jumps only at the observed y_1 values and $\int h_{1n}(y_1) d\hat{Q}_1(y_1) = 0$, and $h_{2n} \in H_{2n}$ is a collection of bounded variation functions with jumps only on observed y_2 values and $\int h_{2n}(y_2) d\hat{Q}_2(y_2) = 0$. Define an inner product on $H_0 \times H_1 \times H_2$ as

$$< (h_0, h_1, h_2), (h_0^*, h_1^*, h_2^*) >= h_0^T h_0^* + \int h_1(y_1) h_1^*(y_1) dQ_{10}(y_1) + \int h_2(y_2) h_2^*(y_2) dQ_{20}(y_2).$$

Note that

$$\begin{aligned} &\sqrt{n} P_n \dot{l}_{(\hat{\gamma},\hat{Q}_1,\hat{Q}_2)}(Y_1, Y_2, Y_3)(h_0, h_{n1}, h_{n2}) \\ &= P_n \dot{l}_{(\hat{\gamma},\hat{Q}_1,\hat{Q}_2)}(Y_1, Y_2, Y_3)\{0, \sqrt{n}(h_{1n} - h_1), \sqrt{n}(h_{2n} - h_2)\} \\ &\quad + \sqrt{n} P_n \dot{l}_{(\hat{\gamma},\hat{Q}_1,\hat{Q}_2)}(Y_1, Y_2, Y_3)(h_0, h_1, h_2), \end{aligned}$$

For any $h_1 \in H_1$ and $h_2 \in H_2$, there are $h_{1n} \in H_1$ and $h_{2n} \in H_2$ such that

$$\sqrt{n}(h_{1n} - h_1) = O_{P_0}(1) \text{ and } \sqrt{n}(h_{2n} - h_2) = O_{P_0}(1).$$

Since $\{\dot{l}_{(\gamma,Q_1,Q_2)}(Y_1, Y_2, Y_3)(h_0, h_1, h_2)\}$ is a Glivenko-Cantelli class for $|\gamma - \gamma_0| < \delta$, $\sup_{y_1} |Q_1(y_1) - Q_{10}(y_1)| < \delta$, $\sup_{y_2} |Q_2(y_2) - Q_{20}(y_2)| < \delta$, and bounded (h_0, h_1, h_2), it follows that

$$\sqrt{n} P_n \dot{l}_{(\hat{\gamma},\hat{Q}_1,\hat{Q}_2)}(Y_1, Y_2, Y_3)(h_0, h_{n1}, h_{n2}) = \sqrt{n} P_n \dot{l}_{(\hat{\gamma},\hat{Q}_1,\hat{Q}_2)}(Y_1, Y_2, Y_3)(h_0, h_1, h_2) + o_{P_0}(1),$$

because $P_0 \dot{l}_{(\gamma_0,Q_{10},Q_{20})}(Y_1, Y_2, Y_3)(h_0^*, h_1^*, h_2^*) = 0$ for any $(h_0^*, h_1^*, h_2^*) \in H_0 \times H_1 \times H_2$. Note that

$$\begin{aligned} o_{P_0}(1) &= \sqrt{n} P_n \dot{l}_{(\hat{\gamma},\hat{Q}_1,\hat{Q}_2)}(Y_1, Y_2, Y_3)(h_0, h_1, h_2) \\ &= -\sqrt{n} P_0 \{\dot{l}_{(\hat{\gamma},\hat{Q}_1,\hat{Q}_2)}(Y_1, Y_2, Y_3) - \dot{l}_{(\gamma_0,Q_{10},Q_{20})}(Y_1, Y_2, Y_3)\}(h_0, h_1, h_2) \\ &\quad + \sqrt{n}(P_n - P_0) \dot{l}_{(\gamma_0,Q_{10},Q_{20})}(Y_1, Y_2, Y_3)(h_0, h_1, h_2) \\ &\quad + \sqrt{n}(P_n - P_0)\{\dot{l}_{(\hat{\gamma},\hat{Q}_1,\hat{Q}_2)}(Y_1, Y_2, Y_3) - \dot{l}_{(\gamma_0,Q_{10},Q_{20})}(Y_1, Y_2, Y_3)\}(h_0, h_1, h_2). \end{aligned}$$

From Lemma 10.2.2 and

$$P_0 \{\dot{l}_{(\hat{\gamma},\hat{Q}_1,\hat{Q}_2)}(Y_1, Y_2, Y_3) - \dot{l}_{(\gamma_0,Q_{10},Q_{20})}(Y_1, Y_2, Y_3)\}^2 (h_0, h_1, h_2) \to 0$$

uniformly, it follows that

$$\sqrt{n}(P_n - P_0)\{\dot{l}_{(\hat{\gamma},\hat{Q}_1,\hat{Q}_2)}(Y_1,Y_2,Y_3) - \dot{l}_{(\gamma_0,Q_{10},Q_{20})}(Y_1,Y_2,Y_3)\}(h_0,h_1,h_2) = o_{P_0}(1).$$

Further, since $\dot{l}_{(\gamma,Q_1,Q_2)}(Y_1,Y_2,Y_3)(h_0,h_1,h_2)$ is Frechet differentiable with respect to (γ,Q_1,Q_2) at (γ_0,Q_{10},Q_{20}), it follows that

$$\begin{aligned}
&-\sqrt{n}P_0\{\dot{l}_{(\hat{\gamma},\hat{Q}_1,\hat{Q}_2)}(Y_1,Y_2,Y_3) - \dot{l}_{(\gamma_0,Q_{10},Q_{20})}(Y_1,Y_2,Y_3)\}(h_0,h_1,h_2)\\
&= \sqrt{n}(\hat{\gamma}-\gamma_0)^T\sigma_0(h_0,h_1,h_2) + \int \sigma_1(h_0,h_1,h_2)d\sqrt{n}(\hat{Q}_1 - Q_{10})\\
&\quad + \int \sigma_2(h_0,h_1,h_2)d\sqrt{n}(\hat{Q}_2 - Q_{20})\\
&\quad + o_{P_0}(|\sqrt{n}(\hat{\beta}-\beta_0)| + ||\sqrt{n}(\hat{Q}_1 - Q_{10})|| + ||\sqrt{n}(\hat{Q}_2 - Q_{20})||),
\end{aligned}$$

where σs are

$$\begin{aligned}
\sigma_0(h_0,h_1,h_2) &= -E_0\left\{\frac{\partial^2 \log\eta}{\partial\gamma^2}h_0 - E_{(\gamma_0,Q_{10},Q_{20})}\left(\frac{\partial^2 \log\eta}{\partial\gamma^2}h_0\middle|Y_3\right)\right\}\\
&\quad + E_0\left[\mathrm{Cov}_{(\gamma_0,Q_{10},Q_{20})}\left\{\frac{\partial\log\eta}{\partial\gamma}, \frac{\partial\log\eta}{\partial\gamma}h_0 + h_1(Y_1) + h_2(Y_2)\middle|Y_3\right\}\right],\\
\sigma_1(h_0,h_1,h_2) &= E_0\left[\left\{E_{(\gamma_0,Q_{10},Q_{20})}\left(\frac{\partial\log\eta}{\partial\gamma}h_0\middle|Y_1 = y_1, Y_3\right)\right.\right.\\
&\quad \left.\left. - E_{(\gamma_0,Q_{10},Q_{20})}\left(\frac{\partial\log\eta}{\partial\gamma}h_0\middle|Y_3\right)\right\}p_1(y_1|Y_3)\right]\\
&\quad + h_1(y_1)E_0\{p_1(y_1|Y_3)\} - E_0[E_{(\gamma_0,Q_{10},Q_{20})}\{h_1(Y_1)|Y_3\}q_1(y_1|Y_3)]\\
&\quad + E_0[E_{(\gamma_0,Q_{10},Q_{20})}\{h_2(Y_2)|Y_1 = y_1, Y_3\}p_1(y_1|Y_3)\\
&\quad - E_{(\gamma_0,Q_{10},Q_{20})}\{h_2(Y_2)|Y_3\}p_1(y_1|Y_3)],\\
\sigma_2(h_0,h_1,h_2) &= E_0\left[\left\{E_{(\gamma_0,Q_{10},Q_{20})}\left(\frac{\partial\log\eta}{\partial\gamma}h_0\middle|Y_2 = y_2, Y_3\right)\right.\right.\\
&\quad \left.\left. - E_{(\gamma_0,Q_{10},Q_{20})}\left(\frac{\partial\log\eta}{\partial\gamma}h_0\middle|Y_3\right)\right\}p_2(y_2|Y_3)\right]\\
&\quad + E_0[E_{(\gamma_0,Q_{10},Q_{20})}\{h_1(Y_1)|Y_2 = y_2, Y_3\}p_2(y_2|Y_3)\\
&\quad - E_{(\gamma_0,Q_{10},Q_{20})}\{h_1(Y_1)|Y_3\}p_2(y_2|Y_3)]\\
&\quad + h_2(y_2)E_0\{p_2(y_2|Y_3)\} - E_0[E_{(\gamma_0,Q_{10},Q_{20})}\{h_2(Y_2)|Y_3\}p_2(y_2|Y_3)].
\end{aligned}$$

where

$$\begin{aligned}
p_1(y_1|y_3) &= \frac{\int \eta(y_1;y_2;y_3|\gamma_0)dQ_{20}(y_2)}{\int\int \eta(y_1;y_2;y_3|\gamma_0)dQ_{10}(y_1)dQ_{20}(y_2)},\\
p_2(y_2|y_3) &= \frac{\int \eta(y_1;y_2;y_3|\gamma_0)dQ_{10}(y_1)}{\int\int \eta(y_1;y_2;y_3)dQ_{10}(y_1)dQ_{20}(y_2)}.
\end{aligned}$$

We show in the following that $\sigma(h_0,h_1,h_2) = (\sigma_0,\sigma_1,\sigma_2)(h_0,h_1,h_2)$ is a continuously invertible operator on $H_0 \times H_1 \times H_2$. Note from assumption 5 that

$$< h, \sigma(h) >= -E_0\{\ddot{l}(h,h)\} \geq 0,$$

and the equality holds only when $h = 0$. This means that σ is a one-to-one map on $H_0 \times H_1 \times H_2$. Next, let

$$\sigma(h_0,h_1,h_2) = A(h_0,h_1,h_2) + B(h_0,h_1,h_2),$$

where $A(h_0, h_1, h_2) = (a_0^T h_0, E_0\{p_1(y_1|Y_3)\}h_1(y_1), E_0\{p_2(y_2|Y_3)\}h_2(y_2))$ and

$$a_0 = E_0\left[Var_{(\gamma_0,Q_{10},Q_{20})}\left\{\frac{\partial \log \eta}{\partial \gamma}(Y_1;Y_2;Y_3|\gamma_0)\bigg|Y_3\right\}\right] > 0.$$

Hence, A is continuously invertible on $H_0 \times H_1 \times H_2$. Further, it can be routinely verified that B is compact by applying Theorem 2 of Kantorovich and Akilov (1982, pages 326–328) which states that the integral operator on $L^2(P_0)$ is compact if its kernel is summable in $L^2(P_0)$. It then follows from that σ is one-to-one on $H_0 \times H_1 \times H_2$ that σ is continuous invertible on $H_0 \times H_1 \times H_2$.

From that

$$\sqrt{n}(P_n - P_0)\dot{l}(Y_1, Y_2; Y_3, \gamma_0)(h_0, h_1, h_2)$$

converges weakly to a tight Gaussian process, it follows that

$$\begin{aligned}\sqrt{n}(\hat{\gamma} - \gamma_0)^T h_0 + \int h_1(y_1)d\sqrt{n}(\hat{Q}_1 - Q_{10}) + \int h_2(y_2)d\sqrt{n}(\hat{Q}_2 - Q_{20})\\ = \sqrt{n}(P_n - P_0)\dot{l}_{(\gamma_0,Q_{10},Q_{20})}(Y_1;Y_2;Y_3)\sigma^{-1}(h_0,h_1,h_2) + o_{P_0}(1),\end{aligned}$$

which converges to a tight Gaussian process with the covariance function

$$P_0\left\{\dot{l}_{(\gamma_0,Q_{10},Q_{20})}(Y_1;Y_2;Y_3)\sigma^{-1}(h_0,h_1,h_2)\dot{l}_{(\gamma_0,Q_{10},Q_{20})}(Y_1;Y_2;Y_3)\sigma^{-1}(h_0^*,h_1^*,h_2^*)\right\}.$$

It follows from setting $(h_1, h_2) = 0$ that

$$\sqrt{n}(\hat{\gamma} - \gamma_0)^T h_0 \to N(0, V)$$

for any bounded $h_0 \in H_0$, where $V = P_0\left\{\dot{l}_{(\gamma_0,Q_{10},Q_{20})}(Y_1;Y_2;Y_3)\sigma^{-1}(h_0,0,0)\right\}^2$.

Proof of consistent variance estimate: Let $\hat{\sigma}$ be defined as σ with P_0 and $(\gamma_0, Q_{10}, Q_{20})$ in the expression respectively replaced by P_n and $(\hat{\gamma}, \hat{Q}_1, \hat{Q}_2)$. To show the consistency of the variance estimate, note first from Lemma 10.2.3 that,

$$\hat{\sigma}(h_0,h_1,h_2)(h_0^*,h_1^*,h_2^*) \to \sigma(h_0,h_1,h_2)(h_0^*,h_1^*,h_2^*)$$

uniformly P_0-almost sure over bounded $h, h^* \in H_0 \times H_1 \times H_2$. it follows from that σ is continuously invertible that $\hat{\sigma}$ is asymptotically continuously invertible almost sure and

$$\hat{\sigma}^{-1}(h_0,h_1,h_2)(h_0^*,h_1^*,h_2^*) \to \sigma^{-1}(h_0,h_1,h_2)(h_0^*,h_1^*,h_2^*)$$

uniformly P_0-almost sure over bounded $h, h^* \in H_0 \times H_1 \times H_2$. Note that

$$\begin{aligned}&P_n\left\{\dot{l}_{(\hat{\gamma},\hat{Q}_1,\hat{Q}_2)}(Y_1;Y_2;Y_3)(h)\right\}^2\\ &= P_0\left\{\dot{l}_{(\gamma_0,Q_{10},Q_{20})}(Y_1;Y_2;Y_3)(h)\right\}^2 + (P_n - P_0)\left\{\dot{l}_{(\hat{\gamma},\hat{Q}_1,\hat{Q}_2)}(Y_1;Y_2;Y_3)(h)\right\}^2\\ &\quad + P_0\left[\left\{\dot{l}_{(\hat{\gamma},\hat{Q}_1,\hat{Q}_2)}(Y_1;Y_2;Y_3)(h)\right\}^2 - \left\{\dot{l}_{(\gamma_0,Q_{10},Q_{20})}(Y_1;Y_2;Y_3)(h)\right\}^2\right].\end{aligned}$$

It follows from the continuity of $P_0\left\{\dot{l}_{(\gamma,Q_1,Q_2)}(Y_1;Y_2;Y_3)(h)\right\}^2$ with respect to (γ, Q_1, Q_2) uniformly over h and Lemma 10.2.3 on the Gilvenko-Cantelli classes that

$$P_n\left\{\dot{l}_{(\hat{\gamma},\hat{Q}_1,\hat{Q}_2)}(Y_1;Y_2;Y_3)(h)\right\}^2 \to P_0\left\{\dot{l}_{(\gamma_0,Q_{10},Q_{20})}(Y_1;Y_2;Y_3)(h)\right\}^2$$

uniformly on $h \in H_0 \times H_1 \times H_2$. It now follows from $\hat{\sigma}^{-1}(h) \to \sigma^{-1}(h)$ that

$$P_n \left\{ \dot{l}_{(\hat{\gamma},\hat{Q}_1,\hat{Q}_2)}(Y_1;Y_2;Y_3)\hat{\sigma}^{-1}(h_0,0,0) \right\}^2 \to P_0 \left\{ \dot{l}_{(\gamma_0,Q_{10},Q_{20})}(Y_1;Y_2;Y_3)\sigma^{-1}(h_0,0,0) \right\}^2.$$

To show the consistency of the numeric differentiation in estimating the asymptotic variance when the odds ratio model is correctly specified, note that, for $\gamma = \hat{\gamma} + h_0/\sqrt{n}$ and $\gamma_1 = \gamma_0 + h_0/\sqrt{n}$,

$$\begin{aligned} \sqrt{n}P_n \left\{ \dot{l}_\gamma(\gamma, Q_{1\gamma}, Q_{2\gamma}) - \dot{l}_\gamma(\hat{\gamma}, \hat{Q}_1, \hat{Q}_2) \right\} &= \sqrt{n}(P_n - P_0) \left\{ \dot{l}_\gamma(\gamma, Q_{1\gamma}, Q_{2\gamma}) - \dot{l}_\gamma(\hat{\gamma}, \hat{Q}_1, \hat{Q}_2) \right\} \\ &+ \sqrt{n}P_0 \left\{ \dot{l}_\gamma(\gamma, Q_{1\gamma}, Q_{2\gamma}) - \dot{l}_\gamma(\hat{\gamma}, \hat{Q}_1, \hat{Q}_2) - \dot{l}_\gamma(\gamma_1, Q_{10\gamma_1}, Q_{20\gamma_1}) + \dot{l}_\gamma(\gamma_0, Q_{10}, Q_{20}) \right\} \\ &+ \sqrt{n}P_0 \left\{ \dot{l}_\gamma(\gamma_1, Q_{10\gamma_1}, Q_{20\gamma_1}) - \dot{l}_\gamma(\gamma_0, Q_{10}, Q_{20}) \right\}, \end{aligned}$$

where $(Q_{10\gamma}, Q_{20\gamma})$ are the maximum likelihood estimators of (Q_1, Q_2) for fixed γ, and $\dot{l}_\gamma$ is the likelihood score for γ. $(Q_{10\gamma_1}, Q_{20\gamma_1})$ are the least favorable model defined by

$$\begin{aligned} dQ_{10\gamma_1}(y_1) &= u(1+ < \sigma_*^{-1}\{\sigma_1(h_0/\sqrt{n},0,0), \sigma_2(h_0/\sqrt{n},0,0)\}, e_1 >)dQ_{10}(y_1), \\ dQ_{20\gamma_1}(y_2) &= v(1+ < \sigma_*^{-1}\{\sigma_1(h_0/\sqrt{n},0,0), \sigma_2(h_0/\sqrt{n},0,0)\}, e_2 >)dQ_{20}(y_2), \end{aligned}$$

where $e_1 = (1,0)$ and $e_2 = (0,1)$, $< a,b >= \sum_k a_k b_k$ for vector $a = (a_k)$ and $b = (b_k)$, $\sigma_*(h_1,h_2) = \sigma(0,h_1,h_2)$, and u and v are taken such that $\int dQ_{10\gamma}(y_1) = 1$ and $\int dQ_{20\gamma}(y_2) = 1$ respectively. The first two terms on the right-hand side of the equation are $o_P(1)$ from Lemma 10.2.3 and the last term converges to $\sigma(h_0,0,0)$ following the arguments of Murphy and van der Vaart (2000)[128].

Corollary 10.2.1. Under regularity conditions 1a, 2–4, and 5a, the maximum likelihood estimator of γ from maximizing (10.9) exists and is almost surely consistent. Furthermore, the maximum likelihood estimator of γ is asymptotically normally distributed. That is,

$$\sqrt{n}(\hat{\gamma} - \gamma_0) \to N(0, V_0),$$

where the asymptotic variance satisfying $h_0^T V_0 h_0 = \sigma^{-1}(h_0,0,0)$ with σ defined in the proof earlier and h_0 being a vector having the same dimension as γ. The variance function σ can be consistently estimated by the sampling version of the inverse of the information operator or by the inverse of the information matrix of the profile likelihood for γ from (10.9).

Proof: This is the direct consequence of Theorem 10.2.1 when the true distribution generated the data is in the semiparametric odds ratio model.

10.4 Properties of the maximum penalized semiparametric likelihood estimator

10.4.1 Settings for the analysis

Let Z be the outcome and X be the covariates. Let the parametric odds ratio function be $\eta(z;x \mid \theta), \theta \in \Theta$. Let x_0 be a fixed point and F denote the distribution function of Z given $X = x_0$. The likelihood for (θ, F) with an independent sample $(Z_i, X_i), i = 1, \cdots, n$ is

$$L^*(\theta, F) = \prod_{i=1}^n \frac{\eta(Z_i; X_i \mid \theta)dF(Z_i \mid x_0)}{\int \eta(z; X_i \mid \theta)dF(z \mid x_0)}.$$

The penalized likelihood approach minimizes the following objective function

$$l_\lambda^*(\theta, F) = l^*(\theta, F) + \lambda_n ||\theta||_1 \tag{10.10}$$

with respect to (θ, F) for some $\lambda_n \geq 0$, where $||\theta||_1 = \sum_{j=1}^r |\theta_j|$ and

$$l^*(\theta, F) = -\frac{1}{n}\sum_{i=1}^n \log \eta(Z_i, X_i|\theta) - \frac{1}{n}\sum_{i=1}^n \log dF(Z_i) + \frac{1}{n}\sum_{i=1}^n \log \int \eta(z; X_i|\theta) dF(z),$$

is the negative log-likelihood given the observed data $(Z_i, X_i), i = 1, \cdots, n$, and x_0 is suppressed. Let $\mathcal{F}$ be the collection of all possible distributions for Z, and $\mathcal{F}_n$ be the collection of distributions for Z with probability masses on the observed Z points. For any (discrete) distribution $F \in \mathcal{F}$, there is an $F_n \in \mathcal{F}_n$ such that $l^*(\theta, F) \geq l^*(\theta, F_n)$. This means that the minimizer, when exists, assigns probability mass to the observed Z values only.

Let P_n and Q_n denote the empirical measures generated by Z and X respectively. Define

$$\mathcal{H}_{P_n} = \left\{ H_n(z) = \log \frac{dF_n(z)}{dP_n(z)} \mid F_n \in \mathcal{F}_n, P_n \ll F_n \ll P_n \right\}.$$

For simplicity, we also use $\mathcal{H}_n$ for $\mathcal{H}_{P_n}$. The minimization problem (10.10) can be transformed into minimizing

$$l_\lambda(\theta, H) = l(\theta, H) + \lambda_n ||\theta||_1,$$

with $(\theta, H) \in \Theta \times \mathcal{H}_n$, where

$$l(\theta, H) = -\frac{1}{n}\sum_{i=1}^n \log \eta(Z_i; X_i|\theta) - \frac{1}{n}\sum_{i=1}^n H(Z_i) + \frac{1}{n}\sum_{i=1}^n \log \int \eta(z; X_i|\theta) e^{H(z)} dP_n(z).$$

For given θ, P, and Q, let $H(z \mid \theta, P, Q)$ denote the minimizer of

$$-\int H(z) dP(z) + \int \left\{ \log \int \eta(z; x \mid \theta) e^{H(z)} dP(z) \right\} dQ(x),$$

in a given set of H when the minimizer exists. For a sufficiently large set of H, the minimizer satisfies

$$H(z \mid \theta, P, Q) = -\log \int \frac{\eta(z; x \mid \theta)}{\int \eta(z; x \mid \theta) \exp\{H(z \mid \theta, P, Q)\} dP(z)} dQ(x), \tag{10.11}$$

where $\int \exp\{H(z \mid \theta, P, Q)\} dP(z) = 1$. Let $H_n(z \mid \theta) = H(z \mid \theta, P_n, Q_n)$, the minimizer for $H \in \mathcal{H}_n$, and $l_{pn}(\theta) = l\{\theta, H_n(\cdot \mid \theta)\}$, the profile likelihood for θ. Let P_0 and Q_0 denote respectively the true marginal probability measures generated by Z and X. Define

$$\mathcal{H}_{P_0} = \left\{ H(z) = \log \frac{dF(z)}{dP_0(z)} \mid F \in \mathcal{F}, P_0 \ll F \ll P_0 \right\}.$$

We also use $\mathcal{H}_0$ for $\mathcal{H}_{P_0}$. Let $H_0(z \mid \theta) = H(z \mid \theta, P_0, Q_0)$.

We make the following assumptions in the study of the penalized likelihood estimator.

1. $\mathcal{Z} \times \mathcal{X} \times \Theta$ is a closed compact set in R^{p+q+r}, where $\mathcal{Z}$, $\mathcal{X}$, and Θ are respectively of dimension p, q, r. Let (Z, X) be random variables defined on probability space $(\mathcal{Z} \times \mathcal{X}, \Omega_{\mathcal{Z}} \times \Omega_{\mathcal{X}}, \mathcal{P})$.

2. $\eta(z; x \mid \theta)$ is an odds ratio function defined on $\mathcal{Z} \times \mathcal{X} \times \Theta$ for z versus x . There exists $M > 1$ such that $M^{-1} \leq \eta(z; x \mid \theta) \leq M$, and $L > 0$ such that, Lipschitz condition holds for $k = 0, 1, 2, 3$,

$$\left| \frac{\partial^k}{\partial^k \theta} \eta(z_1; x_1 \mid \theta_1) - \frac{\partial^k}{\partial^k \theta} \eta(z_2; x_2 \mid \theta_2) \right| \leq L ||(z_1, x_1, \theta_1) - (z_2, x_2, \theta_2)||_2.$$

3. There exists θ_0, an inner point of Θ, such that the density of Z given $X = x$ with respect to a known measure μ, an appropriate product of counting measures and Lebesgue measures, has the form

$$\frac{\eta(z; x \mid \theta_0) f(z \mid x_0)}{\int \eta(z; x \mid \theta_0) f(z \mid x_0) d\mu(z)}$$

where $f(z \mid x_0)$ is the conditional density of Z given $X = x_0$. Let S be a set of indices for θ such that $\theta_{j0} = 0$ for any $j \in S^c$, where $\theta_0 = (\theta_{j0}, j = 1, \cdots, r)$. Let $\theta^S = (\theta_j, j \in S)$, $\theta^{S^c} = (\theta_j, j \in S^c)$, and $\theta = (\theta^S, \theta^{S^c})$. Assume

$$E_0 \left[\mathrm{Var}_0 \left\{ \frac{\partial}{\partial \theta^S} \log \eta(Z; X \mid \theta_0) \middle| X \right\} \right] > 0,$$

where E_0 and Var_0 are mean and variance under the true distribution.

4. Convexity: Assume that Θ is a convex set. For any given probability measure P on $(\mathcal{Z}, \Omega_{\mathcal{Z}})$, let $\mathcal{H}_P^* = \{H(z) \mid \int \exp\{H(z)\} dP(z) < +\infty\}$ which is a convex set by Holder's inequality. For any fixed $(z, x) \in \mathcal{Z} \times \mathcal{X}$, assume that

$$- \log \frac{\eta(z; x \mid \theta) e^{H(z)}}{\int \eta(z; x \mid \theta) e^{H(z)} dP(z)}$$

is a convex function of $(\theta, H) \in \Theta \times \mathcal{H}_P^*$.

5. Minimal eigenvalue: There exists a constant $c_0 > 0$ independent of p_S, q_S, r_S such that

$$e_{min}(B^{SS}) \geq c_0 > 0.$$

6. Irrepresentable condition: There exists $\alpha > 0$ such that

$$|||B^{S^c S}(B^{SS})^{-1}|||_\infty \leq 1 - \alpha,$$

where $|||(a_{jk})|||_\infty = \max_j \sum_k |a_{jk}|$, is the derived matrix norm from the vector supremum norm.

10.4.2 Theoretical properties of the penalized semiparametric likelihood estimator

Lemma 10.4.1. Under Assumptions 1–3, $H_n(z \mid \theta)$ and $H_0(z \mid \theta)$ exist, are unique, and satisfy the score equation (10.11). Both are bounded, differentiable with respect to θ up to the third order. For $k \leq 3$, there exist $C_j > 0, j = 0, 1, 2$, such that

$$P \left\{ \sup_{z \in \mathcal{Z}, \theta \in \Theta} \left| \frac{\partial^k H_n(z|\theta)}{\partial^k \theta} - \frac{\partial^k H_0(z|\theta)}{\partial^k \theta} \right| \geq \delta \right\} \leq C_0 \exp \left[- \frac{n \left\{ \delta - C_1 \sqrt{(p \wedge q + r)/n} \right\}^2}{C_2} \right],$$

for $\delta > C_1\sqrt{(p \wedge q + r)/n}$, where $C_j > 0, j = 0, 1, 2$, are positive constants independent of (p, q, r, n), $p \wedge q = \max(p, q)$, and $\frac{\partial^k}{\partial^k \theta}$ denotes any one of such derivatives.

Proof of Lemma 10.4.1: Existence and uniqueness follow from Lemmas S.1 in the next subsection. Note that Lemmas S.10.4.1–S.10.4.13 in the next subsection are also cited in the proofs of other Lemmas and Theorems. Differentiability follows from Lemma S.10.4.2, and the probability bounds follow from Lemma S.10.4.6.

Lemma 10.4.1 allows us to treat the profile likelihood $l_{pn}(\theta) = l\{\theta, H_n(\cdot \mid \theta)\}$ approximately as the parametric likelihood $l_{p0}(\theta) = l\{\theta, H_0(\cdot \mid \theta)\}$ with high probability. Let

$$A_0\{h(Z)\} \quad = \quad E_0\left[E_0\left\{h(Z) \mid X\right\} \mid Z\right].$$

Under Assumptions 1–3, it is shown in Lemma S.10.4.2 that $(I - A_0)$ is continuously invertible on $\{h(z) \mid \int h(z)e^{H_0(z|\theta_0)}dP_0(z) = 0\}$. Let

$$\begin{aligned} B \quad = \quad & E_0\bigg(\frac{\partial}{\partial\theta}\log\eta(Z;X \mid \theta_0) - E_0\left\{\frac{\partial}{\partial\theta}\log\eta(Z;X \mid \theta_0) \mid X\right\} \\ & -(I - A_0)^{-1}\{h^*(Z)\} + E\left[(I - A)^{-1}\{h^*(Z)\} \mid X\right]\bigg)^2. \end{aligned}$$

where

$$h^*(Z) = E_0\left[\frac{\partial}{\partial\theta}\log\eta(Z;X \mid \theta_0) - E_0\left\{\frac{\partial}{\partial\theta}\log\eta(Z;X \mid \theta_0) \mid X\right\} \mid Z\right],$$

and $B = (B_{ij}, i = 1\cdots, r; j = 1\cdots, r)$. Let $B^{S^cS} = (B_{ij}, i \in S^c, j \in S)$ and $B^{SS} = (B_{ij}, i \in S, j \in S)$.

Lemma 10.4.2. Under Assumptions 1–3, $B^{SS} = E_0\{\ddot{l}_{p0}^{SS}(\theta_0)\}$ and there exists a constant depending on p_S, q_S, r_S only such that

$$e_{\min}(B^{SS}) \geq c(p_S, q_S, r_S) > 0,$$

where $e_{\min}(B^{SS})$ is the minimal eigenvalue of the matrix B^{SS}, and p_S, q_S, and r_S are the reduced dimensions of z, x, and θ^S involved in $\eta\{z; x \mid (\theta^S, 0)\}$.

Proof of Lemma 10.4.2: See Lemma S.10.4.3.

Theorem 10.4.1. Under Assumptions 1–5, there exist constants D, $C_j > 0, j = 0, 1, 2$ such that

$$P\left\{||\tilde{\theta}^S - \theta_0^S||_2 \geq D\sqrt{r_S}\lambda_n\right\} \quad \leq \quad C_0 r_S^2 \exp\left[-\frac{n}{C_2}\left\{\lambda_n - C_1\sqrt{(p_S \wedge q_S + r_S)/n}\right\}^2\right],$$

where $\lambda_n \geq C_1\sqrt{(p_S \wedge q_S + r_S)/n}$.

Proof of Theorem 10.4.1: Let $\delta = c\lambda_n$ in Lemma S.10.4.8 in the next subsection, it follows that

$$P\left\{||\dot{l}_{pn}^S(\theta_0)||_\infty > c\lambda_n\right\} \leq C_0 s \exp\left\{-\frac{n\{\lambda_n - C_1\sqrt{(p_S + q_S + s)/n}\}^2}{C_2}\right\},$$

where $\lambda_n > C_1\sqrt{(p_S + q_S + s)/n}$. Set $\delta = \pi_s/2$ in Lemma S. 9, it follows that

$$P\left\{e_{\min}\{\ddot{l}_{pn}^{SS}(\theta_0)\} \leq \pi_s/2\right\} \leq C_0 s^2 \exp\left\{-\frac{n(\pi_s/s - C_1\sqrt{(p_S + q_S + s)/n})^2}{C_2}\right\},$$

where $\ddot{l}_{pn}^{SS} = \frac{\partial^2 l_{pn}}{\partial^2 \theta^S}$, $\pi_s/s > C_1\sqrt{(p_S+q_S+s)/n}$. There exists $0 < M < \infty$ such that

$$\sup_{||\theta^S-\theta_0^S||_2 \le D\sqrt{s}\lambda_n} ||l_{np}^{(3)SSS}(\theta^S,0)||_\infty \le M,$$

where $l_{pn}^{(3)SSS} = \frac{\partial^3 l_{pn}}{\partial^3 \theta^S}$. From Lemma S.7 and Assumption 5, it follows that

$$\begin{aligned} P\left\{||\tilde{\theta}^S - \theta_0^S||_2 \ge D\sqrt{s}\lambda_n\right\} &\le C_0 s \exp\left[-\frac{n}{C_2}\left\{\lambda_n - C_1\sqrt{(p_S+q_S+s)/n}\right\}^2\right] \\ &\quad + C_0^* s^2 \exp\left[-\frac{n}{C_2^*}\left\{\pi_s/s - C_1^*\sqrt{(p_S+q_S+s)/n}\right\}^2\right]. \end{aligned}$$

The result follows by noting that $\pi_s/s \ge c_0\lambda_n$ and by readjusting the constant C_0, C_1, C_2.

To study maximizing the l_1 penalized likelihood as an approach to selecting the odds ratio model under sparsity restrictions, we start from the first order KKT conditions for the convex programming problem. These conditions are

$$\lambda_n\{||\hat{\theta}||_1 - C(\lambda_n)\} = 0,\ \lambda_n \ge 0,\ ||\hat{\theta}||_1 \le C(\lambda_n), \dot{l}_{pn}(\hat{\theta}) + \lambda_n u = 0,$$

where $u_i = \text{sign}(\theta_i)$ if $\theta_i \ne 0$ and $|u_i| \le 1$ if $\theta_i = 0$. We use the primal-dual witness approach (Ravikumar et al, 2010)[148] to proving sign-consistency of the variable selection. See also Zhao and Yu (2006)[205] and Meinshausen and Buhlmann (2006)[125]. Let

$$u^{S^c} = -\frac{1}{\lambda_n}\dot{l}_{pn}^{S^c}(\tilde{\theta}^S,0) = -\frac{1}{\lambda_n}\frac{\partial}{\partial\theta^{S^c}} l_{pn}(\tilde{\theta}^S,0).$$

Lemma S.12 shows that under Assumptions 1–6, when $\lambda_n \ge C_1\sqrt{(p_S \wedge q_S + r_S)/n}$,

$$P\left\{||u^{S^c}||_\infty < 1\right\} \ge 1 - C_0 r r_S \exp\left[-\frac{n\left\{\lambda_n - C_1\sqrt{(p_S\wedge q_S + r_S)/n}\right\}^2}{C_2}\right],$$

for some constants $C_j > 0, j = 0,1,2$. This allows us to show that $(\tilde{\theta}^S, 0)$ is also a minimizer of (7.16) with high probability. It can be shown that the minimizer of (7.16) is sign-consistent because $\tilde{\theta}^S$ is sign-consistent.

Theorem 10.4.2. Under Assumptions 1-6, if $\theta_{j0} \ge 3D\sqrt{r_S}\lambda_n/2$ for $j \in S_0 \subset S$ and the constant D in Theorem 10.4.1, there exist constants $C_j > 0, j = 0,1,2$ independent of p,q,r and p_S, q_S, r_S such that

$$\begin{aligned} &P\left\{\text{sign}(\hat{\theta}_j) = \text{sign}(\theta_{j0}), j \in S_0;\ \hat{\theta}^{S^c} = 0\right\} \\ &\quad \ge\ 1 - C_0 r r_S \exp\left[-\frac{n\left\{\lambda_n - C_1\sqrt{(p_S\wedge q_S + r_S)/n}\right\}^2}{C_2}\right], \end{aligned}$$

where $\lambda_n \ge C_1\sqrt{(p_S \wedge q_S + r_S)/n}$.

Proof of Theorem 10.4.2: $||u^{S^c}||_\infty < 1$ implies $\hat{\theta}^{S^c} = 0$. When $\hat{\theta}^{S^c} = 0$, $\hat{\theta}^S = \tilde{\theta}^S$. Since

$$||\tilde{\theta}^S - \theta_0^S||_2/\sqrt{s} \le ||\tilde{\theta}^S - \theta_0^S||_\infty \le ||\tilde{\theta}^S - \theta_0^S||_2,$$

when $||\tilde{\theta}^S - \theta_0^S||_2 \le D\sqrt{s}\lambda_n$, it follows that

$$||\tilde{\theta}^S - \theta_0^S||_\infty \le D\sqrt{s}\lambda_n,$$

which implies for $j \in S$

$$\theta_{j0} - D\sqrt{s}\lambda_n \le \tilde{\theta}_j \le \theta_{j0} + D\sqrt{s}\lambda_n.$$

When $|\theta_{j0}| > 3D\sqrt{s}\lambda_n/2$, the sign consistency holds for large n.

To ensure the probability of sign-consistency goes to 1, a sufficient condition is

$$\frac{n\left\{\lambda_n - C_1\sqrt{(p_S \wedge q_S + r_S)/n}\right\}^2}{C_2} - 2\log r \to \infty,$$

when $n \to \infty$. It is sufficient to have

$$\lambda_n \ge C_1^*\sqrt{\frac{p_S \wedge q_S + r_S}{n}} + C_2^*\sqrt{\frac{\log r}{n}},$$

when p_S, q_S, r_S do not change with n, and $r \to \infty$ as $n \to \infty$, where $C_1^* > C_1$ and $C_2^* > \sqrt{2C_2}$.

To distinguish different node clusters in this section, we add a subscription to the variables. Let $\theta_{0k} = (\theta_{j0k}, j = 1, \cdots, r_k)$, for a given node cluster, where $k = 1, \cdots, g$. Let $S_k = \{j \mid \theta_{j0k} \ne 0\}$. Let $\theta_{0k}^{S_k} = (\theta_{j0k}, j \in S_k)$ and $\theta_{0k}^{S_k^c} = (\theta_{j0k}, j \in S_k^c)$.

Theorem 10.4.3. Under Assumptions 1-6, if, for every node cluster k, $\theta_{j0k} > 3D_k\sqrt{r_{S_k}}\lambda_{nk}/2$ for all $j \in S_k$ and constant D_k in Theorem 1 for cluster nodes $k = 1, \cdots, g$, then

$$P(\hat{\mathcal{E}} = \mathcal{E}) \ge 1 - \sum_{k=1}^{g} C_{0k} r r_{S_k} \exp\left[-\frac{n\left\{\lambda_{nk} - C_{1k}\sqrt{(p_{S_k} \wedge q_{S_k} + r_{S_k})/n}\right\}^2}{C_{2k}}\right],$$

where $\hat{\mathcal{E}}$ and $\mathcal{E}$ are respectively estimated and true network edges.

Proof of Theorem 10.4.3: Note that, from Theorem 2,

$$\begin{aligned} P(\hat{\mathcal{E}} = \mathcal{E}) &= P\left[\cap_{k=1}^{g}\left\{\text{sign}(\hat{\theta}_{jk}) = \text{sign}(\theta_{j0k}), j \in S_k;\ \hat{\theta}^{S_k^c} = 0\right\}\right] \\ &\ge 1 - \sum_{k=1}^{g} C_{0k} r_k s_k \exp\left\{-\frac{n(\lambda_{nk} - C_{1k}\sqrt{(p_{Sk} + q_{Sk} + s_k)/n})^2}{C_{2k}}\right\}. \end{aligned}$$

The result follows from (10.14) with the guarantee of the existence of λ_{nk} under conditions (10.12) and (10.13).

Again, to ensure the probability of sign-consistency goes to 1, sufficient conditions are

$$\frac{n\left\{\lambda_{nk} - C_{1k}\sqrt{(p_{S_k} \wedge q_{S_k} + r_{S_k})/n}\right\}^2}{C_{2k}} - 2\log r \to \infty,$$

for all k when $n \to \infty$. It is sufficient to have

$$\lambda_{nk} \ge C_1^*\sqrt{\frac{p_S^* \wedge q_S^* + r_S^*}{n}} + C_2^*\sqrt{\frac{\log r}{n}},$$

when p_S^*, q_S^*, r_S^* do not change with n, and $r \to \infty$ as $n \to \infty$, where $C_1^* > \max_k(C_{1k})$, $C_2^* > \sqrt{2\max_k(C_{2k})}$, $p_S^* = \max_k(p_{S_k})$, $q_S^* = \max_k(q_{S_k})$, $r_S^* = \max_k(r_{S_k})$. The requirement for p_S, q_S, r_S to be constant independent of n can be relaxed.

From the proofs of the Lemmas and Theorems, we can derive more general results that those presented above. In particular, p_S, q_S, r_S may be allowed to depend on n, and π_{S_k} may be depend on p_S, q_S, r_S and thus on n, i.e., Assumption 5 may not hold. Let $p_{S_k} = O(n^{\delta_{1k}})$, $q_{S_k} = O(n^{\delta_{2k}})$, and $r_{S_k} = O(n^{\delta_{3k}})$ where $\delta_{1k}, \delta_{2k}, \delta_{3k} \geq 0$ for $k = 1, \cdots, g$. Let $\pi_{S_k} = \Omega(n^{\gamma_k})$, where, following Ravikumar et al (2010) [146], this notation means there exists $c > 0$ such that $\pi_{S_k} \geq cn^{\gamma_k}$.

From the proofs, it is required that, for some $C_1, C_1^* > 0$,

$$C_1\sqrt{(p_{S_k} \wedge q_{S_k} + r_{S_k})/n} \leq \lambda_{nk} \leq C_1^* \min(\pi_{S_k}/r_{S_k}, \pi_{S_k}^2/\sqrt{r_{S_k}}, 1/r_{S_k}^2).$$

Note that

$$\min(\pi_{S_k}/r_{S_k}, \pi_{S_k}^2/\sqrt{r_{S_k}}, 1/r_{S_k}^2) = \Omega\{\min(n^{\gamma_k - \delta_{3k}}, n^{2\gamma_k - \delta_{3k}/2}, n^{-2\delta_{3k}})\}$$

When $\gamma_k > \delta_{3k}$, the minimum is $n^{\gamma_k - \delta_{3k}}$ which goes to ∞ as $n \to \infty$. This case is less likely to happen and does not restrict λ_n if it does happen. We discuss the case where $\gamma_k \leq \delta_{3k}$ only in the remaining of this section. Assume without loss of generality that $r_k \leq 0 \leq \delta_{3k}$. It follows from $2(\gamma_k - \delta_{3k}) \leq \min\{\gamma_k - \delta_{3k}/2, -2\delta_{3k}\}$ that

$$O\{n^{\{\max(\delta_{1k}, \delta_{2k}, \delta_{3k}) - 1\}/2}\} \leq \lambda_{nk} \leq \Omega\{n^{2(\gamma_k - \delta_{3k})}\}. \tag{10.12}$$

Hence, a requirement is $\{\max(\delta_{1k}, \delta_{2k}, \delta_{3k}) - 1\}/2 < 2(\gamma_k - \delta_{3k})$, which means

$$\gamma_k > \delta_{3k} + \frac{1}{4}\{\max(\delta_{1k}, \delta_{2k}, \delta_{3k}) - 1\}. \tag{10.13}$$

In addition, for a single neighborhood, if

$$\frac{n\left\{\lambda_{nk} - C_{1k}\sqrt{(p_{S_k} \wedge q_{S_k} + r_{S_k})/n}\right\}^2}{C_{2k}} - \log r_k - \log r_{S_k} \to \infty$$

as $n \to \infty$, consistency of the neighborhood selection can be achieved. This requires

$$\lambda_{nk} > C_{1k}\sqrt{(p_{Sk} \wedge q_{S_k} + r_{S_k})/n} + \sqrt{C_{2k}\log(r_k r_{S_k})/n}.$$

For the whole network, if

$$\sum_{k=1}^{g} C_{0k} \exp\left[\frac{-n\left\{\lambda_{nk} - C_{1k}\sqrt{(p_{S_k} \wedge q_{S_k} + r_{S_k})/n}\right\}^2}{C_{2k}} + \log r_k + \log r_{S_k}\right] \to 0 \tag{10.14}$$

as $n \to \infty$, consistency of the whole network selection can be achieved.

A sufficient condition for (10.12), (10.13), and (10.14) to hold simultaneously is

$$\max_k(C_{1k})\sqrt{\frac{n^\delta + \log(\max_k r_k)}{n}} \leq \lambda_n \leq \Omega(n^{2(\gamma - \beta)}),$$

where $\delta = \max_{1 \leq k \leq g} \max(\delta_{1k}, \delta_{2k}, \delta_{3k})$, $\gamma = \min_{1 \leq k \leq g} \gamma_k$, and $\beta = \max_{1 \leq k \leq g} \delta_{3k}$. For example, when p_S, q_S, r_S do not change with n, then $\delta = 0 = \beta$. Then as long as $\gamma > -1/4$ and the size of the network increases polynomially in n, λ_n for yielding a consistent network selection exists. This means that the minimum eigenvalue are allowed to go to 0 as $n \to \infty$. Another example is when p_S, q_S, r_S all changes with n with $\delta = \beta = 1/6$, then λ_n for yielding a consistent network selection exists if $\gamma > -1/24$ and the size of the network increases polynomially in n. When $\gamma = 0$, i.e., the minimal eigenvalue does not go to 0 as $n \to \infty$, $\delta = \beta < 1/5$ and the size of the network increases polynomially in n suffice.

10.4.3 Supplemental lemmas for the proofs of main results

We state and prove a set of lemmas for the proofd of Lemmas and Theorems in the previous subsection. In the following, we assume both $\mathcal{Z}$ and $\mathcal{X}$ are compact subset in Euclidean spaces.

Lemma S.10.4.1. Let P and Q be probability measures on $\mathcal{Z}$ and $\mathcal{X}$ respectively. Assume:

1. There exists $M > 1$ such that $M^{-1} \leq \eta(z; x \mid \theta) \leq M$ for $P \times Q$ almost all $(z, x) \in \mathcal{Z} \times \mathcal{X}$ and any $\theta \in \Theta$,
2. There exists $0 < K < +\infty$ such that, for any $z_1, z_2 \in \mathcal{Z}$, $x \in \mathcal{X}$, and $\theta \in \Theta$,

$$|\eta(z_1; x \mid \theta) - \eta(z_2; x \mid \theta)| \leq K||z_1 - z_2||_2.$$

Then the minimizer $H(z \mid \theta) \in \mathcal{H}_P = \{H(z) = \log \frac{dF}{dP} \mid F \in \mathcal{F}, P \ll F \ll P\}$ of

$$\mathcal{L}(\theta, H) = -\int H(z \mid \theta) dP(z) + \int \log \int \eta(z; x \mid \theta) e^{H(z|\theta)} dP(z) dQ(x)$$

exists and satisfies

$$\dot{\mathcal{L}}_H(\theta, H)(h) = -\int h(z) dP(z) + \int \frac{\int \eta(z; x|\theta) e^{H(z|\theta)} h(z) dP(z)}{\int \eta(z; x|\theta) e^{H(z|\theta)} dP(z)} dQ(x) = 0,$$

for any $h(z) \in \mathcal{H} = \{h(z) \mid \int h(z) e^{H(z|\theta)} dP(z) = 0\}$. The minimizer is unique. If $\eta(z, x \mid \theta)$ is continuous with respect to $\theta \in \Theta$ for any $(z, x) \in \mathcal{Z} \times \mathcal{X}$, then $H(z \mid \theta)$ is continuous with respect to $\theta \in \Theta$ for any $z \in \mathcal{Z}$.

Proof: Let $H(z \mid \theta)$ denote the minimizer in $\mathcal{H}_P$ if it exists. Then

$$\mathcal{L}\{\theta, H(z \mid \theta)\} \leq \mathcal{L}\{\theta, H(z)\}$$

for any $H(z) \in \mathcal{H}_P$. For any $h \in \mathcal{H}$ such that $\int \exp\{H(z \mid \theta) + \epsilon h(z)\} dP(z) = 1$, it follows that $\int h(z) e^{H(z|\theta)} dP(z) = 0$ and that

$$\dot{\mathcal{L}}_H\{\theta, H(z \mid \theta)\}(h) = \lim_{\epsilon \downarrow 0} \frac{1}{\epsilon} [\mathcal{L}\{\theta, H(z \mid \theta) + \epsilon h(z)\} - \mathcal{L}\{\theta, H(z \mid \theta)\}] \geq 0.$$

On the other hand, let $r(\epsilon, h) = \int \exp\{H(z \mid \theta) - \epsilon h(z)\} dP(z)$, then $\int \exp\{H(z \mid \theta) - \epsilon h(z) - \log r(\epsilon, h)\} dP(z) = 1$. Since

$$\mathcal{L}\{\theta, H(z \mid \theta) - \epsilon h(z) - \log r(\epsilon, h)\} = \mathcal{L}\{\theta, H(z \mid \theta) - \epsilon h(z)\},$$

it follows that

$$-\dot{\mathcal{L}}_H\{\theta, H(z \mid \theta)\}(h) = \lim_{\epsilon \downarrow 0} \frac{1}{\epsilon} [\mathcal{L}\{\theta, H(z \mid \theta) - \epsilon h(z) - \log r(\epsilon, h)\} - \mathcal{L}\{\theta, H(z \mid \theta)\}] \geq 0.$$

The score equation follows.

We next prove the existence of $H(z \mid \theta) \in \mathcal{H}_P$. For a fixed $\theta \in \Theta$, consider the mapping

$$\begin{aligned}\varphi(H)(z) &= -\log \int \frac{\eta(z; x \mid \theta)}{\int \eta(z; x \mid \theta) \exp\{H(z)\} dP(z)} dQ(x) \\ &\quad - \log \int \left[\int \frac{\eta(z; x \mid \theta)}{\int \eta(z; x \mid \theta) \exp\{H(z)\} dP(z)} dQ(x)\right]^{-1} dP(z)\end{aligned}$$

where $H \in \mathcal{H}_P$. From the assumption that $M^{-1} \le \eta(z; x \mid \theta) \le M$, it follows that

$$-4 \log M \le \varphi(H)(z) \le 4 \log M$$

and that $\int \exp\{\varphi(H)(z)\} dP(z) = 1$ by the definition of the map. Further, from $|\log x - \log y| \le |x - y| \max(1/x, 1/y)$ for any $x, y > 0$, it follows that

$$|\varphi(H)(z_1) - \varphi(H)(z_2)| \le K^* ||z_1 - z_2||_2,$$

where K^* is a constant depending on K and M only. It follows that $\varphi(\mathcal{H}_P)$ is uniformly bounded and equi-continuous. By Arzela-Ascoli theorem, $\varphi(\mathcal{H}_P)$ is compact in supremum norm. When $H(z \mid \theta)$ exists, it follows from the score equation that

$$\int h(z) \left\{ \int \frac{\eta(z; x \mid \theta)}{\int \eta(z; x \mid \theta) e^{H(z|\theta)} dP(z)} dQ(x) - e^{-H(z|\theta)} \right\} e^{H(z|\theta)} dP(z) = 0,$$

for any $h \in \mathcal{H}$. This implies that

$$\int \frac{\eta(z; x \mid \theta)}{\int \eta(z; x \mid \theta) e^{H(z|\theta)} dP(z)} dQ(x) = e^{-H(z|\theta)}. \tag{10.15}$$

Hence, the minimizer $H(z \mid \theta)$, when exists, satisfies $\varphi\{H(z \mid \theta)\} = H(z \mid \theta)$. It follows that

$$\inf_{H \in \varphi(\mathcal{H}_P)} \mathcal{L}(\theta, H) = \inf_{H \in \mathcal{H}_P} \mathcal{L}(\theta, H).$$

Let $+\infty > A = \inf_{H \in \varphi(\mathcal{H}_P)} \mathcal{L}(\theta, H) > -\infty$. Let $H_m \in \varphi(\mathcal{H}_P)$ such that $\mathcal{L}(\theta, H_m) \le A + 1/m$. Since $\varphi(\mathcal{H})$ is compact in supremum norm, there exists a convergent subsequence of $H_m(z)$ in supremum norm. Without loss of generality, assume $H_m(z)$ converges in supremum norm. Let $H^*(z) = \lim_{m \to \infty} H_m(z)$. It follows that $\sup_{z \in \mathcal{Z}} |H_m(z) - H^*(z)| \to 0$. This implies $\mathcal{L}(\theta, H^*) = A$ and $\int \exp\{H^*(z)\} dP(z) = 1$. Hence $H^* \in \mathcal{H}_P$ is a minimizer and thus satisfies (10.15), which also implies $H^* \in \varphi(\mathcal{H}_P)$.

To prove the uniqueness, let $H_1(z)$ and $H_2(z)$ minimize $\mathcal{L}(\theta, H)$. Then

$$\int \frac{\eta(z; x \mid \theta) e^{H_k(z)}}{\int \eta(z; x \mid \theta) e^{H_k(z)} dP(z)} dQ(x) = 1$$

for $k = 1, 2$. In addition, since $\mathcal{L}\{\theta, H_1(z)\} = \mathcal{L}\{\theta, H_2(z)\}$, it follows that

$$\begin{aligned}
\int \{H_1(z) - H_2(z)\} dP(z) &= \int \log \frac{\int e^{H_1(z) - H_2(z)} \eta(z; x \mid \theta) e^{H_2(z)} dP(z)}{\int \eta(z; x \mid \theta) e^{H_2(z)} dP(z)} dQ(x) \\
&\ge \int \frac{\int \{H_1(z) - H_2(z)\} \eta(z; x \mid \theta) e^{H_2(z)} dP(z)}{\int \eta(z; x \mid \theta) e^{H_2(z)} dP(z)} dQ(x) \\
= \int \{H_1(z) - H_2(z)\} &\int \frac{\eta(z; x \mid \theta) e^{H_2(z)}}{\int \eta(z; x \mid \theta) e^{H_2(z)} dP(z)} dQ(x) dP(z) \\
&= \int \{H_1(z) - H_2(z)\} dP(z).
\end{aligned}$$

The equality holds. Since the equality holds if and only if $H_1(z) - H_2(z) = c$, a constant. From $\int e^{H_1(z)} dP(z) = 1 = \int e^{H_2(z)} dP(z)$, it follows that $c = 0$. Hence, $H_1(z) = H_2(z)$. In the following, to bring in the dependence of H on θ and P and Q, we will use $H(z \mid \theta)$ or $H(z \mid \theta, P, Q)$ to denote the unique solution.

To show that $H(z \mid \theta)$ is continuous with respect to θ, let $H(z \mid \theta + \Delta)$ be the minimizer of $\mathcal{L}(\theta + \Delta, H)$. Let Δ_m be an arbitrary sequence such that $\Delta_m \to 0$ when $m \to \infty$. Hence $H(z \mid \theta + \Delta_m)$ is compact and have convergent subsequence. Without loss of generality, let $H(z \mid \theta + \Delta_m) \to H^*(z)$ in supremum norm. Then by control convergence theorem,

$$H^*(z) = -\log \int \frac{\eta(z; x \mid \theta)}{\int \eta(z; x \mid \theta) e^{H^*(z)} dP(z)} dQ(x).$$

Since $\mathcal{L}\{\theta + \Delta_m, H(z \mid \theta + \Delta_m)\} \leq \mathcal{L}\{\theta + \Delta_m, H(z \mid \theta)\}$, it follows that

$$\mathcal{L}\{\theta, H^*(z)\} \leq \mathcal{L}\{\theta, H(z \mid \theta)\}.$$

By the uniqueness result, we have $H^*(z) = H(z \mid \theta)$. Since any convergent sequence converges to the same limit, the original sequence converges. This implies that $H(z \mid \theta)$ is continuous with respect to $\theta \in \Theta$ for any $z \in \mathcal{Z}$.

Lemma S.10.4.2. Under the conditions of Lemma S.10.4.1, if further that there exists $0 < K < +\infty$ such that, for any $z_1, z_2 \in \mathcal{Z}$, $x \in \mathcal{X}$, and $\theta \in \Theta$,

$$\left| \frac{\partial^k}{\partial^k \theta} \log \eta(z_1; x \mid \theta) - \frac{\partial^k}{\partial^k \theta} \log \eta(z_2; x \mid \theta) \right| \leq K ||z_1 - z_2||_2,$$

for $k = 1, \cdots, K_0$, then $\frac{\partial^k}{\partial^k \theta} H(z \mid \theta)$ exists, $\frac{\partial^{k-1}}{\partial^{k-1} \theta} H(z \mid \theta)$ is equicontinuous with respect to θ for $k = 1, \cdots, K_0$, and

$$\left| \frac{\partial^k}{\partial^k \theta} H(z_1 \mid \theta) - \frac{\partial^k}{\partial^k \theta} H(z_2 \mid \theta) \right| \leq K^* ||z_1 - z_2||_2,$$

for some K^* depending on k, K, and M only.

Proof: From Lemma 10.4.1, $H(z \mid \theta)$ satisfies $\int \exp\{H(z \mid \theta)\} dP(z) = 1$ and

$$H(z \mid \theta) = -\log \int \frac{\eta(z; x \mid \theta)}{\int \eta(z; x \mid \theta) e^{H(z|\theta)} dP(z)} dQ(x),$$

and is continuous with respect to θ. Let

$$s(z \mid \theta, \Delta) = \frac{1}{\Delta} \left\{ e^{H(z|\theta + \alpha\Delta) - H(z|\theta)} - 1 \right\}$$

where α is a direction vector. It follows that $\int s(z \mid \theta, \Delta) e^{H(z|\theta)} dP(z) = 0$ and

$$\begin{aligned} s(z \mid \theta, \Delta) &= \frac{1}{\Delta} \left[\left\{ \int \frac{\eta(z; x \mid \theta + \alpha\Delta) e^{H(z|\theta)}}{\int \eta(z; x \mid \theta + \alpha\Delta) e^{H(z|\theta + \alpha\Delta)} dP(z)} dQ(x) \right\}^{-1} - 1 \right] \\ &= \frac{\int \frac{\eta(z;x|\theta+\alpha\Delta) \int \eta(z;x|\theta+\alpha\Delta) s(z|\theta,\Delta) e^{H(z|\theta)} dP(z)}{\int \eta(z;x|\theta+\alpha\Delta) e^{H(z|\theta+\alpha\Delta)} dP(z) \int \eta(z;x|\theta+\alpha\Delta) e^{H(z|\theta)} dP(z)} dQ(x)}{\int \frac{\eta(z;x|\theta+\alpha\Delta)}{\int \eta(z;x|\theta+\alpha\Delta) e^{H(z|\theta+\alpha\Delta)} dP(z)} dQ(x)} \\ &\quad - \frac{\int \frac{1}{\Delta} \left\{ \frac{\eta(z;x|\theta+\alpha\Delta)}{\int \eta(z;x|\theta+\alpha\Delta) e^{H(z|\theta)} dP(z)} - \frac{\eta(z;x|\theta)}{\int \eta(z;x|\theta) e^{H(z|\theta)} dP(z)} \right\} dQ(x)}{\int \frac{\eta(z;x|\theta+\alpha\Delta)}{\int \eta(z;x|\theta+\alpha\Delta) e^{H(z|\theta+\alpha\Delta)} dP(z)} dQ(x)}. \end{aligned}$$

As $\Delta \to 0$, the first term on the right-hand side of equation is asymptotically equivalent to $A\{s(z)\} + o(||s(z)||_\infty)$, where

$$A(h)(z) = \int \frac{\eta(z; x \mid \theta) e^{H(z|\theta)} \int h(z) \eta(z; x \mid \theta) e^{H(z|\theta)} dP(z)}{\{\int \eta(z; x \mid \theta) e^{H(z|\theta)} dP(z)\}^2} dQ(x).$$

The second term of the right-hand side of the equation converges to

$$-u(z) = \frac{\int \frac{\alpha^T \frac{\partial}{\partial\theta}\eta(z;x|\theta)}{\int \eta(z;x|\theta)e^{H(z|\theta)}dP(z)} - \frac{\eta(z;x|\theta)\int \alpha^T \frac{\partial}{\partial\theta}\eta(z;x|\theta)e^{H(z|\theta)}dP(z)}{\{\int \eta(z;x|\theta)e^{H(z|\theta)}dP(z)\}^2}dQ(x)}{\int \frac{\eta(z;x|\theta)}{\int \eta(z;x|\theta)e^{H(z|\theta)}dP(z)}dQ(x)}$$

$$= \int\left\{\frac{\alpha^T \frac{\partial \log \eta}{\partial\theta}\eta(z;x\mid\theta)e^{H(z|\theta)}}{\int \eta(z;x\mid\theta)e^{H(z|\theta)}dP(z)} - \frac{\eta(z;x\mid\theta)e^{H(z|\theta)}\int \alpha^T \frac{\partial \log\eta}{\partial\theta}\eta(z;x\mid\theta)e^{H(z|\theta)}dP(z)}{\{\int \eta(z;x\mid\theta)e^{H(z|\theta)}dP(z)\}^2}\right\}dQ(x).$$

Hence, the equation can be rewritten as

$$(I-A)\{s(z\mid\theta,\Delta)\}(z) = u(z) + o(1+||s||_\infty),$$

where $\int u(z)dP(z) = 0$. Note that $\int A(h)(z)dP(z) = \int h(z)dP(z)$. Consider

$$(I-A): \left\{h(z)\mid \int h(z)\exp\{H(z\mid\theta)\}dP(z) = 0\right\} \longrightarrow \left\{g(z)\mid \int g(z)dP(z) = 0\right\}.$$

Note first that $(I-A)$ is a one-to-one mapping. This is equivalent to that, for any $h(z)$ such that $\int h(z)\exp\{H(z\mid\theta)\}dP(z) = 0$, if $(I-A)h = 0$ then $h = 0$. Suppose that $Ah = h$. Then,

$$\begin{aligned}
\int h^2(z)dP(z) &= \int h(z)A(h)(z)dP(z) \\
&= \int h(z)\int \frac{\eta(z;x\mid\theta)e^{H(z|\theta)}\int h(z)\eta(z;x\mid\theta)e^{H(z|\theta)}dP(z)}{\{\int \eta(z;x\mid\theta)e^{H(z|\theta)}dP(z)\}^2}dQ(x)dP(z) \\
&= \int\left\{\frac{\int h(z)\eta(z;x\mid\theta)e^{H(z|\theta)}dP(z)}{\int \eta(z;x\mid\theta)e^{H(z|\theta)}dP(z)}\right\}^2 dQ(x) \\
&\le \int \frac{\int h^2(z)\eta(z;x\mid\theta)e^{H(z|\theta)}dP(z)}{\int \eta(z;x\mid\theta)e^{H(z|\theta)}dP(z)}dQ(x) \\
&= \int h^2(z)\int \frac{\eta(z;x\mid\theta)e^{H(z|\theta)}}{\int \eta(z;x\mid\theta)e^{H(z|\theta)}dP(z)}dQ(x)dP(z) \\
&= \int h^2(z)dP(z).
\end{aligned}$$

This means that the equality in the inequality holds, which further means that $h(z) = c$, a constant. Since $\int h(z)e^{H(z)}dP(z) = 0$, this means $c = 0$. Next, it can be directly verified that for any bounded $\{h_m(z)\}$, $\{A(h_m)(z)\}$ is a compact sequence in $||\cdot||_\infty$. This means A is a compact operator in the supremum norm. Hence $(I-A)$ is continuously invertible on $\{g(z)\mid \int g(z)dP(z) = 0\}$ and $(I-A)^{-1}g(z) = \sum_{k=0}^{\infty} A^k g(z)$.

It follows from $\int u(z)dP(z) = 0$ that

$$s(z\mid\theta,\Delta) = (I-A)^{-1}u(z) + o(1+||s(z\mid\theta,\Delta)||_\infty).$$

It can be concluded from the expression first that $||s(z\mid\theta,\Delta)||_\infty = O(1)$ and then that

$$\lim_{\Delta\to 0}||s(z\mid\theta,\Delta) - (I-A)^{-1}(u)(z)||_\infty = 0.$$

The expression also implies that $H(z\mid\theta)$ is equicontinuous with respect to θ. Note that

$$\frac{1}{\Delta}\{H(z\mid\theta+\Delta) - H(z\mid\theta)\} = \frac{1}{\Delta}\log\{1+\Delta s(z\mid\theta,\Delta)\}.$$

It can be concluded that the derivatives of $H(z \mid \theta)$ with respect to θ exists and

$$\alpha^T \frac{\partial}{\partial \theta} H(z \mid \theta) = (I - A)^{-1}(u)(z).$$

By the calculation of implicit function derivative, it can be directly verified that

$$\left| \frac{\partial}{\partial \theta} H(z_1 \mid \theta) - \frac{\partial}{\partial \theta} H(z_2 \mid \theta) \right| \le K^* ||z_1 - z_2||_2.$$

To prove the remaining results, note that

$$0 = \lim_{\Delta \to 0} \int \frac{1}{\Delta} s(z \mid \theta, \Delta) e^{H(z|\theta)} dP(z) = \int \alpha^T \frac{\partial H(z \mid \theta)}{\partial \theta} e^{H(z|\theta)} dP(z).$$

and that

$$(I - A_\theta) \left\{ \alpha^T \frac{\partial}{\partial \theta} H(z \mid \theta) \right\} (z) = u(z \mid \theta),$$

where A_θ is A and $u(z \mid \theta)$ is $u(z)$ when explicit dependence of θ is emphasized. Under the additional condition, it is a routine to show that $u(z \mid \theta)$ is differentiable with respect to θ and has bounded derivatives, and A_θ is differentiable in derived supremum norm and the derived operator, denoted by $\frac{\partial A_\theta}{\partial \theta}$, is bounded linear. Hence,

$$(I - A_\theta)\{s^*(z \mid \theta, \Delta)\} = \beta^T \frac{\partial}{\partial \theta} u(z \mid \theta) + \beta^T \frac{\partial A_\theta}{\partial \theta} \left\{ \alpha^T \frac{\partial}{\partial \theta} H(z \mid \theta) \right\} + o\{\Delta(1 + ||s^*||_\infty)\},$$

where β is another direction vector and

$$s^*(z \mid \theta, \Delta) = \frac{1}{\Delta} \alpha^T \left\{ \frac{\partial H(z \mid \theta + \beta\Delta)}{\partial \theta} - \frac{\partial H(z \mid \theta)}{\partial \theta} \right\}.$$

Since $(I - A_\theta)$ is continuously invertible,

$$s^*(z \mid \theta, \Delta) = (I - A_\theta)^{-1} \left[\beta^T \frac{\partial}{\partial \theta} u(z \mid \theta) + \beta^T \frac{\partial A_\theta}{\partial \theta} \left\{ \alpha^T \frac{\partial}{\partial \theta} H(z \mid \theta) \right\} \right] + o(1 + ||s^*||_\infty).$$

The result follws when $\Delta \to 0$. The results for $k = 3, \cdots, K_0$ follow similar proofs.

Lemma S.10.4.3. Under Assumptions 1–3, $E\ddot{l}^S_{p0}(\theta_0)$ is positive definite, where $l_{p0}(\theta) = l\{\theta, H_0(\cdot \mid \theta)\}$ and $\ddot{l}^S_{p0}(\theta)$ is the second derivative matrix of $l_{p0}(\theta)$ with respect to θ^S.

Proof: From Lemmas S.10.4.1 and S.10.4.2,

$$f(z \mid x, \theta) = \frac{\eta(z; x \mid \theta) e^{H_0(z|\theta)}}{\int \eta(z; x \mid \theta) e^{H_0(z|\theta)} dP_0(z)}$$

is a regular parametric model for $\theta \in \Theta$. It follows that

$$E\ddot{l}^S_{p0}(\theta_0) = E\{\dot{l}^S_{p0}(\theta_0)\}^2,$$

where $a^2 = aa^T$ for a vector a. If $e_{\min}\{E\ddot{l}^S_{p0}(\theta_0)\} = 0$, then there exists $v \neq 0$ such that $v^T E\ddot{l}^S_{p0}(\theta_0) v = 0$, which means $v^T \dot{l}^S_{p0}(\theta_0) = 0$, where

$$\begin{aligned} \dot{l}^S_{p0}(\theta_0) &= -\left[\frac{\partial}{\partial \theta^S} \log \eta(Z; X \mid \theta_0) - E_0 \left\{ \frac{\partial}{\partial \theta^S} \log \eta(Z; X \mid \theta_0) \mid X \right\} \right] \\ &\quad - \left[\frac{\partial}{\partial \theta^S} H_0(Z \mid \theta_0) - E_0 \left\{ \frac{\partial}{\partial \theta^S} H_0(Z \mid \theta_0) \mid X \right\} \right] \end{aligned}$$

Let $Z = z_0$. Then

$$E_0\left\{v^T \frac{\partial}{\partial\theta^S}\log\eta(Z;X\mid\theta_0)\mid X\right\} + E_0\left\{v^T\frac{\partial}{\partial\theta^S}H_0(Z\mid\theta_0)\mid X\right\} = v^T\frac{\partial}{\partial\theta^S}H_0(z_0\mid\theta_0).$$

By plugging in the foregoing equality back to the score, it means that

$$-v^T\frac{\partial}{\partial\theta^S}\log\eta(Z;X\mid\theta_0) = v^T\left\{\frac{\partial}{\partial\theta^S}H_0(Z\mid\theta_0) - \frac{\partial}{\partial\theta^S}H_0(z_0\mid\theta_0)\right\}.$$

By setting $X = x_0$, it follows that

$$v^T\left\{\frac{\partial}{\partial\theta^S}H_0(Z\mid\theta_0) - \frac{\partial}{\partial\theta^S}H_0(z_0\mid\theta_0)\right\} = 0.$$

This further suggests

$$v^T E_0\left[\mathrm{Var}_0\left\{\frac{\partial}{\partial\theta^S}\log\eta(Z;X\mid\theta_0)\mid X\right\}\right]v = 0,$$

which contradicts with Assumption 2.

Lemma S.10.4.4. For a countable class of measurable functions $g \in \mathcal{G}$ and independent random variables $\xi_1,\cdots,\xi_n$, if $a_{ig} \le g(\xi_i) \le b_{ig}$ for all $g \in \mathcal{G}$, then for any $x > 0$,

$$P\left\{\sup_{g\in\mathcal{G}}\sum_{i=1}^n[g(\xi_i) - E\{g(\xi)\}] - E\sup_{g\in\mathcal{G}}\sum_{i=1}^n[g(\xi_i) - E\{g(\xi)\}] \ge x\right\} \le \exp\left(-\frac{2x^2}{L^2}\right),$$

and

$$P\left\{\sup_{g\in\mathcal{G}}\left|\sum_{i=1}^n[g(\xi_i) - E\{g(\xi)\}]\right| - E\sup_{g\in\mathcal{G}}\left|\sum_{i=1}^n[g(\xi_i) - E\{g(\xi)\}]\right| \ge x\right\} \le \exp\left(-\frac{2x^2}{L^2}\right),$$

where $L^2 = \sum_{i=1}^n \sup_{g\in\mathcal{G}}(b_{ig} - a_{ig})^2$.

Proof: These are direct consequence of Theorem 5.3 in Massart (2007).

Lemma S.10.4.5. For a countable class of measurable functions $g \in \mathcal{G}$ and independent random variables $\xi_1,\cdots,\xi_n$, let $P_n(g) = \frac{1}{\sqrt{n}}\sum_{i=1}^n\{g(\xi_i) - Eg(\xi_i)\}$ be a sub-Gaussian process under a semi-metric d on $\mathcal{G}$, i.e.,

$$P\{|P_n(g_1) - P_n(g_2)| > t\} \le 2\exp\left\{-\frac{t^2}{2d^2(g_1,g_2)}\right\},$$

then

$$E\left\{\sup_{g\in\mathcal{G}}|P_n(g)|\right\} \quad\le\quad E\{|P_n(g_0)|\} + C\int_0^{d(\mathcal{G})}\sqrt{\log D(\epsilon,\mathcal{G},d)}\,d\epsilon,$$

where C is a constant, g_0 is a given member in $\mathcal{G}$, $d(\mathcal{G})$ is the supremum of d between any two elements of $\mathcal{G}$, and $D(\epsilon,\mathcal{G},d)$ is ϵ-packing number of $\mathcal{G}$ under semi-metric d, i.e., the maximum number of points in $\mathcal{G}$ that are ϵ-separated from each other in d.

Proof: It is an adaptation of Corollary 2.2.8 on page 101 of van der Vaart and Wellner (1996).

Lemma S.10.4.6. Let $H_n(z \mid \theta) = H(z \mid \theta, P_n, Q_n)$ and $H_0(z \mid \theta) = H(z \mid \theta, P_0, Q_0)$. Under Assumptions 1-4, for $k \leq 3$,

$$P\left\{\sup_{z\in\mathcal{Z},\theta\in\Theta}\left|\frac{\partial^k H_n(z|\theta)}{\partial^k\theta} - \frac{\partial^k H_0(z|\theta)}{\partial^k\theta}\right| \geq \delta\right\} \leq C_0 \exp\left[-\frac{n\left\{\delta - C_1\sqrt{(r+p\wedge q)/n}\right\}^2}{C_2}\right],$$

where $C_j > 0, j = 0,1,2$, are positive constants independent of (p,q,r,n), (p,q,r) are respectively the dimensions of z, x, and θ, (P_n, Q_n) and (P_0, Q_0) are the respectively empirical and true marginal measures of Z and X, and $\frac{\partial^k}{\partial^k\theta}$ denotes any single derivative, and $p \wedge q = \max(p,q)$.

Proof: Note that

$$\begin{aligned}
P\left(\sup_{z\in\mathcal{Z},\theta\in\Theta}|H_n(z\mid\theta) - H_0(z\mid\theta)| \geq \delta\right) &\leq P\left(\sup_{z\in\mathcal{Z},\theta\in\Theta}|H_{n0}(z\mid\theta) - H_0(z\mid\theta)| \geq \delta/2\right) \\
&\quad + P\left(\sup_{z\in\mathcal{Z},\theta\in\Theta}|H_n(z\mid\theta) - H_{n0}(z\mid\theta)| \geq \delta/2\right),
\end{aligned}$$

where

$$\begin{aligned}
H_{n0}(z\mid\theta) &= -\log\int\frac{\eta(z;x\mid\theta)}{\int\eta(z;x\mid\theta)\exp\{H_0(z\mid\theta)\}dP_n(z)}dQ_n(x) \\
&\quad -\log\int\left[\int\frac{\eta(z;x\mid\theta)}{\int\eta(z;x\mid\theta)\exp\{H_0(z\mid\theta)\}dP_n(z)}dQ_n(x)\right]^{-1} dP_n(z).
\end{aligned}$$

It can be seen that $\int\exp\{H_{n0}(z\mid\theta)\}dP_n(z) = 1$.

Let $\mathcal{X}_c, \mathcal{Z}_c, \Theta_c$ be respectively countable dense sets in $\mathcal{X}, \mathcal{Z}, \Theta$. Let

$$\begin{aligned}
R_{n1} &= \sup_{x\in\mathcal{X}_c,\theta\in\Theta_c}\left|\int\eta(z;x\mid\theta)\exp\{H_0(z\mid\theta)\}d(P_n - P_0)(z)\right| \\
R_{n2} &= \sup_{z\in\mathcal{Z}_c,\theta\in\Theta_c}\left|\int\frac{\eta(z;x\mid\theta)}{\int\eta(z;x\mid\theta)\exp\{H_0(z\mid\theta)\}dP_0(z)}d(Q_n - Q_0)(x)\right| \\
R_{n3} &= \sup_{\theta\in\Theta_c}\left|\int\left[\int\frac{\eta(z;x\mid\theta)}{\int\eta(z;x\mid\theta)\exp\{H_0(z\mid\theta)\}dP_0(z)}dQ_0(x)\right]^{-1} d(P_n - P_0)(z)\right|.
\end{aligned}$$

It follows from Lemma S.10.4.4 that

$$\begin{aligned}
P\{R_{n1} - E(R_{n1}) > \delta\} &\leq \exp\left(-\frac{2n\delta^2}{M^6}\right), \\
P\{R_{n2} - E(R_{n2}) > \delta\} &\leq \exp\left(-\frac{2n\delta^2}{M^8}\right), \\
P\{R_{n3} - E(R_{n3}) > \delta\} &\leq \exp\left(-\frac{2n\delta^2}{M^8}\right).
\end{aligned}$$

Note further from Lemma S.20.4.5 that,

$$\begin{aligned}
E(\sqrt{n}R_{n1}) &\leq L\int_0^{D_1}\sqrt{\log D(\mathcal{F}_1,\epsilon,d_{n1})}\,d\epsilon, \\
E(\sqrt{n}R_{n2}) &\leq L\int_0^{D_2}\sqrt{\log D(\mathcal{F}_2,\epsilon,d_{n2})}\,d\epsilon, \\
E(\sqrt{n}R_{n3}) &\leq L_0 + L\int_0^{D_3}\sqrt{\log D(\mathcal{F}_3,\epsilon,d_{n3})}\,d\epsilon,
\end{aligned}$$

where $D(\mathcal{F}_k, \epsilon, d_{nk})$ is the ϵ-packing number of $\mathcal{F}_k$ under the semi-norm $d_{n1}(f) = n^{-1}\sum_{i=1}^n f^2(z_i)$, $d_{n2}(g) = n^{-1}\sum_{i=1}^n g^2(x_i)$, and $d_{n3} = d_{n1}$,

$$\begin{aligned}
\mathcal{F}_1 &= \{\eta(z; x \mid \theta)\exp\{H_0(z \mid \theta)\} \mid x \in \mathcal{X}, \theta \in \Theta\}, \\
\mathcal{F}_2 &= \left\{\frac{\eta(z; x \mid \theta)}{\int \eta(z; x \mid \theta)\exp\{H_0(z \mid \theta)\}dP_0(z)} \mid z \in \mathcal{Z}, \theta \in \Theta\right\}, \\
\mathcal{F}_3 &= \left\{\left[\int \frac{\eta(z; x \mid \theta)}{\int \eta(z; x \mid \theta)\exp\{H_0(z \mid \theta)\}dP_0(z)} dQ_0(x)\right]^{-1} \mid \theta \in \Theta\right\},
\end{aligned}$$

and D_k is the diameter of $\mathcal{F}_k$ under d_{nk} for $k = 1, 2, 3$. Because

$$|\eta(z, x_1 \mid \theta_1)e^{H_0(z|\theta_1)} - \eta(z, x_2 \mid \theta_2)e^{H_0(z|\theta_2)}| \leq K_1(||x_1 - x_2||_2 + ||\theta_1 - \theta_2||_2)$$

for some constant $K_1 > 0$ and the covering number of $\mathcal{X} \times \Theta$ under $||\cdot||_2$ satisfies

$$N(\epsilon, \mathcal{X} \times \Theta, ||\cdot||_2) \leq K\epsilon^{-(q+r)},$$

it follows that $E(\sqrt{n}R_{n1}) \leq L_1\sqrt{q+r}$. This means that, for $\delta > L_1\sqrt{(q+r)/n}$,

$$\begin{aligned}
P\{R_{n1} > \delta\} &\leq \exp\{-2n(\delta - ER_{n1})^2/M^6\} \\
&\leq \exp\left\{-2n\left(\delta - L_1\sqrt{(q+r)/n}\right)^2/M^6\right\}.
\end{aligned}$$

Similarly, it can be shown that $E(\sqrt{n}R_{n1}) \leq L_1\sqrt{p+r}$ and $E(\sqrt{n}R_{n1}) \leq L_0 + L_1\sqrt{r}$. These mean that

$$\begin{aligned}
P\{R_{n2} > \delta\} &\leq \exp\left\{-2n\left(\delta - L_1\sqrt{(p+r)/n}\right)^2/M^8\right\}, \\
P\{R_{n3} > \delta\} &\leq \exp\left\{-2n\left(\delta - L_0/\sqrt{n} - L_1\sqrt{r/n}\right)^2/M^8\right\}.
\end{aligned}$$

It can be concluded that there exist $C_0, C_1, C_2 > 0$ such that

$$P\left\{\sup_{z\in\mathcal{Z},\theta\in\Theta}|H_{n0}(z \mid \theta) - H_0(z \mid \theta)| \geq \delta\right\} \leq C_0\exp\left[-\frac{n}{C_2}\left\{\delta - C_1\sqrt{(r + p \wedge q)/n}\right\}^2\right].$$

Note next that

$$H_n(z \mid \theta) - H_{n0}(z \mid \theta) = -\log\int\frac{\eta(z; x \mid \theta)e^{H_{n0}(z|\theta)}}{\int\eta(z; x \mid \theta)e^{H_n(z|\theta)}dP_n(z)}dQ_n(x).$$

As a result,

$$e^{H_n(z|\theta) - H_{n0}(z|\theta)} - 1 = \left\{\int\frac{\eta(z; x \mid \theta)e^{H_{n0}(z|\theta)}}{\int\eta(z; x \mid \theta)e^{H_n(z|\theta)}dP_n(z)}dQ_n(x)\right\}^{-1} - 1.$$

Let $h_n(z|\theta) = \exp\{H_n(z \mid \theta) - H_{n0}(z \mid \theta)\} - 1$ satisfying $\int h_n(z)\exp\{H_{n0}(z \mid \theta)\}dP_n(z) = 0$. Note that $\exp\{H_n(z \mid \theta)\} = \{1 + h_n(z)\}\exp\{H_{n0}(z)\}$. The expression can be rewritten as

$$h_n(z|\theta) = T_{n1}(z \mid \theta) + T_{n2}(z \mid \theta),$$

where

$$T_{n1}(z|\theta) = \frac{1 - \int \frac{\eta(z;x|\theta)e^{H_{n0}(z|\theta)}}{\int \eta(z;x|\theta)e^{H_{n0}(z|\theta)}dP_n(z)}dQ_n(x)}{\int \frac{\eta(z;x|\theta)e^{H_{n0}(z|\theta)}}{\int \eta(z;x|\theta)\{1+h_n(z|\theta)\}e^{H_{n0}(z|\theta)}dP_n(z)}dQ_n(x)}$$

$$T_{n2}(z \mid \theta) = \frac{\int \frac{\eta(z;x|\theta)e^{H_{n0}(z|\theta)} \int \eta(z;x|\theta)h_n(z|\theta)e^{H_{n0}(z|\theta)}dP_n(z)}{\int \eta(z;x|\theta)\{1+h_n(z|\theta)\}e^{H_{n0}(z|\theta)}dP_n(z) \int \eta(z;x|\theta)e^{H_{n0}(z|\theta)}dP_n(z)}dQ_n(x)}{\int \frac{\eta(z;x|\theta)e^{H_{n0}(z|\theta)}}{\int \eta(z;x|\theta)\{1+h_n(z|\theta)\}e^{H_{n0}(z|\theta)}dP_n(z)}dQ_n(x)}.$$

Let

$$A_{n0}(h_n)(z) = \frac{\int \frac{\eta(z;x|\theta)e^{H_{n0}(z|\theta)} \int \eta(z;x|\theta)h_n(z|\theta)e^{H_{n0}(z|\theta)}dP_n(z)}{\{\int \eta(z;x|\theta)e^{H_{n0}(z|\theta)}dP_n(z)\}^2}dQ_n(x)}{\int \frac{\eta(z;x|\theta)e^{H_{n0}(z|\theta)}}{\int \eta(z;x|\theta)e^{H_{n0}(z|\theta)}dP_n(z)}dQ_n(x)}.$$

Then

$$(I - A_{n0})\{h_n(z \mid \theta)\} = T_{n1}(z \mid \theta) + T_{n2}(z \mid \theta) - A_{n0}\{h_n(z \mid \theta)\}.$$

Note first that $1 + h_n(z) > -1 + \epsilon$ for a constant $\epsilon > 0$, and thus

$$|T_{n1}(z \mid \theta)| \leq M^* \left| 1 - \int \frac{\eta(z; x \mid \theta)e^{H_{n0}(z|\theta)}}{\int \eta(z; x \mid \theta)e^{H_{n0}(z|\theta)}dP_n(z)}dQ_n(x) \right|,$$

for some constant M^* independent of n, h_n or θ. It follows from the result on $|H_{n0}(z) - H_0(z)|$ that, after adjusting for the constants $C_k, k = 0, 1, 2$,

$$P\left\{ \sup_{z\in\mathcal{Z},\theta\in\Theta,h_n\in\mathcal{H}_n} |T_{n1}(z \mid \theta)| > \delta \right\} \leq C_0 \exp\left[-\frac{n}{C_2} \left\{ \delta - C_1\sqrt{(r + p \wedge q)/n} \right\}^2 \right].$$

Next, after similar but more tedious calculation to the foregoing proof, it follows that

$$P\left\{ \sup_{z\in\mathcal{Z},\theta\in\Theta,h_n\in\mathcal{H}_n} |T_{n2}(z \mid \theta) - A_{n0}(h_n)| > \delta \right\} \leq C_0 \exp\left[-\frac{n}{C_2} \left\{ \delta - C_1\sqrt{(r + p \wedge q)/n} \right\}^2 \right].$$

It follows after adjustment of constants that

$$P\left\{ \sup_{z\in\mathcal{Z},\theta\in\Theta,h_n\in\mathcal{H}_n} |(I - A_{n0})(h_n)| > \delta \right\} \leq C_0 \exp\left[-\frac{n}{C_2} \left\{ \delta - C_1\sqrt{(r + p \wedge q)/n} \right\}^2 \right].$$

Further, it can be shown that

$$P\left\{ \sup_{z\in\mathcal{Z},\theta\in\Theta,h_n\in\mathcal{H}_n} |(A_{n0} - A_0)(h_n)| > \delta \right\} \leq C_0 \exp\left[-\frac{n}{C_2} \left\{ \delta - C_1\sqrt{(r + p \wedge q)/n} \right\}^2 \right].$$

These imply

$$P\left\{ \sup_{z\in\mathcal{Z},\theta\in\Theta,h_n\in\mathcal{H}_n} |(I - A_0)(h_n)| > \delta \right\} \leq C_0 \exp\left[-\frac{n}{C_2} \left\{ \delta - C_1\sqrt{(r + p \wedge q)/n} \right\}^2 \right].$$

From the proof of Lemma S.10.4.2, it can also be seen that $(I - A_0)$ is also continuously invertible on the uniformly bounded Lipschitz functions on $\mathcal{Z}$. There exists $\sigma > 0$ such that

$$\sup_{z\in\mathcal{Z},\theta\in\Theta,h_n\in\mathcal{H}_n} |(I - A_0)(h_n)| \geq \sigma \sup_{z\in\mathcal{Z},\theta\in\Theta,h_n\in\mathcal{H}_n} |h_n|.$$

It now follows that

$$
\begin{aligned}
P\left\{\sup_{z\in\mathcal{Z},\theta\in\Theta,h_n\in\mathcal{H}_n}|h_n(z\mid\theta)|>\delta\right\} &\leq P\left\{\sup_{z\in\mathcal{Z},\theta\in\Theta,h_n\in\mathcal{H}_n}|(I-A_0)(h_n)|>\sigma\delta\right\}\\
&\leq C_0\exp\left[-\frac{n}{C_2}\left\{\sigma\delta-C_1\sqrt{(r+p\wedge q)/n}\right\}^2\right].
\end{aligned}
$$

Furthermore, it follows from

$$
M_1|H_n(z\mid\theta)-H_{n0}(z\mid\theta)|\leq|h_n(z\mid\theta)|=|e^{H_n(z|\theta)-H_{n0}(z|\theta)}-1|\leq M_2|H_n(z\mid\theta)-H_{n0}(z|\theta)|,
$$

that, after adjusting the constants $C_j, j=0,1,2$,

$$
P\left\{\sup_{z\in\mathcal{Z},\theta\in\Theta}|H_n(z)-H_{n0}(z)|>\delta\right\}\leq C_0\exp\left[-\frac{n}{C_2}\left\{\delta-C_1\sqrt{(r+p\wedge q)/n}\right\}^2\right].
$$

This combining with

$$
P\left\{\sup_{z\in\mathcal{Z},\theta\in\Theta}|H_{n0}(z)-H_0(z)|>\delta\right\}\leq C_0\exp\left[-\frac{n}{C_2}\left\{\delta-C_1\sqrt{(r+p\wedge q)/n}\right\}^2\right],
$$

implies the result for the case $k=0$.

To prove the result for the case $k=1$, from Lemma S.10.4.2,

$$
\frac{\partial}{\partial\theta}H_n(z\mid\theta)=\{I-A(\theta,H_n,P_n,Q_n)\}^{-1}u(\theta,H_n,P_n,Q_n)
$$

and

$$
\frac{\partial}{\partial\theta}H_0(z\mid\theta)=\{I-A(\theta,H_0,P_0,Q_0)\}^{-1}u(\theta,H_0,P_0,Q_0),
$$

where

$$
\begin{aligned}
u(z\mid\theta,H,P,Q) &= \int\eta(z;x\mid\theta)e^{H(z|\theta)}\left\{\frac{\frac{\partial\log\eta}{\partial\theta}}{\int\eta(z;x\mid\theta)e^{H(z|\theta)}dP(z)}\right.\\
&\quad\left.-\frac{\int\frac{\partial\log\eta}{\partial\theta}\eta(z;x\mid\theta)e^{H(z|\theta)}dP(z)}{\left\{\int\eta(z;x\mid\theta)e^{H(z|\theta)}dP(z)\right\}^2}\right\}dQ(x).
\end{aligned}
$$

By similar arguments, it can be shown that,

$$
\begin{aligned}
&P\left\{\sup_{z\in\mathcal{Z},\theta\in\Theta}|u(z\mid\theta,H_n,P_n,Q_n)-u(z\mid\theta,H_0,P_0,Q_0)|\geq\delta\right\}\\
&\qquad\leq C_0\exp\left(-\frac{n\left\{\delta-C_1\sqrt{(r+p\wedge q)/n}\right\}^2}{C_2}\right)
\end{aligned}
$$

and that

$$
\begin{aligned}
&P\left\{\sup_{z\in\mathcal{Z},\theta\in\Theta,h\in\mathcal{H}_n}|A(\theta,H_n,P_n,Q_n)(h)(z)-A(\theta,H_0,P_0,Q_0)(h)(z)|\geq\delta\right\}\\
&\qquad\leq C_0\exp\left(-\frac{n\{\delta-C_1\sqrt{(r+p\wedge q)/n}\}^2}{C_2}\right).
\end{aligned}
$$

The result for the first derivative of H follows from combining these results and that $A(\theta, H_0, P_0, Q_0)(h)$ is continuously invertible on the uniformly bounded Lipschitz functions on $\mathcal{Z}$. The rest of the results can be proved by similar arguments.

Lemma S.10.4.7. Under Assumptions 1–4,

$$\begin{aligned} P\left\{||\dot{l}_{pn}^S(\theta_0)||_\infty > \delta\right\} &\leq C_0 r_S \exp\left\{-\frac{n\{\delta - C_1\sqrt{(p_S \wedge q_S + r_S)/n}\}^2}{C_2}\right\}, \\ P\left\{||\dot{l}_{pn}^{S^c}(\theta_0)||_\infty \geq \delta\right\} &\leq C_0(r - r_S)\exp\left\{-\frac{n(\delta - C_1\sqrt{(p_S \wedge q_S + r_S)/n})^2}{C_2}\right\}, \end{aligned}$$

where $\delta > C_1\sqrt{(p_S \wedge q_S + r_S)/n}$, and p_S and q_S are the reduced dimensions of z and x in $\eta\{z; x \mid (\theta^S, 0)\}$, and r_S is the dimension of θ^S.

Proof: Note that

$$\begin{aligned} \dot{l}_{pn}(\theta_0) &= -\frac{1}{n}\sum_{i=1}^n \frac{\partial}{\partial\theta}\log\eta(Z_i; X_i \mid \theta_0) - \frac{1}{n}\sum_{i=1}^n \frac{\partial}{\partial\theta}H_n(Z_i \mid \theta_0) \\ &\quad + \frac{1}{n}\sum_{i=1}^n \frac{\partial}{\partial\theta}\log\int \eta(z; X_i \mid \theta_0)e^{H_n(z|\theta_0)}dP_n(z), \end{aligned}$$

and that

$$\begin{aligned} \dot{l}_{p0}(\theta_0) &= -\frac{1}{n}\sum_{i=1}^n \frac{\partial}{\partial\theta}\log\eta(Z_i; X_i \mid \theta_0) - \frac{1}{n}\sum_{i=1}^n \frac{\partial}{\partial\theta}H_0(Z_i \mid \theta_0) \\ &\quad + \frac{1}{n}\sum_{i=1}^n \frac{\partial}{\partial\theta}\log\int \eta(z; X_i \mid \theta_0)e^{H_0(z|\theta_0)}dP_0(z). \end{aligned}$$

By Hoeffding inequality,

$$P\left\{||\dot{l}_{p0}^S(\theta_0)||_\infty \geq c_1\lambda_n\right\} \leq 2r_S\exp\left\{-\frac{nc_1^2\lambda_n^2}{2M^2}\right\}.$$

where M is the uniform bound of $\dot{l}_{p0}^S(\theta_0)$. Note also that

$$\begin{aligned} ||\dot{l}_{pn}^S(\theta_0)||_\infty &\leq ||\dot{l}_{p0}^S(\theta_0)||_\infty + ||\dot{l}_{pn}^S(\theta_0) - \dot{l}_{p0}^S(\theta_0)||_\infty \\ &\leq ||\dot{l}_{p0}^S(\theta_0)||_\infty + M_{2*}\left|\left|\frac{\partial}{\partial\theta^S}H_n(z \mid \theta_0) - \frac{\partial}{\partial\theta^S}H_0(z \mid \theta_0)\right|\right|_\infty \end{aligned}$$

$$+M_{3*}\left|\left|\frac{\int \frac{\partial}{\partial\theta^S}\{\eta(z; x \mid \theta)e^{H_0(z|\theta)}\}_{\theta=\theta_0}dP_n(z)}{\int \eta(z; x \mid \theta_0)e^{H_0(z|\theta_0)}dP_n(z)} - \frac{\int \frac{\partial}{\partial\theta^S}\{\eta(z; x \mid \theta)e^{H_0(z|\theta)}\}_{\theta=\theta_0}dP_0(z)}{\int \eta(z; x \mid \theta_0)e^{H_0(z|\theta_0)}dP_0(z)}\right|\right|_\infty.$$

From Lemma S.10.4.5, it follows that

$$P\{||U_2||_\infty \geq \delta\} \leq C_0 r_S\exp\left[-\frac{n}{C_2}\left\{\delta - C_1\sqrt{(p_S \wedge q_S + r_S)/n}\right\}^2\right],$$

where U_2 is the second term inside the norm. By similar arguments,

$$P(||U_3||_\infty \geq \delta) \leq C_0 r_S\exp\left[-\frac{n}{C_2}\left\{\delta - C_1\sqrt{(p_S \wedge q_S + r_S)/n}\right\}^2\right],$$

where U_3 is the third term inside the norm. Note that those constant $C_j, j = 0, 1, 2$ may be different in different inequalities. But they all independent of n, p_S, q_S, r_S. Hence,

$$\begin{aligned} P\left\{||\dot{l}^S_{pn}(\theta_0)||_\infty > \delta\right\} &\leq P\left\{||\dot{l}^S_{p0}(\theta_0)||_\infty \geq \delta/3\right\} + P\{||U_2||_\infty \geq \delta/(3M_{2*})\} \\ &\quad + P(||U_3||_\infty \geq \delta/(3M_{3*})) \\ &\leq C_0 r_S \exp\left[-\frac{n}{C_2}\left\{\delta - C_1\sqrt{(p_S \wedge q_S + r_S)/n}\right\}^2\right], \end{aligned}$$

for $\delta > C_1\sqrt{(p_S \wedge q_S + s)/n}$. Proof of the other inequality is similar.

Lemma S.10.4.8. Under Assumptions 1–4, there exist $C_0 > 0$, $C_1 > 0$, and $C_2 > 0$ such that

$$P\left\{||\ddot{l}^{SS}_{pn}(\theta_0)\} - E\ddot{l}^{SS}_{p0}(\theta_0)||_\infty \geq \delta\right\} \leq C_0 r_S^2 \exp\left[-\frac{n\left\{\delta - C_1\sqrt{(p_S \wedge q_S + r_S)/n}\right\}^2}{C_2}\right]$$

$$P\left\{||\ddot{l}^{S^cS}_{pn}(\theta_0)\} - E\ddot{l}^{S^cS}_{p0}(\theta_0)||_\infty \geq \delta\right\} \leq C_0(r - r_S)r_S \exp\left[-\frac{n\left\{\delta - C_1\sqrt{(p_S \wedge q_S + r_S)/n}\right\}^2}{C_2}\right]$$

for any $\delta > C_1\sqrt{(p_S \wedge q_S + r_S)/n}$, where $||(a_{jk})||_\infty = \sup_{j,k}|a_{jk}|$.

Proof: Note that

$$\begin{aligned} \ddot{l}_{pn}(\theta_0) &= \ddot{l}_{\theta\theta}\{\theta_0, H_n(\cdot|\theta_0)\} + 2\ddot{l}_{\theta H}\{\theta_0, H_n(\cdot|\theta_0)\}\left\{\frac{\partial H_n(\cdot|\theta_0)}{\partial\theta}\right\} \\ &\quad + \ddot{l}_{HH}\{\theta_0, H_n(\cdot|\theta_0)\}\left\{\frac{\partial H_n(\cdot|\theta_0)}{\partial\theta}\right\}\left\{\frac{\partial H_n(\cdot|\theta_0)}{\partial\theta}\right\} \\ &\quad + \dot{l}_H\{\theta_0, H_n(\cdot|\theta_0)\}\left\{\frac{\partial^2 H_n(\cdot|\theta_0)}{\partial^2\theta}\right\}, \end{aligned}$$

$$\begin{aligned} \ddot{l}_{p0}(\theta_0) &= \ddot{l}_{\theta\theta}\{\theta_0, H_0(\cdot|\theta_0)\} + 2\ddot{l}_{\theta H}\{\theta_0, H_0(\cdot|\theta_0)\}\left\{\frac{\partial H_0(\cdot|\theta_0)}{\partial\theta}\right\} \\ &\quad + \ddot{l}_{HH}\{\theta_0, H_0(\cdot|\theta_0)\}\left\{\frac{\partial H_0(\cdot|\theta_0)}{\partial\theta}\right\}\left\{\frac{\partial H_0(\cdot|\theta_0)}{\partial\theta}\right\} \\ &\quad + \dot{l}_H\{\theta_0, H_0(\cdot|\theta_0)\}\left\{\frac{\partial^2 H_0(\cdot|\theta_0)}{\partial^2\theta}\right\}. \end{aligned}$$

Note that

$$\ddot{l}^{SS}_{pn}(\theta_0) - E\ddot{l}^{SS}_{p0}(\theta_0) = \ddot{l}^{SS}_{pn}(\theta_0) - \ddot{l}^{SS}_{p0}(\theta_0) + \ddot{l}^{SS}_{p0}(\theta_0) - E\ddot{l}^{SS}_{p0}(\theta_0).$$

Since $\ddot{l}^{SS}_{p0}(\theta_0)$ is uniformly bounded, it follows that

$$\begin{aligned} P\left\{||\ddot{l}^{SS}_{p0}(\theta_0) - E_0\ddot{l}^{SS}_{p0}(\theta_0)||_\infty \geq \delta\right\} &= P\left\{\sup_{j\in S, k\in S}|\ddot{l}^{SS}_{p0(jk)}(\theta_0) - E_0\ddot{l}^{SS}_{p0(jk)}(\theta_0)| \geq \delta\right\} \\ &\leq r_S^2 \exp\left\{-\frac{n\delta^2}{C^2}\right\} \end{aligned}$$

where C is the uniform bound. By similar arguments in proving Lemmas S.10.4.6 and

S.10.4.7, there exist constants C_0, C_1, and C_2 such that

$$
\begin{aligned}
P\left\{||\ddot{l}_{pn}^{SS}(\theta_0) - \ddot{l}_{p0}^{SS}(\theta_0)||_\infty \geq \delta\right\} &= P\left\{\sup_{j\in S, k\in S} |\ddot{l}_{pn(jk)}^{SS}(\theta_0) - \ddot{l}_{p0(jk)}^{SS}(\theta_0)| \geq \delta\right\} \\
&\leq C_0 r_S^2 \exp\left[-\frac{n\left\{\delta - C_1\sqrt{(p_S \wedge q_S + r_S)/n}\right\}^2}{C_2}\right],
\end{aligned}
$$

for $\delta > \sqrt{(p_S \wedge q_S + r_S)/n}$. Hence,

$$
P\left\{||\ddot{l}_{pn}^{SS}(\theta_0) - E\ddot{l}_{p0}^{SS}(\theta_0)||_\infty \geq \delta\right\} \leq C_0 r_S^2 \exp\left[-\frac{n\left\{\delta - C_1\sqrt{(p_S \wedge q_S + r_S)/n}\right\}^2}{C_2}\right],
$$

The other part can be proved similarly.

Lemma S.10.4.9. Under Assumptions 1–5, there exist $C_0 > 0$, $C_1 > 0$, and $C_2 > 0$ such that

$$
P\left\{e_{\min}\{\ddot{l}_{pn}^{SS}(\theta_0)\} \leq \pi_S/2\right\} \leq C_0 r_S^2 \exp\left[-\frac{n\left\{\pi_S/r_S - C_1\sqrt{(p_S \wedge q_S + r_S)/n}\right\}^2}{C_2}\right],
$$

for any $\pi_S > C_1 r_S\sqrt{(p_S \wedge q_S + r_S)/n}$ and π_S is the minimum eigenvalue of $E\{\ddot{l}_{p0}^{SS}(\theta_0)\}$.

Proof: Let v_n be the eigenvector of $e_{\min}\{\ddot{l}_{pn}^{S}(\theta_0)\}$ satisfying $||v_n||_2 = 1$. Then,

$$
\begin{aligned}
e_{min}\left\{E\ddot{l}_{p0}^{SS}(\theta_0)\right\} &= \inf_{||v||_2=1} v^T E\ddot{l}_{p0}^{SS}(\theta_0) v \\
&\leq e_{\min}\{\ddot{l}_{pn}^{SS}(\theta_0)\} + v_n^T\left\{E\ddot{l}_{p0}^{S}(\theta_0) - \ddot{l}_{pn}^{SS}(\theta_0)\right\} v_n.
\end{aligned}
$$

It follows that

$$
\begin{aligned}
e_{\min}\{\ddot{l}_{pn}^{SS}(\theta_0)\} &\geq e_{min}\left\{E\ddot{l}_{p0}^{SS}(\theta_0)\right\} - v_n^T\left\{E\ddot{l}_{p0}^{SS}(\theta_0) - \ddot{l}_{pn}^{SS}(\theta_0)\right\} v_n \\
&\geq e_{min}\left\{E\ddot{l}_{p0}^{SS}(\theta_0)\right\} - r_S||E\ddot{l}_{p0}^{SS}(\theta_0) - \ddot{l}_{pn}^{SS}(\theta_0)||_\infty.
\end{aligned}
$$

Note from Lemma S.8. that

$$
P\left\{||\ddot{l}_{pn}^{SS}(\theta_0) - E_0\ddot{l}_{p0}^{SS}(\theta_0)||_\infty \geq \pi_S/(2r_S)\right\} \leq C_0 r_S^2 \exp\left[-\frac{n\left\{\pi_S/r_S - C_1\sqrt{(p_S \wedge q_S + r_S)/n}\right\}^2}{C_2}\right],
$$

for $\pi_S/r_S > C_1\sqrt{(p_S \wedge q_S + r_S)/n}$. It follows by adjusting constant $C_j, j = 1, 2$ that

$$
P\left\{e_{\min}\{\ddot{l}_{pn}^{SS}(\theta_0)\} \leq \pi_S/2\right\} \leq C_0 r_S^2 \exp\left[-\frac{n\left\{\pi_S/r_S - C_1\sqrt{(p_S \wedge q_S + r_S)/n}\right\}^2}{C_2}\right].
$$

Lemma S.10.4.10. Under Assumptions 1–6, there exist $C_0 > 0$, $C_1 > 0$, and $C_2 > 0$ such that

$$\begin{aligned} & P\left\{\left|\left|\left|\ddot{l}_{pn}^{S^cS}(\theta_0^S,0)\left\{\ddot{l}_{pn}^{SS}(\theta_0^S,0)\right\}^{-1} - B^{S^cS}(B^{SS})^{-1}\right|\right|\right|_\infty \geq \delta\right\} \\ \leq\; & C_0 r r_S \exp\left[-\frac{n\left\{\min(\delta/\sqrt{r_S},1)\pi_S/r_S - C_1\sqrt{(p_S\wedge q_S + r_S)/n}\right\}^2}{C_2}\right], \end{aligned}$$

for any $\min(\delta/\sqrt{r_S},1)\pi_S/r_S > C_1\sqrt{(p_S\wedge q_S+r_S)/n}$.

Proof: Note first that $B = E_0\{\dot{l}_{p0}(\theta_0)\}^2 = E_0\{\ddot{l}_{p0}(\theta_0)\}$. Hence $B^{SS} = E_0\{\ddot{l}_{p0}^{SS}(\theta_0)\}$ and $B^{S^cS} = E_0\{\ddot{l}_{p0}^{S^cS}(\theta_0)\}$. It follows that

$$\begin{aligned} \ddot{l}_{pn}^{S^cS}(\theta_0^S,0)\left\{\ddot{l}_{pn}^{SS}(\theta_0^S,0)\right\}^{-1} - B^{S^cS}(B^{SS})^{-1} &= \left[\left\{\ddot{l}_{pn}^{S^cS}(\theta_0^S,0) - E_0\ddot{l}_{p0}^{S^cS}(\theta_0^S,0)\right\}\right. \\ &\left. -B^{S^cS}(B^{SS})^{-1}\left\{\ddot{l}_{pn}^{SS}(\theta_0^S,0) - E_0\ddot{l}_{p0}^{SS}(\theta_0^S,0)\right\}\right]\left\{\ddot{l}_{pn}^{SS}(\theta_0^S,0)\right\}^{-1}. \end{aligned}$$

Note that, for two matrices $A_{a\times b}, B_{b\times c}$,

$$\begin{aligned} |||AB|||_\infty &= \sup_j \sum_l |\sum_k a_{jk}b_{kl}| \leq \sup_j \sum_k |a_{jk}| \sup_k \sum_l |b_{kl}| = |||A|||_\infty |||B|||_\infty, \\ |||A|||_\infty &= \sup_j \sum_{k=1}^b |a_{jk}| \leq b \sup_{j,k} |a_{jk}| = b||A||_\infty, \\ |||A|||_\infty &= \sup_{||u||_\infty=1} \sup_j |(Au)_j| \leq \sup_{||u||_\infty=1} \left\{\sum_j |(Au)_j|^2\right\}^{1/2}, \\ &\leq \sup_{||u||_\infty=1} ||u||_2 |||A|||_2 = \sqrt{b}|||A|||_2, \\ |||A|||_2 &= \sup_{||u||_2=1} \left\{\sum_j (Au)_j^2\right\}^{1/2} = \sup_{||u||_2=1} \left\{u^T A^T A u\right\}^{1/2} \leq e_{\max}^{1/2}(A^T A). \end{aligned}$$

It follows that

$$\begin{aligned} & |||\ddot{l}_{pn}^{S^cS}(\theta_0^S,0)\left\{\ddot{l}_{pn}^{SS}(\theta_0^S,0)\right\}^{-1} - B^{S^cS}(B^{SS})^{-1}|||_\infty \\ \leq\; & |||\left\{\ddot{l}_{pn}^{S^cS}(\theta_0^S,0) - E_0\ddot{l}_{p0}^{S^cS}(\theta_0^S,0)\right\}|||_\infty |||\left\{\ddot{l}_{pn}^{SS}(\theta_0^S,0)\right\}^{-1}|||_\infty \\ & +|||B^{S^cS}(B^{SS})^{-1}|||_\infty |||\ddot{l}_{pn}^{SS}(\theta_0^S,0) - E_0\ddot{l}_{p0}^{SS}(\theta_0^S,0)|||_\infty |||\left\{\ddot{l}_{pn}^{SS}(\theta_0^S,0)\right\}^{-1}|||_\infty \\ \leq\; & r_S||\left\{\ddot{l}_{pn}^{S^cS}(\theta_0^S,0) - E_0\ddot{l}_{p0}^{S^cS}(\theta_0^S,0)\right\}||_\infty \sqrt{r_S} e_{min}^{-1}\left\{\ddot{l}_{pn}^{SS}(\theta_0^S,0)\right\} \\ & +(1-\alpha)r_S||\ddot{l}_{pn}^{SS}(\theta_0^S,0) - E_0\ddot{l}_{p0}^{SS}(\theta_0^S,0)||_\infty \sqrt{r_S} e_{min}^{-1}\left\{\ddot{l}_{pn}^{SS}(\theta_0^S,0)\right\}. \end{aligned}$$

Applying Lemmas S.10.4.8 and S.10.4.9, it follows that

$$\begin{aligned}
& P\left\{\left|\left|\left|\ddot{l}_{pn}^{S^cS}(\theta_0^S,0)\left\{\ddot{l}_{pn}^{SS}(\theta_0^S,0)\right\}^{-1} - B^{S^cS}(B^{SS})^{-1}\right|\right|\right|_\infty \geq \delta\right\} \\
\leq\ & P\left\{||\left\{\ddot{l}_{pn}^{S^cS}(\theta_0^S,0) - E_0\ddot{l}_{p0}^{S^cS}(\theta_0^S,0)\right\}||_\infty \geq e_{min}\left\{\ddot{l}_{pn}^{SS}(\theta_0^S,0)\right\}\delta/(2r_S^{3/2})\right\} \\
& +P\left\{||\ddot{l}_{pn}^{SS}(\theta_0^S,0) - E_0\ddot{l}_{p0}^{SS}(\theta_0^S,0)||_\infty \geq e_{min}\left\{\ddot{l}_{pn}^{SS}(\theta_0^S,0)\right\}\delta/(2r_S^{3/2})\right\} \\
\leq\ & P\left\{||\left\{\ddot{l}_{pn}^{S^cS}(\theta_0^S,0) - E_0\ddot{l}_{p0}^{S^cS}(\theta_0^S,0)\right\}||_\infty \geq \pi_s\delta/(4r_S^{3/2})\right\} \\
& +2P\left\{e_{min}\left\{\ddot{l}_{pn}^{SS}(\theta_0^S,0)\right\} \leq \pi_s/2\right\} \\
& +P\left\{||\ddot{l}_{pn}^{SS}(\theta_0^S,0) - E_0\ddot{l}_{p0}^{SS}(\theta_0^S,0)||_\infty \geq \pi_s\delta/(4r_S^{3/2})\right\} \\
\leq\ & C_0r_S^2\exp\left\{-\frac{n(\pi_S\delta/(4r_S^{3/2}) - C_1\sqrt{(p_S\wedge q_S + r_S)/n})^2}{C_2}\right\} \\
& +2C_0r_S^2\exp\left\{-\frac{n(\pi_S/r_S - C_1\sqrt{(p_S\wedge q_S + r_S)/n})^2}{C_2}\right\} \\
& +C_0(r - r_S)r_S\exp\left\{-\frac{n(\pi_S\delta/(4r_S^{3/2}) - C_1\sqrt{(p_S\wedge q_S + r_S)/n})^2}{C_2}\right\}.
\end{aligned}$$

The result can be obtained by combining the expression and adjusting for the constants.

Let S be the index set of nonzero θ_j components and

$$\tilde{\theta}^S = \arg\min_{\theta^S}\{l_{pn}(\theta^S,0) + \lambda_n||\theta^S||_1\}.$$

It follows from KKT conditions that

$$\dot{l}_{pn}^S(\tilde{\theta}^S,0) + \lambda_n u^S = 0,$$

where $u_j^S = \text{sign}(\theta_j^S)$ if $\theta_j^S \neq 0$ and $|u_j^S| \leq 1$ if $u_j^S = 0$. In the primal-duel witness approach to the proof of selection consistency, define

$$u^{S^c} = -\frac{1}{\lambda_n}\dot{l}_{pn}^{S^c}(\tilde{\theta}^S,0),$$

where $\dot{l}_{pn}^{S^c} = \frac{\partial}{\partial\theta^{S^c}}l_{pn}$. The objective is to show that $||u^{S^c}||_\infty < 1$. For any $a \in R^{r-s}$,

$$a^Tu^{S^c} = -\frac{1}{\lambda_n}a^T\dot{l}_{pn}^{S^c}(\theta_0^S,0) - \frac{1}{\lambda_n}a^T\ddot{l}_{pn}^{S^cS}(\theta_0^S,0)(\tilde{\theta}^S - \theta_0^S) + a^TR_{na},$$

where $R_{na} = (R_{na}(i), i \in S^c)$ and

$$R_{na}(i) = -\frac{1}{\lambda_n}\sum_{j\in S}\sum_{k\in S}\frac{\partial^3}{\partial\theta_i^{S^c}\partial\theta_j^S\partial\theta_k^S}l_{pn}(\theta_*^S,0)(\tilde{\theta}_j^S - \theta_{j0}^S)(\tilde{\theta}_k^S - \theta_{k0}^S),$$

and $\theta_*^S = \theta_0^S + t(\tilde{\theta}^S - \theta_0^S)$ for some $t \in [0,1]$. Furthermore, for any $b \in R^s$,

$$0 = \lambda_n b^Tu^S + b^T\dot{l}_{pn}^S(\theta_0^S,0) + b^T\ddot{l}_{pn}^{SS}(\theta_0^S,0)(\tilde{\theta}^S - \theta_0^S) + b^TR_{nb},$$

where $R_{nb} = (R_{nb}(i), i \in S)$ and

$$R_{nb}(i) = \frac{1}{\lambda_n} \sum_{j\in S} \sum_{k\in S} \frac{\partial^3}{\partial\theta_i^S \partial\theta_j^S \partial\theta_k^S} l_{pn}(\theta_{**}^S, 0)(\tilde{\theta}_j^S - \theta_{j0}^S)(\tilde{\theta}_k^S - \theta_{k0}^S),$$

and $\theta_{**}^S = \theta_0^S + t_1(\tilde{\theta}^S - \theta_0^S)$ for some $t_1 \in [0,1]$. For any $a \in R^{r-s}$, let

$$b^T = a^T \ddot{l}_{pn}^{S^cS}(\theta_0^S, 0) \left\{ \ddot{l}_{pn}^{SS}(\theta_0^S, 0) \right\}^{-1}.$$

It follows from combining the results that

$$\begin{aligned} a^T u^{S^c} &= a^T \ddot{l}_{pn}^{S^cS}(\theta_0^S, 0) \left\{ \ddot{l}_{pn}^{SS}(\theta_0^S, 0) \right\}^{-1} u^S \\ &\quad - \frac{1}{\lambda_n} \left[a^T \dot{l}_{pn}^{S^c}(\theta_0^S, 0) - a^T \ddot{l}_{pn}^{S^cS}(\theta_0^S, 0) \left\{ \ddot{l}_{pn}^{SS}(\theta_0^S, 0) \right\}^{-1} \dot{l}_{pn}^{S}(\theta_0^S, 0) \right] \\ &\quad + a^T \left\{ R_{na} + \ddot{l}_{pn}^{S^cS}(\theta_0^S, 0) \left\{ \ddot{l}_{pn}^{SS}(\theta_0^S, 0) \right\}^{-1} R_{nb} \right\}. \end{aligned}$$

Lemma S.10.4.11. Under Assumptions 1-5, if

$$e_{\min}\{\ddot{l}_{pn}^{SS}(\theta_0)\} \geq 4 \left[\sqrt{r_S} \sup_{\theta^S} ||l_{np}^{(3)SSS}(\theta^S, 0)||_\infty \left\{ ||\dot{l}_{pn}^S(\theta_0)||_\infty + \lambda_n \right\} \right]^{1/2}, \qquad (10.16)$$

then there exists $D \geq 8 \left\{ ||\dot{l}_{pn}^S(\theta_0)||_\infty / \lambda_n + 1 \right\} / e_{\min}\{\ddot{l}_{pn}^{SS}(\theta_0)\}$ such that

$$||\tilde{\theta}^S - \theta_0^S||_2 \leq D\sqrt{r_S}\lambda_n.$$

where $\dot{l}_{pn}^S$, $\ddot{l}_{pn}^{SS}$, and $l_{np}^{(3)SSS}$ are respectively the first,second, and third derivatives of l_{np} restricted on S, and $e_{\min}\{\ddot{l}_{pn}^{SS}(\theta_0)\}$ is the minimal eigenvalue of the matrix $\ddot{l}_{pn}^{SS}(\theta_0)$.

Proof: Let

$$G(u) = l_{pn}(\theta_0 + u) + \lambda_n ||\theta_0 + u||_1 - l_{pn}(\theta_0) - \lambda_n ||\theta_0||_1.$$

Then $G(0) = 0$ and G is a convex function. If we can show that, for any $u = (u^S, 0)$ in a sphere $||u||_2 = \xi > 0$ that $G(u) > 0$, then for any $\theta = (\theta^S, 0)$ such that $||\theta - \theta_0||_2 > \xi$, $G(\theta - \theta_0) > 0$. This is because G is a convex function and thus

$$0 < G\left\{ \frac{\xi}{||\theta - \theta_0||_2}(\theta - \theta_0) \right\} \leq \frac{\xi}{||\theta - \theta_0||_2} G(\theta - \theta_0) + \left\{ 1 - \frac{\xi}{||\theta - \theta_0||_2} \right\} G(0).$$

It follows that $G(\theta - \theta_0) > 0$. This allows us to conclude that $||\tilde{\theta}^S - \theta_0^S||_2 < \xi$ because $G(\tilde{\theta}^S - \theta_0^S, 0) \leq 0$.

Consider $\xi = D\sqrt{r_S}\lambda_n$ where $D > 0$. For $u = (u^S, 0)$, by Taylor expansion,

$$l_{pn}(\theta_0 + u) - l_{pn}(\theta_0) = u^T \dot{l}_{pn}(\theta_0) + \frac{1}{2} u^T \ddot{l}_{pn}(\theta_0 + tu) u,$$

where $0 \leq t \leq 1$. Note that $||\theta_0 + u||_1 - ||\theta_0||_1 \geq -||u||_1 \geq -\sqrt{r_S}||u||_2$. It follows that

$$G(u) \geq u^T \dot{l}_{pn}(\theta_0) + \frac{1}{2} u^T \ddot{l}_{pn}(\theta_0) u - \lambda_n \sqrt{r_S} ||u||_2 + \frac{1}{2} u^T \left\{ \ddot{l}_{pn}(\theta_0 + tu) - \ddot{l}_{pn}(\theta_0) \right\} u.$$

Note first that

$$u^T \dot{l}_{pn}(\theta_0) \geq -||\dot{l}_{pn}^S(\theta_0)||_\infty ||u^S||_1 \geq -||\dot{l}_{pn}^S(\theta_0)||_\infty \sqrt{r_S}||u^S||_2 = -Dr_S\lambda_n||\dot{l}_{pn}^S(\theta_0)||_\infty,$$

where $\dot{l}_{pn}^S(\theta_0)$ is the subvector of $\dot{l}_{pn}(\theta_0)$ with indices in S. Next,

$$u^T \ddot{l}_{pn}(\theta_0)u = D^2 r_S \lambda_n^2 \frac{u^T}{||u||_2}\ddot{l}_{pn}(\theta_0)\frac{u}{||u||_2} \geq D^2 r_S \lambda_n^2 e_{\min}\{\ddot{l}_{pn}^{SS}(\theta_0)\}.$$

Furthermore, $\lambda_n\sqrt{r_S}||u||_2 = Dr_S\lambda_n^2$. Lastly,

$$\begin{aligned} u^T\left\{\ddot{l}_{pn}(\theta_0+tu) - \ddot{l}_{pn}(\theta_0)\right\}u &= \sum_{j\in S}\sum_{k\in S} u_j u_k \left\{\ddot{l}_{np(jk)}^{SS}(\theta_0+tu) - \ddot{l}_{np(jk)}^{SS}(\theta_0)\right\} \\ &= \sum_{j\in S}\sum_{k\in S} u_j u_k \sum_{m\in S} l_{np(jkm)}^{(3)SSS}(\theta_0 + t_{jk}^* u)tu_m \\ &\geq -(D\sqrt{r_S}\lambda_n)^3 \sup_{||\theta^S-\theta_0^S||_2\leq\xi} ||l_{np}^{(3)SSS}(\theta^S,0)||_\infty, \end{aligned}$$

where $\ddot{l}_{np(jk)}^{SS}$ and $l_{np(jkm)}^{(3)SSS}$ are respectively the second and third derivatives. It follows from combining the results that

$$\begin{aligned} G(u) \geq\ & Dr_S\lambda_n^2\Bigg[-\left\{\frac{||\dot{l}_{pn}^S(\theta_0)||_\infty}{\lambda_n}+1\right\} \\ &+\frac{D}{2}\left\{e_{\min}\{\ddot{l}_{pn}^{SS}(\theta_0)\} - D\sqrt{r_S}\lambda_n \sup_{||\theta^S-\theta_0^S||_2\leq\xi}||l_{np}^{(3)SSS}(\theta^S,0)||_\infty\right\}\Bigg]. \end{aligned}$$

If D can be chosen such that

$$\frac{8\left\{||\dot{l}_{pn}^S(\theta_0)||_\infty+\lambda_n\right\}}{\lambda_n e_{\min}\{\ddot{l}_{pn}^{SS}(\theta_0)\}} \leq D \leq \frac{e_{\min}\{\ddot{l}_{pn}^{SS}(\theta_0)\}}{2\sqrt{r_S}\lambda_n \sup_{||\theta^S-\theta_0^S||_2\leq\xi}||l_{np}^{(3)SSS}(\theta^S,0)||_\infty},$$

then

$$G(u) \geq \frac{1}{8}D^2 r_S\lambda_n^2 e_{\min}\{\ddot{l}_{pn}^{SS}(\theta_0)\} > 0.$$

Such a choice of D is possible when

$$e_{\min}\{\ddot{l}_{pn}^{SS}(\theta_0)\} \geq 4\left[\sqrt{r_S}\sup_{||\theta^S-\theta_0^S||_2\leq\xi}||l_{np}^{(3)SSS}(\theta^S,0)||_\infty\left\{||\dot{l}_{pn}^S(\theta_0)||_\infty+\lambda_n\right\}\right]^{1/2} > 0.$$

Since ξ depends on D, to avoid circular argument, let $\Delta = \sup_{\theta^S}||l_{np}^{(3)SSS}(\theta^S,0)||_\infty$. Replace $\sup_{||\theta^S-\theta_0^S||_2\leq\xi}||l_{np}^{(3)SSS}(\theta^S,0)||_\infty$ by Δ in the foregoing arguments, the same conclusion can be drawn. It now follows that $||\tilde{\theta}^S-\theta_0^S||_2 \leq D\sqrt{r_S}\lambda_n$, when

$$e_{\min}\{\ddot{l}_{pn}^{SS}(\theta_0)\} \geq 4\left[\sqrt{r_S}\Delta\left\{||\dot{l}_{pn}^S(\theta_0)||_\infty+\lambda_n\right\}\right]^{1/2} > 0,$$

and

$$\frac{8\left\{||\dot{l}_{pn}^S(\theta_0)||_\infty+\lambda_n\right\}}{\lambda_n e_{\min}\{\ddot{l}_{pn}^{SS}(\theta_0)\}} \leq D \leq \frac{e_{\min}\{\ddot{l}_{pn}^{SS}(\theta_0)\}}{2\sqrt{r_S}\lambda_n\Delta}.$$

Lemma S.10.4.12. Under Assumptions 1–6,

$$P\left\{||u^{S^c}||_\infty < 1\right\} \geq 1 - C_0 r r_S \exp\left\{-\frac{n(\lambda_n - C_1\sqrt{(p_S \wedge q_S + r_S)/n})^2}{C_2}\right\},$$

where $\lambda_n \geq C_1\sqrt{(p_S \wedge q_S + r_S)/n}$, $c_0\lambda_n < \pi_S/r_S$ for some constant $c_0 > 0$, and $\lambda_n \leq \alpha/\{2(4-\alpha)MD^2r_S^2\}$.

Proof: Note first from the expression for $R_{na}(i)$ that

$$\begin{aligned}\sup_{i\in S^c}|R_{na}(i)| &\leq \frac{M}{\lambda_n}\sum_{j\in S}\sum_{k\in S}|(\tilde{\theta}_j^S - \theta_{j0}^S)(\tilde{\theta}_k^S - \theta_{k0}^S)| \\ &\leq \frac{M}{\lambda_n}r_S||\tilde{\theta}^S - \theta_0^S||_2^2 \leq MD^2r_S^2\lambda_n,\end{aligned}$$

when $||\tilde{\theta}^S - \theta_0^S||_2 \leq D\sqrt{r_S}\lambda_n$. Similar, it can be shown that

$$\sup_{i\in S}|R_{nb}(i)| \leq MD^2r_S^2\lambda_n.$$

Note next that

$$\begin{aligned}||u^{S^c}||_\infty &= \sup_{||a||_1=1}|a^Tu^{S^c}| \\ &\leq |||\ddot{l}_{pn}^{S^cS}(\theta_0^S,0)\left\{\ddot{l}_{pn}^{SS}(\theta_0^S,0)\right\}^{-1}|||_\infty||u^S||_\infty \\ &\quad+\frac{1}{\lambda_n}\left[||\dot{l}_{pn}^{S^c}(\theta_0^S,0)||_\infty + |||\ddot{l}_{pn}^{S^cS}(\theta_0^S,0)\left\{\ddot{l}_{pn}^{SS}(\theta_0^S,0)\right\}^{-1}|||_\infty||\dot{l}_{pn}^S(\theta_0^S,0)||_\infty\right] \\ &\quad+\sup_{i\in S^c}|R_{na}(i)| + |||\ddot{l}_{pn}^{S^cS}(\theta_0^S,0)\left\{\ddot{l}_{pn}^{SS}(\theta_0^S,0)\right\}^{-1}|||_\infty\sup_{i\in S}|R_{nb}(i)|.\end{aligned}$$

When

$$\begin{aligned}|||\ddot{l}_{pn}^{S^cS}(\theta_0^S,0)\left\{\ddot{l}_{pn}^{SS}(\theta_0^S,0)\right\}^{-1}|||_\infty &\leq 1-\alpha/4, \\ ||\dot{l}_{np}(\theta_0^S,0)||_\infty &\leq c\lambda_n, \\ ||\tilde{\theta}^S - \theta_0^S||_2 &\leq D\sqrt{r_S}\lambda_n,\end{aligned}$$

then

$$||u^{S^c}||_\infty \leq 1-\alpha/2 + c(2-\alpha/2) + (2-\alpha/2)MD^2r_S^2\lambda_n.$$

If we choose $c(2-\alpha/2) \leq \alpha/4$ and $(2-\alpha/2)MD^2r_S^2\lambda_n \leq \alpha/4$, then $||u^{S^c}||_\infty < 1$. This yields $c \leq \alpha/(8-2\alpha)$ and $\lambda_n \leq \alpha/\{(8-2\alpha)MD^2r_S^2\}$.

From Lemma S.10.4.10, it follows by choosing $\delta = \alpha/2$ that

$$\begin{aligned}&P\left\{|||\ddot{l}_{pn}^{S^cS}(\theta_0^S,0)\left\{\ddot{l}_{pn}^{SS}(\theta_0^S,0)\right\}^{-1}|||_\infty \leq 1-3\alpha/4\right\} \\ &\geq P\left\{|||\ddot{l}_{pn}^{S^cS}(\theta_0^S,0)\left\{\ddot{l}_{pn}^{SS}(\theta_0^S,0)\right\}^{-1} - B^{S^cS}(B^{SS})^{-1}|||_\infty \leq \alpha/4\right\} \\ &\geq 1 - C_0 r r_S\exp\left[-\frac{n\left\{\min(\alpha/(4\sqrt{r_S}),1)\pi_S/r_S - C_1\sqrt{(p_S\wedge q_S+r_S)/n}\right\}^2}{C_2}\right],\end{aligned}$$

where $\min(\alpha/(4\sqrt{r_S}),1)\pi_S/r_S \geq C_1\sqrt{(p_S \wedge q_S + r_S)/n}$. From Lemma S.10.4.8, it follows by taking $\delta = \lambda_n\alpha/\{2(4-\alpha)\}$ and adjusting constants that

$$P\left\{||\dot{l}_{pn}(\theta_0)||_\infty \leq \lambda_n\alpha/\{2(4-\alpha)\}\right\} \geq 1 - C_0 r \exp\left\{-\frac{n\{\lambda_n - C_1\sqrt{(p_S \wedge q_S + r_S)/n}\}^2}{C_2}\right\},$$

for $\lambda_n \geq C_1\sqrt{(p_S \wedge q_S + r_S)/n}$. From Theorem 1,

$$P\left\{||\tilde{\theta}^S - \theta_0^S||_2 \geq D\sqrt{s}\lambda_n\right\} \quad \geq \quad 1 - C_0 s^2 \exp\left[-\frac{n}{C_2}\left\{\lambda_n - C_1\sqrt{(p_S \wedge q_S + r_S)/n}\right\}^2\right],$$

where $\lambda_n \geq C_1\sqrt{(p_S \wedge q_S + r_S)/n}$ and $\lambda_n \leq \alpha/\{2(4-\alpha)MD^2s^2\}$. It follows from combining the results that

$$\begin{aligned}
P\left\{||u^{S^c}||_\infty < 1\right\} \geq 1 - C_0 r r_S \exp&\left\{-\frac{n(\min(\alpha/(4\sqrt{r_S}),1)\pi_S/r_S - C_1\sqrt{(p_S \wedge q_S + r_S)/n})^2}{C_2}\right\} \\
- C_0 r \exp&\left\{-\frac{n\{\lambda_n - C_1\sqrt{(p_S \wedge q_S + r_S)/n}\}^2}{C_2}\right\} \\
- C_0 r_S^2 \exp&\left[-\frac{n}{C_2}\left\{\lambda_n - C_1\sqrt{(p_S \wedge q_S + r_S)/n}\right\}^2\right].
\end{aligned}$$

The result of the Lemma is obtained by combining terms and adjusting constants.

Lemma 13. Under assumptions 1-4, if $\hat{\theta}_A$ and $\hat{\theta}_B$ are two minimizers of (7.16) with $\lambda_n > 0$, then $||\hat{\theta}_A||_1 = ||\hat{\theta}_B||_1$, and if $\hat{\theta}_{jB} = 0$ and $|u_{jB}| < 1$ then $\hat{\theta}_{jA} = 0$. If $\hat{\theta}_{jA} \neq 0$ and $\hat{\theta}_{jB} \neq 0$, then $\text{sign}(\hat{\theta}_{jA}) = \text{sign}(\hat{\theta}_{jB})$.

Proof: For any two solutions to the penalized likelihood problem $\hat{\theta}_A$ and $\hat{\theta}_B$,

$$\begin{aligned}
\lambda_n\{||\hat{\theta}_A||_1 - C(\lambda_n)\} = 0,\ \lambda_n \geq 0,\ ||\hat{\theta}_A||_1 \leq C(\lambda_n), \dot{l}_{pn}(\hat{\theta}_A) + \lambda_n u_A = 0, \\
\lambda_n\{||\hat{\theta}_B||_1 - C(\lambda_n)\} = 0,\ \lambda_n \geq 0,\ ||\hat{\theta}_B||_1 \leq C(\lambda_n), \dot{l}_{pn}\{\hat{\theta}_B\} + \lambda_n u_B = 0,
\end{aligned}$$

When $\lambda_n > 0$, it follows from the complementary slackness that $||\hat{\theta}_A||_1 = ||\hat{\theta}_B||_1$. It follows from the convexity of the objective function that

$$0 = l_{pn}(\hat{\theta}_B) - l_{pn}(\hat{\theta}_A) \geq \dot{l}_{pn}(\hat{\theta}_A)(\hat{\theta}_B - \hat{\theta}_A) = \lambda_n(||\hat{\theta}_A||_1 - < u_A, \hat{\theta}_B >).$$

Hence, $< u_A, \hat{\theta}_B > \geq ||\hat{\theta}_A||_1 = ||\hat{\theta}_B||_1$. Since $||u_A||_\infty \leq 1$, it follows that, if $u_{jA} < 1$, then $\hat{\theta}_{jB} = 0$. If $\hat{\theta}_{jA} \neq 0$ and $\hat{\theta}_{jB} \neq 0$, it follows from $u_{jA} = \text{sign}(\hat{\theta}_{jA})$ and $u_{jB} = \text{sign}(\hat{\theta}_{jB})$ that $u_{jA} = u_{jB}$ and the signs of $\hat{\theta}_{jA}$ and $\hat{\theta}_{jB}$ are the same.

10.5 Bounds for the permutation distribution

Let $(Y_1, \cdots, Y_G)$ be the variables of a multivariate data set. We model the data by the semiparametric odds ratio model as

$$f(y_1, \cdots, y_G) = \frac{\eta(y_1; \cdots; y_G \mid \theta)\prod_{k=1}^G dF_k(y_k)}{\int \cdots \int \eta(y_1; \cdots; y_G \mid \theta)\prod_{k=1}^G dF_k(y_k)}.$$

Let the observed data be $(Y_{i1}, \cdots, Y_{iG}), i = 1, \cdots, n$. The likelihood for parameters $(\theta, F_1, \cdots, F_G)$ is

$$\mathcal{L}(\theta, F_1, \cdots, F_G) = \prod_{i=1}^{n} \frac{\eta(Y_{i1}; \cdots; Y_{iG} \mid \theta) \prod_{k=1}^{G} dF_k(Y_{ik})}{\int \cdots \int \eta(y_1; \cdots; y_G \mid \theta) \prod_{k=1}^{G} dF_k(y_k)}.$$

The permutation likelihood eliminating all the baseline functions appears as

$$L(\theta) = \frac{\prod_{i=1}^{n} \eta(Y_{i1}; \cdots; Y_{iG} \mid \theta)}{\sum_{\tau \in \mathcal{P}} \prod_{i=1}^{n} \eta(Y_{\tau_1(i)1}; \cdots; Y_{\tau_G(i)G} \mid \theta)}, \tag{10.17}$$

where $\tau = (\tau_1, \cdots, \tau_G) \in \mathcal{P}$ and $\mathcal{P}$ is the collection of all the such permutations. Define a probability mass function on all the permutations $\mathcal{P}$ as

$$P^*(\tau) = \frac{\prod_{i=1}^{n} \eta(Y_{\tau_1(i)1}; \cdots; Y_{\tau_G(i)G} \mid \theta)}{\sum_{\tau \in \mathcal{P}} \prod_{i=1}^{n} \eta(Y_{\tau_1(i)1}; \cdots; Y_{\tau_G(i)G} \mid \theta)}, \tag{10.18}$$

The probability measure P^* is defined on all the permutations $\mathcal{P}$ acting on the indices of the observed data. Let $\sigma = (\sigma_1, \cdots, \sigma_G)$ be the vector of permutations such that $(Y_{\sigma_k^{-1}(1)k}, \cdots, Y_{\sigma_k^{-1}(n)k}) = (Y_{(1)k}, \cdots, Y_{(n)k})$, i.e., σ_k^{-1} permutes the indices of the observed values $Y_{1k}, \cdots, Y_{nk}$ into the indices of the ordered values from the smallest to the largest. If we are given the order statistics, then σ transforms the order statistic into the observed data. A probability measure P defined on all the permutations $\mathcal{P}$ acting on the indices of the order statistic has

$$P(\tau) = P^*(\tau\sigma^{-1}).$$

It follows that

$$P(\sigma) = P^*(I) = L(\theta),$$

where I is the identity permutation on the indices.

For simplicity of presentation, consider the case with $G = 2$. Note that, for any given index i,

$$\mathcal{P}_n = \{\tau_n \mid \tau_n = c(i)\} \cup_{i_2 \neq i} \{\tau_n \mid \tau_n = c(i, i_2)\} \cdots \cup_{i_2, \cdots, i_n \neq i} \{\tau_n \mid \tau_n = c(i, i_2, \cdots, i_n)\},$$

where $i, i_2, \cdots, i_n$ are all different. Since no two sets in the foregoing expression have non-empty intersection, it follows that

$$1 = P\{\tau_n = c(i)\} + \sum_{i_2 \neq i} P\{\tau_n = c(i, i_2)\} + \cdots + \sum_{i_2, \cdots, i_n \neq i} P\{\tau_n = c(i, i_2, \cdots, i_n)\}. \tag{10.19}$$

Let

$$D_n(Y_1, \cdots, Y_n) = \sum_{\tau_n \in \mathcal{P}_n} \prod_{k=1}^{n} \eta(Y_{\tau_n(k)1}, Y_{k2}),$$

and

$$Y_{-(i_1, \cdots, i_j)} = \{Y_k \mid k \in \{1, \cdots, n\} - \{i_1, \cdots, i_j\}\}.$$

It follows that, for a fixed i with $i_1 = i$,

$$\begin{aligned} D_n(Y) \;&=\; \eta(Y_{i1}; Y_{i2}) D_{n-1}(Y_{-i}) + \cdots + \sum_{i_2, \cdots, i_j \neq i} \prod_{k=1}^{j} \eta(Y_{i_{k+1}1}; Y_{i_k 2}) D_{n-j}(Y_{-(i, i_2, \cdots, i_j)}) \\ &\quad + \cdots + \sum_{i_2, \cdots, i_n \neq i} \prod_{k=1}^{n} \eta(Y_{i_{k+1}1}; Y_{i_k 2}) D_0, \end{aligned}$$

where $D_0 = 1$ and with some abuse of notation $i_{j+1} = i$ in each product item of the form $\prod_{k=1}^{j}$. It follows that

$$\begin{aligned} D_n(Y) &= D_{n-1}(Y_{-i})\Bigg[\eta(Y_{i1};Y_{i2}) + \cdots \\ &+ \sum_{i_2,\cdots,i_j\neq i} \frac{\prod_{k=1}^{j}\eta(Y_{i_{k+1}1};Y_{i_k2})}{\prod_{k=2}^{j}\eta(Y_{i^*_{k+1}1};Y_{i_k2})} \frac{\prod_{k=2}^{j}\eta(Y_{i^*_{k+1}1};Y_{i_k2})D_{n-j}(Y_{-S_i})}{D_{n-1}(Y_{-i})} \\ &+ \cdots + \sum_{i_2,\cdots,i_n\neq i} \frac{\prod_{k=1}^{n}\eta(Y_{i_{k+1}1};Y_{i_k2})}{\prod_{k=2}^{n}\eta(Y_{i^*_{k+1}1};Y_{i_k2})} \frac{\prod_{k=2}^{n}\eta(Y_{i^*_{k+1}1};Y_{i_k2})D_0}{D_{n-1}(Y_{-i})}\Bigg], \end{aligned}$$

where, with some abuse of notation, $i^*_{j+1} = i_2$ and $i^*_k = i_k$ for $k < j$ in each product item of the form $\prod_{k=2}^{j}$, and $i_1 = i$. Note also that, for any $j \geq 2$,

$$\frac{\prod_{k=1}^{j}\eta(Y_{i_{k+1}1};Y_{i_k2})}{\eta(Y_{i1};Y_{i2})\prod_{k=2}^{j}\eta(Y_{i^*_{k+1}1};Y_{i_k2})} = \frac{\eta(Y_{i_21};Y_{i2})\eta(Y_{i1};Y_{i_j2})}{\eta(Y_{i1};Y_{i2})\eta(Y_{i_21};Y_{i_j2})}.$$

It follows that

$$\begin{aligned} D_n(Y) &= D_{n-1}(Y_{-i})\eta(Y_{i1};Y_{i2})\Bigg[1 + \cdots \\ &+ \sum_{i_2,\cdots,i_j\neq i} \frac{\eta(Y_{i_21};Y_{i2})\eta(Y_{i1};Y_{i_j2})}{\eta(Y_{i1};Y_{i2})\eta(Y_{i_21};Y_{i_j2})} \frac{\prod_{k=2}^{j}\eta(Y_{i^*_{k+1}1};Y_{i_k2})D_{n-j}(Y_{-S_i})}{D_{n-1}(Y_{-i})} \\ &+ \cdots + \sum_{i_2,\cdots,i_n\neq i} \frac{\eta(Y_{i_21};Y_{i2})\eta(Y_{i1};Y_{i_n2})}{\eta(Y_{i1};Y_{i2})\eta(Y_{i_21};Y_{i_n2})} \frac{\prod_{k=2}^{n}\eta(Y_{i^*_{k+1}1};Y_{i_k2})D_0}{D_{n-1}(Y_{-i})}\Bigg]. \end{aligned}$$

For a given index set S_i where the subscript i means the index set including i in it,

$$P\{\tau_n = c(S_i)\} = \frac{\sum_{\tau_n=c(S_i)}\prod_{k=1}^{n}\eta(Y_{\tau_n(k)1},Y_{k2})}{\sum_{\tau_n\in\mathcal{P}_n}\prod_{k=1}^{n}\eta(Y_{\tau_n(k)1},Y_{k2})}.$$

It follows that

$$\begin{aligned} P_n\{\tau_n = c(S_i)\} &= \frac{\eta(Y_{i_21};Y_{i2})\eta(Y_{i1};Y_{i_{|S_i|}2})}{\eta(Y_{i1};Y_{i2})\eta(Y_{i_21};Y_{i_{|S_i|}2})} P_{n-1}\{\tau_{n-1}^{-(i)} = c(S_i^{-i})\}\Bigg[1 + \cdots \\ &+ \sum_{i_2,\cdots,i_j\neq i} \frac{\eta(Y_{i_21};Y_{i2})\eta(Y_{i1};Y_{i_j2})}{\eta(Y_{i1};Y_{i2})\eta(Y_{i_21};Y_{i_j2})} P_{n-1}\{\tau_{n-1}^{-(i)} = c(i_2,\cdots,i_j)\} \\ &+ \cdots + \sum_{i_2,\cdots,i_n\neq i} \frac{\eta(Y_{i_21};Y_{i2})\eta(Y_{i1};Y_{i_n2})}{\eta(Y_{i1};Y_{i2})\eta(Y_{i_21};Y_{i_n2})} P_{n-1}\{\tau_{n-1}^{-(i)} = c(i_2,\cdots,i_n)\}\Bigg]^{-1}, \end{aligned}$$

where i_1 is fixed at i, $\tau_{n-1}^{-(i)}$ is a permutation on the index set $\{1,\cdots,n\}\backslash\{i\}$, and S_i^{-i} is S_i with i removed. In particular,

$$\begin{aligned} P_n\{\tau_n = c(i)\} &= \Bigg[1 + \cdots + \sum_{i_2,\cdots,i_j\neq i} \frac{\eta(Y_{i_21};Y_{i2})\eta(Y_{i1};Y_{i_j2})}{\eta(Y_{i1};Y_{i2})\eta(Y_{i_21};Y_{i_j2})} P_{n-1}\{\tau_{n-1}^{-(i)} = c(i_2,\cdots,i_j)\} \\ &+ \cdots + \sum_{i_2,\cdots,i_n\neq i} \frac{\eta(Y_{i_21};Y_{i2})\eta(Y_{i1};Y_{i_n2})}{\eta(Y_{i1};Y_{i2})\eta(Y_{i_21};Y_{i_n2})} P_{n-1}\{\tau_{n-1}^{-(i)} = c(i_2,\cdots,i_n)\}\Bigg]^{-1}. \end{aligned}$$

Lemma 10.5.1. Assume that, $0 < C_1 \le 1$ and $1 \le C_2$, and

$$C_1 \le \max_{i,j,k} \frac{\eta(Y_{j1};Y_{i2})\eta(Y_{i1};Y_{k2})}{\eta(Y_{i1};Y_{i2})\eta(Y_{j1};Y_{k2})} \le C_2. \tag{10.20}$$

Then

$$\frac{(C_1/C_2)^k}{n!/(n-k)!} \le P_n\{\tau_n = c(i_1,\cdots,i_k)\} \le \frac{(C_2/C_1)^k}{n!/(n-k)!}.$$

Proof: Note that, for $i_2 \ne i$,

$$P_{n-1}\{\tau_{n-1}^{-(i)} = c(i_2)\} + \cdots + \sum_{i_3,\cdots,i_n \ne i} P_{n-1}\{\tau_{n-1}^{-(i)} = c(i_2,\cdots,i_n)\} = 1.$$

It follows that

$$\frac{C_1}{1+C_2(n-1)} P_{n-1}\{\tau_{n-1}^{-(i)} = c(S_i^{-i})\} \le P_n\{\tau_n = c(S_i)\} \le \frac{C_2}{1+C_1(n-1)} P_{n-1}\{\tau_{n-1}^{-(i)} = c(S_i^{-i})\}. \tag{10.21}$$

It follows by induction that

$$\frac{(C_1/C_2)^j}{n!/(n-j)!} \le \prod_{k=1}^{j} \frac{C_1}{1+C_2(n-k)} \le P_n\{\tau_n = c(i_1,\cdots,i_j)\} \le \prod_{k=1}^{j} \frac{C_2}{1+C_1(n-k)} \le \frac{(C_2/C_1)^j}{n!/(n-j)!}.$$

Lemma 10.5.2. Under condition (10.20), for distinctive index sets $(i_1,\cdots,i_k)$ and $(j_1,\cdots,j_k)$,

$$\frac{(C_1/C_2)^k}{n!/(n-k)!} \le P_n\{\tau_n(i_1,\cdots,i_k) = (j_1,\cdots,j_k)\} \le \frac{(C_2/C_1)^k}{n!/(n-k)!}.$$

Proof: We prove this by induction. For $k=1$, let $w_{ij}^{(n)} = P_n\{\tau_n(i) = j\}$. The result holds for $i=j$ from Lemma 10.5.1. If $i \ne j$, then

$$\begin{aligned} w_{ij}^{(n)} &= P_n\{\tau_n = c(i,j)\} + \cdots + \sum_{(i_3,\cdots,i_n)} P_n\{\tau_n = c(i,j,i_3\cdots,i_n)\} \\ &\le \frac{C_2}{1+C_1(n-1)} \Big[P_{n-1}\{\tau_{n-1}^{-(i)} = c(j)\} + \sum_{i_3} P_{n-1}\{\tau_{n-1}^{-(i)} = c(j,i_3)\} \\ &\quad + \cdots + \sum_{(i_3,\cdots,i_n)} P_{n-1}\{\tau_{n-1}^{-(i)} = c(j,i_3,\cdots,i_n)\} \Big] = \frac{C_2}{1+C_1(n-1)}. \end{aligned}$$

Applying similar inequality to the other direction, it follows that

$$\frac{C_1/C_2}{n} \le \frac{C_1}{1+C_2(n-1)} \le w_{ij}^{(n)} \le \frac{C_2}{1+C_1(n-1)} \le \frac{C_2/C_1}{n}.$$

Assume that $k \le t$ holds. For $k = t+1$, if reordering $(j_1,\cdots,j_{t+1})$ into $(l_1,\cdots,l_{t+1})$ results in $\tau = c(l_1,\cdots,l_{t+1})$, the result holds from Lemma 3. If there is an embedded sub-cycle in $\tau_n(i_1,\cdots,i_{t+1}) = (j_1,\cdots,j_{t+1})$, i.e., there exists $(l_1,\cdots,l_s) \subset \{i_1,\cdots,i_{t+1}\} \cap \{j_1,\cdots,j_{t+1})$ with $1 \le s \le t$ such that $\tau_n\{l_1,\cdots,l_s\} = \{l_1,\cdots,l_s,l_1\}$, then

$$\begin{aligned} P_n\{\tau_n(i_1,\cdots,i_{t+1}) = (j_1,\cdots,j_{t+1})\} &= P_n\{\tau_n = c(l_1,\cdots,l_s)\} \\ \times P_{n-s}\Big[\tau_{n-s}^{-(l_1,\cdots,l_s)}(\{i_1,\cdots,i_{t+1}\}\backslash\{l_1,\cdots,l_s\}) &= \{j_1,\cdots,j_{t+1}\}\backslash\{l_1,\cdots,l_s\}\Big] \\ &\le \frac{(C_2/C_1)^s}{n!/(n-s)!} \frac{(C_2/C_1)^{t+1-s}}{(n-s)!/(n-t-1)!}. \end{aligned}$$

If there is no such an embedded cycle, let $IC_1^M, \cdots, IC_r^M$ be all the non-overlapped maximum incomplete cycles found in $\tau_n(i_1, \cdots, i_{t+1}) = (j_1, \cdots, j_{t+1})$, where an incomplete cycle with indices $\{l_1, \cdots, l_s\}$ for $\tau_n(i_1, \cdots, i_{t+1}) = (j_1, \cdots, j_{t+1})$ means $(l_2, \cdots, l_s) \subset \{i_1, \cdots, i_{t+1}\} \cup \{j_1, \cdots, j_{t+1}\}$, $\tau_n(l_1, \cdots, l_{s-1}) = (l_2, \cdots, l_s)$, and $l_s \neq l_1$, and the maximum incomplete cycles of $\tau_n(i_1, \cdots, i_{t+1}) = (j_1, \cdots, j_{t+1})$ are all incomplete cycles with the maximum lengths. Let $IC^M(l_1, \cdots, l_s)$ denote a maximum incomplete cycle with index set $\{l_1, \cdots, l_s\}$. Let

$$\{\tau_n \mid \tau_n(i_1, \cdots, i_{t+1}) = (j_1, \cdots, j_{t+1})\} = \cup_{k=1}^r IC_k^M(l_{k1}, \cdots, l_{ks_k}).$$

Note also that

$$IC(l_1, \cdots, l_s) = c(l_1, \cdots, l_s) \cup_{u=1}^{n-s} \cup_{\{k_1, \cdots, k_u\} \cap \{l_1, \cdots, l_s\} = \emptyset} c(l_1, \cdots, l_s, k_1, \cdots, k_u).$$

It follows that

$$\begin{aligned}
P_n\{\tau_n = IC(l_1, \cdots, l_s)\} &= P_n\{\tau_n = c(l_1, \cdots, l_s)\} \\
&\quad + \sum_{u=1}^{n-s} \sum_{\{k_1, \cdots, k_u\} \cap \{l_1, \cdots, l_s\} = \emptyset} P_n\{\tau_n = c(l_1, \cdots, l_s, k_1, \cdots, k_u)\} \\
&\leq \frac{(C_2/C_1)^{s-1}}{n!/(n-s+1)!} \Bigg[P_{n-s-t+2}\left\{\tau_{n-s+1}^{-(l_1, \cdots, l_{s-1})} = c(l_s)\right\} + \\
&\quad + \sum_{u=1}^{n-s} \sum_{\{k_1, \cdots, k_u\} \cap \{l_s\} = \emptyset} P_{n-s+1}\left\{\tau_{n-s+1}^{-(l_1, \cdots, l_{s-1})} = c(l_s, k_1, \cdots, k_u)\right\} \Bigg] \\
&= \frac{(C_2/C_1)^{s-1}}{n!/(n-s+1)!}.
\end{aligned}$$

Furthermore,

$$\begin{aligned}
&P_n\{\tau_n = IC(l_1, \cdots, l_s) IC(j_1, \cdots, j_t)\} \\
&= \sum_{u,v} \sum_{(i_1, \cdots, i_u) \cup (k_1, \cdots, k_v)} P_n\{\tau_n = c(l_1, \cdots, l_s, i_1, \cdots, i_u) c(j_1, \cdots, j_t, k_1, \cdots, k_v)\} \\
&\quad + \sum_{u,v} \sum_{(i_1, \cdots, i_u) \cup (k_1, \cdots, k_v)} P_n\{\tau_n = c(l_1, \cdots, l_s, i_1, \cdots, i_u, j_1, \cdots, j_t, k_1, \cdots, k_v)\} \\
&\leq \frac{(C_2/C_1)^{s+t-2}}{n!/(n-s-t+2)!} \Bigg[\\
&\quad \sum_{(i_1, \cdots, i_u) \cup (k_1, \cdots, k_v)} P_{n-s-t+2}\left\{\tau_{n-s-t+2}^{-(l_1, \cdots, l_{s-1}, j_1, \cdots, j_{t-1})} = c(l_s, i_1, \cdots, i_u) c(j_t, k_1, \cdots, k_u)\right\} \\
&\quad + \sum_{(i_1, \cdots, i_u) \cup (k_1, \cdots, k_v)} P_{n-s-t+2}\left\{\tau_{n-s-t+2}^{-(l_1, \cdots, l_{s-1}, j_1, \cdots, j_{t-1})} = c(l_s, i_1, \cdots, i_u, j_t, k_1, \cdots, k_u)\right\} \Bigg] \\
&= \frac{(C_2/C_1)^{s+t-2}}{n!/(n-s-t+2)!}.
\end{aligned}$$

By similar arguments, it can be shown that, for any r,

$$\begin{aligned}
P_n\left\{\tau_n = \cup_{k=1}^r IC_k^M(l_{k1}, \cdots, l_{ks_k})\right\} &\leq \prod_{k=1}^r \frac{(C_2/C_1)^{s_k - 1}}{\left(n - \sum_{j=1}^{k-1}(s_j - 1)\right)! / \left(n - \sum_{j=1}^k (s_k - 1)\right)!} \\
&= \frac{(C_2/C_1)^{\sum_{k=1}^r (s_k - 1)}}{n! / \left(n - \sum_{k=1}^r (s_k - 1)\right)!}.
\end{aligned}$$

This concludes the proof.

Lemma 10.5.3. Under condition (10.20), for indices in $i, j_1, j_2 \in \{1, \cdots, n\}$ and $j_1 \neq i \neq j_2$, we have

$$E_0\left[u(Y_{i1}, Y_{i2})h(Y_{j_1 1}, Y_{j_2 2})P\{\tau_n(j_2) = j_1\}\right] = O(n^{-2}),$$

if $E_0\{u(Y_{i1}, Y_{i2})\} = 0$ and $E_0[E_\tau\{h(Y_{\tau(i)1}, Y_{i2})\}] = 0$.

Proof: Let $i, i_2, \cdots, i_n$ be distinctive indices in $\{1, \cdots, n\}$ and $j_1 \neq i \neq j_2$. Then

$$\begin{aligned}\{\tau_n(j_2) = j_1\} &= (\{\tau_n(j_2) = j_1\} \cap \{\tau_n \mid \tau_n = c(i)\}) \\ &\quad \cup (\{\tau_n(j_2) = j_1\} \cap (\cup_{i_2 \neq i}\{\tau_n \mid \tau_n = c(i, i_2)\})) \cdots \\ &\quad \cap (\{\tau_n(j_2) = j_1\} \cap (\cup_{i_2, \cdots, i_n \neq i}\{\tau_n \mid \tau_n = c(i, i_2, \cdots, i_n)\}))\,.\end{aligned}$$

It follows that

$$\begin{aligned}P\{\tau_n(j_2) = j_1\} &= P[\{\tau_n(j_2) = j_1\} \mid \{\tau_n = c(i)\}]\, P[\{\tau_n = c(i)\}] \\ &\quad + \sum_{i_2 \neq i} P[\{\tau_n(j_2) = j_1\} \mid \{\tau_n = c(i, i_2)\}]\, P[\{\tau_n = c(i, i_2)\}] + \cdots \\ &\quad + \sum_{i_2, \cdots, i_n \neq i} P[\{\tau_n(j_2) = j_1\} \mid \{\tau_n = c(i, i_2 \cdots, i_n))\}]\, P[\{\tau_n = c(i, i_2 \cdots, i_n)\}]\,.\end{aligned}$$

Note that, for $j_1 \neq i \neq j_2$,

$$\begin{aligned}P[\{\tau_n(j_2) = j_1\} \mid \{\tau_n = c(i)\}] &= P\{\tau_{n-1}^{-(i)}(j_2) = j_1\}, \\ P[\{\tau_n(j_2) = j_1\} \mid \{\tau_n = c(i, i_2)\}] &= P\{\tau_{n-2}^{-(i,i_2)}(j_2) = j_1\}.\end{aligned}$$

It follows that

$$\begin{aligned}&P\{\tau_n(j_2) = j_1\} - P\{\tau_{n-1}^{-(i)}(j_2) = j_1\} \\ &\quad = \sum_{i_2 \neq i} \left[P\{\tau_{n-2}^{-(i,i_2)}(j_2) = j_1\} - P\{\tau_{n-1}^{-(i)}(j_2) = j_1\}\right] P\{\tau_n = c(i, i_2)\} + \cdots \\ &\quad\quad + \sum_{i_2, \cdots, i_n \neq i} \left[P\{\tau_n(j_2) = j_1 \mid \tau_n = c(i, i_2, \cdots, i_n)\} - P\left\{\tau_{n-1}^{-(i)}(j_2) = j_1\right\}\right] \\ &\quad\quad \times P\{\tau_n = c(i, i_2, \cdots, i_n)\}.\end{aligned}$$

The above expression can be rewritten as

$$
\begin{aligned}
&P\{\tau_n(j_2) = j_1\} - P\{\tau_{n-1}^{-(i)}(j_2) = j_1\} \\
&\quad = \sum_{i_2 \neq i} \left[P\{\tau_{n-2}^{-(i,i_2)}(j_2) = j_1\} - P\{\tau_{n-1}^{-(i)}(j_2) = j_1\}\right] \Bigg[P\{\tau_n = c(i, i_2)\} + \cdots \\
&\qquad + \sum_{i_3, \cdots, i_n \neq i} P\{\tau_n = c(i, i_2, \cdots, i_n)\}\Bigg] \\
&\qquad + \sum_{i_2, i_3 \neq i} \left[P\{\tau_n(j_2) = j_1 \mid \tau_n = c(i, i_2, i_3)\} - P\left\{\tau_{n-2}^{-(i,i_2)}(j_2) = j_1\right\}\right] \Bigg[P\{\tau_n = c(i, i_2, i_3) \\
&\qquad + \cdots + \sum_{i_4, \cdots, i_n \neq i} P\{\tau_n = c(i, i_2, \cdots, i_n)\}\Bigg] \\
&\qquad + \sum_{k=3}^{n-1} \sum_{i_2, \cdots, i_{k+1} \neq i} \\
&\qquad \times \Bigg[P\{\tau_n(j_2) = j_1 \mid \tau_n = c(i, i_2, \cdots, i_{k+1})\} - P\{\tau_n(j_2) = j_1 \mid \tau_n = c(i, i_2, \cdots, i_k)\}\Bigg] \\
&\qquad \times \Bigg[P\{\tau_n = c(i, i_2, \cdots, i_{k+1})\} + \cdots + \sum_{i_{k+2}, \cdots, i_n \neq i} P\{\tau_n = c(i, i_2, \cdots, i_n)\}\Bigg].
\end{aligned}
$$

Note also that, if $j_1 \neq j_2 \in \{i, i_2, \cdots, i_k\}$ for any $k \geq 3$, then

$$
\begin{aligned}
&P\{\tau_n(j_2) = j_1 \mid \tau_n = c(i, i_2, \cdots, i_{k+1}) - P\{\tau_n(j_2) = j_1 \mid \tau_n = c(i, i_2, \cdots, i_k) = 0, \\
&\sum_{k=2}^{n-1} \sum_{i_2, \cdots, i_{k+1} \neq i} \Bigg[P\{\tau_n = c(i, i_2, \cdots, i_{k+1})\} + \cdots + \sum_{i_{k+2}, \cdots, i_n \neq i} P\{\tau_n = c(i, i_2, \cdots, i_n)\}\Bigg] \\
&= O(n^{-2}).
\end{aligned}
$$

It follows that

$$
\begin{aligned}
P\{\tau_n(j_2) = j_1\} - P\{\tau_{n-1}^{-(i)}(j_2) = j_1\} &= O_p(n^{-2}) \\
&+ \sum_{k=1}^{n-3} \sum_{i_2, \cdots, i_{k+1} \neq i, j_1, j_2} \Bigg[P\{\tau_{n-k-1}^{-(i,i_2,\cdots,i_{k+1})}(j_2) = j_1\} - P\left\{\tau_{n-k}^{-(i,i_2,\cdots,i_k)}(j_2) = j_1\right\}\Bigg] \\
&\times \Bigg[P\{\tau_n = c(i, i_2, \cdots, i_{k+1})\} + \cdots + \sum_{i_{k+2}, \cdots, i_n \neq i} P\{\tau_n = c(i, i_2, \cdots, i_n)\}\Bigg].
\end{aligned}
$$

Furthermore, by similar arguments that the additional terms are of $O_p(n^{-2})$, we have

$$
\begin{aligned}
P\{\tau_n(j_2) = j_1\} - P\{\tau_{n-1}^{-(i)}(j_2) = j_1\} &= O_p(n^{-2}) \\
&+ \sum_{k=1}^{n-3} \sum_{i_2, \cdots, i_{k+1} \neq i, j_1, j_2} \Bigg[P\{\tau_{n-k-1}^{-(i,i_2,\cdots,i_{k+1})}(j_2) = j_1\} - P\left\{\tau_{n-k}^{-(i,i_2,\cdots,i_k)}(j_2) = j_1\right\}\Bigg] \\
&\times \Bigg[P\{\tau_n = c(i, i_2, \cdots, i_{k+1})\} + \cdots + \sum_{i_{k+2}, \cdots, i_n \neq i, j_1, j_2} P\{\tau_n = c(i, i_2, \cdots, i_n)\}\Bigg].
\end{aligned}
$$

It now follows that, for $j_1 \neq i \neq j_2$, apart frrom an $O(n^{-2})$ term,

$$E_0\left[u(Y_{i1}, Y_{i2})h(Y_{j_11}, Y_{j_22})P\{\tau_n(j_2) = j_1\}\right]$$

$$= E_0\{u(Y_{i1}, Y_{i2})\}\, E_0\left[h(Y_{j_11}, Y_{j_22})P\{\tau_{n-1}^{-(i)}(j_2) = j_1\}\right] + \sum_{k=1}^{n-3} \sum_{i_2,\cdots,i_{k+1}\neq i,j_1,j_2}$$

$$\times E_0\Big(h(Y_{j_11}, Y_{j_22})\left[P\{\tau_{n-k-1}^{-(i,i_2,\cdots,i_{k+1})}(j_2) = j_1\} - P\left\{\tau_{n-k}^{-(i,i_2,\cdots,i_k)}(j_2) = j_1\right\}\right]\Big)$$

$$\times E_0\left(u(Y_{i1}, Y_{i2})\left[P\{\tau_n = c(i, i_2, \cdots, i_{k+1})\} + \cdots + \sum_{i_{k+2},\cdots,i_n\neq i,j_1,j_2} P\{\tau_n = c(i, i_2, \cdots, i_n)\}\right]\right).$$

Note from $E_0\{u(Y_{i1}, Y_{i2})\} = 0$ and $E_0[E_\tau\{h(Y_{\tau(i)1}, Y_{i2})\}] = 0$ that

$$E_0\Big(h(Y_{j_11}, Y_{j_22})\left[P\{\tau_{n-k-1}^{-(i,i_2,\cdots,i_{k+1})}(j_2) = j_1\} - P\left\{\tau_{n-k}^{-(i,i_2,\cdots,i_k)}(j_2) = j_1\right\}\right]\Big) = O\left\{(n-k)^{-2}\right\},$$

It can be concluded that, for $j_1 \neq i \neq j_2$,

$$E_0\left[u(Y_{i1}, Y_{i2})h(Y_{j_11}, Y_{j_22})P\{\tau_n(j_2) = j_1\}\right] = O(n^{-2}).$$

Bibliography

[1] P. S. Albert, D. Ratnasinghe, J. Tangrea, and S. Wacholder. Limitations of the case-only design for identifying gene-environment interactions. *American Journal of Epidemiology*, 154:687–693, 2001.

[2] B. C. Arnold, E. Castillo, and J. M. Sarabia. Conditionally specified distributions: An introduction (with discussion). *Statistical Science*, 16:249–274, 2001.

[3] B. C. Arnold and S. J. Press. Compatibility conditional distributions. *Journal of the American Statistical Association*, 84:152–156, 1989.

[4] J. B. Aubin and S. Leoni-Aubin. Projection density estimation under a m-sample semiparametric model. *Computational Statistics and Data Analysis*, 52:2451–2468, 2008.

[5] T. B. Berrett, Y. Wang, R. F. Barber, and R. J. Samworth. The conditional permutation test for independence while controlling for confounders. *Journal of the Royal Statistical Society, Ser. B*, 82:175–197, 2020.

[6] P. Berti, E. Dreassi, and P. Rigo. Compatibility results for conditional distributions. *Journal of Multivariate Analysis*, 125:190–203, 2014.

[7] J. Besag. Spatial interaction and the statistical analysis of lattice systems (with discussion). *Journal of the Royal Statistical Society, Ser. B*, 36:192–236, 1974.

[8] J. Besag. On statistical analysis of dirty pictures (with discussion). *Journal of the Royal Statistical Society, Ser. B*, 48:192–236, 1996.

[9] J. Besag, A. Gelman, T. E. Raghunathan, H. Joe, B. C. Arnold, E. Castillo, and J. M. Sarabia. Conditionally specified distributions: An introduction-Discussion. *Statistical Science*, 16:265–274, 2001.

[10] J. Besag and C. Kooperberg. On conditional and intrinsic autoregressions. *Biometrika*, 82:733–746, 1995.

[11] P. J. Bickel, C. A. J. Klaassen, Y. Ritov, and J. A. Wellner. *Efficient and Adaptive Estimation for Semiparametric Models.* John Hopkins University Press, Baltimore, 1993.

[12] S. Boyd and L. Vanderberghe. *Convex Optimization.* Cambridge University Press, 2004.

[13] N. Breslow, B. McNeney, and J. A. Wellner. Large sample theory for semiparametric regression model with two-stage, outcome-dependent sampling. *Annals of Statistics*, 31:2014–2042, 2012.

[14] N. E. Breslow. Statistics in Epidemiology: the case-control study. *Journal of the American Statistical Association*, 91:14–28, 1996.

[15] N. E. Breslow, N. E. Day, K. T. Halvorsen, R. L. Prentice, and C. Sabai. Estimation of multiple relative risk functions in matched case-control studies. *American Journal of Epidemiology*, 108:299–307, 1978.

[16] D. Brook. On the distinction between the conditional probability and the joint probability approaches in the specification of the nearest neighbor systems. *Biometrika*, 51:481–483, 1964.

[17] P. Buhlmann, M. Kalisch, and M.H. Maathuis. Variable selection in high-dimensional linear models: partial faithful distributions and the PC-simple algorithm. *Biometrika*, 97:261–278, 2010.

[18] S. P. J. Byrne and A. P. Dawid. Retrospective-prospective symmetry in the likelihood and Bayesian analysis of case-control studies. *Biometrika*, 101:189–204, 2014.

[19] T. Cai, H. Li, W. Liu, and J. Xie. Covariate adjusted precision matrix estimation with an application in genetical genomics. *Biometrika*, 100:139–156, 2013.

[20] T. Cai and M. Yuan. Adaptive covariance matrix estimation through block thresholding. *Annals of Statistics*, 40:2014–2042, 2012.

[21] R. J. Carroll, S. Wang, and C. Y. Wang. Prospective analysis of logistic case-control studies. *Journal of the American Statistical Association*, 90:157–169, 1995.

[22] K. C. G. Chan. Nuisance parameter elimination for proportional likelihood ratio models with nonignorable missingness and random truncation. *Biometrika*, 100:269–276, 2013.

[23] N. Chatterjee and R. J. Carroll. Semiparametric maximum likelihood estimation exploiting gene-environment independence in case-control studies. *Biometrika*, 92:399–418, 2005.

[24] S. Chatterjee. A new coefficient of correlation. *Journal of the American Statistical Association*, Published online 2020.

[25] H. Y. Chen. Double nonparametric likelihood method for the Cox regression model with missing covariates. *Journal of the American Statistical Association*, 97:565–576, 2002.

[26] H. Y. Chen. A note on prospective analysis of outcome dependent samples. *Journal of the Royal Statistical Society, Ser. B*, 65:575–584, 2003.

[27] H. Y. Chen. Nonparametric and semiparametric models for missing covariates in parametric regression. *Journal of the American Statistical Association*, 99:1176–1189, 2004.

[28] H. Y. Chen. A semiparametric odds ratio model for measuring association. *Biometrics*, 63:413–421, 2007.

[29] H. Y. Chen. Compatibility of conditionally specified models. *Statistics & Probability Letters*, 80:670–677, 2010.

[30] H. Y. Chen. A unified framework for parameter identification and estimation in biased sampling designs. *Biometrika*, 98:163–175, 2011.

[31] H. Y. Chen. A note on the convergence of an iterative algorithm for the semiparametric odds ratio model. *Biometrika*, 102:747–751, 2015.

[32] H. Y. Chen and J. Chen. On information coded in gene-environment independence in case-control design. *American Journal of Epidemiology*, 174:736–743, 2011.

[33] H. Y. Chen and M. Li. Improving power for detecting genetic association with extreme-value sampling design. *Genetic Epidemiology*, 35:823–830, 2011.

[34] H. Y. Chen and R. J. A. Little. Proportional hazards regression with missing covariates. *Journal of the American Statistical Association*, 94:896–908, 1999.

[35] H. Y. Chen, D. E. Rader, and M. Li. Likelihood inference on semiparametric odds ratio Model. *Journal of the American Statistical Association*, 110:1125–1135, 2015.

[36] H. Y. Chen, M. P. Reilly, and M. Li. Semiparametric odds ratio model for case-control and matched case-control designs. *Statistics in Medicine*, 32:3126–3142, 2013.

[37] H. Y. Chen, H. Xie, and Y. Qian. Multiple imputation for missing data through conditional semiparametric odds ratio models. *Biometrics*, 67:799–809, 2011.

[38] H. Y. Chen, W. Zhang, and R. Kittles. Bias correction to secondary trait analysis with case-control design. *Statistics in Medicine*, 32:1494–1508, 2013.

[39] J. Chen and Z. Chen. Extended Bayesian information criteria for model selection with large model spaces. *Biometrika*, 95:759–771, 2008.

[40] J. Chen and Z. Chen. Extended BIC for small-n-large-p sparse GLM. *Statistica Sinica*, 22:555–574, 2012.

[41] J. S. Chen, G. R. Terrell, I. Kim, and M. L. Daviglus. Proportional odds model with log-concave density estimation. *Statistica Sinica*, 30:877–901, 2020.

[42] Y. H. Chen, N. Chatterjee, and R. J. Carroll. Retrospective analysis of haplotype based case-control under a flexible model for gene-environment association. *Biostatistics*, 9:81–99, 2008.

[43] K. F. Cheng and C. K. Chu. Semiparametric density estimation under a two sample density ratio model. *Bernoulli*, 10:583–604, 2004.

[44] W. G. Cochran. Some methods for strengthening the common χ^2 tests. *Biometrics*, 10:417–451, 1954.

[45] P. Cole. The evolving case-control study. *Journal of Chronic Diseases*, 32:15–27, 1979.

[46] J Cornfield. A method of estimating comparative rates from clinical data, applications to cancer of the lung, breast, and cervix. *Journal of the National Cancer Institute*, 11:1269–1275, 1951.

[47] S. R. Cosslett. Maximum likelihood estimator for choice-based samples. *Econometrics*, 49:1289–1316, 1981.

[48] P. Danaher, P. Wang, and D. M. Witten. The joint graphical lasso for inverse covariance estimation across multiple classes. *Journal of the Royal Statistical Society, Ser. B*, 76:273–397, 2014.

[49] L. J. Davis. Generalization of the Mantel-Haenszel estimator to nonconstant odds ratios. *Biometrics*, 41:487–495, 1985.

[50] A. P. Dawid. Conditional independence in statistical theory. *Journal of the Royal Statistical Society, Ser. B*, 41:1–31, 1979.

[51] A. P. Dawid. Some misleading arguments involving conditional independence. *Journal of the Royal Statistical Society, Ser. B*, 41:249–252, 1979.

[52] A. P. Dawid. Conditional independence for statistical operations. *Annals of Statistics*, 8:598–617, 1980.

[53] M. de Carvalho and A. C. Davison. Spectral density ratio models for multivariate extremes. *Journal of the American Statistical Association*, 109:764–776, 2014.

[54] A. P. Dempster. Covariance selection. *Biometrics*, 28:157–175, 1972.

[55] P. Diaconis and B. Sturmfels. Algebraic algorithms for sampling from conditional distributions. *Annals of Statistics*, 26:363–379, 1998.

[56] V. Didelez, S. Kreiner, and N. Keiding. Graphical models for inference under outcome-dependent sampling. *Statistical Science*, 25:368–387, 2010.

[57] S. Duane, A. D. Kennedy, B. J. Pendleton, and D. Roweth. Hybrid Monet Carlo. *Physics Letters B*, 195:216–222, 1987.

[58] A. C. Edmondson, P. S. Braund, I. M. Stylianou, A. V. Khera, C. P. Nelson, M. L. Wolfe, S. L. Derohannessian, B. J. Keating, L. Qu, J. He, M. D. Tobin, M. Tomaszewski, J. Baumert, N. Klopp, A. Doering, B. Thorand, M. Li, M. P. Reilly, W. Koenig, N. J. Samani, and D. J. Rader. Dense genotyping of candidate gene loci identifies variants associated with high-density lipoprotein cholesterol. *Circulation Cardivascular Genetics*, 4:145–155, 2011.

[59] D. M. Edwards. *Introduction to Graphical Modelling*. Springer, New York, 2000.

[60] M. D. Ernst. Permutation methods: A basis for exact inference. *Statistical Science*, 19:676–685, 2004.

[61] B. Fellinghauer, P. Buhlmann, M. Ryffel, M. Von Rhein, and J. D. Reinhardt. Stable graphical model estimation with random forests for discrete, continuous, and mixed variables. *Computational Statistics and Data Analysis*, 64:132–152, 2013.

[62] K. Fokianos. Merging information for semiparametric density estimation. *Journal of the Royal Statistical Society, Ser. B*, 66:941–958, 2004.

[63] K. Fokianos and A Savvides. On comparing several densities. *Technometrics*, 50:317–331, 2008.

[64] J. Friedman, T. Hastie, and R Tibshirani. Sparse inverse covariance estimation with the graphical lasso. *Biostatistics*, 3:432–441, 2007.

[65] W. J. Fu. Penalized regression: The Bridge versus the Lasso. *Journal of Computational and Graphical Statistics*, 7:397–416, 1998.

[66] S. Fuchino and S. Plewik. On a theorem of E. Helly. *Proceedings of the American Mathematical Society*, 127:491–497, 1999.

[67] K Fukumizu, A Gretton, X Sun, and B. Scholkopf. Kernel measures of conditional dependence. *Adv Neur In*, 20:489–496, 2008.

[68] G. Geenens and P. L. de Micheaux. The Hellinger correlation. *Journal of the American Statistical Association*, Published online, 2020.

[69] C. J. Geyer and E. A. Thompson. Constrained Monte Carlo Maximum likelihood for dependent data. *Journal of the Royal Statistical Society, Ser. B*, 54:657–699, 1992.

[70] P. B. Gilbert. Large sample theory of maximum likelihood estimates in semiparametric biased sampling models. *Annals of Statistics*, 28:151–194, 2000.

[71] P. B. Gilbert, S. R. Lele, and Y. Vardi. Maximum likelihood estimation in semiparametric selection bias models with application to AIDS vaccine trials. *Biometrika*, 86:27–43, 1999.

[72] M Girolami and B. Calderhead. Riemann manifold Langevin and Hamiltonian Monte Carlo methodsConstrained Monte Carlo. *Journal of the Royal Statistical Society, Ser. B*, 73:123–214, 2011.

[73] A. Gretton, O. Bousquet, A. Smola, and B. Scholkopf. Measuring statistical dependence with Hilbert-Schmidt norms. *Algorithmic Learning Theory*, 3734:63–77, 2005.

[74] M. L. Hazelton and B. A. Turlach. Semiparametric density deconvolution. *Scandinavian Journal of Statistics*, 37:91–108, 2010.

[75] A. H. Herring, J. G. Ibrahim, and S. R. Lipsitz. Likelihood-based methods for missing covariates in the Cox proportional hazards model. *Journal of the American Statistical Association*, 96:292–302, 2001.

[76] J. P. Hobert and G. Casella. Functional compatibility, Markov chains, and Gibbs sampling with improper posteriors. *Journal of Computational and Graphical Statistics*, 7:42–60, 1998.

[77] W. Hoeffding. A class of statistics with asymptotically normal distributions. *Annals of Mathematical Statistics*, 19:293–325, 1948.

[78] W. Hoeffding. A combinatorial central limit theorem. *Annals of Mathematical Statistics*, 22:558–566, 1948.

[79] W. Hoeffding. *The strong law of large numbers for U-statistics. Institute of Statistics Mimeo Series No. 302.* University of North Carolina, Chapel Hill, 1961.

[80] M. D. Hoffman and A. Gelman. The No-U-Turn sampler: adaptive setting path lengths in Hamiltonian Monte Carlo. *Journal of Machine Learning Research*, 15:1593–1623, 2014.

[81] A. Huang and P. J. Rathouz. Proportional likelihood ratio models for mean regression. *Biometrika*, 99:223–229, 2012.

[82] J. G. Ibrahim, M. H. Chen, and D. Sinha. *Bayesian Survival Analysis.* Springer, New York, 2001.

[83] Y. N. Jiang, A. J. Scott, and C. J. Wild. Secondary analysis of case-control data. *Statistics in Medicine*, 25:1323–1339, 2006.

[84] H. Joe. *Multivariate Models and Dependence Concepts.* Chapman and Hall, London, 1997.

[85] M. C. Jones. Kernel density estimation for length-biased data. *Biometrika*, 78:511–519, 1991.

[86] J. Kalbfleisch. Likelihood methods and Nonparametric tests. *Journal of the American Statistical Association*, 73:167–170, 1978.

[87] J. D. Kalbfleisch. Nonparametric Bayesian analysis of survival time data. *Journal of the Royal Statistical Society, Ser. B*, 40:214–221, 1978.

[88] L. V. Kantorovich and G. P. Akilov. *Functional Analysis*, 2nd edition. Pergamon Press: Oxford, 1982.

[89] I. Kojadinovic and Holmes. M. Tests of independence among continuous random vectors based on Cramer-von Mises functionals of empirical copula process. *Journal of Multivariate Analysis*, 100:1137–1154, 2009.

[90] K. L. Kuo and Y. J. Wang. Simulating conditionally specified models. *Journal of Multivariate Analysis*, 167:171–180, 2018.

[91] C. Lam and J. Fan. Sparsistency and rates of convergence in large covariance matrices estimation. *Annals of Statistics*, 37:4254–4278, 2009.

[92] S. Lauritzen. *Graphical Models.* Clarendon Press, Oxford, UK, 1996.

[93] J. F. Lawless, J. D. Kalbfleisch, and C. J. Wild. Semiparametric methods for response-selective and missing data problems in regression. *Journal of the Royal Statistical Society, Ser. B*, 61:413–438, 1999.

[94] A. J. Lee. *U-Statistics.* Marcel Dekker Inc., New York, 1990.

[95] A. J. Lee, L. McMurchy, and A. J. Scott. Re-using data from case-control studies. *Statistics in Medicine*, 16:1377–1389, 1997.

[96] B. Li, H. Chun, and H. Zhao. Sparse estimation of conditional graphical models with application to gene networks. *Journal of the American Statistical Association*, 107:152–167, 2012.

[97] H. Li and J. Gui. Gradient directed regularization for sparse Gaussian concentration graphs, with applications to inference of genetic networks. *Biostatistics*, 7:302–317, 2006.

[98] F. Liang, Q. Song, and P. Qiu. An equivalent measure of partial correlation coefficients for high-dimensional Gaussian graphical models. *Journal of the American Statistical Association*, 110:1248–1265, 2015.

[99] K. Y. Liang. On information and ancillarity in the presence of a nuisance parameter. *Biometrika*, 70:607–612, 1983.

[100] K. Y. Liang. The asymptotic efficiency of conditional likelihood methods. *Biometrika*, 71:305–313, 1984.

[101] K. Y. Liang. Odds ratio inference with dependent data. *Biometrika*, 72:678–682, 1985.

[102] K. Y. Liang. Extended Mantel-Haenszel estimating procedure for multivariate regression models. *Biometrics*, 43:289–299, 1987.

[103] K.Y. Liang and J. Qin. Regression analysis under non-standard situations: a pairwise pseudolikelihood approach. *Journal of the Royal Statistical Society, Ser. B*, 62:773–786, 2000.

[104] D. Y. Lin and Z. L. Ying. Cox regression with incomplete covariate measurements. *Journal of the American Statistical Association*, 88:1341–1349, 1993.

[105] D. Y. Lin and D. Zeng. Proper analysis of secondary phenotype data in case-control studies. *Genetic Epidemiology*, 33:256–265, 2009.

[106] S. R. Lipsitz and J. G. Ibrahim. A conditional model for incomplete covariates in parametric regression models. *Biometrika*, 83:916–922, 1996.

[107] S. R. Lipsitz, J. G. Ibrahim, and L. Zhao. A weighted estimating equation for missing covariate data with properties similar to maximum likelihood. *Journal of the American Statistical Association*, 94:1147–1160, 1999.

[108] R. J. A. Little. Regression with missing X's: A review. *Journal of the American Statistical Association*, 87:1127–1137, 1992.

[109] R. J. A. Little and D. B. Rubin. *Statistical Analysis with Missing Values*, 2nd edition. Wiley: New York, 2002.

[110] H. Liu, F. Han, M. Yuan, J. Lafferty, and L. Wasserman. High dimensional semiparametric Gaussian copula graphical models. *Annals of Statistics*, 40:2293–2326, 2012.

[111] J. Liu. *Monte Carlo Strategies in Scientific Computing.* Springer series in Statistics, New York, 2001.

[112] J. Liu, A. Gelman, J. Hill, Y. Su, and J. Kropko. On the stationary distribution of iterative imputations. *Biometrika*, 101:155–173, 2014.

[113] S. Liu, T. Suzuki, R. Relator, J. Sese, Sugiyama, and K. Fukumizu. Support consistency of direct sparse-change learning in Markov networks. *Annals of Statistics*, 45:959–990, 2017.

[114] C. R. Loader. Local likelihood density estimation. *Annals of Statistics*, 24:1602–1608, 1996.

[115] P. L. Loh and M. J. Wainwright. Structure estimation for discrete graphical models: Generalized covariance matrices and their inverses. *Annals of Statistics*, 41:3022–3049, 2013.

[116] X. Luo and W. Y. Tsai. A proportional likelihood ratio model. *Biometrika*, 99:211–222, 2012.

[117] Y. Ma and R. Carroll. Semiparametric estimation in the secondary analysis of case-control studies. *Journal of the Royal Statistical Society, Ser. B*, 78:127–151, 2016.

[118] D. Malinsky, I. Shpitser, and E. J. Tchetgen Tchetgen. Semiparametric inference for non-monotone missing-not-atrandom data: the no self-censoring mode. *Journal of the American Statistical Association*, Published online, 2021.

[119] C. F. Manski and S. R. Lerman. The estimation of choice probabilities from choice based samples. *Econometrics*, 45:1977–1988, 1977.

[120] N. Mantel. Chi-square tests with one degree of freedom; Extension of the Mantel-Haenszel procedure. *Journal of the American Statistical Association*, 58:690–700, 1963.

[121] N. Mantel and W. Haenszel. Statistical analysis of data from retrospective studies of disease. *Journal of the National Cancer Institute*, 22:719–748, 1959.

[122] T. Martinussen. Cox regression with incomplete covariate measurements using the EM-algorithm. *Scandinavian Journal of Statistics*, 26:479–491, 1999.

[123] R. Mazumder and T. Hastie. Exact covariance thresholding into connected components for large-scale graphical lasso. *Journal of Machine Learning Research*, 13:781–794, 2012.

[124] C. R. Mehta, N. R. Patel, and P. Senchaudhuri. Efficient Monte Carlo Methods for conditional logistic regression. *Journal of the American Statistical Association*, 95:99–108, 2000.

[125] N. Meinshausen and P. Buhlmann. High-dimensional graphs and variable selection with the Lasso. *Annals of Statistics*, 34:1436–1462, 2006.

[126] W. Miao and E. J. Tchetgen Tchetgen. On varieties of doubly robust estimators under missing not at random with a shadow variable. *Biometrika*, 103:475–482, 2016.

[127] J. Moussouris. Gibbs and Markov random systems with constraints. *Journal of Statistical Physics*, 10:11–33, 1974.

[128] S. Murphy and A. W. van der Vaart. On profile likelihood. *Journal of the American Statistical Association*, 95:449–465, 2000.

[129] R. M. Neal. *Handbook of Markov Chain Monte Carlo, chapter 5: MCMC using Hamiltoian Dynamics.* CRC Press, 2011.

[130] Y. Ning, T. Zhao, and H. Liu. A likelihood ratio framework for high-dimensional semiparametric regression. *Annals of Statistics*, 45:2299–2327, 2017.

[131] G. Osius. The association between two random elements: A complete characterization and odds ratio models. *Metrika*, 60:261–277, 2005.

[132] G. Osius. Asymptotic inference for semiparametric association models. *Annals of Statistics*, 37:459–489, 2009.

[133] M. C. Paik. Quasi-likelihood regression models with missing covariates. *Biometrika*, 83:825–834, 1997.

[134] B. Park, W. C. Kim, and M. C. Jones. On local likelihood density estimation. *Annals of Statistics*, 30:1480–1495, 2002.

[135] M. S. Pepe and T. R. Fleming. A nonparametric method for dealing with mismeasured covariate data. *Journal of the American Statistical Association*, 86:108–113, 1991.

[136] L. Pesarin, F. Salmaso. *Permutation Tests for Complex Data: Theory, Applications and Software.* John Wiley, UK, 1989.

[137] P. H. Peskun. Optimum Monte Carlo sampling using Markov chains. *Biometrika*, 60:607–612, 1973.

[138] N. Pfister, P. Buhlmann, B. Scholkopf, and J. Peters. Kernel based tests for joint independence. *Journal of the Royal Statistical Society, Ser. B*, 80:5–31, 2018.

[139] A. Philips and P. W. Holland. Estimators of the variance of the Mantel-Haenszel log-odds-ratio estimate. *Biometrics*, 43:425–431, 1987.

[140] W. W. Piegorsch, C. R. Weinberg, and J. A. Taylor. Non-hierarchical logistic models and case-only designs for assessing susceptibility in population-based case-control studies. *Statistics in Medicine*, 13:153–162, 1994.

[141] R. L. Prentice and R. Pyke. Logistic disease incidence models and case-control studies. *Biometrika*, 66:403–411, 1979.

[142] J. Qin. Inference for case-control and semiparametric two-sample density ratio models . *Biometrika*, 85:619–630, 1998.

[143] J. Qin and B. Zhang. A goodness-of-fit test for logistic regression models based on case-control data . *Biometrika*, 84:609–618, 1997.

[144] J. Qin and B. Zhang. Using logistic regression procedures for estimating receiver operating characteristic curves . *Biometrika*, 90:585–596, 2003.

[145] J. Qin and B. Zhang. Density estimation under a two-sample semiparametric model. *Journal of Nonparametric Statistics*, 17:665–683, 2005.

[146] D. Rabinowitz. A note on efficient estimation from case-control data. *Biometrika*, 84:486–488, 1997.

[147] P. J. Rathouz and L. P. Gao. Generalized linear models with unspecified reference distribution. *Biostatistics*, 10:205–218, 2009.

[148] P. Ravikumar, M. J. Wainwright, and J. Lafferty. High-dimensional Ising model selection using l_1-regularized logistic regression. *Annals of Statistics*, 38:1287–1319, 2010.

[149] M. Reilly and M. S. Pepe. A mean score method for missing and auxiliary covariate data in regression model. *Biometrika*, 82:299–314, 1995.

[150] J. Robins, N. Breslow, and S. Greenland. Estimators of the Mantel-Haenszel variance consistent in both sparse data and large strata models. *Biometrics*, 42:311–323, 1986.

[151] J. Robins, S. Green, and N. Breslow. A general estimator for the variance of the Mantel-Haenszel odds ratio. *American Journal of Epidemiology*, 124:719–723, 1986.

[152] J. M. Robins, F. Hsieh, and W. Newey. Semiparametric estimation of a conditional density with missing or mismeasured covariates. *Journal of the Royal Statistical Society, Ser. B*, 57:409–424, 1995.

[153] J. M. Robins, A Rotnitzky, and L. Zhao. Estimation of regression coefficients when some regressors are not always observed. *Journal of the American Statistical Association*, 89:477–485, 1994.

[154] P. R. Rosenbaum. Conditional permutation tests and the propensity score in observational studies. *Journal of the American Statistical Association*, 79:565–574, 1984.

[155] J. S. Rosenthal. Minorization conditions and convergence rates for Markov chain Monte Carlo. *Journal of the American Statistical Association*, 90:558–566, 1995.

[156] D. B. Rubin. Inference and missing data. *Biometrika*, 63:581–592, 1976.

[157] M. Sadinle and J. Reiter. Itemwise conditionally independent nonresponse modelling for incomplete multivariate data. *Biometrika*, 104:207–220, 2017.

[158] G. Satten and L. Kupper. Inference about exposure-disease associations using probability-of-exposure information. *Journal of the American Statistical Association*, 88:200–208, 1993.

[159] G. A. Satten and R. J. Carroll. Conditional and unconditional categorical regression models with missing covariates. *Biometrics*, 56:384–388, 2000.

[160] S. Schmidt and D. J. Schaid. Potential misinterpretation of the case-only study to assess gene-environment interaction. *American Journal of Epidemiology*, 150:878–885, 1999.

[161] A. J. Scott and C. J. Wild. Fitting logistic models under case-control or choice based sampling. *Journal of the Royal Statistical Society, Ser. B*, 48:170–182, 1986.

[162] A. J. Scott and C. J. Wild. Fitting logistic regression models in stratified case-control studies. *Biometrics*, 47:497–510, 1991.

[163] A. J. Scott and C. J. Wild. Fitting regression models to case-control data by maximum likelihood. *Biometrika*, 84:57–71, 1997.

[164] R. D. Shah and J. Peters. The hardness of conditional independence testing and the generalized covariance measure. *Annals of Statistics*, 48:1514–1538.

[165] J. Shao and L. Wang. Semiparametric inverse propensity weighting for nonignorable missing data. *Biometrika*, 84:175–187, 2016.

[166] R. Simon. Length biased sampling in etiologic studies. *American Journal of Epidemiology*, 111:444–452, 1980.

[167] Y. Slaoui. Bias reduction in kernel density estimation. *Journal of Nonparametric Statistics*, 30:505–522, 2018.

[168] K. Song. Testing conditional independence via Rosenblatt transforms. *Annals of Statistics*, 37:4011–4045.

[169] W. Stute. Conditional U-statistics. *Annals of Probability*, 19:1110–1139, 2003.

[170] L. Su and H. White. A consistent characteristic function-based test for conditional independence. *Journal of Econometrics*, 141:807–834, 2007.

[171] L. Su and H. White. Testing conditional independence via empirical likelihood. *Journal of Econometrics*, 182:27–44, 2014.

[172] B. Sun, L. Liu, W. Miao, K. Wirth, J. Robins, and E. J. Tchetgen Tchetgen. Semiparametric estimation with data missing not at random using an instrumental variable. *Statistica Sinica*, 28:1965–1983, 2018.

[173] G. J. Szekely and M. L. Rizzo. Partial distance correlation with methods for dissimilarities. *Annals of Statistics*, 42:2382–2412, 2014.

[174] G. J. Szekely, M. L. Rizzo, and N. K. Bakirov. Measuring and testing dependence by correlation of distances. *Annals of Statistics*, 35:2769–2794, 2007.

[175] Z. Tan. A note on profile likelihood for exponential tilt mixture models. *Biometrika*, 96:229–236, 2009.

[176] Z. Tan. On doubly robust estimation for logistic partially linear models. *Statistics & Probability Letters*, 155:108, 2019.

[177] G. Tang, R. J. A. Little, and T. E. Raghunathan. Analysis of multivariate missing data with nonignorbale nonresponse. *Biometrika*, 90:747–764, 2003.

[178] Tchetgen Tchetgen and J. M. E. Robins. The semiparametric case-only estimator. *Biometrics*, 67:380–387, 2010.

[179] E. Tchetgen Tchetgen, J. M. Robins, and A. Rotnitsky. On doubly robust estimation in a semiparametric odds ratio model. *Biometrika*, 97:171–180, 2010.

[180] E. J. Tchetgen Tchetgen. On a closed-form doubly robust estimator of the adjusted odds ratio for a binary exposure. *American Journal of Epidemiology*, 177:1314–1316, 2013.

[181] E. J. Tchetgen Tchetgen and A. Rotnitzky. Double-robust estimation of an exposure-outcome odds ratio adjusting for confounding in cohort and case-control studies. *Statistics in Medicine*, 32:335–347, 2011.

[182] G. R. Terrell and D. W. Scott. Variable kernel density estimation. *Annals of Statistics*, 20:1236–1265, 1992.

[183] H. Tjelmeland and J. Besag. Markov random fields with higher-order interactions. *Scandinavian Journal of Statistics*, 25:413–433, 1998.

[184] P. Tseng. Convergence of a block coordinate descent method for nondifferentiable minimization. *Journal of Optimization Theory and Applications*, 109:475–494, 2001.

[185] P. Tseng and S. Yun. A coordinate gradient descent method for nonsmooth separable minimization. *Mathematical Programming*, 117:387–423, 2009.

[186] D. M. Umbach and C. M. Weinberg. Designing and analyzing case-control studies to exploit independence of genotype and exposure. *Statistics in Medicine*, 16:1731–1743, 1997.

[187] G. J. G. Upton. Conditional independence, Mantel-Haenszel test, and the Yates correction. *American Statistician*, 54:112–115–244, 2000.

[188] S. Van Buuren, J. P. L. Brand, C. G. M. Groothuis-Oudshoorn, and D. B. Rubin. Fully conditional specification in multivariate imputation. *Journal of Statistical Computation and Simulation*, 76:1049–1064, 2006.

[189] D. A. van Dyk and T. Park. Partially collapsed Gibbs samplers: Theory and methods. *Journal of the American Statistical Association*, 103:790–796, 2008.

[190] T. J. VanderWeele, S. Hermandez-Diaz, and M. A. Hernan. Case-only gene-environment interaction studies: When does association imply mechanistic interaction? *Genetic Epidemiology*, 34:327–334, 2010.

[191] Y. Vardi. Empirical distributions in selection bias models. *Annals of Statistics*, 13:178–203, 1985.

[192] A. Voorman, A. Shojaie, and D. Witten. Graph estimation with joint additive models. *Biometrika*, 101:85–101, 2014.

[193] M. C. Wang. Hazards regression analysis for length-biased data. *Biometrika*, 83:343–354, 1996.

[194] S. Wang, J. Shao, and J. K. Kim. An instrumental variable approach for identification and estimation with nonignorable nonrespeonse. *Statistical Science*, 24:1097–1116, 2014.

[195] Y. J. Wang and E. H. Ip. Conditionally specified continuous distributions. *Biometrika*, 95:735–746, 2008.

[196] M. A. Weaver and H. B. Zhou. An estimated likelihood method for continuous outcome regression model with outcome-dependent sampling. *Journal of the American Statistical Association*, 100:459–469, 2005.

[197] C. R. Weinberg and S. Wacholder. Prospective analysis of case-control data under general multiplicative-intercept risk models. *Biometrika*, 80:461–465, 1993.

[198] J. E. White. A two-stage design for the study of the relationship between a rare exposure and a rare disease. *American Journal of Epidemiology*, 115:119–128, 1982.

[199] T. Wu and K. Lange. Coordinate descent procedures for lasso penalized regression. *Annals of Applied Statistics*, 2:224–244, 2008.

[200] L. Xue and H. Zou. Regularized rank-based estimation of high-dimensional nonparanormal graphical models. *Annals of Statistics*, 40:2541–2571, 2012.

[201] Y. Yang. Penalized semiparametric density estimation. *Statistics and Computing*, 19:355–366, 2009.

[202] M. Yuan and Y. Lin. Model selection and estimation in the Gaussian graphical model. *Biometrika*, 94:19–35, 2007.

[203] J. Zetterqvist, K. Vermeulen, S. Vansteelandt, and A. Sjolander. Doubly robust conditional logistic regression. *Statistics in Medicine*, 38:4749–4760, 2019.

[204] J. Zhao and J. Shao. Semiparametric pseudo-likelihoods in generalized linear models with nonignorable missing data. *Journal of the American Statistical Association*, 110:1577–1590, 2015.

[205] P. Zhao and B. Yu. On model selection consistency of Lasso. *Journal of Machine Learning Research*, 7:2541–2563, 2006.

[206] H. B. Zhou and M. S. Pepe. Auxiliary covariate data in failure time regression. *Biometrika*, 82:139–149, 1995.

Index